PARALLELOGRAM

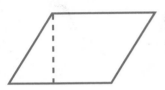

Area: $A = bh$

TRAPEZOID

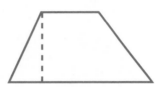

Area: $A = \frac{1}{2}h(b_1 + b_2)$

RECTANGULAR SOLID

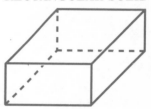

Volume: $V = LWH$

CIRCLE

Area: $A = \pi r^2$
Circumference: $C = \pi D$

SPHERE

Volume: $V = \frac{4}{3}\pi r^3$
Surface area: $S = 4\pi r^2$

RIGHT CIRCULAR CYLINDER

Volume: $V = \pi r^2 h$
Lateral surface area: $S = 2\pi rh$

CONE

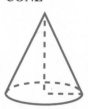

Volume: $V = \frac{1}{3}\pi r^2 h$
Surface area: $S = \pi r\sqrt{r^2 + h^2}$

Other Formulas

Distance: $D = RT$ $\qquad R$ = rate, T = time

Percent: $p = br$ $\qquad p$ = percentage, b = base, r = rate

Temperature: $F = \frac{9}{5}C + 32$ $\qquad C = \frac{5}{9}(F - 32)$

Simple interest: $I = Prt$ $\qquad P$ = principal, r = rate or percent, t = time in years

Amount: $A = P + Prt = P(1 + rt)$

Sale price: $S = L - rL$ $\qquad S$ = sale price, L = list price, r = rate of discount

Intermediate Algebra

MARK DUGOPOLSKI

Southeastern Louisiana University

Addison-Wesley Publishing Company

Reading, Massachusetts • Menlo Park, California • New York
Don Mills, Ontario • Wokingham, England • Amsterdam • Bonn
Sydney • Singapore • Tokyo • Madrid • San Juan

or	Charles B. Glaser
or	Stephanie Botvin
Managing Editor	Karen M. Guardino
Production Supervisor	Marion E. Howe
Copy Editor	Barbara Willette
Proofreader	Laura Michaels
Text Designer	Geri Davis, Quadrata, Inc.
Design Consultant	Meredith Nightingale
Layout Artist	Lorraine Hodsdon
Cover Designer	Marshall Henrichs
Art Consultant	Loretta M. Bailey
Illustrator	Tech-Graphics
Photo Researcher	Susan Van Etten
Manufacturing Manager	Roy Logan
Marketing Manager	Melissa Acuña
Production Services Manager	Herbert Nolan
Compositor, Color Separator	York Graphic Services, Inc.
Printer	R.R. Donnelley & Sons

Library of Congress Cataloging-in-Publication Data

Dugopolski, Mark.
    Intermediate algebra / by Mark Dugopolski.
      p. cm.
    Includes index.
    ISBN 0-201-16894-4
    1. Algebra.   I. Title.
    QA154.2.D84   1991
    512.9—dc20   90-39641
Reprinted with corrections, June 1991   CIP

To my wife and daughters,
Cheryl, Sarah, and Alisha

Preface

This text is designed for a one term course in intermediate algebra, with an emphasis on the basics in the textual material as well as in the exercises. Although a complete development for each topic is provided, it is assumed that the student has had some prior training in algebra. (The text *Elementary Algebra* in this series is available for students with no prior experience in algebra.) The unifying theme of this text is first the development of the skills necessary for solving equations and inequalities and then application of those skills to solving applied problems.

My primary goal in writing this book was to write one that students can read, understand, and enjoy, while gaining confidence in their ability to use mathematics. Toward this end, I have endeavored to design a comprehensive, yet flexible, presentation of intermediate algebra that will satisfy the individual needs of instructors and a variety of students. I have taken what I refer to as a "common-ground" approach. Whenever possible, the text reminds students of ideas they are already familiar with and then builds on those ideas. This approach allows the instructor to meet the students on a common ground, before advancing to a new level.

Key Features

▶ Every chapter begins with a **Chapter Opener,** which features an application of an idea developed in that chapter. This gives the student a concrete idea of what will be accomplished in that chapter.

▶ Every section begins with a list of topics that tells what is **In This Section.** The sections are divided into subsections and the subsection titles correspond to the topics from the In-This-Section list.

▶ Important ideas are set apart in boxes for quick reference. These boxes are used for **definitions, rules, summaries, and strategies.**

▶ Simple **fractions and decimals** are used throughout the text. This feature helps to reinforce the basic arithmetic skills that are necessary for success in algebra.

▶ One of the main goals of the text is the development of the skills necessary for solving equations and inequalities, and then application of those skills to solving **applied problems.** For this reason, word problems occur wherever possible. This text contains over 400 word problems and they occur in over 50% of the exercise sets.

▶ **Inequalities** are used in applications as well as equations. This helps students to see the need for solving inequalities. There are word problems involving linear inequalities, absolute value inequalities, and quadratic inequalities.

▶ There are numerous **geometric word problems** throughout the text, which are designed to review basic geometric facts and figures. A summary of common geometric facts can be found inside the front cover.

▶ Each exercise set is preceded by **Warm-ups,** a set of 10 simple statements that are to be marked either true or false. These exercises are designed to bridge the gap between the lecture and the exercise sets and to stimulate discussion of the concepts presented in the section. Many of the false statements point out examples of common student errors. The answers to all of these exercises are given in the answer section.

▶ The exercise sets throughout the text contain **keyed and non-keyed exercises.** The exercises at the end of each section follow the same order as the textual material and they are keyed to the examples. This organization allows instructors the flexibility of being able to cover only part of a section and easily see which exercises are appropriate to assign. The keyed exercises always give the student a place to start and build confidence. Wherever appropriate, non-keyed exercises follow the keyed exercises and are designed to bring all of the ideas together. Answers to all odd-numbered exercises are in the answer section.

▶ **Calculator exercises** are included whenever appropriate. These exercises may be omitted by instructors who do not require calculators in their courses. However, to attain the maximum benefit from this course a calculator is essential. Topics such as quadratic equations, compound interest, and logarithms are more understandable when students can find approximate decimal answers as well as the exact answers.

▶ **Mental Exercises** occur at the end of many exercise sets and in the chapter review exercises. These exercises consist mainly of simple equations that can be solved in one or two steps, or simple expressions that can be simplified mentally. These exercises are meant to be done (perhaps more than once) only after the written exercises. Mental exercises point out to all students what many of them would discover on their own.

▶ **Challenging Exercises** are included for instructors wishing to assign more difficult problems to the more able students. These optional exercises are indicated by an asterisk.

▶ **Logarithmic and exponential functions** are covered from a modern point of view. A scientific calculator should be required when discussing logarithmic and exponential functions. Calculating with logarithms is not discussed. However, a common logarithm table with instructions on how to use it is given in Appendix B for those instructors who use tables. The applications of logarithmic and exponential functions are stressed, as well as their properties and graphs.

▶ Each chapter contains a **Math at Work** feature that illustrates how an idea from that chapter occurs in a real-life situation. The applications are drawn from a variety of fields to maximize student interest. Each Math at Work feature contains an optional exercise pertaining to the application presented.

▶ Each chapter ends with a four part **Wrap-up.**

The **Chapter Summary** is a summary of the important concepts from the chapter along with brief illustrative examples.

The **Review Exercises** are intended to provide a review of each section of the chapter. The Review Exercises contain exercises that are keyed to the sections of the chapter, as well as miscellaneous exercises. Answers to the odd-numbered Review Exercises are in the answer section.

The **Chapter Test** is designed to help the student test his or her readiness for a chapter test. To help the student to be independent of the examples and sections, the Chapter Test has no keyed exercises. All answers for each Chapter Test are given in the answer section.

Tying It All Together exercises are designed to help students fit the current ideas in with ideas from the previous chapters. They may also contain exercises that will help in understanding the upcoming chapter. These exercises could be worked after a chapter test, but prior to starting the next chapter. All answers for these exercises are given in the answer section.

Ancillary Materials

▶ An **Annotated Instructor's Edition** is available to instructors. In addition to the material presented in the student edition, the AIE contains marginal notes to the instructor, in-text answers to all numerical exercises, and answers to all even-numbered graphing exercises in the instructor's answer section, which follows the To the Instructor introductory material.

▶ The **Instructor's Solution Manual** contains a detailed solution to each exercise in the text.

▶ The **Student's Solution Manual** contains detailed solutions to all of the odd-numbered exercises as well as solutions to all exercises from the Warm-ups, Chapter Tests, and Tying It All Together sections.

▶ The **Student's Study Guide** includes diagnostic tests for each chapter as well as detailed solutions and explanations for each test question.

▶ There are several options available for **Computerized Testing.**

  AWTest (Apple) is an algorithm-based testing system keyed to the text. AWTest will generate multiple choice, open-ended, and true-false questions.

  Omni Test (IBM) is a powerful new testing system developed exclusively for Addison-Wesley. It is an algorithm-driven system that allows the user to create up to 99 perfectly parallel forms of any test effortlessly. It also allows the user to add his or her own test items and edit existing items with an easy to use, on screen "What-You-See-Is-What-You-Get" text editor.

  Mac Test (Macintosh) is a test item bank containing ten questions for each objective drawn from the text. Questions are in a multiple choice format.

▶ The **Printed Test Bank** contains three alternate forms of tests for each chapter, as well as cumulative review tests and final exams.

▶ **Tutorial Software** is available in several forms.

  IMPACT: An Interactive Mathematics Tutorial (IBM & Mac) is a tutorial package correlated to each section of the text. This package will generate exercises similar to those in the text exercise sets. If the student gets an exercise wrong, he or she can see either an example or a solution. The program will lead students interactively through step-by-step solutions to the exercises so that they can easily pinpoint specific trouble areas if they make an error.

  Students using **The Math Lab** (IBM & Apple II) pick the topic area, level of difficulty, and number of problems. If they get a wrong answer, the program will prompt them with the first step of the solution. This program keeps detailed records of students' results, which can be stored on the disk.

  In **Professor Weissman's Software** (IBM), students pick the topic area and level of difficulty. The program generates problems and gives step-by-step solutions if they get a wrong answer. Problems increase in difficulty as students are successful. This program also keeps records of students' scores.

  For the **Algebra Problem Solver** (IBM), students choose a topic and can either request an exercise from the computer or enter their own homework exercise. The computer will show them a step-by-step solution to the problem.

▶ An extensive set of **video tapes** is available free upon adoption. These videos cover every topic presented in the text. Particular attention has been paid to working examples and giving complete explanations of the steps involved.

▶ **The Math Hotline** is a new idea in teaching and learning packages created especially for this text. Suppose a student is at home studying and can't get a problem started. Now, simply by calling the Math Hotline from a touch-tone phone and responding to a few questions with the keypad, a student can receive a hint for any odd-numbered exercise. This means thousands of hints are as close as the phone. The Math Hotline will provide hints 24-hours a day, 365 days a year.

Acknowledgments

I would like to express my appreciation to the reviewers whose comments and suggestions were invaluable in writing this book:

Wayne Andrepont
University of Southwestern Louisiana

Laura Clark
Milwaukee Area Technical College

Johanna Ellner
New York City Technical College

Paul Finster
El Paso Community College

Susan L. Friedman
CUNY, Baruch College

Laura Holden
Indiana University Bloomington

Steven Kahn
Anne Arundel Community College

Cynthia Mansour
Hillsborough Community College

Steven P. Marsden
Glendale Community College

Richard Marshall
Eastern Michigan University

Cynthia Miller
Georgia State University

Karla Neal
Louisiana State University

Allan R. Newhart
Parkersburg Community College

Stephen B. Rodi
Austin Community College

Anna Jo Ruddel
CUNY, Baruch College

Helen Santiz
University of Michigan

Richard Semmler
Northern Virginia Community College

Thomas A. Verrecchia
Community College of Rhode Island
Jon D. Weerts
Triton College

I thank Beth Gray, Southeastern Louisiana University, for working all of the problems, and Tina Golding, Southeastern Louisiana University, for class-testing the manuscript. I thank James Morgan, Southeastern Louisiana University, and Wayne Andrepont, University of Southwestern Louisiana, for error checking.

I thank the staff at Addison-Wesley for all of their help and encouragement throughout this project, especially Chuck Glaser, Betsy Burr, Stuart Johnson, and Stephanie Botvin.

I also want to express my sincere appreciation to my wife, Cheryl, and my daughters, Sarah and Alisha, for their patience and support. I could not have completed the project without them.

Hammond, Louisiana   M.D.

To the Student

This text was written to help you learn algebra. I have made the explanations as clear as possible and I have provided plenty of examples to illustrate the ideas. This text was written with you in mind. You should be aware of the many features that were designed into this text to make the text easy to use.

The exercises follow the same order as the textual material and they are keyed to the examples in the text. This feature is to help you get started. Don't try to understand an entire section before beginning the exercises. Get started by doing exercises. If you don't understand how to do an exercise, then refer to the example that corresponds to that exercise. If you miss class and you need to work on your own, read the section in pieces. Read until you get through an example, then work the corresponding exercises.

To learn algebra you must do algebra and doing algebra means working problems. I have written an abundance of problems for this text. I have tried to avoid overly complicated exercises to give you plenty of practice with the basics. The answers to all odd-numbered exercises are given in the back of the book. These answers should be consulted only after you have worked a problem and checked your work.

To help you bring each chapter together, a four-part Wrap-up is included at the end of each chapter. In each Wrap-up you will find a Chapter Summary that lists the key ideas in each chapter along with illustrative examples. You will also find Review Exercises that are keyed to the sections of the chapter, along with miscellaneous exercises from the chapter. A Chapter Test is included for each chapter to let you check your progress. Complete answers for chapter tests are given in the back of the book. The exercises in the Chapter Tests are not keyed to the section from which they come. After each Chapter Test (starting with Chapter 2) I have included some exercises called Tying It All Together. These exercises are designed to help you fit current ideas in with ideas from previous chapters. Sometimes these exercises include problems that will be helpful to review before starting the next chapter. The Tying It All Together exercises could be worked after an in-class test but prior to the next class meeting.

If you need additional help with the exercises, there is a Student's Solution Manual available that includes complete solutions to those problems that have answers in the book. There is also a Student's Study Guide available to help you to prepare for tests. I have also prepared a computerized help line that you can call from a touch tone phone and get a brief message on how to work a problem. Help is available 24 hours a day for those exercises that have answers in the answer section of the book. Relax and do your homework faithfully, and I am sure that you will be successful in algebra.

Mark Dugopolski

Contents

6   Quadratic Equations and Inequalities   284

7   Linear Equations in Two Variables   328

8   Functions   371

Basics

Why can you put your groceries on the checkout counter in any order and still get the same total?

Do we get the same amount of tax if the sales tax is computed on each item and then totaled, or if the items are totaled first and then the tax computed?

Why is $8 divided among 4 people not the same as $4 divided among 8 people?

What do these questions have to do with algebra? The answers to these questions are found in the fundamental properties of arithmetic. Success in algebra strongly depends on a solid foundation of arithmetic. The computational skills developed in arithmetic are essential in algebra. However, in algebra we must understand the underlying concepts of arithmetic as well. In this chapter we will review some of the concepts of arithmetic, learn the basic terminology of algebra, and discover the answers to the above questions (and many others).

1.1 Sets

IN THIS SECTION:
- **Set Notation**
- **Union**
- **Intersection**
- **Subsets**
- **Parentheses**

Every subject has its own terminology, and algebra is no different. In this section we will learn the basic terms and facts about sets.

Set Notation

A **set** is a collection of objects. We may speak of a set of dishes or a set of steak knives. In algebra we generally discuss sets of numbers. For example, we refer to the numbers 1, 2, 3, 4, 5, and so on, as the set of **counting numbers.** Of course, these are the numbers that we use for counting.

The objects or numbers in a set are called **elements** or **members** of the set. To describe sets with a convenient notation, we use braces, { }, and name the sets with capital letters. For example,

$$A = \{1, 2, 3\}$$

means that set A is the set whose members are the numbers 1, 2, and 3.

A set may have a **finite** number of elements such as {1, 2, 3}, or a set may have an **infinite** number of elements such as the set of counting numbers. To indicate a continuing pattern, we use a series of three dots. The set of counting numbers is written

$$\{1, 2, 3, \ldots \}.$$

The set of counting numbers between 4 and 40 is written

$$\{5, 6, 7, 8, \ldots , 39\}.$$

Note that since the members of this set are *between* 4 and 40, it does not include 4 or 40.

Set-builder notation is another method of describing sets. In this notation we use a variable to represent the numbers in the set. A **variable** is a letter that is used to stand for some numbers. The set is then built from the variable and a description of the numbers that the variable represents. For example, the set

$$B = \{1, 2, 3, \ldots, 49\}$$

is written in set-builder notation as

$$B = \{x \mid x \text{ is a counting number smaller than } 50\}.$$

The set of ⌐⌐ Such that

This notation is read as "*B* is the set of numbers *x*, such that *x* is a counting number smaller than 50." Notice that the number 50 is not a member of set *B*.

To indicate that a certain element is a member of a set, we use the symbol $\in$. To indicate that a certain element is not a member of the set, we use the symbol $\notin$. If we let

$$A = \{1, 2, 3\},$$

then the statement $1 \in A$ is read as "1 is an element of *A*," "1 belongs to *A*," or "1 is a member of *A*." The statement $4 \notin A$ is read as "4 is not an element of *A*," "4 does not belong to *A*," or "4 is not a member of *A*."

Two sets are **equal** if they contain exactly the same members. Otherwise, they are said to be not equal. To indicate equal sets, we use the symbol $=$. For sets that are not equal, we use the symbol $\neq$. For example,

$$\{3, 4, 7\} = \{3, 4, 7\},$$
$$\{2, 4, 1\} = \{1, 2, 4\},$$
$$\{3, 5, 6\} \neq \{3, 5, 7\}.$$

Note that the elements in two equal sets do not need to be written in the same order.

EXAMPLE 1 Let $A = \{1, 2, 3, 5\}$ and $B = \{x \mid x \text{ is an even counting number smaller than } 10\}$. Determine whether each of the following statements is true or false.

a) $3 \in A$ **b)** $5 \in B$ **c)** $4 \notin A$ **d)** $A = B$
e) $A = \{x \mid x \text{ is a counting number smaller than } 6\}$
f) $B = \{2, 4, 6, 8\}$

Solution

a) True, because 3 is a member of set *A*.
b) False, because 5 is not an even counting number.
c) True, because 4 is not a member of set *A*.
d) False, because 4 is a member of *B* but not a member of *A*.
e) False, because 4 is a counting number smaller than 6, and 4 is not a member of *A*.
f) True, because 2, 4, 6, and 8 are precisely the even counting numbers smaller than 10.   ◄

Union

For any two sets A and B, a new set can be formed using the following definition.

> If A and B are sets, the **union** of A and B, denoted $A \cup B$, is the set whose members are either elements of A, elements of B, or elements of both. In symbols,
>
> $$A \cup B = \{x \mid x \in A \text{ or } x \in B\}.$$

In mathematics, the word "or" is always used in an inclusive manner (allowing the possibility of both alternatives).

Intersection

Another way to form a new set from two known sets is given in the following definition.

> If A and B are sets, the **intersection** of A and B, denoted $A \cap B$, is the set whose elements are members of both A and B. In symbols,
>
> $$A \cap B = \{x \mid x \in A \text{ and } x \in B\}.$$

It is possible for two sets to have no elements in common. A set with no members is called the **empty set** and is denoted by the symbol $\varnothing$. Note that $\{0\}$ is not the empty set. This set has one member, the number 0. Note also that $A \cup \varnothing = A$ and $A \cap \varnothing = \varnothing$ for any set A.

EXAMPLE 2   Let $A = \{0, 2, 3\}$, $B = \{2, 3, 7\}$, and $C = \{7, 8\}$. List the elements in each of the following sets.

a) $A \cup B$ **b)** $A \cup C$ **c)** $A \cap B$ **d)** $B \cap C$ **e)** $A \cap C$

Solution

a) $A \cup B = \{0, 2, 3, 7\}$ **b)** $A \cup C = \{0, 2, 3, 7, 8\}$
c) $A \cap B = \{2, 3\}$ **d)** $B \cap C = \{7\}$ **e)** $A \cap C = \varnothing$ ◄

EXAMPLE 3   Let $A = \{1, 2, 3, 5\}$, $B = \{2, 3, 7, 8\}$, and $C = \{6, 7, 8, 9\}$. Place one of the symbols $=$, $\neq$, $\in$, or $\notin$ in the blank to make each statement correct.

a) 5 _____ $A \cup B$ **b)** 5 _____ $A \cap B$
c) $A \cup B$ _____ $\{1, 2, 3, 5, 7, 8\}$ **d)** $A \cap B$ _____ $\{2\}$
e) $A \cap C$ _____ $\varnothing$ **f)** 8 _____ $A \cup C$

Solution

a) $5 \in A \cup B$, because 5 is a member of A.

b) $5 \notin A \cap B$, because 5 must belong to both A and B in order to be a member of $A \cap B$.

c) $A \cup B = \{1, 2, 3, 5, 7, 8\}$, because the elements of A together with those of B are listed. Note that 3 and 5 are members of both sets but are listed only once.

d) $A \cap B \neq \{2\}$, because both 2 and 3 are members of both sets.

e) $A \cap C = \varnothing$ because A and C have no elements in common.

f) $8 \in A \cup C$, because 8 is a member of C.   ◄

Subsets

A set A is a **subset** of set B, written $A \subseteq B$, if every member of A is also a member of B. For example,

$$\{2, 3\} \subseteq \{2, 3, 4\}.$$

⌐ Is a subset of

A set A that is not a subset of B is written as $A \nsubseteq B$. To claim that $A \nsubseteq B$, there *must* be an element of A that does *not* belong to B. For example,

$$\{1, 2\} \nsubseteq \{2, 3, 4\},$$

because the number 1 is a member of the first set but not the second.

Consider the empty set $\varnothing$ and $\{1, 2\}$. Is $\varnothing$ a subset of $\{1, 2\}$? If the empty set is *not* a subset of $\{1, 2\}$, then there must be an element of $\varnothing$ that does not belong to $\{1, 2\}$. This is impossible, because $\varnothing$ is empty. Thus the empty set $\varnothing$ is a subset of $\{1, 2\}$. In fact, the empty set is a subset of every set.

EXAMPLE 4 Determine whether each of the following statements is true or false.

a) $\{1, 2, 3\}$ is a subset of the set of counting numbers.

b) The set of counting numbers is not a subset of $\{1, 2, 3\}$.

c) $\{1, 2, 3\} \nsubseteq \{2, 4, 6, 8\}$

d) $\{2, 6\} \subseteq \{1, 2, 3, 4, 5\}$

e) $\varnothing \subseteq \{2, 4, 6\}$

Solution

a) True.

b) True, because 5 is a counting number and $5 \notin \{1, 2, 3\}$.

c) True, because 1 is in the first set, but not the second.

d) False, because 6 is not in the second set.

e) True, because we cannot find anything in $\varnothing$ that fails to be in $\{2, 4, 6\}$.   ◄

Parentheses

Union and intersection are operations that we perform on a pair of sets to get a new set. To combine three or more sets using the operations of union and intersection, we use parentheses to indicate which pair of sets to combine first. In the following example, notice that different results are obtained from different placements of the parentheses.

EXAMPLE 5 Let $A = \{1, 2, 3, 4\}$, $B = \{2, 5, 6, 8\}$, and $C = \{4, 5, 7\}$. List the elements of each of the following sets.

a) $(A \cup B) \cap C$ **b)** $A \cup (B \cap C)$

Solution

a) The parentheses indicate that the union of A and B is to be found first, and then that set is to be intersected with C.

$$A \cup B = \{1, 2, 3, 4, 5, 6, 8\}$$

The only numbers that are members of this set and C are 4 and 5. Thus

$$(A \cup B) \cap C = \{4, 5\}.$$

b) The parentheses now indicate that the intersection of B and C is to be found first. The only number that is a member of both B and C is 5. Thus $B \cap C = \{5\}$. The union of this set and set A consists of all numbers that are members of one set or the other. Thus

$$A \cup (B \cap C) = \{1, 2, 3, 4, 5\}. \qquad \blacktriangleleft$$

Warm-ups

Let $A = \{1, 2, 3, 4\}$, $B = \{3, 4, 5\}$, and $C = \{3, 4\}$. Determine whether each statement is true or false.

1. $A = \{x \mid x \text{ is a counting number}\}$

2. The set B has an infinite number of elements.

3. The set of counting numbers less than 50 million is an infinite set.

4. $1 \in A \cap B$

5. $3 \in A \cup B$

6. $A \cap B = C$

7. $C \subseteq B$

8. $A \subseteq B$

9. $\varnothing \subseteq C$

10. $A \nsubseteq C$

1.1 EXERCISES

For Exercises 1–40, let

$A = \{x \mid x \text{ is an odd counting number smaller than } 10\}$,
$B = \{2, 4, 6, 8\}$, and
$C = \{1, 2, 3, 4, 5\}$.

Determine whether each statement is true or false. See Example 1.

1. $6 \in A$

2. $8 \in A$

3. $A = \{1, 3, 5, 7, 9, 11, \ldots\}$

4. $A \neq B$

5. $3 \in C$

6. $4 \notin B$

7. $B \neq C$

8. $A = \{1, 3, 7, 9\}$

List the elements in the following sets. See Example 2.

9. $A \cup B$

10. $A \cap B$

11. $A \cap C$

12. $A \cup C$

13. $B \cup C$

14. $B \cap C$

15. $A \cup \emptyset$

16. $B \cup \emptyset$

17. $A \cap \emptyset$

18. $B \cap \emptyset$

Use one of the symbols $\in$, $\notin$, $=$, $\neq$, $\cup$, or $\cap$ in the blank of each statement to make it correct. See Example 3.

19. $A \cap B$ _____ $\emptyset$

20. $A \cap C$ _____ $\emptyset$

21. A _____ $B = \{1, 2, 3, 4, 5, 6, 7, 8, 9\}$

22. A _____ $B = \emptyset$

23. B _____ $C = \{2, 4\}$

24. B _____ $C = \{1, 2, 3, 4, 5, 6, 8\}$

25. 3 _____ $A \cap B$

26. 3 _____ $A \cap C$

27. 4 _____ $B \cap C$

28. 8 _____ $B \cup C$

Determine whether each statement is true or false. See Example 4.

29. Set A is a subset of the set of counting numbers.

30. Set B is a subset of the set of counting numbers.

31. $\{2, 3\} \subseteq C$

32. $C \subseteq A$

33. $B \not\subseteq C$

34. $C \not\subseteq A$

35. $\emptyset \subseteq B$

36. $\emptyset \subseteq C$

37. $A \subseteq \emptyset$

38. $B \subseteq \emptyset$

39. $A \cap B \subseteq C$

40. $B \cap C \subseteq \{2, 4, 6, 8\}$

For Exercises 41–66, let

$A = \{3, 5, 7\}$,
$B = \{2, 4, 6, 8\}$, and
$C = \{1, 2, 3, 4, 5\}$.

List the elements in the following sets. See Example 5.

41. $A \cup B$

42. $A \cap B$

43. $A \cap C$

44. $A \cup C$

45. $B \cup C$

46. $B \cap C$

47. $(A \cup B) \cap C$

48. $(A \cup C) \cap B$

49. $A \cup (B \cap C)$

50. $A \cup (C \cap B)$

51. $(A \cap C) \cup (B \cap C)$

52. $(A \cap B) \cup (C \cap B)$

53. $(A \cup B) \cap (A \cup C)$

54. $(A \cup C) \cap (A \cup B)$

Use one of the symbols $\in$, $\subseteq$, $=$, $\cup$, *or* $\cap$ *in the blank of each statement to make it correct.*

55. A _____ $\{x \mid x$ is an odd counting number$\}$

56. B _____ $\{x \mid x$ is an even counting number smaller than 9$\}$

57. 3 _____ A **58.** $\{3\}$ _____ A **59.** A _____ $B = \varnothing$

60. $A \cap B$ _____ A **61.** $A \cap C$ _____ C **62.** $3 \notin B$ _____ C

63. $B \nsubseteq B$ _____ C **64.** $B \subseteq B$ _____ C **65.** A _____ $C = C \cup A$

66. B _____ $C = C \cap B$

1.2 The Real Numbers

THIS SECTION:

- Rational Numbers
- The Number Line
- Irrational Numbers
- Summary

The set of real numbers is the basic set of numbers used in algebra. There are many different types of real numbers. To better understand the set of real numbers, we will study some of the subsets of numbers that make up this set.

Rational Numbers

We have already introduced the most fundamental set of numbers, the **counting numbers.** These numbers are also called the **natural numbers.** The letter N is used to represent this set. The set of natural numbers is written in symbols as follows:

The Natural Numbers

$$\{1, 2, 3, \ldots\}$$

The set of natural numbers together with the number 0 is called the set of **whole numbers.** We use the letter W to represent the whole numbers. The set of whole numbers is written as follows:

The Whole Numbers

$$\{0, 1, 2, 3, \ldots\}$$

The idea of a negative number is used to represent a loss or a debt. A debt of $10 would be expressed by the negative number -10. The whole numbers together with the negatives of the counting numbers form the set of **integers.** We use the letter I to represent the integers, and we write this set as follows:

The Integers

$$\{ \ldots, \ -3, \ -2, \ -1, \ 0, \ 1, \ 2, \ 3, \ \ldots \}$$

To represent ratios or quotients, we use **rational numbers.** For example, if 4 out of 5 dentists recommend sugarless gum, then we say that 4/5 of the dentists recommend sugarless gum. We use the letter Q to represent the set of rational numbers, and we write this set as follows:

The Rational Numbers

$$\{a/b \mid a \text{ and } b \text{ are integers, with } b \neq 0\}.$$

Examples of rational numbers are

$$7, \quad 9/4, \quad -\frac{17}{10}, \quad 0, \quad \frac{0}{4}, \quad 3/1, \quad -\frac{47}{3}, \quad \text{and} \quad -2/-6.$$

Note that the rational numbers are the numbers that can be expressed as a ratio (or quotient) of integers. The integer 7 is a rational number, because it can be expressed as $7/1$.

Another way to describe rational numbers is by using their decimal form. To obtain the decimal form, we divide the denominator into the numerator. For some rational numbers the division terminates, and for others it continues indefinitely. The following examples show some rational numbers and their equivalent decimal forms.

$$\frac{26}{100} = .26 \qquad \textbf{Terminating decimal}$$

$$\frac{4}{1} = 4.0 \qquad \textbf{Terminating decimal}$$

$$\frac{1}{4} = .25 \qquad \textbf{Terminating decimal}$$

$$\frac{2}{3} = .6666 \ldots \qquad \textbf{6 repeats.}$$

$$\frac{25}{99} = .252525 \ldots \qquad \textbf{The pair of digits 25 repeats.}$$

$$\frac{4177}{990} = 4.2191919 \ldots \qquad \textbf{The pair 19 repeats.}$$

The fact that the decimal numbers either repeat or terminate can be used to describe the rational numbers. The rational numbers are precisely those decimal numbers that either repeat or terminate.

EXAMPLE 1 Determine whether each statement is true or false.

a) $0 \in W$ **b)** $N \subseteq I$ **c)** $.75 \in I$ **d)** $I \subseteq Q$

Solution

a) True, because 0 is a whole number.
b) True, because every natural number is also a member of the set of integers.
c) False, because the rational number .75 is not an integer.
d) True, because the rational numbers include the integers. ◀

The Number Line

To construct a number line, we draw a straight line and label any convenient point with the number 0. Now we choose any convenient length and use it to locate points to the right of 0 as points corresponding to the positive integers, and points to the left of 0 as points corresponding to the negative integers (see Fig. 1.1). The numbers corresponding to the points on the line are called the **coordinates** of the points. The distance between two consecutive integers is called a **unit,** and is the same for any two consecutive integers. The point with coordinate 0 is called the **origin.** The numbers on the number line increase in size from left to right. When we compare the size of any two numbers, the larger number lies to the right of the smaller on the number line.

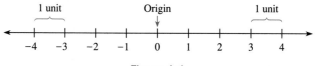

Figure 1.1

It is often convenient to illustrate sets of numbers on a number line. The set of integers, I, would be illustrated or graphed as in Fig. 1.2. The three dots to the right and left on the number line indicate that the numbers go on indefinitely in both directions.

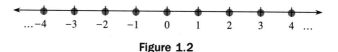

Figure 1.2

EXAMPLE 2 List the elements of each of the following sets and graph each set on a number line.

a) $\{x \mid x \text{ is a whole number less than 4}\}$
b) $\{a \mid a \text{ is an integer between 3 and 9}\}$
c) $\{y \mid y \text{ is an integer greater than } -3\}$

Solution

a) The whole numbers less than 4 are 0, 1, 2, and 3. This set is shown in Fig. 1.3.

Figure 1.3

b) The integers between 3 and 9 are 4, 5, 6, 7, and 8. The graph is shown in Fig. 1.4.

Figure 1.4

c) The integers greater than -3 are -2, -1, 0, 1, and so on. To indicate the continuing pattern we use a series of dots on the graph in Fig. 1.5.   ◄

Figure 1.5

For every rational number there is a point on the number line. For example, the number $1/2$ corresponds to a point halfway between 0 and 1 on the number line, and $3/2$ corresponds to a point halfway between 1 and 2 (see Fig. 1.6). However, there are also points on the number line that do not correspond to rational numbers. The set of numbers that corresponds to *all* points on a number line is called the set of **real numbers.** Those real numbers that are not rational are called **irrational.** We will let the letter R stand for the set of real numbers and M stand for the set of irrational numbers.

Figure 1.6

Irrational Numbers

Irrational numbers are those real numbers that cannot be expressed as a ratio of integers. This definition makes it difficult to identify irrational numbers. One source of irrational numbers is the idea of square root. The square root of 2, written $\sqrt{2}$, is a number that we multiply by itself to get 2. Thus

$$\sqrt{2} \cdot \sqrt{2} = 2.$$ The raised dot "$\cdot$" is used as a multiplication symbol.

The numbers that we deal with most often are rational numbers. Is there a rational number that we can multiply by itself to get 2? A calculator or the square root table of Appendix A will tell us that $\sqrt{2}$ is 1.414. This is a terminating decimal and is a rational number. If we multiply 1.414 by itself, we get

$$(1.414) \cdot (1.414) = 1.999396.$$

So 1.414 is not really the square root of 2; it is the square root of 1.999396. The number 1.414 is only an approximate value for $\sqrt{2}$. There is no terminating or repeating decimal that will give us exactly 2 when we multiply it by itself. It can be shown that other numbers involving square roots, such as $\sqrt{3}$, $\sqrt{5}$, and $\sqrt{7}$, cannot be expressed as a ratio of integers. (Note that a square root such as $\sqrt{4}$ is a rational number.)

Irrational numbers can also be described as those decimal numbers that neither repeat nor terminate. This description makes it easy to write down some irrational numbers in decimal form. The following decimal numbers are neither repeating nor terminating decimal numbers and therefore must be irrational.

$$.606000600000600000006 \ldots$$

$$.15115111511115 \ldots$$

$$3.12345678910111213 \ldots$$

Each of these numbers has a pattern, but it neither repeats nor terminates.

Note that the rational and the irrational numbers have no elements in common, and together they form the set of real numbers.

Because we generally work with rational numbers, the irrational numbers may seem to be unnecessary. However, irrational numbers occur in some very real situations. Over 2000 years ago, people in the Orient and Egypt observed that the ratio between the circumference and the diameter is the same for any circle. This constant value was proven to be an irrational number by Johann Heinrich Lambert in 1767. Like other irrational numbers, it does not have any convenient representation as a decimal number. This number has been given the name π (Greek letter pi) (see Fig. 1.7). The value of π rounded to 9 decimal places is 3.141592654. When using π in computations, we frequently use the rational number 3.14 as a approximate value for π.

$$\pi = \frac{\text{Circumference}}{\text{Diameter}} \qquad \pi = \frac{C}{D}$$

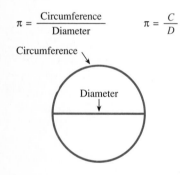

Figure 1.7

EXAMPLE 3 Determine which elements of the set

$$\{-\sqrt{7}, \ -1/4, \ 0, \ \sqrt{5}, \ \pi, \ 4.16, \ 12\}$$

are members of each of the following sets.

a) Real numbers   **b)** Rational numbers   **c)** Integers

Solution

a) All of the numbers in the set are real numbers.

b) The numbers $-1/4$, 0, 4.16, and 12 are rational numbers.

c) The only integers in this set are 0 and 12.   ◄

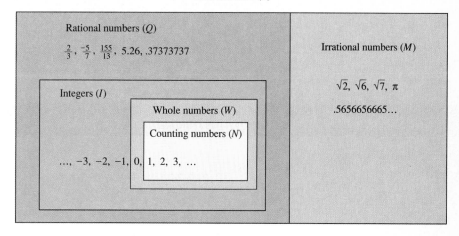

Figure 1.8

Summary

A summary of these important sets of numbers is given in Fig. 1.8.

EXAMPLE 4   Refer to Fig. 1.8 to determine whether each of the following statements is true or false.

a) $\sqrt{7} \in Q$ **b)** $I \subseteq W$ **c)** $M \cap Q = \varnothing$ **d)** $-3 \in N$
e) $I \cap M = \varnothing$ **f)** $Q \subseteq R$ **g)** $R \subseteq N$ **h)** $\pi \in R$

Solution

a) F   **b)** F   **c)** T   **d)** F
e) T   **f)** T   **g)** F   **h)** T   ◀

Warm-ups

True or false?

1. The number π is a rational number.
2. The set of rational numbers is a subset of the set of real numbers.
3. The number 0 is the only number that is in both the set of rational numbers and the set of irrational numbers.
4. The set of real numbers is a subset of the set of irrational numbers.
5. The decimal number .44444 . . . is a rational number.
6. The decimal number 4.212112111211112 . . . is a rational number.
7. The intersection of the set of rational numbers and the set of irrational numbers is the empty set.

8. The union of the set of irrational numbers with the set of rational numbers is the set of real numbers.

9. The set of counting numbers from 1 through 5 trillion is an infinite set.

10. There are infinitely many rational numbers.

1.2 EXERCISES

For this set of exercises,

N represents the natural numbers,
W the whole numbers,
I the integers,
Q the rational numbers,
M the irrational numbers, and
R the real numbers.

Determine whether each statement is true or false. See Example 1.

1. $-6 \in Q$
2. $2/7 \in Q$
3. $0 \notin Q$
4. $0 \notin N$
5. $.6666 \ldots \in Q$
6. $.00976 \notin Q$
7. $N \subseteq Q$
8. $Q \subseteq I$

List the elements in each set and graph each set on a number line. See Example 2.

9. $\{x \mid x \text{ is a whole number smaller than } 6\}$
10. $\{x \mid x \text{ is a natural number less than } 7\}$
11. $\{a \mid a \text{ is an integer greater than } -5\}$
12. $\{z \mid z \text{ is an integer between } 2 \text{ and } 12\}$
13. $\{w \mid w \text{ is a counting number between } 0 \text{ and } 5\}$
14. $\{y \mid y \text{ is a whole number greater than } 0\}$
15. $\{x \mid x \text{ is an integer between } -3 \text{ and } 5\}$
16. $\{y \mid y \text{ is an integer between } -4 \text{ and } 7\}$

Determine which elements of the set

$$A = \{-\sqrt{10}, -3, -5/2, -.025, 0, \sqrt{2}, 3\frac{1}{2}, 8/2\}$$

are members of the following sets. See Example 3.

17. Real numbers
18. Natural numbers
19. Whole numbers
20. Integers
21. Rational numbers
22. Irrational numbers

Determine whether each statement is true or false. See Example 4.

23. $Q \subseteq R$
24. $M \subseteq Q$
25. $M \cap Q = \{0\}$
26. $I \subseteq Q$
27. $M \cup Q = R$
28. $I \cap Q = \varnothing$
29. $.2121121112 \ldots \in Q$
30. $.3333 \ldots \in Q$

31. $3.252525 \ldots \in M$ **32.** $3.1010010001 \ldots \in M$ **33.** $.999 \ldots \in M$

34. $.666 \ldots \in Q$ **35.** $\pi \in M$ **36.** $\pi \in Q$

Place one of the symbols $\subseteq$, $\not\subseteq$, $\in$, *or* $\notin$ *in each blank to make the statements true.*

37. N _____ W **38.** I _____ Q **39.** I _____ N **40.** Q _____ W

41. Q _____ R **42.** M _____ R **43.** $\varnothing$ _____ M **44.** $\varnothing$ _____ Q

45. N _____ R **46.** W _____ R **47.** 5 _____ I **48.** -6 _____ I

49. 7 _____ Q **50.** 8 _____ Q **51.** $\sqrt{2}$ _____ R **52.** $\sqrt{2}$ _____ M

53. 0 _____ M **54.** 0 _____ Q **55.** $\{2, 3\}$ _____ Q **56.** $\{0, 1\}$ _____ N

57. $\{3, \sqrt{2}\}$ _____ R **58.** $\{3, \sqrt{2}\}$ _____ Q

1.3 Operations on the Set of Real Numbers

IN THIS SECTION:
- **Absolute Value**
- **Addition**
- **Subtraction**
- **Multiplication**
- **Division**
- **Division by Zero**

Computations in algebra are performed with positive and negative numbers. In this section we will extend the basic operations of arithmetic to the negative numbers.

Absolute Value

The real numbers are the coordinates of the points on the number line. However, to simplify matters we generally refer to the points as numbers. The numbers 5 and -5 are both 5 units away from 0 on the number line (see Fig. 1.9). A number's distance from 0 on the number line is called the **absolute value** of the number. We write $|a|$ for "the absolute value of a." Therefore $|5| = 5$ and $|-5| = 5$.

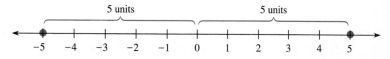

Figure 1.9

EXAMPLE 1 Determine the values of the following.

a) $|4|$ **b)** $|-4|$ **c)** $|0|$

Solution

a) $|4| = 4$, because 4 is 4 units away from 0.
b) $|-4| = 4$, because -4 is 4 units away from 0.
c) $|0| = 0$, because 0 is 0 units away from 0.   ◄

Note that $|a|$ represents distance, and distance is never negative. For any number a, $|a|$ is greater than or equal to zero.

Two numbers that are located on opposite sides of zero and have the same absolute value are called **opposites** of each other. The opposite of zero is zero. Every number has a unique opposite. The numbers 9 and -9 are opposites of one another. The minus sign, $-$, is used to signify "opposite" in addition to "negative." When the minus sign is used in front of a number, it is read as "negative." When it is used in front of parentheses or a variable, it is read as "opposite." For example,

$-(9) = -9$ means "the opposite of 9 is negative 9," and

$-(-9) = 9$ means "the opposite of negative 9 is 9."

In general, $-a$ means "the opposite of a." If a is positive, $-a$ is negative. If a is negative, $-a$ is positive. Opposites have the following property.

PROPERTY
Opposite of an Opposite

For any number a, $-(-a) = a$.

Remember that we have defined $|a|$ to be the distance between 0 and a on the number line. Using opposites, we can give a symbolic definition of absolute value.

DEFINITION
Absolute Value

$$|a| = \begin{cases} a & \text{if } a \text{ is positive or zero,} \\ -a & \text{if } a \text{ is negative.} \end{cases}$$

Using this definition, we would write

$$|7| = 7,$$

because 7 is positive. To find the absolute value of -7, we would use the second line of the definition and write

$$|-7| = -(-7) = 7.$$

Addition

A good way to understand positive and negative numbers is to think of the *positive numbers as assets* and the *negative numbers as debts*. For this illustration we can think of assets simply as cash. Think of debts as unpaid bills, such as the electric bill, the phone bill, etc. If you have debts of $6 and $7, then your total debt is $13. This can be expressed symbolically as

$$(-6) \quad + \quad (-7) \quad = \quad -13$$

$6 debt   plus $7 debt $13 debt

Think of this addition as adding the absolute values of -6 and -7 ($6 + 7 = 13$), and then putting a negative sign on that result to get -13. This illustrates the following rule.

RULE
Sum of Two Negative Numbers

> To find the sum of two negative numbers, add their absolute values and then put a negative sign on the result.

If you have a debt of $5 and have only $5 in cash, then your debts equal your assets (in absolute value), and your **net worth** is $0. Net worth is the total of debts and assets. Symbolically,

$$-5 + 5 = 0.$$

Debt of $5   Net worth
Asset of $5

For any number a, a and its opposite $-a$ have a sum of zero. For this reason, a and $-a$ are called **additive inverses** of one another. Note that the words "negative," "opposite," and "additive inverse" are often used interchangeably.

PROPERTY
Additive Inverse

> For any number a, $a + (-a) = 0$ and $(-a) + a = 0$.

To understand the sum of a positive and a negative number, consider the following situation. If you have a debt of $7 and $10 in cash, you may have $10 in hand, but your net worth is only $3. Your assets exceed your debts (in absolute value), and you have a positive net worth. In symbols,

$$-7 + 10 = 3.$$

Note that to get 3, we actually subtract 7 from 10.

If you have a debt of $8 but have only $5 in cash, then your debts exceed your assets (in absolute value). You have a net worth of $-\$3$. In symbols,

$$-8 + 5 = -3.$$

Note that to get the 3 in the answer we subtract 5 from 8.

As you can see from these examples, the sum of a positive number and a negative number (with different absolute values) may be either positive or negative. These examples help us to understand the rule for adding numbers with opposite signs and different absolute values.

Sum of Two Numbers with Opposite Signs (and Different Absolute Values)

If a and b have opposite signs, the sum of a and b is found by subtracting the absolute values of the numbers.

The answer is positive if the number with the larger absolute value is positive.

The answer is negative if the number with the larger absolute value is negative.

EXAMPLE 2 Evaluate the following sums.

a) $-6 + 13$ **b)** $-9 + (-7)$ **c)** $2 + (-2)$

d) $-5.4 + 2.3$ **e)** $-7 + .05$ **f)** $\dfrac{1}{2} + \left(-\dfrac{3}{4}\right)$

Solution

a) The absolute values of these numbers are 6 and 13. Subtract them to get 7. Because the number with the larger absolute value is 13 and it is positive, the result is 7.

b) -16

c) 0

d) Line up the decimal points and subtract 2.3 from 5.4. Because 5.4 is larger than 2.3, and 5.4 has a negative sign, the sign of the answer is negative.

$$-5.4 + 2.3 = -3.1$$

e) Line up the decimal points and subtract .05 from 7.00. Because 7.00 is larger than .05, and 7.00 has the negative sign, the sign of the answer is negative.

$$-7 + .05 = -6.95$$

f) $\dfrac{1}{2} + \left(-\dfrac{3}{4}\right) = \dfrac{2}{4} + \left(-\dfrac{3}{4}\right) = -\dfrac{1}{4}$ ◄

Subtraction

Think of subtraction as removing debts or assets, and addition as receiving debts or assets. For example, if you have $10 in cash and $4 is taken from you, your resulting net worth is the same as if you have $10 and a water bill for $4 arrives in the mail. In symbols,

$$10 - 4 = 10 + (-4).$$

Remove — Cash
Debt — Receive

Suppose you have $17 in cash, but owe $7 in library fines. Your net worth is $10. If the debt of $7 is cancelled or forgiven, your net worth will increase to $17, the same as if you received $7 in cash. In symbols,

$$10 - (-7) = 10 + 7.$$

Remove — Debt
Cash — Receive

Removing a debt is equivalent to receiving cash.

Notice that each subtraction problem above is equivalent to an addition problem where we add the opposite of what we were going to subtract. These examples illustrate the definition of subtraction.

DEFINITION
Subtraction of Real Numbers

For any real numbers a and b,

$$a - b = a + (-b).$$

EXAMPLE 3 Perform each subtraction.

a) $7 - 3$ b) $-7 - 3$ c) $-7 - (-3)$

d) $\dfrac{1}{3} - \left(-\dfrac{1}{6}\right)$ e) $-3.6 - (-7)$ f) $.02 - 7$

Solution To do *any* subtraction, we can change the operation to addition of the opposite. (Note that in part (a) it is certainly not necessary to do so.)

Subtract   Positive 3   Add   Negative 3

a) $7 \quad - \quad 3 \quad = \quad 7 \quad + \quad (-3) \quad = \quad 4$

b) $-7 - 3 = -7 + (-3) = -10$

Subtract   Negative 3   Add   Positive 3

c) $-7 \quad - \quad (-3) \quad = \quad -7 \quad + \quad 3 \quad = \quad -4$

d) $\dfrac{1}{3} - \left(-\dfrac{1}{6}\right) = \dfrac{2}{6} - \left(-\dfrac{1}{6}\right) = \dfrac{2}{6} + \dfrac{1}{6} = \dfrac{3}{6} = \dfrac{1}{2}$

e) $-3.6 - (-7) = -3.6 + 7 = 3.4$

f) $.02 - 7 = .02 + (-7) = -6.98$ ◀

Multiplication

The result of multiplying two numbers is referred to as the **product** of the numbers. The numbers multiplied are referred to as **factors.** In algebra we use a raised dot "·" between the factors to indicate multiplication, or we place symbols next to one another to indicate multiplication. Thus ab or $a \cdot b$ are both referred to as the product of a and b. When multiplying numbers, we may enclose them in parentheses to make the meaning clear. To write 4 times 3, we may write $4 \cdot 3$, $4(3)$, $(4)3$, or $(4)(3)$. In multiplying a number and a variable, no sign is used between them. Thus $4x$ is used to represent the product of 4 and x.

Multiplication is just a short way to do repeated additions. Adding 2's together five times gives

$$2 + 2 + 2 + 2 + 2 = 10.$$

So we have the multiplication fact $5 \cdot 2 = 10$. Adding together five negative twos gives

$$(-2) + (-2) + (-2) + (-2) + (-2) = -10.$$

So we must have $5(-2) = -10$. We can think of $5(-2) = -10$ as saying that taking on 5 debts of \$2 each is equivalent to a debt of \$10. Losing 5 debts of \$2 each is equivalent to gaining \$10, so we must have $-5(-2) = 10$.

The rules for multiplying signed numbers are easy to state and remember.

RULE
Product of Signed Numbers

The product of two numbers with the same sign is equal to the product of their absolute values.

The product of two numbers with opposite signs is a negative number, found by multiplying their absolute values and then putting a negative sign on the result.

For example, to multiply -4 and -5, we multiply their absolute values ($4 \cdot 5 = 20$). Thus $(-4)(-5) = 20$. To multiply -6 and 3, we multiply their absolute values ($6 \cdot 3 = 18$) and then put a negative sign on the result. Thus, $-6 \cdot 3 = -18$. This definition can be simply remembered as

$$\text{same sign} \longleftrightarrow \text{positive result,}$$

$$\text{opposite signs} \longleftrightarrow \text{negative result.}$$

EXAMPLE 4   Evaluate the following products.

a) $(-3)(-6)$ **b)** $(3)(5)$ **c)** $(-4)(10)$

d) $\left(\dfrac{2}{3}\right)\left(-\dfrac{1}{5}\right)$ **e)** $(-.01)(.02)$ **f)** $(-200)(-.03)$

Solution

a) First find the product of the absolute values.

$$|-3| \cdot |-6| = 18$$

Because -3 and -6 have the same sign, we get $(-3)(-6) = 18$.

b) 15 Same signs, positive result.

c) -40 Opposite signs, negative result.

d) $\left(\dfrac{2}{3}\right)\left(-\dfrac{1}{5}\right) = -\dfrac{2}{15}$

e) When multiplying decimals, we total the number of decimal places used in the numbers multiplied to get the number of decimal places in the answer. Thus, $(-.01)(.02) = -.0002$.

f) $(-200)(-.03) = 6$ ◀

Division

We say that $8 \div 2 = 4$ because $4 \cdot 2 = 8$. This illustrates how division can be defined in terms of multiplication.

DEFINITION
Division of Real Numbers

If a, b, and c are real numbers with $b \neq 0$, then

$$a \div b = c \text{ provided that } c \cdot b = a.$$

The number c in the above definition is called the **quotient** of a and b. We also refer to $a \div b$ and a/b as the quotient of a and b.

From this definition we get

$$8 \div (-2) = -4, \text{ because } (-4)(-2) = 8;$$
$$-8 \div 2 = -4, \text{ because } (-4)(2) = -8;$$

and

$$-8 \div (-2) = 4, \text{ because } (4)(-2) = -8.$$

These examples show us that the rules for dividing signed numbers are similar to those for multiplying signed numbers.

RULE
Division of Signed Numbers

> If a and b ($b \neq 0$) have the same sign, the quotient of a divided by b is the same as the quotient of their absolute values.
>
> To find the quotient a and b ($b \neq 0$) when they have opposite signs, find the quotient of their absolute values and put a negative sign on that result.

This rule can be simply remembered as

$$\text{same sign} \longleftrightarrow \text{positive result,}$$
$$\text{opposite signs} \longleftrightarrow \text{negative result.}$$

EXAMPLE 5   Evaluate.

a) $(-6) \div (-2)$ b) $(-6) \div 6$ c) $6 \div \left(-\dfrac{1}{3}\right)$

d) $(-6) \div (-.03)$ e) $(-1.5) \div (.03)$

Solution

a) $(-6) \div (-2) = 3$ b) $(-6) \div 6 = -1$

c) $6 \div \left(-\dfrac{1}{3}\right) = 6\left(-\dfrac{3}{1}\right) = -18$ d) $(-6) \div (-.03) = 200$

e) $(-1.5) \div (.03) = -50$ ◀

We use the same rules for division when division is indicated by a fraction bar. For example,

$$\frac{-6}{3} = -2, \qquad \frac{6}{-3} = -2, \qquad \frac{-1}{3} = \frac{1}{-3} = -\frac{1}{3}, \qquad \text{and} \qquad \frac{-6}{-3} = 2.$$

Note that if one negative sign appears in a fraction, the fraction has the same value whether the negative sign is in the numerator, in the denominator, or in front of the fraction. If the numerator and denominator of a fraction are both negative, then the fraction has a positive value.

Division by Zero

Why do we omit division by zero from the definition of division? If we write $10 \div 0 = c$, we need to find a number c such that $c \cdot 0 = 10$. This is impossible. If we write $0 \div 0 = c$, we need to find a number c such that $c \cdot 0 = 0$. Now, $c \cdot 0 = 0$ is true for any value of c. Having $0 \div 0$ equal to any number would be confusing in doing computations. Thus $a \div b$ is only defined for $b \neq 0$. Quotients such as

$$5 \div 0, \qquad 0 \div 0, \qquad \frac{7}{0}, \qquad \text{and} \qquad \frac{0}{0}$$

are said to be **undefined.**

Warm-ups

True or false?
1. The additive inverse of -6 is 6.
2. The opposite of negative 5 is positive 5.
3. The absolute value of 6 is -6.
4. The result of a subtracted from b is the same as b plus the opposite of a.
5. The product of a positive number and a negative number is always a negative number.
6. The sum of a positive number and a negative number is always a negative number.
7. $(-3)^2 = -6$
8. $6 \div (-1/2) = -3$
9. $5 - (-2) = 3$
10. $0 \div (-7) = 0$

1.3 EXERCISES

Determine the values of the following. See Example 1.

1. $|-34|$
2. $|17|$
3. $|34|$
4. $|-15|$
5. $|0|$
6. $|8| - |8|$
7. $-|-9|$
8. $-|-3|$
9. $|2| + |-2|$
10. $|-6| - |-6|$

Evaluate the following. See Example 2.

11. $-5 + 9$
12. $-3 + 10$
13. $-4 + (-3)$
14. $(-15) + (-11)$
15. $-6 + 4$
16. $5 + (-15)$
17. $7 + (-17)$
18. $-8 + 13$
19. $(-11) + (-15)$
20. $-18 + 18$
21. $18 + (-20)$
22. $7 + (-19)$
23. $-14 + 9$
24. $-6 + (-7)$
25. $-4 + 4 \quad 0$
26. $-7 + 9$
27. $-\dfrac{1}{10} + \dfrac{1}{5}$
28. $-\dfrac{1}{8} + \left(-\dfrac{1}{8}\right)$
29. $\dfrac{1}{2} + \left(-\dfrac{2}{3}\right)$
30. $-\dfrac{3}{4} + \dfrac{1}{2}$
31. $-15 + .02$
32. $.45 + (-1.3)$
33. $-2.7 + (-.01)$
34. $.8 + (-1)$

35. $47.39 + (-44.587)$

36. $-.65357 + (-2.375)$

37. $.2351 + (-.5)$

38. $-1.234 + (-4.756)$

Evaluate. See Example 3.

39. $7 - 10$

40. $8 - 19$

41. $-4 - 7$

42. $-5 - 12$

43. $7 - (-6)$

44. $3 - (-9)$

45. $-1 - 5$

46. $-4 - 6$

47. $-12 - (-3)$

48. $-15 - (-6)$

49. $20 - (-3)$

50. $50 - (-70)$

51. $\dfrac{9}{10} - \left(-\dfrac{1}{10}\right)$

52. $\dfrac{1}{8} - \dfrac{1}{4}$

53. $1 - \dfrac{3}{2}$

54. $-\dfrac{1}{2} - \left(-\dfrac{1}{3}\right)$

55. $2 - .03$

56. $.02 - 3$

57. $5.3 - (-2)$

58. $-4.1 - .13$

59. $-2.44 - 48.29$

60. $8.8 - 9.164$

61. $-3.89 - 5.16$

62. $0 - (-3.5)$

Evaluate. See Examples 4 and 5.

63. $(25)(-3)$

64. $(5)(-7)$

65. $(-3)(-8)$

66. $(-51) \div (-17)$

67. $-6 \div 3$

68. $84 \div (-2)$

69. $30 \div (-5)$

70. $(-9)(-6)$

71. $(.3)(-.3)$

72. $(-.1)(-.5)$

73. $(-.02)(-10)$

74. $(.05)(-2.5)$

75. $(-.8) \div (.1)$

76. $7 \div (-.5)$

77. $(-.1) \div (-.4)$

78. $(-18) \div (-.9)$

79. $9 \div \left(-\dfrac{1}{2}\right)$

80. $-\dfrac{1}{3} \div \left(-\dfrac{1}{2}\right)$

81. $\left(-\dfrac{2}{3}\right)\left(-\dfrac{9}{10}\right)$

82. $\left(\dfrac{1}{2}\right)\left(-\dfrac{2}{5}\right)$

83. $(.25)(-365)$

84. $7.5 \div (-.15)$

85. $(-51) \div (-.003)$

86. $(-2.8)(5.9)$

Perform the following computations.

87. $-6 + 2$

88. $-8 + 3$

89. $-3 - (-2)$

90. $-7 - (-1)$

91. $|-5|$

92. $-|-5|$

93. $(1/2)(-6)$

94. $(1/3)(-12)$

95. $\dfrac{1}{2} - \left(-\dfrac{1}{4}\right)$

96. $\dfrac{1}{8} - \left(-\dfrac{1}{4}\right)$

97. $\left(-\dfrac{1}{3}\right)(-9)$

98. $\left(-\dfrac{1}{3}\right)(-3)$

99. $-.2 + 1$

100. $-.6 + 2$

101. $(-.3) + (-.03)$

102. $-.02 + (-.5)$

103. $(-8) \div 2$

104. $0 \div (-3)$

105. $|-7| + |-3|$

106. $6 - |-12|$

107. $0 \div (-.5)$

108. $(-20) \div \left(-\dfrac{1}{2}\right)$

109. $2 \div (-.5)$

110. $3 \div (-.2)$

111. $-\dfrac{1}{3} + \dfrac{1}{6}$

112. $-\dfrac{2}{3} + \dfrac{1}{6}$

113. $-6 + |8|$

114. $|-3| - 2$

115. $-\dfrac{1}{2} + \left(-\dfrac{1}{2}\right)$

116. $\left(-\dfrac{2}{3}\right) + \left(-\dfrac{2}{3}\right)$

117. $-\dfrac{1}{2} - 1$

118. $-\dfrac{1}{3} - 2$

119. $2 - (.01)$

120. $5 - (.1)$

121. $-2 - (.3)$

122. $-.2 - .3$

123. $(-2)(.3)$

124. $(-3)(.1)$

125. $(-10)(-.2)$

126. $(-1/2)(-5)$

1.4 Evaluating Expressions

IN THIS SECTION:
- Exponential Expressions
- Arithmetic Expressions
- Algebraic Expressions

In algebra we will learn to deal with letters, numbers, formulas, and equations. We can become experts at manipulating symbols, but often there is nothing more important than a numerical answer. This section is concerned with finding values of expressions (computation).

Exponential Expressions

We use the notation of exponents to simplify the writing of repeated multiplication. For example, the product $2 \cdot 2 \cdot 2$ is written as 2^3. The product $5 \cdot 5$ is written as 5^2. The exponent indicates the number of times the factor occurs in the product.

DEFINITION
Exponents

For any counting number n,
$$a^n = \underbrace{a \cdot a \cdot a \cdot \ldots \cdot a}_{n \text{ factors of } a}.$$

We call a the **base,** n the **exponent,** and a^n an **exponential expression.**

We read a^n as "a to the nth power." For 3^5 and 10^6 we would say "3 to the 5th power" and "10 to the 6th power." We can also use the words "squared" and "cubed" for the second and third powers. For example, 5^2 and 2^3 would be read as "5 squared" and "2 cubed," respectively.

EXAMPLE 1 Evaluate the following exponential expressions.

a) 2^3 **b)** $(-2)^4$ **c)** $\left(-\dfrac{1}{2}\right)^5$ **d)** $(.5)^2$

Solution
a) $2^3 = 2 \cdot 2 \cdot 2 = 8$
b) $(-2)^4 = (-2)(-2)(-2)(-2) = 16$

c) $\left(-\dfrac{1}{2}\right)^5 = \left(-\dfrac{1}{2}\right)\left(-\dfrac{1}{2}\right)\left(-\dfrac{1}{2}\right)\left(-\dfrac{1}{2}\right)\left(-\dfrac{1}{2}\right) = -\dfrac{1}{32}$

d) $(.5)^2 = (.5)(.5) = .25$ ◀

Arithmetic Expressions

The result of writing numbers in meaningful combination with the ordinary operations of arithmetic is called an **arithmetic expression,** or simply an **expression.** Consider the expression

$$5 + (2 \cdot 3).$$

The parentheses in this expression are used as **grouping symbols.** The parentheses indicate which operation to perform first. To evaluate this expression, we find the product of 2 and 3, and then add that result to 5:

$$5 + (2 \cdot 3) = 5 + 6 = 11.$$

If we move the parentheses, we get a different value.

$$(5 + 2)3 = 7 \cdot 3 = 21$$

To simplify the writing of expressions, parentheses are often omitted. If we saw the expression

$$5 + 2 \cdot 3$$

written without parentheses, we would not know how to evaluate it unless we had some way to tell which operation to perform first. To evaluate such expressions consistently, there is an accepted **order of operations.** When no parentheses or absolute value symbols are present, the accepted order is:

RULE
Order of Operations

> **1.** Evaluate each exponential expression (in order from left to right).
> **2.** Perform multiplication and division (in order from left to right).
> **3.** Perform addition and subtraction (in order from left to right).

When parentheses or absolute value symbols are involved, we first evaluate expressions within each set of parentheses or absolute value symbols using the above order. When an expression involves a fraction bar, the numerator and denominator are each treated as if they are in parentheses. The use of the order of operations is illustrated in the following examples.

EXAMPLE 2   Evaluate each expression.

a) $5 + 2 \cdot 3$ **b)** $2^3 \cdot 3^2$ **c)** $(6 - 4^2)^2$

Solution

a) $5 + 2 \cdot 3 = 5 + 6$ Multiply first.

$\qquad\qquad = 11$ Then add.

b) $2^3 \cdot 3^2 = 8 \cdot 9$ Evaluate exponential expressions first.

$\qquad\qquad = 72$ Then multiply.

c) $(6 - 4^2)^2 = (6 - 16)^2$ Evaluate within parentheses first.

$\qquad\qquad\quad = (-10)^2$

$\qquad\qquad\quad = 100$ ◄

Consider the expression $-1 \cdot 2^4$. By the order of operations, the exponential expression is evaluated first. Thus

$$-1 \cdot 2^4 = -1 \cdot 16 = -16.$$

Math at Work

Everywhere you look, people are walking, running, jumping, and riding their way to fitness. Exercise has become the nation's number-one leisure activity, and Americans of all ages are joining in. Health professionals have found that exercise has the power to prolong your life, decrease your appetite, and improve the way you look and feel.

While many sports can help you stay fit, experts have found that seven activities provide the most fitness benefit: Aerobic exercising, calisthenics, cycling, exercise-walking, exercising with equipment, running, and swimming. Whatever activity you choose, athletic trainers recommend that you set realistic goals and work your way toward them slowly. Overexerting yourself at the start can cause pain and may discourage you from continuing your program. Stretch your muscles before each session to avoid injuries, and cool down slowly after exercising. To reach maximum health benefits, experts suggest that you exercise at least three times a week.

The best measure of your exercise program is your heart rate. After finding your pulse at your neck or wrist, count the number of pulsations for 10 seconds and multiply that number by 6. As you get into better condition, your heart rate after exercise will be closer to your heart rate at rest. To find your target heart rate for beneficial exercise, evaluate the formula below:

Target heart rate for men: $.65(220 - \text{Age})$

Target heart rate for women: $.65(225 - \text{Age})$

Now consider the expression -2^4. This expression is read as "the opposite of two to the fourth power." The negative sign in front of the 2 is understood to mean "the opposite of." Because taking an opposite and multiplying by -1 give the same result, taking an opposite is done in the same order as multiplication. The expression -2^4 has the same value as $-1 \cdot 2^4$. Thus

$$-2^4 = -16.$$

Note that if parentheses are used as in Example 1(b) above, we get

$$(-2)^4 = 16.$$

EXAMPLE 3   Evaluate each expression.

a) -5^2 b) $(-5)^2$ c) $-(3 - 5)^2$ d) $-(5^2 - 4 \cdot 7)^2$

Solution

a) $-5^2 = -25$ Square first, then take the opposite.
b) $(-5)^2 = (-5)(-5) = 25$
c) The negative sign in front of the parentheses indicates that we are to take an opposite. Thus the negative sign is given the same order in this computation as multiplication.

$$-(3 - 5)^2 = -(-2)^2 \quad \text{Evaluate within parentheses first.}$$
$$= -4 \quad \text{Square } -2 \text{ to get 4, then take the opposite of 4 to get } -4.$$

d) $-(5^2 - 4 \cdot 7)^2 = -(25 - 28)^2$ Evaluate within parentheses first.
$$= -(-3)^2$$
$$= -9 \quad \text{Square } -3 \text{ to get 9, then take the opposite of 9 to get } -9.$$ ◄

The next example illustrates how the fraction bar groups the numerator and denominator.

EXAMPLE 4   Evaluate each expression.

a) $\dfrac{10 - 8}{6 - 8}$ b) $\dfrac{-6^2 + 2 \cdot 7}{4 - 3 \cdot 2}$

Solution

a) $\dfrac{10 - 8}{6 - 8} = \dfrac{2}{-2}$ Evaluate numerator and denominator first.
$$= -1 \quad \text{Then divide.}$$
b) $\dfrac{-6^2 + 2 \cdot 7}{4 - 3 \cdot 2} = \dfrac{-36 + 14}{4 - 6}$ Evaluate numerator and denominator first.
$$= \dfrac{-22}{-2}$$
$$= 11$$ ◄

The following example shows how grouping symbols can occur within grouping symbols. When this happens, we evaluate within the innermost grouping symbol first. Brackets, [ ], may be used as parentheses when grouping occurs within grouping.

EXAMPLE 5 Evaluate each expression.

a) $2 - 3[2 - (3 - 5)]$ **b)** $7^2 - 2|3 - (9 - 5)|$

Solution

a) $2 - 3[2 - (3 - 5)] = 2 - 3[2 - (-2)]$ Evaluate within innermost
                                                     parentheses first.

$$= 2 - 3[4]$$
$$= 2 - 12 \quad \text{Multiply.}$$
$$= -10 \quad \text{Subtract.}$$

b) $7^2 - 2|3 - (9 - 5)| = 7^2 - 2|3 - (4)|$ First evaluate within the absolute value.

$$= 7^2 - 2|-1|$$
$$= 7^2 - 2 \quad |-1| = 1$$
$$= 49 - 2$$
$$= 47$$ ◀

Algebraic Expressions

The result of combining numbers and variables with the ordinary operations of arithmetic (in some meaningful way) is called an **algebraic expression.** For example,

$$2x - 5y, \qquad -5x^2, \qquad (x - 3)(x + 2), \qquad b^2 - 4ac, \qquad 5, \qquad \text{and} \qquad \frac{x}{2}$$

are algebraic expressions, or simply **expressions.**

An expression such as $2x - 5y$ has no definite value unless we assign values to x and y. For example, if $x = 3$ and $y = 4$, then the value of $2x - 5y$ is found by replacing x with 3 and y with 4 and evaluating:

$$2x - 5y = 2(3) - 5(4) = 6 - 20 = -14.$$

Note the importance of the order of operations in evaluating an algebraic expression.

To find the value of $2x - 5y$ when $x = -2$ and $y = -3$, replace x and y by -2 and -3 respectively, and then evaluate.

$$2x - 5y = 2(-2) - 5(-3) = -4 - (-15) = -4 + 15 = 11$$

EXAMPLE 6 If $a = 2$, $b = -3$, and $c = 4$, evaluate each expression.

a) $a - c^2$ **b)** $a - b^2$ **c)** $b^2 - 4ac$ **d)** $\dfrac{a - b}{c - b}$

Solution

a) $a - c^2 = 2 - 4^2 = 2 - 16 = -14$
b) $a - b^2 = 2 - (-3)^2 = 2 - 9 = -7$
c) $b^2 - 4ac = (-3)^2 - 4(2)(4) = 9 - 32 = -23$
d) $\dfrac{a - b}{c - b} = \dfrac{2 - (-3)}{4 - (-3)} = \dfrac{5}{7}$ ◄

A symbol such as y_1 is treated like any other variable. We read y_1 as "y one" or "y sub one." The 1 is called a **subscript.** We can think of y_1 as the "first y" and y_2 as the "second y." The subscript notation is used in the following example.

EXAMPLE 7   Let $y_1 = -12$, $y_2 = -5$, $x_1 = -3$, and $x_2 = 4$. Find the value of

$$\frac{y_1 - y_2}{x_1 - x_2}.$$

Solution   Substitute the appropriate values into the expression.

$$\frac{y_1 - y_2}{x_1 - x_2} = \frac{-12 - (-5)}{-3 - 4} = \frac{-7}{-7} = 1$$ ◄

Warm-ups

True or false?

1. $2^3 = 6$
2. $-1 \cdot 2^2 = -4$
3. $-2^2 = -4$
4. $6 + 3 \cdot 2 = 18$
5. $(6 + 3) \cdot 2 = 81$
6. $(6 + 3)^2 = 18$
7. $6 + 3^2 = 15$
8. $(-3)^3 = -3^3$
9. $|-3 - (-2)| = 5$
10. $|7 - 8| = |7| - |8|$

1.4 EXERCISES

Evaluate each exponential expression. See Example 1.

1. 2^5
2. 3^4
3. $(-1)^4$
4. $(-1)^5$
5. $(-4)^2$
6. $\left(-\dfrac{1}{2}\right)^6$
7. $\left(-\dfrac{3}{4}\right)^3$
8. $\left(-\dfrac{1}{2}\right)^7$
9. $(.1)^3$
10. $(.1)^4$

Evaluate each expression. See Examples 2 and 3.

11. $4 - 6 \cdot 2$
12. $8 - 3 \cdot 9$
13. $5 - 6(3 - 5)$
14. $8 - 3(4 - 6)$
15. $(5 - 6)(3 - 5)$
16. $(8 - 3)(4 - 6)$
17. $-7^2 + 9$
18. $-6^2 + (-3)^3$
19. $-(2 + 3)^2$
20. $-(1 - 3 \cdot 2)^3$

21. $-5^2 \cdot 2^3$

22. $8 - 2^3$

23. $5 \cdot 2^3$

24. $(8 - 2)^3$

25. $-(3^2 - 4)^2$

26. $-(6 - 2^3)^4$

Evaluate each expression. See Example 4.

27. $\dfrac{2 - 6}{9 - 7}$

28. $\dfrac{9 - 12}{4 - 5}$

29. $\dfrac{-3 - 5}{6 - (-2)}$

30. $\dfrac{-14 - (-2)}{-3 - 3}$

31. $\dfrac{4 + 2 \cdot 7}{3 \cdot 2 - 9}$

32. $\dfrac{11 - 2(-3)}{8 - 3(-3)}$

33. $\dfrac{-3^2 - 5}{2 - 3^2}$

34. $\dfrac{-2^4 - 5}{3^2 - 2^4}$

Evaluate each expression. See Example 5.

35. $3 - 7[4 - (2 - 5)]$

36. $9 - 2[3 - (4 + 6)]$

37. $3 - 4(2 - |4 - 6|)$

38. $3 - (|-4| - |-5|)$

39. $4[2 - (5 - |-3|)^2]$

40. $[5 - (-3)]^2 + [4 - (-2)]^2$

Evaluate each expression for $a = -1$, $b = 3$, and $c = -4$. See Example 6.

41. $b^2 - 4ac$

42. $a^2 - 4bc$

43. $\dfrac{a - b}{a - c}$

44. $\dfrac{b - c}{b - a}$

45. $(a - b)(a + b)$

46. $(a - c)(a + c)$

47. $c^2 - 2c + 1$

48. $b^2 - 2b - 3$

49. $\dfrac{2}{a} + \dfrac{b}{c} - \dfrac{1}{c}$

50. $\dfrac{c}{a} + \dfrac{c}{b} - \dfrac{a}{b}$

51. $|a - b|$

52. $|b + c|$

Find the value of $\dfrac{y_1 - y_2}{x_1 - x_2}$ for each choice of y_1, y_2, x_1, and x_2. See Example 7.

53. $y_1 = 2$, $y_2 = -3$, $x_1 = 5$, $x_2 = -3$

54. $y_1 = 4$, $y_2 = -6$, $x_1 = 2$, $x_2 = -7$

55. $y_1 = -1$, $y_2 = 2$, $x_1 = -3$, $x_2 = 1$

56. $y_1 = -3$, $y_2 = -3$, $x_1 = 4$, $x_2 = -5$

57. $y_1 = -6$, $y_2 = -6$, $x_1 = 3$, $x_2 = -5$

58. $y_1 = -2$, $y_2 = 5$, $x_1 = 2$, $x_2 = 6$

59. $y_1 = 2.4$, $y_2 = 5.6$, $x_1 = 5.9$, $x_2 = 4.7$

60. $y_1 = -5.7$, $y_2 = 6.9$, $x_1 = 3.5$, $x_2 = 4.2$

Find the value of $b^2 - 4ac$ for each of the following choices of a, b, and c.

61. $a = -1$, $b = -2$, $c = 3$

62. $a = 1$, $b = 2$, $c = 3$

63. $a = 2$, $b = 3$, $c = -5$

64. $a = 2$, $b = -3$, $c = -4$

65. $a = 4$, $b = 2$, $c = 3$

66. $a = 3$, $b = 1$, $c = 2$

67. $a = 1$, $b = -3$, $c = -5$

68. $a = 1$, $b = -3$, $c = -4$

69. $a = -3$, $b = 4$, $c = -2$

70. $a = -3$, $b = 5$, $c = -6$

71. $a = .5$, $b = 8$, $c = 2$

72. $a = .2$, $b = 3$, $c = 5$

73. $a = 4.2$, $b = 6.7$, $c = 1.8$

74. $a = -3.5$, $b = 9.1$, $c = 3.6$

75. $a = -1.2$, $b = 3.2$, $c = 5.6$

76. $a = 2.4$, $b = -8.5$, $c = -5.8$

Evaluate the following expressions.

77. $-2^2 + 5(3)^2$

78. $-3^2 + 3(6)^2$

79. $(-2 + 5)3^2$

80. $(-3 + 3)6^2$

81. $3 + 2^3$

82. $5 - 2^3$

83. $(3 + 2)^3$

84. $(5 - 2)^3$

85. $-\dfrac{6}{3} + 2$

86. $-\dfrac{10}{2} + 5$

87. $(-2)^2 - 4(-1)(5)$

88. $(-3)^2 - 4(2)(-3)$

89. $3^2 - 4(2)(-3)$

90. $5^2 - 4(-1)(6)$

91. $(-12)^2 - 4(6)(0)$

92. $(-11)^2 - 4(7)(0)$

93. $5^2 - 4(1)(6)$

94. $6^2 - 4(2)(3)$

95. $[13 + 2(-5)]^2$

96. $[6 + 2(-4)]^2$

97. $\dfrac{4 - (-1)}{-3 - 2}$

98. $\dfrac{2 - (-3)}{3 - 5}$

99. $3(-2)^2 - 5(-2) + 4$

100. $3(-1)^2 + 5(-1) - 6$

101. $-4\left(\dfrac{1}{2}\right)^2 + 3\left(\dfrac{1}{2}\right) - 2$

102. $8\left(\dfrac{1}{2}\right)^2 - 6\left(\dfrac{1}{2}\right) + 1$

103. $4 - 3|2 - 9|$

104. $3 - 5|7 - 9|$

105. $-\dfrac{1}{2}|6 - 2|$

106. $-\dfrac{1}{3}|9 - 6|$

107. $|6 - 3 \cdot 7| + |7 - 5|$

108. $|12 - 4| - |3 - 4 \cdot 5|$

1.5 Properties

IN THIS SECTION:

- The Commutative Properties
- The Associative Properties
- The Distributive Properties
- Identity Properties
- Inverse Properties
- Multiplication Property of Zero

You know that the price of a hamburger plus the price of a Coke is the same as the price of a Coke plus the price of a hamburger. But do you know that this illustrates the commutative property of addition? In arithmetic we may be unaware that we are using the properties, but in algebra we need a better understanding of the properties. In this section we will study the properties of the basic operations on the set of real numbers.

The Commutative Properties

We get the same result whether we evaluate $3 + 7$ or $7 + 3$. This fact illustrates the commutative property of addition. The fact that $4 \cdot 5$ and $5 \cdot 4$ are both 20 is an illustration of the commutative property of multiplication.

Commutative Properties

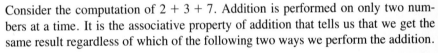

For any real numbers a and b,

$$a + b = b + a$$

and

$$a \cdot b = b \cdot a.$$

When writing the product of a number and a variable, it is customary to write the number first. We write $3x$ rather than $x3$. When writing the product of two variables, it is customary to write them in alphabetical order. We write cd rather than dc.

Addition and multiplication are commutative operations, but what about subtraction and division? Because $7 - 3 = 4$ and $3 - 7 = -4$, subtraction is not commutative. To see that division is not commutative, consider dividing $8 among 4 people and $4 among 8 people.

The Associative Properties

Consider the computation of $2 + 3 + 7$. Addition is performed on only two numbers at a time. It is the associative property of addition that tells us that we get the same result regardless of which of the following two ways we perform the addition.

$$(2 + 3) + 7 = 5 + 7 = 12$$
$$2 + (3 + 7) = 2 + 10 = 12$$

This is why we can leave out the parentheses when writing $2 + 3 + 7$. The associative and commutative properties of addition also allow us to place our groceries on the checkout counter in any order and get the same total.

For multiplication we also have an associative property. Consider the following.

$$(2 \cdot 3) \cdot 5 = 6 \cdot 5 = 30$$
$$2 \cdot (3 \cdot 5) = 2 \cdot 15 = 30$$

We get the same result for either order of multiplication. This property allows us to write $2 \cdot 3 \cdot 5$ without parentheses.

Associative Properties

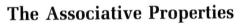

For any real numbers a, b, and c,

$$(a + b) + c = a + (b + c)$$

and

$$(ab)c = a(bc).$$

Consider the expression

$$4 - 9 + 8 - 5 - 8 + 6 - 13.$$

According to the accepted order of operations, we could evaluate this by computing from left to right. However, if we use the definition of subtraction, we can rewrite this expression as

$$4 + (-9) + 8 + (-5) + (-8) + 6 + (-13).$$

The commutative and associative properties of addition allow us to add these in any order we choose. A good way to add these is to add the positive numbers, add the negative numbers, and then combine those two totals.

$$4 + 8 + 6 + (-9) + (-5) + (-8) + (-13) = 18 + (-35) = -17$$

There is no need to rewrite this expression as we have done here. We can sum the positive numbers and the negative numbers from the original expression, and then combine their totals.

EXAMPLE 1   Evaluate.

a) $4 - 7 + 10 - 5$ **b)** $6 - 5 - 9 + 7 - 2 + 5 - 8$

Solution

a) $4 - 7 + 10 - 5 = 14 + (-12) = 2$

Sum of positive numbers┘   └Sum of negative numbers

b) $6 - 5 - 9 + 7 - 2 + 5 - 8 = 18 + (-24) = -6$ ◄

Note that subtraction and division are not associative operations. For example,

$$(8 - 4) - 1 \neq 8 - (4 - 1)$$ Evaluate each side.

and

$$(8 \div 4) \div 2 \neq 8 \div (4 \div 2).$$

The Distributive Properties

Consider the following.

$$3(4 + 6) = 3 \cdot 10 = 30$$
$$3 \cdot 4 + 3 \cdot 6 = 12 + 18 = 30$$

Therefore

$$3(4 + 6) = 3 \cdot 4 + 3 \cdot 6.$$

We say that multiplication is **distributed** over the addition. It is the distributive property that gives us the same amount of tax whether we compute the tax on each item or total the items and then multiply by the tax rate.

Consider the following expressions involving multiplication and subtraction.

$$3(6 - 4) = 3 \cdot 2 = 6$$
$$3 \cdot 6 - 3 \cdot 4 = 18 - 12 = 6$$

Therefore

$$3(6 - 4) = 3 \cdot 6 - 3 \cdot 4.$$

Notice that the multiplication by 3 is distributed over each term in the parentheses. These examples illustrate the distributive properties.

Distributive Properties

For any real numbers a, b, and c,

$$a(b + c) = ab + ac$$

and

$$a(b - c) = ab - ac.$$

If we start with $5(x + 4)$ and write

$$5(x + 4) = 5x + 20,$$

we are using the distributive property to remove the parentheses. We are multiplying 5 and $x + 4$ to get $5x + 20$.

If we start with $5x - 15$ and write

$$5x - 15 = 5(x - 3),$$

we are using the distributive property to **factor** $5x - 15$. We are factoring out the common factor 5.

EXAMPLE 2   Use the distributive property to rewrite each of the following. Either factor out the common factor or multiply, whichever is appropriate.

a) $3x - 21$ **b)** $b(2 - a)$ **c)** $3a + ac$ **d)** $-2(x - 3)$

Solution

a) Factor: $3x - 21 = 3(x - 7)$
b) Multiply: $b(2 - a) = 2b - ab$ Note that $b \cdot 2 = 2b$ by the commutative property.
c) Factor: $3a + ac = a(3 + c)$
d) Multiply: $-2(x - 3) = -2x - (-2)(3)$
$$= -2x - (-6)$$
$$= -2x + 6$$

Identity Properties

The numbers 0 and 1 have special properties. Multiplication of a number by 1 does not change the number, and addition of 0 to a number does not change the number. For this reason, 1 is called the **multiplicative identity,** and 0 is called the **additive identity.**

Identity Properties

For any real number a,

$$a \cdot 1 = 1 \cdot a = a$$

and

$$a + 0 = 0 + a = a.$$

Inverse Properties

The idea of additive inverses was introduced in Section 1.3. Every real number a has a unique additive inverse or opposite $-a$ such that $a + (-a) = 0$. Every nonzero real number a also has a unique **multiplicative inverse,** written $1/a$, such that $a(1/a) = 1$. For rational numbers, the multiplicative inverse is easy to find. For example, the multiplicative inverse of $2/5$ is $5/2$, because

$$\frac{2}{5} \cdot \frac{5}{2} = \frac{10}{10} = 1.$$

Inverse Properties

For any real number a, there is a unique number $-a$, such that

$$a + (-a) = 0.$$

For any nonzero real number a, there is a unique number $1/a$ such that

$$a \cdot (1/a) = 1.$$

EXAMPLE 3   Find the multiplicative inverse of each of the following numbers.

a) 4   **b)** .7   **c)** $-\dfrac{3}{5}$ **d)** 1.9

Solution

a) The multiplicative inverse of 4 is $1/4$, because

$$4\left(\frac{1}{4}\right) = 1.$$

b) To find the inverse of .7 we must write it as a ratio of integers:

$$.7 = \frac{7}{10}.$$

The multiplicative inverse of .7 is 10/7, because

$$\frac{7}{10} \cdot \frac{10}{7} = \frac{70}{70} = 1.$$

c) The multiplicative inverse of $-3/5$ is $-5/3$, because

$$\left(-\frac{3}{5}\right)\left(-\frac{5}{3}\right) = 1.$$

d) First convert 1.9 to a ratio of integers:

$$1.9 = 1\frac{9}{10} = \frac{19}{10}.$$

Thus, the multiplicative inverse is 10/19. ◀

Multiplication Property of Zero

Zero has a property that no other number has. Multiplication involving zero always results in zero.

Multiplication Property of Zero

For any real number a, $0 \cdot a = 0$ and $a \cdot 0 = 0$.

EXAMPLE 4 Identify the property that justifies each equality.

a) $5 \cdot 9 = 9 \cdot 5$ **b)** $(3)(1/3) = 1$ **c)** $1 \cdot 865 = 865$
d) $3 + (5 + a) = (3 + 5) + a$ **e)** $4x + 6x = (4 + 6)x$
f) $7 + (x + 3) = 7 + (3 + x)$ **g)** $4567 \cdot 0 = 0$
h) $239 + 0 = 239$ **i)** $-8 + 8 = 0$
j) $-4(x - 5) = -4x + 20$

Solution

a) Commutative **b)** Inverse **c)** Identity
d) Associative **e)** Distributive **f)** Commutative
g) Multiplication property of zero **h)** Identity
i) Inverse **j)** Distributive ◀

Warm-ups

True or false?

1. Addition is a commutative operation.
2. $8 \div (4 \div 2) = (8 \div 4) \div 2$
3. $10 \div 2 = 2 \div 10$
4. $5 - 3 = 3 - 5$
5. $10 - (7 - 3) = (10 - 7) - 3$
6. $4(6 \div 2) = (4 \cdot 6) \div (4 \cdot 2)$
7. The multiplicative inverse of .02 is 50.
8. Division is not an associative operation.
9. $3 + 2x = 5x$ for any value of x.
10. $(2x - 3) - (5x - 2) = -3x - 1$ for any value of x.

1.5 EXERCISES

Evaluate by first finding the sum of the positive numbers and then the sum of the negative numbers. See Example 1.

1. $9 - 4 + 6 - 10$
2. $-3 + 4 - 12 + 9$
3. $6 - 10 + 5 - 8 - 7$
4. $5 - 11 + 6 - 9 + 12 - 2$
5. $-4 - 11 + 6 - 8 + 13 - 20$
6. $-8 + 12 - 9 - 15 + 6 - 22 + 3$
7. $-3.2 + 1.4 - 2.8 + 4.5 - 1.6$
8. $4.4 - 5.1 + 3.6 - 2.3 + 8.1$
9. $3.27 - 11.41 + 5.7 - 12.36 - 5$
10. $4.89 - 2.1 + 7.58 - 9.06 - 5.34$

Use the distributive property to either multiply or factor, whichever is appropriate. See Example 2.

11. $4(x - 6)$
12. $5(a - 1)$
13. $2m + 10$
14. $3y + 9$
15. $a(3 + t)$
16. $b(y + w)$
17. $-2(w - 5)$
18. $-4(m - 7)$
19. $-2(3 - y)$
20. $-5(4 - p)$
21. $5x - 5$
22. $3y + 3$
23. $-1(x - 3)$
24. $-1(y - 9)$
25. $-1(w + 5)$
26. $-1(x + 6)$
27. $3y - 15$
28. $5x + 10$
29. $3a + 9$
30. $7b - 49$
31. $\frac{1}{2}(4x + 8)$
32. $\frac{1}{3}(3x + 6)$
33. $-\frac{1}{2}(2x - 4)$
34. $-\frac{1}{3}(9x - 3)$

Give the multiplicative inverse (reciprocal) of each of the following numbers. See Example 3.

35. $\frac{1}{2}$
36. $\frac{1}{3}$
37. -3
38. -5
39. 6
40. 8

41. .25   **42.** .75   **43.** .7   **44.** −.9   **45.** 1.5   **46.** 2.5

47. 1   **48.** −1

Name the property that justifies each equality. See Example 4.

49. $3 + x = x + 3$ **50.** $x \cdot 5 = 5x$ **51.** $5(x - 7) = 5x - 35$

52. $a(3b) = (a \cdot 3)b$ **53.** $3(xy) = (3x)y$ **54.** $3(x - 1) = 3x - 3$

55. $4(.25) = 1$ **56.** $.3 + 9 = 9 + .3$ **57.** $y^3x = xy^3$

58. $0 \cdot 52 = 0$ **59.** $1 \cdot x = x$

60. $(.1)(10) = 1$ **61.** $2x + 3x = (2 + 3)x$

62. $8 + 0 = 8$ **63.** $7 + 0 = 7$

64. $1 \cdot y = y$ **65.** $(36 + 79)0 = 0$

66. $5x + 5 = 5(x + 1)$ **67.** $xy + x = x(y + 1)$

68. $ab + 3ac = a(b + 3c)$

Complete each statement using the property named.

69. $5 + w = \underline{\qquad}$, commutative   **70.** $2x + 2 = \underline{\qquad}$, distributive

71. $5(xy) = \underline{\qquad}$, associative   **72.** $x + \dfrac{1}{2} = \underline{\qquad}$, commutative

73. $\dfrac{1}{2}x - \dfrac{1}{2} = \underline{\qquad}$, distributive   **74.** $3(x - 7) = \underline{\qquad}$, distributive

75. $6x + 9 = \underline{\qquad}$, distributive   **76.** $(x + 7) + 3 = \underline{\qquad}$, associative

77. $8(.125) = \underline{\qquad}$, inverse property   **78.** $-1(a - 3) = \underline{\qquad}$, distributive

79. $0 = 5(\underline{\qquad})$, multiplication property of zero   **80.** $8 \cdot (\underline{\qquad}) = 8$, identity property

81. $.25(\underline{\qquad}) = 1$, inverse property   **82.** $45(1) = \underline{\qquad}$, identity property

1.6 Using the Properties

IN THIS SECTION:

- Using the Properties in Computation
- Like Terms
- Combining Like Terms
- Products and Quotients
- Removing Parentheses

The properties of the real numbers can be helpful when we are doing computations. In this section we will see how the properties can be applied in arithmetic and algebra.

Using the Properties in Computation

Consider how to find the product of 36 and 200:

$$
\begin{array}{r}
200 \\
\times\ \ 36 \\
\hline
1200 \\
600 \\
\hline
7200
\end{array}
$$

Using the associative property, we can write

$$(36)(200) = (36 \cdot 2)(100).$$

To find this product mentally, first multiply 36 by 2 to get 72, then multiply 72 by 100 to get 7200.

We can use the commutative property to simplify the addition of a negative and a positive number, such as $-9 + 25$. By the commutative property,

$$-9 + 25 = 25 + (-9) = 25 - 9.$$

The commutative property has allowed us to see $-9 + 25$ as a simple subtraction problem, $25 - 9$. Thus

$$-9 + 25 = 25 - 9 = 16.$$

EXAMPLE 1   Make use of the appropriate property to perform these computations mentally.

a) $536 + 25 + 75$ **b)** $5 \cdot 426 \cdot \dfrac{1}{5}$

c) $6 \cdot 29 + 4 \cdot 29$ **d)** $-18 + 99$

Solution

a) To perform this addition mentally, the associative property can be applied as follows.

$$536 + (25 + 75) = 536 + 100 = 636$$

b) Use the commutative and associative properties to mentally rearrange this product.

$$
\begin{aligned}
5 \cdot 426 \cdot \frac{1}{5} &= 426\left(5 \cdot \frac{1}{5}\right) \\
&= 426 \cdot 1 \qquad \text{\small Inverse property} \\
&= 426
\end{aligned}
$$

c) We can use the distributive property to rewrite this problem and do it mentally.

$$6 \cdot 29 + 4 \cdot 29 = (6 + 4)29 = 10 \cdot 29 = 290$$

d) Use the commutative property to rewrite this problem as a simple subtraction.

$$-18 + 99 = 99 - 18 = 81 \qquad \blacktriangleleft$$

Like Terms

The properties of the real numbers are also used with algebraic expressions. An expression containing a number or the product of a number and one or more variables is called a **term.** For example,

$$-2, \qquad 4x, \qquad -5x^2y, \qquad b, \qquad \text{and} \qquad -abc$$

are terms. The number preceding the variables in a term is called the **coefficient.** In the term $4x$, the coefficient of x is 4. In the term $-5x^2y$, the coefficient of x^2y is -5. In the term b, the coefficient of b is 1, and in the term $-abc$ the coefficient of abc is -1. If two terms contain the same variables with the same powers, they are called **like terms.** For example, $3x^2$ and $-5x^2$ are like terms.

Combining Like Terms

Using the distributive property on an expression involving the sum of like terms allows us to combine them. For example,

$$2x + 5x = (2 + 5)x \qquad \text{Distributive property}$$
$$= 7x \qquad \text{Add 2 and 5.}$$

The distributive property says that $a(b + c) = ab + ac$ for any values of a, b, and c. In using the distributive property above, x could be any number. Thus $2x + 5x = 7x$ no matter what number is used for x.

We can also use the distributive property on a difference of like terms to combine them. For example,

$$-3xy - (-2xy) = [-3 - (-2)]xy \qquad \text{Distributive property}$$
$$= -1xy \qquad \text{Subtract.}$$
$$= -xy \qquad \text{Multiplying by } -1 \text{ is the same as taking the opposite.}$$

Of course, we do not want to write out these steps every time we combine like terms. We can combine like terms as easily as we can add or subtract their coefficients.

EXAMPLE 2 Combine the like terms.

a) $b + 3b$ $\qquad\qquad$ **b)** $5x^2 + 7x^2$ $\qquad\qquad$ **c)** $-3x + 8x$

d) $5xy - (-13xy)$ $\qquad$ **e)** $-2a + (-9a)$

Solution

a) $b + 3b = 1b + 3b = 4b$

b) $5x^2 + 7x^2 = 12x^2$

c) $-3x + 8x = 5x$

d) $5xy - (-13xy) = 18xy$

e) $-2a + (-9a) = -11a$ ◄

Expressions such as

$$3x + 5, \qquad 7xy + 9x, \qquad 5b + 6a, \qquad \text{and} \qquad 6x^2 + 7x$$

do not involve like terms, so their terms cannot be combined.

Products and Quotients

We can use the associative property of multiplication to simplify the product of two terms. For example,

$$4(7x) = (4 \cdot 7)x \qquad \text{Associative property}$$
$$= (28)x$$
$$= 28x \qquad \text{Remove unnecessary parentheses.}$$

Notice the difference between $4(7x)$ and $4(7 + x)$. In $4(7 + x)$ the distributive property is used to distribute the 4 over the addition. The 7 and the x are both multiplied by the 4.

$$4(7 + x) = 28 + 4x$$

In the next example, we use the fact that dividing by 3 is equivalent to multiplying by $1/3$, the reciprocal of 3.

$$3\left(\frac{x}{3}\right) = 3\left(\frac{1}{3} \cdot x\right)$$
$$= \left(3 \cdot \frac{1}{3}\right)x \qquad \text{Associative property}$$
$$= 1 \cdot x \qquad 3 \cdot \frac{1}{3} = 1$$
$$= x \qquad \text{The multiplicative identity is 1.}$$

To find the product $(3x)(5x)$, we use both the commutative and associative properties. For example,

$$(3x)(5x) = (3x \cdot 5)x \qquad \text{Associative property}$$
$$= (3 \cdot 5x)x \qquad \text{Commutative property}$$
$$= (3 \cdot 5)(x \cdot x) \qquad \text{Associative property}$$
$$= (15)(x^2) \qquad \text{Simplify.}$$
$$= 15x^2 \qquad \text{Remove unnecessary parentheses.}$$

We do not want to spend time going through all of these steps every time we want to find the result of $(3x)(5x)$. We must be able to write only the answer.

EXAMPLE 3 Find the following products.

a) $(-5)(6x)$ **b)** $(-3a)(-8a)$ **c)** $(-4y)(-6)$ **d)** $(-5a)\left(\dfrac{b}{5}\right)$

Solution

a) $-30x$ **b)** $24a^2$ **c)** $24y$ **d)** $-ab$ ◄

EXAMPLE 4 Simplify each quotient.

a) $\dfrac{5x}{5}$ **b)** $\dfrac{4x + 8}{2}$

Solution

a) Since dividing by 5 is equivalent to multiplying by 1/5, we can use the associative property to write

$$\frac{5x}{5} = \frac{1}{5}(5x) = \left(\frac{1}{5} \cdot 5\right)x = 1 \cdot x = x.$$

All that we have really done is divide 5 by 5 to get 1.

b) Since dividing by 2 is equivalent to multiplying by 1/2, we can use the distributive property to write.

$$\frac{4x + 8}{2} = \frac{1}{2}(4x + 8) = 2x + 4 \qquad \text{Note that both 4 and 8 are divided by 2.}$$

We could also write this as

$$\frac{4x + 8}{2} = \frac{4x}{2} + \frac{8}{2} = 2x + 4. \qquad\qquad ◄$$

Removing Parentheses

Multiplying a number by -1 merely changes the sign of the number. For example,

$$(-1)(7) = -7 \qquad \text{and} \qquad (-1)(-8) = 8.$$

Thus -1 times a number is the *opposite* of the number. In symbols we write

$$(-1)x = -x \qquad \text{or} \qquad -1(a + 2) = -(a + 2).$$

When a minus sign appears in front of a sum or a difference, we can think of it as multiplication by -1 and use the distributive property. For example,

$$-(a + 2) = -1(a + 2) = (-1)a + (-1)2 = -a + (-2) = -a - 2,$$
$$-(x - 5) = -1(x - 5) = (-1)x - (-1)5 = -x + 5.$$

Note that a minus sign in front of a set of parentheses is distributed over each term in the parentheses, changing the sign of each term.

EXAMPLE 5   Simplify the following.

a) $6 - (x + 8)$ **b)** $4x - 6 - (7x - 4)$ **c)** $3x - (-x + 7)$

Solution

a) $6 - (x + 8) = 6 - x - 8$ Change sign of each term in parentheses.
$$= 6 - 8 - x$$ Rearrange the terms.
$$= -2 - x$$ Combine like terms.

b) $4x - 6 - (7x - 4) = 4x - 6 - 7x + 4$ Remove parentheses.
$$= 4x - 7x - 6 + 4$$ Rearrange the terms.
$$= -3x - 2$$ Combine like terms.

c) $3x - (-x + 7) = 3x + x - 7$ Remove parentheses.
$$= 4x - 7$$ Then combine like terms.   ◄

The commutative and associative properties of addition allow us to rearrange the terms so that we may combine the like terms. However, it is not necessary actually to write down the rearrangement. We can identify the like terms and combine them without rearranging.

EXAMPLE 6   Simplify these algebraic expressions.

a) $(-5x + 7) + (2x - 9)$ **b)** $-4x + 7x + 3(2 - 5x)$
c) $-3x(4x - 9) - (x - 5)$ **d)** $x - .03(x + 300)$

Solution

a) $(-5x + 7) + (2x - 9) = -3x - 2$ Combine like terms.

b) $-4x + 7x + 3(2 - 5x) = -4x + 7x + 6 - 15x$ Distributive property
$$= -12x + 6$$ Combine like terms.

c) $-3x(4x - 9) - (x - 5) = -12x^2 + 27x - x + 5$ Remove parentheses.
$$= -12x^2 + 26x + 5$$ Combine like terms.

d) $x - .03(x + 300) = 1x - .03x - 9$ $(-.03)(300) = -9$
$$= .97x - 9$$ $1 - .03 = .97$ ◄

Warm-ups

True or false?

A statement involving variables should be marked true only if it is true for all values of the variable.

1. $5(x + 7) = 5x + 35$ **2.** $-4x + 8 = -4(x + 8)$
3. $-1(a - 3) = -(a - 3)$ **4.** $5y + 4y = 9y$
5. $(2x)(5x) = 10x$ **6.** $-2t(5t - 3) = -10t^2 + 6t$
7. $a + a = a^2$ **8.** $b \cdot b = 2b$
9. $1 + 7x = 8x$ **10.** $(3x - 4) - (8x - 1) = -5x - 3$

1.6 EXERCISES

Perform each computation mentally, making use of the appropriate properties. See Example 1.

1. $45(200)$

2. $25(300)$

3. $\frac{4}{3}(.75)$

4. $5(.2)$

5. $(427 + 68) + 32$

6. $(194 + 78) + 22$

7. $47 \cdot 4 + 47 \cdot 6$

8. $53 \cdot 3 + 53 \cdot 7$

9. $19 \cdot 5 \cdot 2 \cdot \frac{1}{5}$

10. $17 \cdot 4 \cdot 2 \cdot \frac{1}{4}$

11. $(120)(400)$

12. $150 \cdot 300$

13. $13 \cdot 377(-5 + 5)$

14. $(456 \cdot 8)\frac{1}{8}$

15. $(348 + 5) + 45$

16. $(135 + 38) + 12$

17. $354 + 7 + 8 + 3 + 2$

18. $564 + 35 + 65 + 72 + 28$

19. $\frac{2}{3}(1.5)$

20. $(1.25)(.8)$

21. $-11 + 66$

22. $-15 + 65$

23. $8 - 16 + 76$

24. $6 - 19 + 49$

25. $(-19 + 25)(-4 + 9)$

26. $(-17 + 48)200$

27. $(567 + 874)(-2 \cdot 4 + 8)$

28. $(567^2 + 48)[3(-5) + 15]$

Combine like terms where possible. See Example 2.

29. $-4n + 6n$

30. $-3a + 15a$

31. $3w - (-4w)$

32. $3b - (-7b)$

33. $4mw^2 - 15mw^2$

34. $2xb^2 - 16xb^2$

35. $-5x - (-2x)$

36. $-11 - 7t$

37. $-4 - 7z$

38. $-19m - (-3m)$

39. $4t^2 + 5t^2$

40. $5a + 4a^2$

41. $-4a + 3a^2$

42. $-7x^2 + 5x^2$

43. $9n - n$

44. $3m - m$

45. $y - 3y$

46. $t - 5t$

47. $-k - k$

48. $w - w$

Simplify each product or quotient. See Examples 3 and 4.

49. $4(7t)$

50. $-3(4r)$

51. $(-2x)(-5x)$

52. $(-3h)(-7h)$

53. $-7a(3b)$

54. $-8w(2x)$

55. $-5a(1 + b)$

56. $-6x(5 + y)$

57. $(-h)(-h)$

58. $x(-x)$

59. $7w(-4)$

60. $-5t(-1)$

61. $-x(1 - x)$

62. $-p(p - 1)$

63. $(5k)(5k)$

64. $(-4y)(-4y)$

65. $3\left(\frac{y}{3}\right)$

66. $5z\left(\frac{z}{5}\right)$

67. $9\left(\frac{2y}{9}\right)$

68. $8\left(\frac{y}{8}\right)$

69. $\frac{6x}{2}$

70. $\frac{-8x}{4}$

71. $\frac{3x + 15}{3}$

72. $\frac{6x - 8}{2}$

73. $\frac{2x - 4}{-2}$

74. $\frac{-6x - 9}{-3}$

75. $\frac{-x + 10}{-2}$

76. $\frac{-2x + 8}{-4}$

Simplify the following. See Example 5.

77. $a - (4a - 1)$ **78.** $5x - (2x - 7)$ **79.** $6 - (x - 4)$

80. $9 - (w - 5)$ **81.** $4m + 6 - (m + 5)$ **82.** $5 - 6t - (3t + 4)$

83. $-5 - (-a + 7)$ **84.** $-4x - (-7x + 2)$ **85.** $t - 5 - (-2 - t)$

86. $n - 6 - (-n - 2)$ **87.** $x - (x - y - z)$ **88.** $5 - (6 - 3y - z)$

Simplify the following expressions by combining like terms. See Example 6.

89. $2x + 7x + 3 + 5$ **90.** $3x + 4x + 5 + 12$ **91.** $-3x + 4 + 5x - 6$

92. $-4x + 11 + 6x - 8$ **93.** $4a - 5 - (6a - 7)$ **94.** $3x - 4 - (x - 5)$

95. $5(t - 3) - 2(-3 - t)$ **96.** $6(y + 2) - 5(-y - 1)$

97. $-7m + 3(m - 4) + 5m$ **98.** $-6m + 4(m - 3) + 7m$

99. $8 - 7(k + 3) - 4$ **100.** $6 + 5(k - 2) - k + 5$

101. $x - .04(x + 50)$ **102.** $x - .03(x + 500)$

103. $.1(x + 5) - .04(x + 50)$ **104.** $.06x + .14(x + 200)$

105. $3k + 5 - 2(3k - 4) - k + 3$ **106.** $5w - 2 + 4(w - 3) - 6(w - 1)$

107. $5.7 - 4.5(x - 3.9) - 5.42$ **108.** $.04(5.6x - 4.9) + .07(7.3x - 34)$

Simplify each expression as much as possible. —

109. $3 - (x - 5)$ **110.** $(36x + 27x) + 64x$ **111.** $3w \cdot 5w$

112. $3w + 5w$ **113.** $3w - 5w$ **114.** $3(w + 5w)$

115. $3 - 2(x - 3)$ **116.** $-5m - 5m$ **117.** $-5m(-5m)$

118. $\dfrac{4w}{4}$ **119.** $\dfrac{-6x}{-3}$ **120.** $\dfrac{20a + 5}{5}$

121. $\dfrac{-8t + 2}{-2}$ **122.** $\dfrac{7x - 5x}{-2}$

Wrap-up

CHAPTER 1

SUMMARY

	Sets	Examples
Set	A collection of objects. The objects (or numbers) are elements of the set.	$C = \{1, 2, 3\}$ $D = \{3, 4\}$
Variable	A letter used to stand for some numbers	

Set-builder notation	Notation for describing a set using variables	$C = \{x \mid x$ is a counting number smaller than $4\}$
Membership	The symbol $\in$ means "is an element of."	$1 \in C \quad 4 \notin C$
Union	$A \cup B = \{x \mid x \in A \text{ or } x \in B\}$	$C \cup D = \{1, 2, 3, 4\}$
Intersection	$A \cap B = \{x \mid x \in A \text{ and } x \in B\}$	$C \cap D = \{3\}$
Subset	A is a subset of B if every element of A is also an element of B.	$\{1, 2\} \subseteq C$
	$A \nsubseteq B$ if there is an element of A that fails to belong to B.	$D \nsubseteq C$
	$\varnothing \subseteq A$ for any set A.	$\varnothing \subseteq C, \varnothing \subseteq D$

Sets of Numbers — **Examples**

Counting, or natural, numbers	$N = \{1, 2, 3, \ldots\}$	
Whole numbers	$W = \{0, 1, 2, 3, \ldots\}$	
Integers	$I = \{\ldots, -3, -2, -1, 0, 1, 2, 3, \ldots\}$	
Rational numbers	$Q = \{a/b \mid a \text{ and } b \text{ are integers with } b \neq 0\}$	$3/2, 5, -6, 0, .25252525 \ldots$
Irrational numbers	$M = \{x \mid x \text{ is a real number that is not rational}\}$	$\sqrt{2}, \sqrt{3}, \pi, .1515515551 \ldots$
Real numbers	$R = \{x \mid x \text{ is the coordinate of a point on the number line}\}$	$R = Q \cup M$

Operations with Real Numbers — **Examples**

Absolute value	$	a	= \begin{cases} a & \text{if } a \text{ is positive or zero} \\ -a & \text{if } a \text{ is negative} \end{cases}$	$	6	= 6,	0	= 0$ $	-6	= 6$
Sum of two negative numbers	Add their absolute values and put a negative sign on the result.	$-2 + (-7) = -9$
Sum of two numbers with opposite signs	Subtract their absolute values. The answer is positive if the number with the larger absolute value is positive, and negative if the number with the larger absolute value is negative.	$-6 + 9 = 3$ $-9 + 6 = -3$
Definition of subtraction	For any real numbers a and b, $a - b = a + (-b)$. (Change sign and add.)	$4 - 7 = 4 + (-7)$ $5 - (-3) = 5 + 3$
Definition of division	If a, b, and c are real numbers, with $b \neq 0$, then $a \div b = c$, provided $b \cdot c = a$. (Division by zero is undefined.)	$8 \div (-2) = -4$ because $(-4)(-2) = 8$
Multiplication and division of signed numbers	Same sign $\longleftrightarrow$ Positive result Opposite signs $\longleftrightarrow$ Negative result	$(-4)(-2) = 8$ $(-4)(2) = -8$ $(-8) \div (-2) = 4$ $(-8) \div (2) = -4$

Definition of exponents	For any counting number n, $a^n = a \cdot a \cdot a \cdot \cdots \cdot a$ (n factors of a). In the exponential expression a^n, a is the base and n is the exponent.	$2^3 = 2 \cdot 2 \cdot 2 = 8$

Order of
operations
	Without parentheses or absolute value:
	  1. Evaluate exponential expressions
	  2. Perform multiplication and division
	  3. Perform addition and subtraction
	With parentheses or absolute value:
	  First evaluate within each set of parentheses
	  or absolute value, using the above order.
		$7 + 2^3 = 15$
		$3 + 4 \cdot 6 = 27$
		$5 + 4 \cdot 3^2 = 41$
		
		$(2 + 4)(5 - 9) = -24$
		$3 + 4\lvert 2 - 3\rvert = 7$

	Properties	**Examples**
Commutative properties	$a + b = b + a$ $a \cdot b = b \cdot a$	$3 + 7 = 7 + 3$ $4 \cdot 3 = 3 \cdot 4$
Associative properties	$a + (b + c) = (a + b) + c$ $a \cdot (b \cdot c) = (a \cdot b) \cdot c$	$1 + (3 + 5) = (1 + 3) + 5$ $3 \cdot (5 \cdot 7) = (3 \cdot 5) \cdot 7$
Distributive properties	$a(b + c) = ab + ac$ $a(b - c) = ab - ac$	$3(4 + x) = 12 + 3x$ $5x - 10 = 5(x - 2)$
Identity properties	$a + 0 = a$ and $0 + a = a$. Zero is the additive identity. $1 \cdot a = a$ and $a \cdot 1 = a$. The multiplicative identity is 1.	$6 + 0 = 0 + 6 = 6$ $6 \cdot 1 = 1 \cdot 6 = 6$
Inverse properties	For any real number a, there is a unique number $-a$ (additive inverse of a) such that $a + (-a) = 0$ and $(-a) + a = 0$. For any nonzero real number a, there is a unique number $1/a$ (multiplicative inverse of a) such that $a \cdot (1/a) = 1$ and $(1/a) \cdot a = 1$.	$8 + (-8) = 0$ $-8 + 8 = 0$ $(8)(1/8) = 1$ $(-8)(-1/8) = 1$
Multiplication property of 0	$a \cdot 0 = 0$ and $0 \cdot a = 0$.	$9 \cdot 0 = 0$ $(0)(-4) = 0$

	Other Important Facts	**Examples**
Algebraic expressions	Any meaningful combination of numbers, variables, and operations	$x^2 + y^2$, $-5abc$
Term	An expression containing a number or the product of a number and one or more variables	$3x^2$, $-7x^2y$, 8
Coefficient	The number preceding the variable in a term	$3x^2$ (3 is the coefficient.)
Like terms	Identical variable parts	$-7x + 9x = 2x$ $4bc - 8bc = -4bc$

REVIEW EXERCISES

1.1 *Let* $A = \{1, 2, 3\}$, $B = \{3, 4, 5\}$, $C = \{1, 2, 3, 4, 5\}$, $D = \{3\}$, *and* $E = \{4, 5\}$. *Determine whether each of the following statements is true or false.*

1. $A \cap B = D$
2. $A \cap B = E$
3. $A \cup B = E$
4. $A \cup B = C$
5. $B \cup C = C$
6. $A \cap C = B$
7. $A \cap \varnothing = A$
8. $A \cup \varnothing = \varnothing$
9. $(A \cap B) \cup E = B$
10. $(C \cap B) \cap A = D$
11. $B \subseteq C$
12. $A \subseteq E$
13. $A = B$
14. $B = C$
15. $3 \in D$
16. $5 \notin A$
17. $0 \in E$
18. $D \subseteq \varnothing$
19. $\varnothing \subseteq E$
20. $1 \in A$

1.2 *Which elements of the set*

$$\{-\sqrt{2}, -1, 0, 1, 1.732, \sqrt{3}, \pi, 22/7, 31\}$$

are members of the following sets?

21. Whole numbers
22. Natural numbers
23. Integers
24. Rational numbers
25. Irrational numbers
26. Real numbers

True or false?

27. The set of whole numbers is a subset of the set of counting numbers.
28. Zero is not a real number.
29. The set of counting numbers larger than -3 is $\{-2, -1, 0, 1, 2, 3, \ldots \}$.
30. The set of rational numbers is finite.
31. The ratio of the circumference to the diameter of any circle is exactly 3.14.
32. Every terminating decimal number is an integer.
33. Every repeating decimal number is a rational number.
34. The number $\sqrt{9}$ is an irrational number.
35. The number zero is both rational and irrational.
36. The irrational numbers have no decimal representation.

1.3 *Evaluate the following.*

37. $-4 + 9$
38. $-3 + (-5)$
39. $25 - 37$
40. $-6 - 10$
41. $(-4)(6)$
42. $(-7)(-6)$
43. $(-8) \div (-4)$
44. $40 \div (-8)$
45. $-\dfrac{1}{4} + \dfrac{1}{12}$
46. $\dfrac{1}{3} - \left(-\dfrac{1}{12}\right)$
47. $\dfrac{-20}{-2}$
48. $\dfrac{30}{-6}$
49. $-.04 + 10$
50. $-.05 + (-3)$
51. $-6 - (-2)$
52. $-.2 - (-.04)$
53. $-.5 + .5$
54. $-.04 \div .2$
55. $3.2 \div (-.8)$
56. $(.2)(-.9)$
57. $0 \div (-.3545)$
58. $(-6)(-.5)$
59. $(1/4)(-12)$
60. $-7 - (-9)$

1.4 *Evaluate the following expressions.*

61. $4 + 7(5)$

62. $(4 + 7)5$

63. $(4 + 7)^2$

64. $4 + 7^2$

65. $5 + 3 \cdot |6 - 4 \cdot 3|$

66. $6 - (7 - 8)$

67. $(6 - 8) - (5 - 9)$

68. $5 - 6 - 8 - 10$

69. $-3 - 5(6 - 2 \cdot 5)$

70. $4^2 - 9 + 3^2$

71. $5^2 - (6 + 5)^2$

72. $|3 - 4 \cdot 2| - |5 - 8|$

73. $\dfrac{6}{2} + 2$

74. $\dfrac{6 + 3}{3}$

75. $\dfrac{-4 - 5}{7 - (-2)}$

76. $\dfrac{5 - 9}{2 - 4}$

77. $1 - (.8)(.3)$

78. $5 - (.2)(.1)$

79. $(-3)^2 - (4)(-1)(-2)$

80. $3^2 - 4(1)(-3)$

81. $3 - 2|4 - 7|$

82. $3|4 - 6| + 1$

83. $\dfrac{-5 - 3}{-4 + 2}$

84. $\dfrac{-6 + 3}{3 - 9}$

Let $a = -2$, $b = 3$, and $c = -1$. Find the value of each algebraic expression.

85. $b^2 - 4ac$

86. $a^2 - 4b$

87. $(c - b)(c + b)$

88. $(a + b)(a - b)$

89. $a^2 + 2ab + b^2$

90. $a^2 - 2ab + b^2$

91. $a^3 - b^3$

92. $a^3 + b^3$

93. $\dfrac{b + c}{a + b}$

94. $\dfrac{b - c}{2b - a}$

95. $|a - b|$

96. $|b - a|$

97. $(a + b)c$

98. $ac + bc$

1.5 *Name the property that justifies each statement.*

99. $a + x = x + a$

100. $0 \cdot 5 = 0$

101. $3(x - 1) = 3x - 3$

102. $10 + (-10) = 0$

103. $5(2x) = (5 \cdot 2)x$

104. $w + y = y + w$

105. $1 \cdot y = y$

106. $4(1/4) = 1$

107. $5(.2) = 1$

108. $2 + (3 + 4) = (2 + 3) + 4$

109. $18 + 0 = 18$

110. $3 \cdot 1 = 3$

111. $-5 + 5 = 0$

112. $2w + 2m = 2(w + m)$

113. $12 \cdot 0 = 0$

114. $x + 1 = 1 + x$

Use the distributive property to either multiply or factor, whichever is appropriate.

115. $3x - 3a$

116. $5x - 5y$

117. $3(w + 1)$

118. $2(m + 14)$

119. $7x + 7$

120. $3w + 3$

121. $5(x - 5)$

122. $13(b - 3)$

123. $-3(2x - 5)$

124. $-2(5 - 4x)$

125. $p - pt$

126. $ab + b$

1.6 *Simplify by combining like terms.*

127. $3a + 7 + 4a - 5$

128. $2m + 6 + m - 2$

129. $2(a - 6) + 3a - 8$

130. $3(a - 5) + 5(a + 2)$

131. $5(t - 4) - 3(2t - 6)$

132. $2(x - 3) + 2(3 - x)$

133. $-(a - 2) + 2 - a$

134. $-(w - y) + 3(y - w)$

135. $5 - 3(x - 2) + 7(x + 4) - 6$

136. $7 - 2(x - 7) + 7 - x$

137. $.2(x + .1) - (x + .5)$

138. $.1(x - .2) - (x + .1)$

139. $.05(x + 3) - .1(x + 20)$

140. $.02(x + 100) - .2(x + 50)$

141. $\frac{1}{2}(x + 4) - \frac{1}{4}(x - 8)$

142. $\frac{1}{2}(2x - 1) + \frac{1}{4}(x + 1)$

Miscellaneous

Evaluate the following expressions mentally for w = 24, x = −6, y = 6, and z = 4. Name the property or properties used.

143. $32z(x + y)$

144. $(wz)\dfrac{1}{w}$

145. $768z + 768y$

146. $28z + 28y$

147. $(12z + x) + y$

148. $(42 + x) + y$

149. $752x + 752y$

150. $37y + 37x$

151. $(47y)\dfrac{z}{w}$

152. $3w + 3y$

153. $(xw)\dfrac{1}{y}$

154. $(xz)\dfrac{1}{x}$

155. $5(x + y)(z + w)$

156. $(4x + 7y)(w + xz)$

CHAPTER 1 TEST

Let A = {2, 4, 6, 8, 10}, B = {3, 4, 5, 6, 7}, and C = {6, 7, 8, 9, 10}. List the elements in each of the following sets.

1. $A \cup B$

2. $B \cap C$

3. $A \cap (B \cup C)$

Which elements of $\{-4, -\sqrt{3}, -1/2, 0, 1.65, \sqrt{5}, \pi, 8\}$ are members of the following sets?

4. Whole numbers

5. Integers

6. Rational numbers

7. Irrational numbers

Evaluate the following expressions.

8. $6 + 3(-5)$

9. $(-2)^2 - 4(2)(-5)$

10. $\dfrac{-3 - (-7)}{3 - 5}$

11. $\dfrac{-6 - 2}{4 - 2}$

12. $-5 + 6 - 12$

13. $.02 - 2$

14. $(5 - 9)(6 - 10)$

15. $(498 + 89) + 11$

16. $|3 - 5(2)|$

17. $5 - 2|6 - 10|$

18. $(452 + 695)[2(-4) + 8]$

19. $478(8) + 478(2)$

Identify the property that justifies each equality.

20. $2(5 + 7) = 10 + 14$

21. $57 \cdot 4 = 4 \cdot 57$

22. $2 + (6 + x) = (2 + 6) + x$

23. $-6 + 6 = 0$

24. $1 \cdot (-6) = (-6) \cdot 1$

25. $0 \cdot 28 = 0$

Simplify the following by combining like terms.

26. $3x - 5 + 2(x - 6)$

27. $x + 3 - .05(x + 2)$

28. $3(m - 5) - 4(-2m - 3)$

29. $6 - 4(3 - t) - 5$

30. $\dfrac{1}{2}(x - 4) + \dfrac{1}{4}(x + 3)$

Use the distributive property to factor each of the following.

31. $5x - 40$

32. $8t - 8$

Linear Equations and Inequalities

A certain lawyer likes to tell the story of his friend Albert and Albert's peculiar Last Will and Testament. Upon his death, Albert, who had never married, wanted his estate divided among his three nephews, Daniel, Raymond, and Brian. Daniel was a good lad, but it was Brian who really helped Uncle Albert throughout his old age. So, Albert decided that Daniel's share of the estate should be only one-half as large as Brian's share. Raymond was the black sheep of the family, so Albert wanted him to receive only one-third as much as Brian. Then Albert remembered the $1000 that Raymond had borrowed and never repaid. So he told the lawyer to give Raymond $1000 less than one-third of Brian's share. Uncle Albert died before the lawyer had a chance to convince him to make his will simpler. After Albert's property was sold, $25,400 was left to be divided among the nephews. How much did each one inherit?

The amounts that two of the nephews received were based on an unknown quantity, Brian's share. Unknowns are what algebra thrives on. This chapter focuses on solving equations, inequalities, and problems involving unknowns. We will find that a problem such as this can be solved using an equation with one unknown, or variable. When we see this problem as Exercise 45 in Section 2.4, we will be able to solve it easily.

2.1 Linear Equations in One Variable

IN THIS SECTION:
- Basic Ideas and Definitions
- Solving Equations
- Identities
- Conditional Equations
- Inconsistent Equations

The applications of algebra often lead to equations. The skills that we learned in Chapter 1, such as combining like terms and performing operations with algebraic expressions, will now be used to solve equations.

Basic Ideas and Definitions

An **equation** is a sentence that expresses the equality of two algebraic expressions. Consider the equation

$$2x + 1 = 7.$$

Because $2(3) + 1 = 7$ is true, we say that the number 3 is a **solution** to the equation, or that 3 **satisfies** the equation. No other number in place of x will make the statement $2x + 1 = 7$ true. Any number that satisfies an equation is called a **solution** or a **root** to the equation.

DEFINITION
Solution Set

> The set of all solutions to an equation is called the **solution set** to that equation.

To **solve** an equation means to find its solution set. The solution set to the equation $2x + 1 = 7$ is $\{3\}$.

EXAMPLE 1   Determine whether each number is in the solution set to the equation that follows it.

a) -5, $3x + 7 = -8$ **b)** 4, $2(x - 1) = 2x + 3$

Solution

a) Replace x by -5 and evaluate each side of the equation.

$$3x + 7 = -8$$
$$3(-5) + 7 = -8$$
$$-15 + 7 = -8$$
$$-8 = -8 \quad \text{Correct.}$$

Because both sides have a value of -8, -5 is a member of the solution set to the equation.

b) Replace x by 4 and evaluate each side of the equation.

$$2(x - 1) = 2x + 3$$
$$2(4 - 1) = 2(4) + 3$$
$$2(3) = 8 + 3$$
$$6 = 11 \quad \text{Incorrect.}$$

Because the two sides of the equation have different values when $x = 4$, 4 is *not* a member of the solution set to the equation. ◀

Solving Equations

Consider the equations

$$2x + 1 = 7, \qquad 2x = 6, \qquad \text{and} \qquad x = 3.$$

If we replace x by 3 in each one of these equations, we get the following true statements:

$$2(3) + 1 = 7, \qquad 2(3) = 6, \qquad \text{and} \qquad 3 = 3.$$

The solution set to each of these equations is $\{3\}$. Equations that have the same solution set are called **equivalent** equations.

A closer look at these equivalent equations shows that each equation can be obtained from the one before it. Subtracting 1 from each side of $2x + 1 = 7$ gives us $2x = 6$, and dividing each side of $2x = 6$ by 2 gives us $x = 3$.

Note that the same expression can be subtracted from each side of an equation without changing the solution set, and that each side of an equation can be divided by the same nonzero number without changing the solution set. In fact, we can perform any of the four basic operations of arithmetic on each side of an equation without changing the solution set. These important properties of equality are stated formally in the following box.

Properties of Equality

If A, B, and C are algebraic expressions, then the equation $A = B$ is equivalent to the following equations:

$A + C = B + C$	Addition Property of Equality
$A - C = B - C$	Subtraction Property of Equality
$CA = CB \quad (C \neq 0)$	Multiplication Property of Equality
$\dfrac{A}{C} = \dfrac{B}{C} \quad (C \neq 0)$	Division Property of Equality

Note that we do not allow multiplication of each side of an equation by 0. If we were to multiply the equation $x = 3$ by 0 on each side, we would get $0 = 0$, which is not equivalent to $x = 3$.

The following example illustrates how these properties of equality can be used in a systematic way to solve an equation.

EXAMPLE 2   Solve the equation $5x - 7 = -22$.

Solution   We want to use the properties of equality to obtain an equivalent equation with only a single x on one side and a number on the other side. We can eliminate -7 from the left side by adding 7 to each side.

$$5x - 7 = -22$$
$$5x - 7 + 7 = -22 + 7 \qquad \text{Add 7 to each side of the equation.}$$
$$5x = -15 \qquad \text{Simplify.}$$
$$\frac{5x}{5} = \frac{-15}{5} \qquad \text{Divide each side by 5 to get a single } x.$$
$$x = -3 \qquad \frac{5x}{5} = x$$

All of these equations are equivalent. The solution set to the equation $x = -3$ is $\{-3\}$, and so the solution set to the original equation is also $\{-3\}$. To check the solution, we replace x by -3 in the original equation.

$$5x - 7 = -22$$
$$5(-3) - 7 = -22 \qquad \text{Replace } x \text{ by } -3.$$
$$-15 - 7 = -22 \qquad \text{Simplify.}$$
$$-22 = -22 \qquad \text{Correct.}$$

We can now be certain that $\{-3\}$ is the correct solution set.   ◄

In each of the following examples we use the properties of equality to isolate the variable on the left side of the equation. Because all of the equations along the way are equivalent, we know that the solution to the original equation is the same as the solution to the last simple equation.

EXAMPLE 3   Solve the equation $6 - 3x = 8 - 2x$.

Solution   To isolate x on the left side, subtract 6 from each side.

$$6 - 3x = 8 - 2x$$
$$6 - 3x - 6 = 8 - 2x - 6 \qquad \text{Subtract 6 from each side.}$$
$$-3x = 2 - 2x \qquad \text{Simplify.}$$
$$-3x + 2x = 2 - 2x + 2x \qquad \text{Add } 2x \text{ to each side.}$$
$$-x = 2 \qquad \text{Simplify.}$$
$$(-1)(-x) = -1 \cdot 2 \qquad \text{Multiply each side by } -1.$$
$$x = -2$$

Note that instead of multiplying each side by -1, we could have realized that $-x = 2$ is equivalent to $x = -2$. (If the opposite of x is 2, then x is -2.) All of these equations are equivalent. Because the solution set for the last equation is $\{-2\}$, that is also the solution set to the original equation. Checking this solution in the original equation gives us the correct statement $6 - 3(-2) = 8 - 2(-2)$.   ◄

EXAMPLE 4   Solve the equation $2(x - 4) + 5x = 34$.

Solution

$$2x - 8 + 5x = 34 \qquad \text{Use distributive property to remove parentheses.}$$
$$7x - 8 = 34 \qquad \text{Combine like terms.}$$
$$7x - 8 + 8 = 34 + 8 \qquad \text{Add 8 to each side.}$$
$$7x = 42 \qquad \text{Simplify.}$$
$$\frac{7x}{7} = \frac{42}{7} \qquad \text{Divide each side by 7 to get a single } x \text{ on the left side.}$$
$$x = 6 \qquad \frac{7x}{7} = \frac{x}{1} = x$$

All of these equations are equivalent. The solution set to $x = 6$ is $\{6\}$, and so the solution set to the original equation is also $\{6\}$. To check the solution, we replace x by 6 in the original equation and simplify.

$$2(6 - 4) + 5(6) = 34$$
$$2 \cdot 2 + 30 = 34$$
$$34 = 34$$

This check verifies that $\{6\}$ is the correct solution set.   ◄

When an equation involves fractions, it is often helpful to multiply each side by a number that is evenly divisible by all of the denominators. The smallest such number is called the **least common denominator (LCD)**. Multiplying each side of the equation by the LCD will eliminate all of the fractions.

EXAMPLE 5   Find the solution set for the equation

$$\frac{x}{2} - \frac{1}{3} = \frac{x}{3} + \frac{5}{6}.$$

Solution   To solve this equation we multiply each side by 6, the smallest number that is evenly divisible by the denominators 2, 3, and 6.

$$6\left(\frac{x}{2} - \frac{1}{3}\right) = 6\left(\frac{x}{3} + \frac{5}{6}\right) \qquad \text{Multiply each side by 6.}$$

$$\frac{6x}{2} - \frac{6}{3} = \frac{6x}{3} + \frac{30}{6} \qquad \begin{array}{l}\text{Use the distributive property}\\\text{to remove parentheses.}\end{array}$$

$$3x - 2 = 2x + 5 \qquad \text{Divide to eliminate the fractions.}$$

$$3x - 2 - 2x = 2x + 5 - 2x \qquad \text{Subtract } 2x \text{ from each side.}$$

$$x - 2 = 5 \qquad \text{Combine like terms.}$$

$$x - 2 + 2 = 5 + 2 \qquad \text{Add 2 to each side.}$$

$$x = 7 \qquad \text{Combine like terms.}$$

The solution set is {7}. We leave it to the reader to check.   ◄

EXAMPLE 6   Solve: $\dfrac{y}{2} - \dfrac{y-4}{5} = \dfrac{23}{10}.$

Solution   We want to multiply each side of the equation by 10, the LCD. We do not have to write down that step, however; we can simply use the distributive property to multiply each term of the equation by 10.

$$\overset{5}{\cancel{10}}\left(\frac{y}{\cancel{2}}\right) - \overset{2}{\cancel{10}}\left(\frac{y-4}{\cancel{5}}\right) = \cancel{10}\left(\frac{23}{\cancel{10}}\right) \qquad \text{Multiply each side by 10.}$$

$$5y - 2(y - 4) = 23 \qquad \begin{array}{l}\text{Divide each denominator into}\\\text{10 to eliminate fractions.}\end{array}$$

$$5y - 2y + 8 = 23 \qquad \begin{array}{l}\text{Be careful with}\\-2(y - 4) = -2y + 8.\end{array}$$

$$3y + 8 = 23 \qquad \text{Combine like terms.}$$

$$3y + 8 - 8 = 23 - 8 \qquad \text{Subtract 8 from each side.}$$

$$3y = 15 \qquad \text{Simplify.}$$

$$\frac{3y}{3} = \frac{15}{3} \qquad \text{Divide each side by 3.}$$

$$y = 5$$

The solution set is {5}. You should check that 5 solves the original equation.   ◄

Equations that involve decimal numbers can be solved like equations involving fractions. If we multiply a decimal number by 10, 100, or 1000, the decimal point is moved 1, 2, or 3 places to the right, respectively. If the decimal points are all moved far enough to the right, the decimal numbers will be replaced by whole numbers. The next example shows how to use the multiplication property of equality to eliminate decimal numbers in an equation.

EXAMPLE 7   Solve: $x - .1x = .75x + 4.5$.

Solution   Because the number with the most decimal places in this equation is .75 (75 hundredths), multiplying by 100 will eliminate all decimals.

$$100(x - .1x) = 100(.75x + 4.5) \qquad \text{Multiply each side by 100.}$$
$$100x - 10x = 75x + 450 \qquad \begin{array}{l}\text{Use the distributive property} \\ \text{to remove parentheses.}\end{array}$$
$$90x = 75x + 450 \qquad \text{Combine like terms.}$$
$$90x - 75x = 75x + 450 - 75x \qquad \text{Subtract 75x from each side.}$$
$$15x = 450 \qquad \text{Combine like terms.}$$
$$\frac{15x}{15} = \frac{450}{15} \qquad \text{Divide each side by 15.}$$
$$x = 30$$

The solution set is {30}. Check to see that 30 solves the original equation.   ◄

Identities

Some equations are true no matter what we choose for the value of the variable. Consider the equations

$$x + 1 = x + 1, \qquad x \div 2 = \frac{1}{2}x, \qquad \text{and } x + x = 2x.$$

The first equation expresses the equality of identical expressions, and the last two express the equality of different forms of the same expression. Each of these equations is true for any value of x. These equations are called **identities.** The solution set to each of these equations is the set of real numbers, R. The equation

$$\frac{x}{x} = 1$$

is true for all values of x except 0, because 0/0 is undefined. This equation is also considered an identity.

DEFINITION
Identity

An equation that is satisfied by every number for which both sides are defined is called an **identity.**

It is easy to recognize identities like those just listed. The following example shows an identity that is not as easily recognized.

EXAMPLE 8   Solve the equation

$$8 - 3(x - 5) + 7 = 3 - (x - 5) - 2(x - 11).$$

Solution

$$8 - 3x + 15 + 7 = 3 - x + 5 - 2x + 22 \qquad \text{Distributive property}$$
$$30 - 3x = 30 - 3x \qquad \text{Combine like terms.}$$

This last equation is satisfied by any value of x because the two sides are identical. Because the last equation is equivalent to the original equation, the original equation is satisfied by any value of x and is an identity. The solution is R, the set of all real numbers.   ◄

Conditional Equations

The statement $2x + 1 = 7$ is true only on condition that we choose $x = 3$. For this reason it is called a **conditional equation.**

DEFINITION
Conditional Equation

> A **conditional equation** is an equation that is satisfied by at least one real number but is not an identity.

The equations solved in Examples 2 through 7 are conditional equations.

Inconsistent Equations

It is easy to write equations that are false no matter what value we use to replace the variable. If we replace x by any real number in the equations

$$x = x + 1, \qquad x - x = 4, \qquad \text{and } 0 \cdot x + 5 = 3,$$

we get false statements. (Try some.) The solution set to each of these equations is the empty set, $\varnothing$.

DEFINITION
Inconsistent Equation

> An equation whose solution set is the empty set is called an **inconsistent equation.**

The next example shows an inconsistent equation that is not as easily recognized as the ones just listed.

EXAMPLE 9   Solve the equation $5 - 3(x - 6) = 4(x - 9) - 7x$.

Solution

$$5 - 3x + 18 = 4x - 36 - 7x \qquad \text{Distributive property}$$

$$23 - 3x = -36 - 3x \qquad \text{Combine like terms.}$$

$$23 - 3x + 3x = -36 - 3x + 3x \qquad \text{Add } 3x \text{ to each side.}$$

$$23 = -36 \qquad \text{Combine like terms.}$$

This last equation is false for any choice of x. Because these equations are all equivalent, the original equation is also false for any choice of x. The solution set is the empty set, $\varnothing$, and the equation is an inconsistent equation.   ◄

In Examples 2–7 of this section we solved linear equations in one variable. Even though all of these equations look different, they could all be rewritten in a standard form. In Chapter 7 we will see a relationship between these equations and straight lines.

DEFINITION
Linear Equation in
One Variable

> A **linear equation in one variable** x is any equation that can be written in the form $ax + b = 0$, where a and b are real numbers, with $a \neq 0$.

The equations in Examples 8 and 9 appear to be linear equations, but because they cannot be written in the form $ax + b = 0$, with $a \neq 0$, they are not linear equations. A linear equation has exactly one solution. It is certainly not necessary to write a linear equation in the form of the definition in order to solve it. You should try to solve equations with a minimum number of steps. The strategy we use for solving linear equations is summarized in the following box.

STRATEGY
Solving a Linear Equation

> 1. If fractions are present, multiply each side by the LCD to eliminate the fractions.
> 2. Use the distributive property to remove parentheses.
> 3. Combine any like terms.
> 4. Use the addition and subtraction properties to get all variables on one side and numbers on the other.
> 5. Use the multiplication or division property to get a single variable on one side.
> 6. Check your work by replacing the variable in the original equation with your solution.

Not all equations require the use of all steps, and the steps are not always performed in the same order.

Warm-ups

True or false?

1. The equation $-2x + 3 = 8$ is equivalent to $-2x = 11$.
2. The equation $x - (x - 3) = 5x$ is equivalent to $3 = 5x$.
3. To solve $\frac{3}{4}x = 12$, we should multiply each side by 3/4.
4. If $2x + 5 = 0$, then $x = 5/2$.
5. When an equation involves fractions, it is usually a good idea to multiply each side by the LCD of all of the fractions.
6. The solution set to $x - x = 3$ is the empty set.
7. The equation $3x - 7 = 0$ is an inconsistent equation.
8. The equation $4(x + 3) = x + 3$ is a conditional equation.
9. The equation $x - .2x = .8x$ is an identity.
10. The equation $3x - 5 = 7$ is a linear equation.

2.1 EXERCISES

Determine whether or not each number is in the solution set to the equation following it. See Example 1.

1. 4, $3x - 7 = 5$
2. 6, $-3x + 5 = -13$
3. -2, $3(x - 4) = x + 5$
4. -3, $1 - 2(x + 3) = x + 2$
5. 7, $2x + 1 = 2x + 3$
6. 5, $x + 3 = x - 1$
7. -2, $x + x = 2x$
8. 5, $3(x - 4) = 3x - 12$

Solve each of these linear equations. Show your work and check your answer. See Examples 2, 3, and 4.

9. $2x - 3 = 0$
10. $5x + 7 = 0$
11. $-2x + 5 = 7$
12. $-3x - 4 = 11$
13. $-7 - x = 5$
14. $5 - x = -9$
15. $-3(x - 6) = 2 - x$
16. $-2(x + 5) = 3 - x$
17. $-2x - 5 = 7$
18. $-3x + 7 = -2$
19. $2(x + 1) - x = 36$
20. $3(x - 1) - x = 9$
21. $12 = 4x + 3$
22. $14 = -5x - 21$
23. $-3x - 1 = 5 - 2x$
24. $-5x + 2 = -5 - 4x$

Solve each equation. See Examples 5 and 6.

25. $\frac{2}{3}x + 5 = -\frac{1}{3}x + 17$
26. $\frac{1}{4}x - 6 = -\frac{3}{4}x + 14$
27. $-\frac{3}{7}x = 4$
28. $\frac{5}{6}x = -2$
29. $-\frac{5}{7}x - 1 = 3$
30. $4 - \frac{3}{5}x = -6$

31. $\frac{1}{2}x + \frac{1}{4} = \frac{1}{4}(x - 6)$ **32.** $\frac{1}{3}(x - 2) = \frac{2}{3}x - \frac{13}{3}$ **33.** $\frac{x}{3} + \frac{1}{2} = \frac{7}{6}$

34. $\frac{x}{6} - \frac{1}{3} = 1$ **35.** $\frac{1}{8} + \frac{x}{6} = \frac{x}{4}$ **36.** $\frac{1}{4} + \frac{1}{5} = \frac{x}{2}$

37. $\frac{y - 3}{2} = \frac{1}{4}$ **38.** $\frac{y + 5}{3} = 7$ **39.** $8 - \frac{x - 2}{2} = \frac{x}{4}$

40. $\frac{x}{3} - \frac{x - 5}{5} = 3$ **41.** $\frac{y - 3}{3} - \frac{y - 2}{2} = -1$ **42.** $\frac{x - 2}{2} - \frac{x - 3}{4} = \frac{7}{4}$

Solve each equation. See Example 7.

43. $x - .2x = 72$ **44.** $x - .1x = 63$

45. $.03(x + 200) + .05x = 86$ **46.** $.02(x - 100) + .06x = 62$

47. $.1x + .05(x - 300) = 105$ **48.** $.2x - .05(x - 100) = 35$

Solve each equation. Identify each as a conditional equation, an inconsistent equation, or an identity. See Examples 8 and 9.

49. $2(x + 1) = 2(x + 3)$ **50.** $2x + 3x = 6x$ **51.** $x + x = 2x$

52. $4x - 3x = x$ **53.** $x + x = 2$ **54.** $4x - 3x = 5$

55. $2(x + 3) - 7 = 5(5 - x) + 7(x + 1)$ **56.** $2(x + 4) - 8 = 2x + 1$

57. $2(x + 3) - 7 = 3(x + 1) - (x + 4)$ **58.** $2(x + 4) - 8 = 2x$

59. $2(x + 3) - 7 = 3(x + 1)$ **60.** $2(x + 4) - 8 = 3x - 7$

61. $\frac{4x}{4} = x$ **62.** $5x \div 5 = x$

63. $x \cdot x = x^2$ **64.** $\frac{x}{x} = 1$

For each equation find the value of c that makes the given equation equivalent to the equation $x = -3$.

65. $x = c$ **66.** $x = -c$ **67.** $2x = c$ **68.** $3x = c$

69. $2x + 1 = x + c$ **70.** $3x - 1 = x + 2c$ **71.** $-x + c = 3$ **72.** $x - c = 5$

Solve each equation.

73. $\frac{P + 7}{3} - \frac{P - 2}{5} = \frac{7}{3} - \frac{P}{15}$ **74.** $\frac{w - 3}{8} - \frac{5 - w}{4} = \frac{4w - 1}{8} - 1$

75. $x - .06x = 50,000$ **76.** $x - .05x = 800$

*77. $5 - [4 - (t - 6)] = 17 + 2[t + 3(t - 6)]$ *78. $[-3 - 5(2 - a)] + 6a = -6 - 2[-1 + 4(a - 6)]$

*79. $.3 - [4 - (t - .2)] = [t - (2t + 6)] + 3 - t$ *80. $x - (.3 + x) = -10.9 - 2[x - 6(1 - x)]$

81. $2.365x + 3.694 = 14.8095$ **82.** $-3.48x + 6.981 = 4.329x - 6.851$

83. $5.39 - [4.21 - (x - 3.52)] = 17.6 + 2(x + 13.581)$ **84.** $5.024 - 3.67(x - 1.09) = 12.5x - 3(1.4x + 10.72)$

2.2 Mental Algebra

IN THIS SECTION:

- Combining Steps
- Reading from Right to Left
- Solving Equations Mentally

We have been solving complicated equations by reducing them to simple equations in a step-by-step procedure. Although this method is essential for success in algebra, writing down every step of the procedure is not always necessary. Solving an equation is often part of a larger problem, and anything that we can do to make the process more efficient will make solving the entire problem faster and easier. In this section we will learn how to solve equations more efficiently.

Combining Steps

Consider the following two solutions to the equation $4x - 5 = 23$.

Combining Steps

$$4x - 5 = 23$$
$$4x = 28 \quad \text{Add 5 to each side.}$$
$$x = 7 \quad \text{Divide each side by 4.}$$

Writing Every Step

$$4x - 5 = 23$$
$$4x - 5 + 5 = 23 + 5$$
$$4x = 28$$
$$\frac{4x}{4} = \frac{28}{4}$$
$$x = 7$$

The solution on the left combines two steps into one. The addition of 5 to each side is done mentally, and the division by 4 is also done mentally. In fact, all of algebra is really done mentally—what we write on paper is just a reminder of what is taking place in our minds. Combining steps takes a little practice but is a valuable skill that can be used throughout our work in algebra. The following examples also illustrate how to combine some steps.

EXAMPLE 1   Solve the equation $4x + 3(x - 5) = 2x + 10$.

Solution

$$7x - 15 = 2x + 10 \quad \text{Remove parentheses and combine like terms.}$$
$$7x = 2x + 25 \quad \text{Add 15 to each side.}$$
$$5x = 25 \quad \text{Subtract } 2x \text{ from each side.}$$
$$x = 5 \quad \text{Divide each side by 5.}$$

The solution set is {5}. Check that 5 solves the original equation.   ◀

EXAMPLE 2 Solve the equations:

a) $8 - x = 3$ **b)** $-3 - x = 6$

Solution

a) $8 - x = 3$

$\quad\quad -x = -5$ Subtract 8 from each side.

$\quad\quad\quad x = 5$ If the opposite of x is -5, then x is 5.

The solution set is $\{5\}$. Check.

b) $-3 - x = 6$

$\quad\quad -x = 9$ Add 3 to each side.

$\quad\quad\quad x = -9$ If the opposite of x is 9, then x is -9.

The solution set is $\{-9\}$. Check.   ◀

Reading from Right to Left

The equation $7 = x$ is equivalent to the equation $x = 7$, but there is no need to rewrite it as $x = 7$. Instead, read it from right to left. Thus $7 = x$ can be read as "x equals 7." In solving an equation, it is often simpler to leave the variable on the right, as in the following example.

EXAMPLE 3 Solve the equation $5 = 2x - 13$.

Solution

$$18 = 2x \quad\quad \text{Add 13 to each side.}$$

$$9 = x \quad\quad \text{Divide each side by 2.}$$

The solution set is $\{9\}$. Check.   ◀

In the next example we could end up with x on the left or right side of the equation. We choose to get x on the right to save one step.

EXAMPLE 4 Solve the equation $3(x + 3) = 4x - 5$.

Solution

$$3x + 9 = 4x - 5 \quad\quad \text{Remove the parentheses.}$$

$$14 = x \quad\quad \text{Subtract } 3x \text{ from each side and add 5 to each side.}$$

The solution set is $\{14\}$. Check it. Notice that isolating x on the right side simplified the arithmetic by avoiding negative numbers. Try solving this same equation by isolating x on the left.   ◀

Solving Equations Mentally

After a little practice, equations that require only one or two steps in their solution can often be solved mentally. It is not a requirement of algebra that equations be solved mentally, but this can be a labor-saving procedure that will be useful when we solve more complicated problems. Solving equations mentally should be attempted only after becoming proficient at solving equations with pencil and paper. The next example shows the types of equations that we can learn to solve mentally.

EXAMPLE 5   Solve the following equations mentally.

a) $2x - 3 = 7$ b) $3x + 4 = 0$ c) $2x + 3 = x + 8$

d) $x - 3 = 2x + 1$ e) $\dfrac{x}{5} = -2$

Solution   Each of these equations can be solved by using two or fewer steps. To solve them mentally, we perform the same steps as we would in a written problem, except that we do not write down the intermediate results.

a) The two steps to solve this equation are adding 3 to each side and then dividing each side by 2. To solve it mentally, add 3 and 7 mentally to get 10, then divide by 2 to get 5. The solution set is {5}. Check mentally: 2 times 5 is 10, and 10 minus 3 is 7.

b) Mentally subtract 4 from 0 to get -4, then divide -4 by 3. The solution set is $\{-4/3\}$. Check mentally.

c) Isolate x on the left side of the equation. Subtract x from $2x$ to get x, and then subtract 3 from 8 to get 5. The solution set is {5}. Check.

d) In this equation, isolate x on the right side. Subtract x from $2x$ to get x on the right, then subtract 1 from -3 to get -4. The solution set is $\{-4\}$. Check.

e) Multiplying by 5 gives us the solution set $\{-10\}$. Check.   ◀

We can also learn to recognize simple identities and inconsistent equations without doing any work.

EXAMPLE 6   Solve the following equations mentally.

a) $4(x - 1) = 4x + 5$ b) $4(x - 1) = 4x - 4$

Solution

a) If we subtract $4x$ from each side of this equation, we will get $-4 = 5$. So this equation is inconsistent. The solution set is the empty set, $\varnothing$.

b) If we multiply to remove the parentheses on the left side, the two sides will be identical. This equation is an identity. The solution set is R.   ◀

Warm-ups

True or false?

1. The equation $5x - 7 = 3x + 9$ is equivalent to $2x - 7 = 16$.
2. The equation $-3x = 12$ is equivalent to $x = -4$.
3. The equation $-x = -6$ is equivalent to $x = 6$.
4. The equation $-13 - x = 0$ is equivalent to $-13 = x$.

> **5.** The equation $2x + 1 = 0$ is equivalent to $x = -1/2$.
> **6.** The solution set to $3x + 5 = 7$ is $\{2/3\}$.
> **7.** The equation $-3(x - 4) = 5 - 3x$ is an identity.
> **8.** The equation $x + 3x = 5x$ is an inconsistent equation.
> **9.** The equation $2(x + 4) = 2x + 9$ is an inconsistent equation.
> **10.** The equation $2(x + 4) = 3x + 8$ is an inconsistent equation.

2.2 EXERCISES

Solve each of the following equations. Practice combining some steps and look for more efficient ways to solve each equation. See Examples 1–4.

1. $3x - 9 = 0$　　　**2.** $5x + 1 = 0$　　　**3.** $3x - 2 = 6$　　　**4.** $-2x + 3 = 7$

5. $7 - x = -9$　　　**6.** $-3 - x = 3$　　　**7.** $-x + 4 = 9$　　　**8.** $-x - 7 = -8$

9. $7 = 5 - 4x$　　　**10.** $4 = 7 - 6x$　　　**11.** $3x + 5 = 4x - 1$

12. $2x - 7 = 3x + 1$　　　**13.** $-3x - 6 = -2x + 5$　　　**14.** $-6x - 3 = -5x + 7$

15. $5x - 8 = 3x + 1$　　　**16.** $6x - 5 = 9x - 3$

17. $2(x + 4) = 3(x - 5)$　　　**18.** $1 - 5(x - 2) = -4(x - 3)$

19. $5x + 10(x + 2) = 110$　　　**20.** $1 - 3(x - 2) = 4(x - 1) - 3$

21. $\dfrac{x}{2} - \dfrac{1}{3} = \dfrac{1}{6}$　　　**22.** $\dfrac{x}{3} - \dfrac{1}{4} = \dfrac{5}{12}$　　　**23.** $\dfrac{x - 3}{6} + 1 = 5$

24. $5 - \dfrac{x - 1}{4} = 2$　　　**25.** $.05x + 75 = .2x$　　　**26.** $.06x - 16 = .02x$

Solve each of the following equations mentally. If there are equations that you cannot solve mentally, solve them by writing out every step, then try again later. See Examples 5 and 6.

27. $2x = 6$　　　**28.** $3x = 15$　　　**29.** $3x = 4$　　　**30.** $2x = 5$

31. $-x = 7$　　　**32.** $-x = -13$　　　**33.** $x - 6 = -8$　　　**34.** $x + 4 = -9$

35. $x - 7 = 0$　　　**36.** $x + 9 = 0$　　　**37.** $5 - x = 0$　　　**38.** $-x - 19 = 0$

39. $\dfrac{1}{3}x = 6$　　　**40.** $\dfrac{1}{2}x = 10$　　　**41.** $-\dfrac{1}{4}x = -5$　　　**42.** $-\dfrac{1}{3}x = -5$

43. $\dfrac{x}{5} = -\dfrac{1}{3}$　　　**44.** $\dfrac{x}{2} = -\dfrac{2}{3}$　　　**45.** $\dfrac{1}{2}x + 1 = 15$　　　**46.** $\dfrac{1}{3}x - 1 = 8$

47. $3x - 1 = 0$　　　**48.** $5x - 2 = 0$　　　**49.** $-2x + 1 = 0$　　　**50.** $-3x + 7 = 0$

51. $3x - 2 = 7$　　　**52.** $-5x + 1 = 11$　　　**53.** $-4x + 2 = 1$　　　**54.** $6x - 7 = 11$

55. $1 - x = 7$　　　**56.** $3 - x = -5$　　　**57.** $5 - x = -3$　　　**58.** $4 - x = 6$

59. $\dfrac{1}{x} = \dfrac{1}{3}$

60. $\dfrac{1}{x} = -\dfrac{1}{5}$

61. $\dfrac{2}{3}x = 5$

62. $\dfrac{3}{2}x = 2$

63. $\dfrac{x-5}{3} = 2$

64. $\dfrac{x+6}{5} = 3$

65. $\dfrac{x-1}{4} = -3$

66. $\dfrac{x-4}{6} = -1$

67. $2x - 3 = x + 5$

68. $2x - 6 = x + 7$

69. $3x - 4 = 4x + 1$

70. $6x - 9 = 7x - 3$

71. $3(x + 1) = 3x + 3$

72. $2(x + 1) = 2x + 1$

73. $3(x + 1) = 3x + 1$

74. $2(x + 1) = 2x + 2$

75. $x + x = 3x$

76. $x + x = x$

77. $.1x = 5$

78. $.1x = 3$

79. $.03x = 6$

80. $.05x = 10$

81. $.02x = 8$

82. $.01x = 40$

83. $x \div \dfrac{1}{2} = 4$

84. $x \div \dfrac{1}{3} = 6$

85. $x \div \dfrac{1}{5} = 10$

86. $x \div \dfrac{1}{2} = 3$

2.3   Formulas

IN THIS SECTION:

- Solving for a Variable
- Finding the Value of a Variable

A real-life problem may involve many variable quantities that are related to each other. The relationship between these variables may be expressed by a formula. In this section we will combine our knowledge of evaluating expressions from Chapter 1 with the equation-solving skills of the last two sections in order to work with formulas.

Solving for a Variable

A formula is an equation involving two or more variables. The formula

$$A = LW$$

expresses the relationship between the length (L), width (W), and area (A) of a rectangle. The formula

$$C = \frac{5}{9}(F - 32)$$

expresses the relationship between the Fahrenheit and Celsius measurements of temperature. For example, if the Fahrenheit temperature (F) is 95° we can use the formula to find the Celsius temperature (C) as follows:

$$C = \frac{5}{9}(95 - 32) = \frac{5}{9}(63) = 35.$$

A temperature of 95°F is equivalent to 35°C.

In real-life problems, it is sometimes necessary to solve a formula for one variable in terms of the others without substituting numbers for the variables. We will use the same steps in solving for a particular variable as we did in solving linear equations.

EXAMPLE 1 Solve the formula $C = \frac{5}{9}(F - 32)$ for F.

Solution To solve the formula for F we isolate F on one side of the equation. Because there is a fraction in the equation, we would usually multiply by the LCD, 9. However, we can eliminate both the 9 and the 5 from the right side of the equation by multiplying by 9/5, the reciprocal of 5/9.

$$C = \frac{5}{9}(F - 32)$$

$$\frac{9}{5}C = \frac{9}{5} \cdot \frac{5}{9}(F - 32) \qquad \text{Multiply each side by the reciprocal of } \frac{5}{9}.$$

$$\frac{9}{5}C = F - 32 \qquad \frac{9}{5} \cdot \frac{5}{9} = 1$$

$$\frac{9}{5}C + 32 = F - 32 + 32 \qquad \text{Add 32 to each side.}$$

$$\frac{9}{5}C + 32 = F$$

This formula is usually written as $F = \frac{9}{5}C + 32$. Both this formula and the original formula express the relationship between C and F. Note that if we substitute 35 for C in this formula, we get

$$F = \frac{9}{5}(35) + 32 = 63 + 32 = 95. \qquad \blacktriangleleft$$

In the formula $S = P + Prt$, the variable P occurs twice. To solve for P we must isolate P on one side of the equation and have no occurrence of P on the other side. If we subtract Prt from each side, we get

$$S - Prt = P.$$

This formula is *not* solved for P. The following example shows how to convert the two occurrences of P into one by using the distributive property.

EXAMPLE 2 Solve $S = P + Prt$ for P.

Solution

$$S = P(1 + rt) \qquad \text{Factor out } P, \text{ using the distributive property.}$$

$$\frac{S}{(1 + rt)} = \frac{P(1 + rt)}{(1 + rt)} \qquad \text{Divide each side by the expression } (1 + rt).$$

$$\frac{S}{1 + rt} = P$$

Note that parentheses are really not needed around the expression $1 + rt$ in the denominator, because the fraction bar acts as a grouping symbol. ◄

When the variable for which we are solving occurs on opposite sides of the equation, we must move all terms involving that variable to the same side and then apply the distributive property as in the next example.

EXAMPLE 3 Solve $3x + 7 = -5xy + y$ for x.

Solution

$$3x + 5xy + 7 = y \qquad \text{Get all terms involving } x \text{ to the left side.}$$

$$3x + 5xy = y - 7 \qquad \text{Subtract 7 from each side.}$$

$$x(3 + 5y) = y - 7 \qquad \text{Factor } x \text{ out of } 3x + 5xy.$$

$$\frac{x(3 + 5y)}{3 + 5y} = \frac{y - 7}{3 + 5y} \qquad \text{Divide each side by } (3 + 5y).$$

$$x = \frac{y - 7}{3 + 5y}$$

◄

In Chapter 7 we will frequently have to solve equations involving x and y for one of the variables.

EXAMPLE 4 Solve $3x - 2y = 6$ for y.

Solution

$$-2y = -3x + 6 \qquad \text{Subtract } 3x \text{ from each side.}$$

$$\frac{-2y}{-2} = \frac{-3x + 6}{-2} \qquad \text{Divide each side by } -2.$$

$$\frac{-2y}{-2} = \frac{-3x}{-2} + \frac{6}{-2}$$

$$y = \frac{3x}{2} - 3$$

◄

Finding the Value of a Variable

In many situations, we know the values of all the variables in a formula except one. We can use the formula to determine the unknown value. A list of common formulas and their meanings is given inside the front cover. This list may be helpful for doing the exercises at the end of this section.

EXAMPLE 5 If the simple interest is $50, the principal is $500, and the time is 2 years, find the rate.

Solution We can calculate simple interest by using the formula $I = Prt$, found inside the front cover. First we solve the formula for r (rate), then insert the values of P (principal), I (interest), and t (time).

$$Prt = I$$

$$\frac{Prt}{Pt} = \frac{I}{Pt} \qquad \text{Divide each side by } Pt.$$

$$r = \frac{I}{Pt}$$

$$r = \frac{50}{500(2)} \qquad \text{Substitute values for } I, P, \text{ and } t.$$

$$r = .05 = 5\% \qquad \text{A rate is usually written as a percent.} \qquad \blacktriangleleft$$

In Example 5 we solved the formula for r and then inserted the values of the other variables. We could also have inserted the values of I, P, and t into the original formula and then solved for r. If we had to find the interest rate for many different loans, it would be simpler to solve for r first. To illustrate the two different approaches, we will work the next example by first inserting the values of the variables into the formula.

EXAMPLE 6 A trapezoid with area 30 square meters has a lower base of 10 meters and a height of 5 meters. Find the length of the upper base.

Solution The formula for the area of a trapezoid can be found inside the front cover. When solving a geometric problem, it is helpful to draw a diagram. First, put the given information on a diagram, as shown in Fig. 2.1. Substitute the given values into the formula for the area of a trapezoid and then solve for b_2.

$$A = \frac{1}{2}h\left(b_1 + b_2\right)$$

$$30 = \frac{1}{2}(5)(10 + b_2) \qquad \text{Substitute given values into the formula for the area of a trapezoid.}$$

$$60 = 5(10 + b_2) \qquad \text{Multiply each side by 2.}$$

$$12 = 10 + b_2 \qquad \text{Divide each side by 5.}$$

$$2 = b_2 \qquad \text{Subtract 10 from each side.}$$

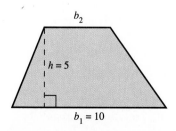

Figure 2.1

The upper base is 2 meters. $\qquad \blacktriangleleft$

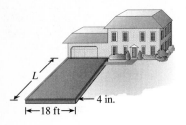

L

4 in.

|←18 ft→|

Figure 2.2

EXAMPLE 7   Millie has just completed pouring 14 cubic yards of concrete to construct a rectangular driveway. If the concrete is 4 inches thick and the driveway is 18 feet wide, then how long is her driveway?

Solution   First draw a diagram, as in Fig. 2.2. The driveway is a rectangular solid. From the front end papers, we can get the formula for volume of a rectangular solid, $V = LWH$. The volume of a rectangular solid is the product of its length, width, and height. Before we insert the values of the variables into the formula, we must convert all of them to the same unit of measurement. We will convert feet and inches to yards.

$$4 \text{ inches} = 4 \text{ in.} \cdot \frac{1 \text{ yd}}{36 \text{ in.}} = \frac{1}{9} \text{ yard}$$

$$18 \text{ feet} = 18 \text{ ft} \cdot \frac{1 \text{ yd}}{3 \text{ ft}} = 6 \text{ yards}$$

Now replace W, H, and V by the appropriate values.

$$V = LWH$$

$$14 = L \cdot 6 \cdot \frac{1}{9}$$

$$\frac{9}{6} \cdot 14 = L \qquad \text{Multiply each side by } \frac{9}{6}.$$

$$21 = L$$

The length of the driveway is 21 yards, or 63 feet.   ◄

Warm-ups

True or false?

1. If we solve $S = P + Prt$ for P, we get $P = S - Prt$.
2. In solving $S = P + Prt$ for P, we do not need the distributive property.
3. Solving $I = Prt$ for t give us $t = I - Pr$.
4. If $a = (bh)/2$, $b = 5$, and $h = 6$, then $a = 15$.
5. The perimeter of a rectangle is found by multiplying its length and width.
6. The volume of a rectangular box is found by multiplying its length, width, and height.
7. The length of an NFL football field, excluding the end zones, is 99 yards.
8. Solving $x - y = 5$ for y gives us $y = x - 5$.
9. If $x = -3$ and $y = -2x - 4$, then $y = 2$.
10. The area of a rectangle is a measurement of the total distance around the outside edge.

2.3 EXERCISES

Solve each formula for the specified variable. Many of these formulas are from the list of formulas inside the front cover. See Examples 1–3.

1. $I = Prt$ for t

2. $d = rt$ for r

3. $F = \dfrac{9}{5}C + 32$ for C

4. $A = \dfrac{1}{2}bh$ for h

5. $A = LW$ for W

6. $C = 2\pi r$ for r

7. $A = \dfrac{1}{2}\left(b_1 + b_2\right)$ for b_1

8. $A = \dfrac{1}{2}\left(b_1 + b_2\right)$ for b_2

9. $S = P + Prt$ for t

10. $S = P + Prt$ for r

11. $ab + a = 1$ for a

12. $y - wy = m$ for y

13. $xy + 5 = y - 7$ for y

14. $xy + 5 = x + 7$ for x

15. $xy^2 + xz^2 = xw^2 - 6$ for x

16. $xz^2 + xw^2 = xy^2 + 5$ for x

*17. $\dfrac{1}{R} = \dfrac{1}{R_1} + \dfrac{1}{R_2}$ for R_1

*18. $\dfrac{1}{a} + \dfrac{1}{b} = \dfrac{1}{2}$ for a

Solve for y. See Example 4.

19. $2x + 3y = 9$

20. $4y + 5x = 8$

21. $x - y = 4$

22. $y - x = 6$

23. $4x - 2y = 6$

24. $2y + 4x = -8$

25. $x + 3y = -6$

26. $3x - 2y = 12$

27. $x = 2y + 5$

28. $x = 4y - 8$

29. $x = \dfrac{1}{2}y + 3$

30. $x = \dfrac{1}{3}y + 2$

31. $\dfrac{1}{2}x - \dfrac{1}{3}y = 2$

32. $\dfrac{1}{3}x - \dfrac{1}{4}y = 1$

33. $y - 2 = \dfrac{1}{2}(x - 3)$

34. $y - 3 = \dfrac{1}{3}(x - 4)$

35. $y + 5 = 4(x - 6)$

36. $y + 2 = -3(x + 6)$

37. $.1725x + .0575y = 56.58$

38. $.0875x - .04375y = 87.5$

Find y, given that x = 3. See Examples 5 and 6.

39. $2x - 3y = 5$

40. $-3x - 4y = 4$

41. $-4x + 2y = 1$

42. $x - y = 7$

43. $y = -2x + 5$

44. $y = -3x - 6$

45. $-x + 2y = 5$

46. $-x - 3y = 6$

47. $y - 1.046 = 2.63(x - 5.09)$

48. $y - 2.895 = -1.07(x - 2.89)$

In each formula, find x, given that y = 2, z = -3, and w = 4. See Examples 5 and 6.

49. $wxy = 5$

50. $wxz = 4$

51. $x + xz = 7$

52. $xw - x = 3$

53. $w(x - z) = y(x - 4)$

54. $z(x - y) = y(x + 5)$

55. $w = \dfrac{1}{2}xz$

56. $y = \dfrac{1}{2}wx$

57. $\dfrac{1}{w} + \dfrac{1}{x} = \dfrac{1}{y}$

58. $\dfrac{1}{w} + \dfrac{1}{y} = \dfrac{1}{x}$

Solve each of the following problems. Refer to the endpapers inside the front cover for a list of common formulas. See Examples 6 and 7.

59. The area of a rectangle is 23 square yards. The width is 4 yards. Find the length.

60. The area of a rectangle is 55 square meters. The length is 7 meters. Find the width.

61. The volume of a rectangular box is 36 cubic feet. The bottom is 2 feet by 2.5 feet. Find the height of the box.

62. A shipping box has a volume of 2.5 cubic meters. The box measures one meter high by 1.25 meters wide. How long is the box?

63. The volume of a rectangular aquarium is 900 gallons. The bottom is 4 feet by 6 feet. Find the height of the aquarium. (Hint: There are 7.5 gallons per cubic foot.)

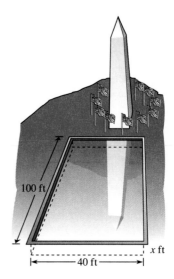

Figure for Exercise 64

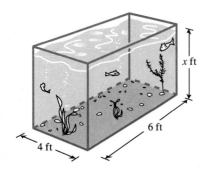

Figure for Exercise 63

68. The area of triangle is 40 square meters. If the height is 10 meters, then what is the length of the base?

69. The area of a trapezoid is 300 square inches. If the height is 20 inches and the lower base is 16 inches, then what is the length of the upper base?

64. A rectangular reflecting pool with a horizontal bottom holds 60,000 gallons of water. If the pool is 40 feet by 100 feet, how deep is the water?

65. If the simple interest on $1000 for 2 years is $300, then what is the rate?

66. If the simple interest on $2000 at 18% is $180, then what is the time?

67. The area of a triangle is 30 square feet. If the base is 4 feet, then what is the height?

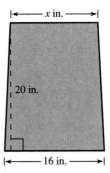

Figure for Exercise 69

70. The area of a trapezoid is 200 square centimeters. The bases are 16 centimeters and 24 centimeters. Find the height.

71. If it takes 600 feet of fence to enclose a rectangular lot that is 132 feet wide, then how deep is the lot?

72. The perimeter of a football field in the NFL, excluding the end zones, is $306 \frac{2}{3}$ yards. How wide is the field?

Figure for Exercise 72

73. If the circumference of the Earth is 25,000 miles, then what is the radius?

74. If a satellite travels 26,000 miles in each circular orbit of the Earth, then how high above the Earth is the satellite orbiting? (See Exercise 73.)

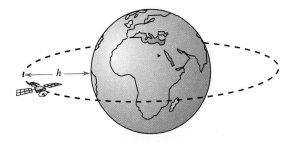

Figure for Exercise 74

75. If the circumference of a circle is 3π meters, then what is the radius?

76. If the circumference of a circle is 12π inches, then what is the diameter?

77. If the volume of a can is 30 cubic inches and the diameter of the top is 2 inches, then what is the height of the can?

78. If the volume of a cylinder is 6.3 cubic meters and the diameter of the lid is 1.2 meters, then what is the height of the cylinder?

*79. Harold Johnson lives on a four-sided, 12,000-square-foot lot that is bounded on two sides by parallel streets. The city has assessed him $320 for curb repair, $2 for each foot of property bordering these two streets. How far apart are the streets?

*80. Harold's sister, Maude, lives next door on a triangular lot of 15,000 square feet that also extends from street to street, but has frontage on only one street. What will her assessment be? (See Exercise 79.)

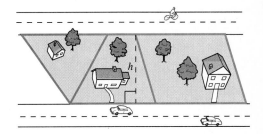

Figure for Exercise 79–81

*81. Harold's other sister, Juniper, lives on the other side of him on a lot of 45,000 square feet in the shape of a parallelogram. What will her assessment be? (See Exercise 79.)

*82. Harold's mother, who lives across the street, is pouring a concrete driveway, 12 feet wide and 4 inches thick, from the street straight to her house. This is too much work for Harold to do in one day, so his mother has agreed to buy 4 cubic yards of concrete each Saturday for 3 consecutive Saturdays. How far is it from the street to her house?

2.4 Applications

We are often faced with problems that can be most easily solved if we translate them into algebraic equations. Sometimes we can use formulas such as those inside the front cover. More often, we have to set up a new equation that describes the problem. We begin with translating verbal expressions into algebraic expressions.

Writing Algebraic Expressions

Consider the three consecutive integers 5, 6, and 7. Note that each integer is 1 larger than the previous integer. We can represent three *unknown* consecutive integers as follows:

Three Consecutive Integers

Let x = the first integer,

$x + 1$ = the second integer,

and $x + 2$ = the third integer.

Consider the three consecutive odd integers 7, 9, and 11. Note that each odd integer is 2 larger than the previous odd integer. We can represent three *unknown* consecutive odd integers as follows:

Three Consecutive Odd Integers

Let x = the first odd integer,

$x + 2$ = the second odd integer,

and $x + 4$ = the third odd integer.

Note that consecutive even integers as well as consecutive odd integers differ by 2.

How would we represent two numbers that have a sum of 8? If one of the numbers is 2, the other is certainly $8 - 2 = 6$. Thus if x is one of the numbers, then $8 - x$ is the other. The expressions

$$x \quad \text{and} \quad 8 - x$$

have a sum of 8 for any value of x.

EXAMPLE 1 Write algebraic expressions to represent the following.

a) Two numbers that differ by 12

b) The length of a rectangle, if the width is x meters and the perimeter is 10 meters

Solution

a) The expressions x and $x + 12$ differ by 12. Note that we could also use x and $x - 12$ for two numbers that differ by 12.

b) Because the perimeter is 10 and $P = 2L + 2W$, the sum of the length and width is 5. Because the width is x, the length is $5 - x$. ◄

Many verbal phrases occur repeatedly in applications. The following is a list of some frequently occurring verbal phrases and their translations into algebraic expressions.

Translating Words into Algebra

Verbal Phrase	*Algebraic Expression*
Addition	
The sum of a number and 8	$x + 8$
Five is added to a number	$x + 5$
Two more than a number	$x + 2$
A number increased by 3	$x + 3$
Subtraction	
Four is subtracted from a number	$x - 4$
Three less than a number	$x - 3$
The difference between 7 and a number	$7 - x$
Some number decreased by 2	$x - 2$
A number less 5	$x - 5$
Multiplication	
The product of 5 and a number	$5x$
Seven times a number	$7x$
Twice a number	$2x$
One-half of a number	$\frac{1}{2}x$

	Division
The ratio of a number to 6	$x/6$
The quotient of 5 and a number	$5/x$
Three divided by some number	$3/x$

Solving Problems

We will now see how algebraic expressions can be used to form an equation. If the equation correctly describes a problem, then we may be able to solve the equation to get the solution to the problem. Some problems in this section could be solved without using algebra. However, the purpose of this section is to gain experience in setting up equations and using algebra to solve problems. We will show a complete solution to each problem to gain the experience needed to solve more complex problems. We begin with a simple number problem.

EXAMPLE 2   The sum of three consecutive integers is 228. Find the integers.

Solution   We first represent the unknown quantities with variables. The unknown quantities are the three consecutive integers. Let

$$x = \text{the first integer,}$$
$$x + 1 = \text{the second integer,}$$
and
$$x + 2 = \text{the third integer.}$$

The sum of these three expressions for the consecutive integers is 228. We now write that fact as an equation, and then we solve the equation.

$$x + (x + 1) + (x + 2) = 228 \qquad \text{The sum of the integers is 228.}$$
$$3x + 3 = 228$$
$$3x = 225$$
$$x = 75$$
$$x + 1 = 76 \qquad \text{Identify the other unknown quantities.}$$
$$x + 2 = 77$$

To verify that these values are the correct integers, we compute $75 + 76 + 77 = 228$. The three consecutive integers that add up to 228 are 75, 76, and 77.   ◄

General Strategy for Problem Solving

The steps to follow in providing a complete solution to a verbal problem can be stated as follows.

STRATEGY
Solving Word Problems

1. Read the problem.
2. If possible, draw a diagram to illustrate the problem.
3. Choose a variable and write down what it represents.
4. Represent any other unknowns in terms of that variable.
5. Write an equation that fits the situation.
6. Solve the equation.
7. Be sure that your solution answers the question posed in the original problem.
8. Check your answer by using it to solve the original problem (not the equation).

We will now see how this strategy can be applied to various types of problems.

Geometric Problems

$2x + 1$

Figure 2.3

EXAMPLE 3 The length of a rectangular piece of property is 1 foot more than twice the width. If the perimeter is 302 feet, find the length and width.

Solution Let x = the width. Then, because the length is 1 foot more than twice the width, we get length = $2x + 1$ (see Fig. 2.3). The formula for the perimeter of a rectangle is $2L + 2W = P$. Replacing $2x + 1$ for L and x for W in this formula yields the equation below.

$$
\begin{array}{cc}
\quad L \qquad W \qquad P & \\
2(2x + 1) + 2(x) = 302 & \\
4x + 2 + 2x = 302 & \text{Remove the parentheses.} \\
6x = 300 & \\
x = 50 & \text{This is the width.} \\
2x + 1 = 101 & \text{Since } 2(50) + 1 = 101
\end{array}
$$

The length is 101 feet, and the width is 50 feet. Because $P = 2(101) + 2(50) = 302$, and 101 is 1 more than twice 50, we can be sure that the answer is correct.   ◄

Investment Problems

EXAMPLE 4 Greg Smith invested some money in a certificate of deposit with an annual yield of 9%. He invested twice as much money in a mutual fund with an annual yield of 12%. His interest from the two investments at the end of the year was $396. How much money was invested at each rate?

Solution Recall the formula $I = Prt$. In this problem the time is 1 year, so $I = Pr$. If we let x represent the amount invested at the 9% rate, then $2x$ is the amount invested at 12%. The interest on these investments is the principal times the rate, or $.09x$ and $.12(2x)$. It is often helpful to make a table for the unknown quantities.

Interest rate	9%	12%
Amount invested	x	$2x$
Amount of interest	$.09x$	$.12(2x)$

The fact that the total interest from the investments was $396 is expressed in the following equation.

$$.09x + .12(2x) = 396$$
$$.09x + .24x = 396$$ Note that we could multiply each side by 100 to eliminate the decimals.
$$.33x = 396$$
$$x = \frac{396}{.33}$$
$$x = \$1200 \qquad \text{Amount invested at 9\%}$$
$$2x = \$2400 \qquad \text{Amount invested at 12\%}$$

Greg invested $1200 at 9% and $2400 at 12%. To check this, we find $.09(\$1200) = \$108$ and $.12(\$2400) = \$288$. Now $\$108 + \$288 = \$396$. ◀

Figure 2.4

Mixture Problems

EXAMPLE 5 How many gallons of milk containing 5% butterfat must be mixed with 90 gallons of 1% milk to obtain 2% milk?

Solution If x represents the number of gallons of 5% milk, then $.05x$ represents the amount of fat in that milk. If we mix x gallons of 5% milk with 90 gallons of 1% milk, we will have $x + 90$ gallons of 2% milk (see Fig. 2.4). We can make a table to classify all of the unknown quantities.

Percentage of fat	5%	1%	2%
Gallons of milk	x	90	$x + 90$
Amount of fat	$.05x$	$.01(90)$	$.02(x + 90)$

In mixture problems we always write an equation that accounts for one of the

ingredients in the process. In this case, we write an equation to express the fact that the total fat from the first two milks is the same as the fat in the mixture.

$$.05x + .01(90) = .02(x + 90)$$

$$.05x + .9 = .02x + 1.8 \quad \text{Remove parentheses.}$$

$$.03x = .9 \quad \begin{array}{l}\text{Note that we chose to work with the decimals} \\ \text{rather than eliminate them.}\end{array}$$

$$x = 30$$

We should use 30 gallons of 5% milk. There are 1.5 gallons of fat in the 30 gallons of 5% milk. The 1% milk will contribute .9 gallons of fat, and there will be 2.4 gallons of fat in 120 gallons of 2% milk. Because $1.5 + .9 = 2.4$, we have the correct solution. ◄

EXAMPLE 6 A dealer has 10,000 gallons of unleaded gasoline. He wants to add just enough ethanol to make the fuel a 10% ethanol mixture. How many gallons of ethanol should be added?

Solution Let x represent the number of gallons of ethanol that should be added. Note that the original gasoline has no ethanol in it and that the ethanol is pure ethanol (100% ethanol). We can classify all of this information in a table.

	Gasoline	Ethanol	Mixture
Percentage of ethanol	0%	100%	10%
Number of gallons	10,000	x	$10,000 + x$
Amount of ethanol	0	x	$.10(10,000 + x)$

The equation expresses the fact that the amount of ethanol in the gasoline plus the amount of ethanol added is equal to the amount of ethanol in the final mixture. This accounts for all of the ethanol in the process.

$$0 + x = .10(10,000 + x)$$

$$x = 1000 + .1x \quad \text{Remove parentheses.}$$

$$.9x = 1000 \quad \text{Subtract .1x from each side of the equation.}$$

$$x = \frac{1000}{.9} \quad \text{Divide each side by .9.}$$

$$x = 1,111.1$$

The amount of ethanol has been rounded to the nearest tenth of a gallon, so we cannot expect this to check exactly. If we combine 10,000 gallons of gasoline with 1,111.1 gallons of ethanol, we obtain 11,111.1 gallons of fuel. If we take 10% of 11,111.1, we get 1,111.11 gallons of ethanol in the mixture. Note that we did not take 10% of 10,000 gallons to get the amount of ethanol to be added. The amount of ethanol is 10% of the total mixture, 10% of $(10,000 + x)$. ◄

Uniform-motion Problems

Problems that involve motion at a constant rate are referred to as **uniform-motion problems.**

EXAMPLE 7   Jennifer drove her car for 3 hours in a dust storm. When the skies cleared, she increased her speed by 30 miles per hour and drove for 4 more hours, completing her 295-mile trip. How fast did she travel during the dust storm?

Solution   Let x be Jennifer's speed during the dust storm and $x + 30$ her speed under clear skies. For problems involving motion, we need the formula $D = RT$ (distance equals rate times time). It is again helpful to make a table to classify the information given.

	Rate	Time	Distance
Dust storm	x	3	$3x$
Clear skies	$x + 30$	4	$4(x + 30)$

The equation expresses the fact that the total distance traveled was 295 miles.

$$3x + 4(x + 30) = 295$$
$$3x + 4x + 120 = 295 \qquad \text{Remove parentheses.}$$
$$7x = 175$$
$$x = 25 \text{ miles per hour}$$

Jennifer traveled 25 mph during the storm. Check this answer in the original problem.   ◄

Commission Problems

EXAMPLE 8   Sonia is selling her house through a real estate agent whose commission is 6% of the selling price. What should the selling price be so that Sonia can get $42,300.

Solution   Let x be the selling price. The commission is 6% of x, not 6% of 42,300. Sonia receives the selling price less the sales commission.

$$\text{Selling price} - \text{commission} = \text{Sonia's share}$$

$$x - .06x = \$42{,}300$$
$$.94x = 42{,}300$$
$$x = \frac{42{,}300}{.94} = \$45{,}000$$

The house should sell for $45,000. The commission is .06(45,000) = $2,700, and $45,000 − $2,700 = $42,300 is Sonia's share.   ◄

Warm-ups

True or false?

1. The first step in solving a word problem is to write the equation.
2. When solving word problems, always write what the variable stands for.
3. If we solve the equation correctly, then we must have the correct solution to the problem.
4. To represent two consecutive odd integers, we use x and $x + 1$.
5. We can represent two numbers that have a sum of 6 by x and $6 - x$.
6. Two numbers that differ by 7 can be represented by x and $x + 7$.
7. If $5x$ miles is 2 miles more than $3(x + 20)$ miles, then $5x + 2 = 3(x + 20)$.
8. If x is the selling price and the commission is 8% of the selling price, then the commission is $.08x$.
9. If you need $40,000 for your house, and the agent gets 10% of the selling price, then the agent gets $4,000, and the house sells for $44,000.
10. When we mix a 10% acid solution with a 14% acid solution, we can get as high as 24% acid.

2.4 EXERCISES

Find algebraic expressions for each of the following. See Example 1.

1. Two consecutive even integers
2. Two consecutive odd integers
3. Two numbers with a sum of 10
4. Two numbers with a difference of 3
5. Eighty-five percent of the selling price
6. The product of a number and 3
7. The distance traveled in 3 hours at x miles per hour
8. The time it takes to travel 100 miles at $x + 5$ miles per hour
9. The perimeter of a rectangle, if the width is x feet and the length is 5 feet longer than the width
10. The width of a rectangle, if the length is x meters and the perimeter is 20 meters

Show a complete solution for each problem.

Number problems. See Example 2.

11. The sum of three consecutive integers is 84. Find the integers.
12. Find three consecutive integers whose sum is 171.
13. Find three consecutive even integers whose sum is 252.
14. Find three consecutive even integers whose sum is 84.
15. Two consecutive odd integers have a sum of 128. What are the integers?
16. Four consecutive odd integers have a sum of 56. What are the integers?

Geometric problems. See Example 3.

17. If the perimeter of a rectangle is 278 meters and the length is one meter longer than twice the width, then what are the length and width?

18. A frame maker made a large picture frame using 10 feet of frame molding. If the length of the finished frame was 2 feet more than the width, then what were the dimensions of the frame?

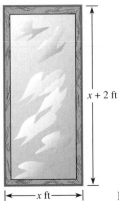

$x + 2$ ft

|←——x ft——→| Figure for Exercise 18

19. Having finished fencing the perimeter of a triangular piece of land, Lance observed that the second side was just 10 feet short of being twice as long as the first side, and the third side was exactly 50 feet longer than the first side. If he used 684 feet of fencing, what are the lengths of the three sides?

20. A navigational flag in the shape of an isosceles triangle has a base that is 3.5 inches shorter than either of the equal sides. If the perimeter of the flag is 49 inches, what is the length of the equal sides?

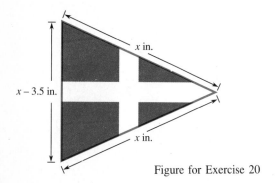

x in.

$x - 3.5$ in.

x in.

Figure for Exercise 20

21. Farmer Hodges has 50 feet of fencing to make a rectangular hog pen beside a very large red barn. He needs to fence only three sides, because the barn will form the fourth side. Studies have shown that under those conditions, the side parallel to the barn should be 5 feet longer than twice the width. If Farmer Hodges uses all of the fencing, what should the dimensions be?

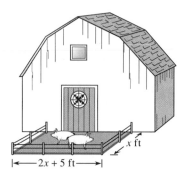

x ft

|←——$2x + 5$ ft——→|

Figure for Exercise 21

22. A carpenter made a doorway that is 1 foot taller than twice the width. If she used 3 pieces of door edge molding with a total length of 17 feet, then what are the approximate dimensions of the doorway?

Investment problems. See Example 4.

23. Ahmed invested some money at 6% simple interest and some money at 10% simple interest. He put $1000 more in the second investment than he put in the first. If the income from both investments was $340, then how much did he invest at each rate?

24. Samantha lent her brother some money at 9% simple interest and her sister one-half as much money at 16% simple interest. If she received a total of 34 cents in interest, then how much did she lend to each one?

25. Norman invested one-half of his inheritance in a certificate of deposit that had a 10% yield. He lent one-quarter of his inheritance to his brother-in-law at 12% simple interest. His income from these two investments was $6400. How much was the inheritance?

26. Gary invested one-third of his insurance settlement in a certificate of deposit that yielded 12%. He also invested one-third in Tara's computer business. Tara paid Gary 15% on this investment. If Gary's total income from these investments was $10,800, what was his insurance settlement?

Mixture problems. See Examples 5 and 6.

27. How many gallons of a 5% acid solution should be mixed with 20 gallons of a 10% acid solution to obtain an 8% acid solution?

28. How many liters of a 10% alcohol solution should be mixed with 12 liters of a 20% alcohol solution to obtain a 14% alcohol solution?

29. A gallon of vinegar is labeled 5% acidity. How many ounces of pure acid must be added to get 6% acidity.

30. A gallon of bleach is labeled 5.25% sodium hypochlorite by weight. If a gallon of bleach weighs 8.3 pounds, then how many ounces of sodium hypochlorite must be added so that the bleach will be 6% sodium hypochlorite?

Uniform-motion problems. See Example 7.

31. Carlo drove for 3 hours in a fog, then increased his speed by 30 miles per hour and drove 6 more hours. If his total trip was 540 miles, then what was his speed in the fog?

32. Louise walked for 2 hours then ran for 1 1/2 hours. If she runs twice as fast as she walks and the total trip was 20 miles, then how fast does she run?

33. A commuter bus takes 2 hours to get downtown, while an express bus, averaging 25 mph faster, takes 45 minutes to cover the same route. What is the average speed for the commuter bus?

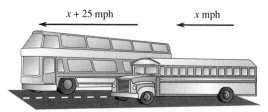

$x + 25$ mph x mph

Figure for Exercise 33

34. A freight train takes 1 1/4 hours to get to the city, while a passenger train averaging 40 mph faster takes only 45 minutes to cover the same distance. What is the average speed of the passenger train?

Commission problems. See Example 8.

35. Kwong wants to get $80,000 for his house. The real estate agent charges 8% of the selling price for selling the house. What should the selling price be?

36. Martha sells hot tamales at a sidewalk stand. Her total receipts, including the 5% sales tax, were $915.60. What amount of sales tax did she collect?

37. Olinda bought a new car. The selling price plus the 7% state sales tax was $9,041.50. What was the selling price?

38. Hector is selling his car through a broker. Hector wants to get $3000 for himself, but the broker gets a commission of 10% of the selling price. What should the selling price be?

Miscellaneous.

39. Nicodemo blends 30 kilograms of premium Brazilian coffee with 50 kilograms of standard Colombian coffee. If the Brazilian coffee sells for $10 per kilogram and the Colombian coffee sells for $8 per kilogram, then what should the price per kilogram be for the blended coffee?

40. Cheryl's Famous Pumpkin Pie Seasoning consists of a blend of cinnamon, nutmeg, and cloves. When Cheryl mixes up a batch, she uses 200 ounces of cinnamon, 100 ounces of nutmeg, and 100 ounces of cloves. If cinnamon sells for $1.80 per ounce, nutmeg sells for $1.60 per ounce, and cloves sell for $1.40 per ounce, what should be the price per ounce of the mixture?

41. Dried bananas sell for $.80 per quarter-pound and dried apricots sell for $1.00 per quarter-pound. How many pounds of apricots should be mixed with 10 pounds of bananas to get a mixture that sells for $.95 per quarter-pound?

Photo for Exercise 41

42. Cashews sell for $1.20 per quarter-pound and Brazil nuts sell for $1.50 per quarter-pound. How many pounds of cashews should be mixed with 20 pounds of Brazil nuts to get a mix that sells for $1.30 per quarter-pound?

43. A mechanic finds that a car with a 20-quart radiator has a mixture containing 30% antifreeze in it. How much of this mixture would he have to drain out and replace with pure antifreeze to get a 50% antifreeze mixture?

44. A mechanic has found that a car with a 16-quart radiator has a 40% antifreeze mixture in the radiator. She has on hand a 70% antifreeze solution. How much of the 40% solution would she have to replace with the 70% solution to get the solution in the radiator up to 50%?

45. Uncle Albert's estate is to be divided among his three nephews. The will specifies that Daniel receive one-half of the amount that Brian receives and Raymond receive $1000 less than one-third of the amount that Brian receives. If the estate amounts to $25,400, then how much does each inherit?

46. Mary Hall's will specifies that her lawyer is to liquidate her assets and divide the proceeds among her three sisters. Lena's share is to be one-half of Lisa's, and Lisa's share is to be one-half of Lauren's. If the lawyer has agreed to a fee that is equal to 10% of the largest share and the proceeds amount to $164,428, then how much does each person get?

47. If the larger of 2 consecutive integers is subtracted from twice the smaller, the result is 21. Find the integers.

48. If the smaller of 2 consecutive odd integers is subtracted from twice the larger, the result is 13. Find the integers.

49. Stacy has 70 meters of fencing and plans to make a square pen. In one side she is going to leave an opening that is one-half the length of the side. If she uses all 70 meters of fencing, how big can the square be?

50. Shawn is building a tool shed with a square foundation and has enough siding to cover 32 linear feet of walls. If he leaves a 4-foot space for a door, then what size foundation would use up all of his siding?

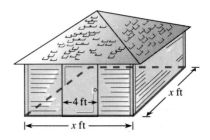

Figure for Exercise 50

51. Joan had $3000 to invest. She invested part of it in an investment paying 8% and the remainder in an investment paying 10%. If the total income on these investments was $290, then how much did she invest at each rate?

52. Dorothy had $8000 to invest. She invested part of it in an investment paying 6% and the rest in an investment paying 9%. If the total income from these investments was $690, then how much did she invest at each rate?

53. Rosalina has two solutions available in the laboratory, one with 5% alcohol and the other with 10% alcohol. How much of each should she mix together to obtain 5 gallons of an 8% solution?

54. Tanisha has a solution containing 12% alcohol. How much of this solution and how much water would she have to use to get 6 liters of a solution containing 10% alcohol?

*55. In 6 years, Todd will be twice as old as Darla was when they met 6 years ago. If their ages total 78 years, then how old are they now?

*56. The three Hoffman brothers advertise that together they have a century of plumbing experience. Bart has twice the experience of Al, and in 3 years, Carl will have twice the experience that Al had a year ago. How many years of experience does each of them have?

2.5 Inequalities

IN THIS SECTION:

- Basic Ideas
- Graphing Inequalities
- Solving Linear Inequalities
- Applications

So far in this chapter we have been working with equations. Equations express the equality of two algebraic expressions. But we are often concerned with two algebraic expressions that are not equal, one expression being greater than or less than the other. In this section we will begin our study of inequalities.

Basic Ideas

Statements that express the inequality of algebraic expressions are called **inequalities.** The symbols that we use to express inequality are given below with their meanings.

**DEFINITION
Inequality Symbols**

Symbol	Meaning
$<$	Is less than
$\leq$	Is less than or equal to
$>$	Is greater than
$\geq$	Is greater than or equal to

It seems clear that 5 is less than 10, but how do we compare -5 and -10? If we think of negative numbers as debts, we would say that -10 is the larger debt. However, in algebra the size of a number is determined only by its position on the number line. For two numbers a and b, we say that a is *less* than b if and only if a is to the *left* of b on the number line. To compare -5 and -10, we locate each point on the number line in Fig. 2.5. Because -10 is to the left of -5 on the number

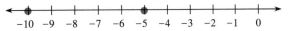

Figure 2.5

line, we say that -10 is less than -5. In symbols,

$$-10 < -5.$$

We say that a is greater than b if and only if a is to the *right* of b on the number line. Thus we can also write

$$-5 > -10.$$

The statement $a \leq b$ is true if a is less than b or if a is equal to b. The statement $a \geq b$ is true if a is greater than b or if a equals b. For example, the statement $3 \leq 5$ is true, and so is the statement $5 \leq 5$.

EXAMPLE 1 Determine whether each statement is true or false.

a) $-5 < 3$ **b)** $-9 > -6$ **c)** $-3 \leq 2$ **d)** $4 \geq 4$

Solution

a) True, since -5 is to the left of 3 on the number line. In fact, any negative number is less than any positive number.
b) False, because -9 is not located to the right of -6.
c) True, because -3 is less than 2.
d) True, because $4 = 4$ is true. ◄

Graphing Inequalities

The statement $x < 3$ means that x is a number to the left of 3 on the number line. The numbers 1, 0, and -2 are all to the left of 3 on the number line, and so the statements

$$1 < 3, \qquad 0 < 3, \qquad \text{and} \qquad -2 < 3$$

are true statements. The set of all numbers that give a true statement when used as a replacement for x is called the **solution set** to the inequality. The solution set to the inequality $x < 3$ is the set of all real numbers to the left of 3 on the number line, in symbols, $\{x \mid x < 3\}$.

When we indicate the solution set on the number line, we are **graphing** the solution set, or graphing the inequality. We can gain a better understanding of inequalities by graphing their solution sets. The graph of $\{x \mid x < 3\}$, the solution set to $x < 3$, is shown in Fig. 2.6. We shade the region to the left of 3, but we use an open circle at 3 to indicate that 3 is not a solution to the inequality $x < 3$.

Figure 2.6

The solution set to an inequality can be written using **interval notation.** For example, $\{x \mid x < 3\}$ is written as $(-\infty, 3)$ and is referred to as the interval of real numbers between $-\infty$ (negative infinity) and 3. We use parentheses to indicate that a number is not included in an interval, and brackets to indicate that a number is included in an interval. For example, $\{x \mid x \geq 1\} = [1, +\infty)$.

EXAMPLE 2   Graph the solution set to each inequality on the number line.

a) $x > -5$ b) $x \leq 2$

Solution

a) The solution set to the inequality $x > -5$, $\{x \mid x > -5\}$, includes all numbers to the right of -5 on the number line. This set is written in interval notation as $(-5, +\infty)$, and is graphed in Fig. 2.7.

Figure 2.7

b) The solution set to the inequality $x \leq 2$ includes 2 and all numbers to the left of 2. To indicate that 2 is included, we draw a solid dot at 2. The graph of $\{x \mid x \leq 2\}$ is shown in Fig. 2.8. The interval notation for this set is $(-\infty, 2]$.

Figure 2.8   ◄

Solving Linear Inequalities

In Section 2.1 we defined a linear equation as any equation that could be written in the form $ax + b = 0$. If we replace the equality symbol in a linear equation with an inequality symbol, we have a **linear inequality.**

DEFINITION
Linear Inequality

> A linear inequality in one variable x is any inequality that can be written in the form $ax + b < 0$, where a and b are real numbers, with $a \neq 0$.

We stated the definition using the symbol $<$, but we can also use any of the other inequality symbols in place of it.

Consider what happens to an inequality when we perform the same operation on each side. If we start with the inequality $2 < 6$ and add 2 to each side, we get the true statement $4 < 8$. Examine the results of performing the same operation on each side of $2 < 6$.

Perform these operations on each side.

		Add 2	Subtract 2	Multiply by 2	Divide by 2
Start with	$2 < 6$	$4 < 8$	$0 < 4$	$4 < 12$	$1 < 3$

All of the resulting inequalities are correct. Now if we repeat these operations using -2, we get the results shown below.

		Add -2	Subtract -2	Multiply by -2	Divide by -2
Start with	$2 < 6$	$0 < 4$	$4 < 8$	$-4 > -12$	$-1 > -3$

Notice that the direction of the inequality is the same for all of the results except the last two. When we multiplied each side by -2 and when we divided each side by -2, we had to reverse the inequality symbol to get a correct result. In general, *when we multiply or divide each side on an inequality by a negative number, the inequality symbol must be reversed.*

EXAMPLE 3   Solve the inequality $5 - 3x < 11$. State and graph the solution set.

Solution   We proceed exactly as we do when solving an equation.

$5 - 3x < 11$

$-3x < 6$ Subtract 5 from each side.

$x > -2$ Divide each side by -3 and reverse the inequality symbol.

All of these inequalities are equivalent, so the solution set to the original is the same as the solution set to the last. The solution set is written in set notation as $\{x \mid x > -2\}$ and in interval notation as $(-2, +\infty)$. The graph is shown in Fig. 2.9.

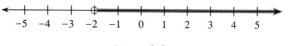

Figure 2.9   ◄

While a linear equation has only one solution, a linear inequality has infinitely many solutions. We find the solution to a linear inequality by converting it into an equivalent inequality with an obvious solution set. **Equivalent inequalities** are inequalities with the same solution set. We have rules for obtaining equivalent inequalities just like the rules for obtaining equivalent equations. We now state these properties of inequality formally.

Properties of Inequality

If A and B are algebraic expressions and C is a real number, then the inequality $A < B$ is equivalent to

$$A + C < B + C$$ **Addition Property of Inequality**

$$A - C < B - C$$ **Subtraction Property of Inequality**

$$CA < CB \ (C \text{ positive})$$ **Multiplication Property of Inequality**

$$CA > CB \ (C \text{ negative})$$

$$\frac{A}{C} < \frac{B}{C} \ (C \text{ positive})$$ **Division Property of Inequality**

$$\frac{A}{C} > \frac{B}{C} \ (C \text{ negative})$$

We stated the above properties only for $A < B$. Similar properties hold for $\leq$, $>$, and $\geq$. These rules tell us that we may perform operations on each side of an inequality just as we do when solving equations. However, *when we multiply or divide by a negative number, we must reverse the direction of the inequality symbol.*

EXAMPLE 4   Solve $\dfrac{8 + 3x}{-5} \geq -4$. State and graph the solution set.

Solution

$$8 + 3x \leq 20$$ Multiply each side by -5 and reverse the inequality.

$$3x \leq 12$$ Subtract 8 from each side.

$$x \leq 4$$ Divide each side by 3.

The solution set is $\{x \mid x \leq 4\}$. Its graph is shown in Fig. 2.10.

$$
\begin{array}{ccccccccccc}
-5 & -4 & -3 & -2 & -1 & 0 & 1 & 2 & 3 & 4 & 5
\end{array}
$$

Figure 2.10   ◀

EXAMPLE 5   Solve $\dfrac{1}{2}x - \dfrac{2}{3} \leq x + \dfrac{4}{3}$. State and graph the solution set.

Solution

$$6 \cdot \frac{1}{2}x - 6 \cdot \frac{2}{3} \leq 6x + 6 \cdot \frac{4}{3}$$ Multiply each side by 6, the LCD.

$$3x - 4 \leq 6x + 8$$ The inequality symbol is not reversed with multiplication by a positive number.

$$3x \leq 6x + 12$$ Add 4 to each side.

$$-3x \leq 12$$ Subtract 6x from each side.

$$x \geq -4$$ Divide each side by -3 and reverse the inequality symbol.

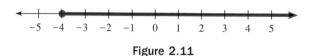

Figure 2.11

The solution set is $\{x \mid x \geq -4\}$. Its graph is shown in Fig. 2.11.   ◀

Applications

Inequalities occur in applications just as equations do. To use inequalities, we must be able to translate a verbal problem into an algebraic inequality. Inequality can be expressed verbally in a variety of ways.

Two phrases that can be misunderstood are "at most" and "at least." If a boat has a capacity of *at most* 4 people, then it will hold 4 people or fewer. If you must be *at least* 18 to vote, then you can vote if you are 18 years old or older. The phrase

"at most"   means   "less than or equal to,"

and the phrase

"at least"   means   "greater than or equal to."

Consider the statement "Chris will pay no more than $200 for a suit." If the price Chris will pay is *not greater than* $200, then it is *less than or equal to* $200. This example illustrates an important property of the real numbers, the **trichotomy property.**

Trichotomy Property

For any two real numbers a and b, exactly one of the following is true:
$$a < b, \qquad a = b, \qquad \text{or } a > b.$$

The trichotomy property expresses the fact that there are only three options. We can use the trichotomy property to understand the phrase "no fewer than 10 programmers." If the number of programmers is not less than 10, then by the trichotomy property it must be equal to or greater than 10.

We basically follow the same steps to solve problems involving inequalities as we do to solve problems involving equations.

EXAMPLE 6 Lois plans to spend less than $500 on an electric dryer, including the 9% sales tax and a $64 set-up charge. In what range is the selling price of the dryer that she can afford?

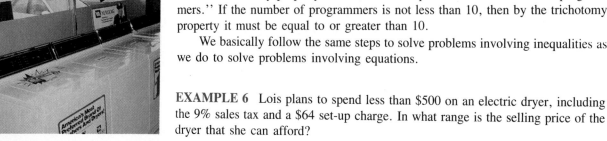

Solution   If we let x represent the selling price of the dryer, then the amount of sales tax is $.09x$. To keep the total cost less than $500, x must satisfy the inequality

$$x + .09x + 64 < 500.$$

$$1.09x < 436 \qquad \text{Subtract 64 from each side.}$$

$$x < \frac{436}{1.09} \qquad \text{Divide each side by 1.09.}$$

$$x < 400$$

The selling price of the dryer must be less than $400.   ◄

Note that if we had written the equation $x + .09x + 64 = 500$ for the last example, we would have gotten $x = 400$. We could then have concluded that the selling price must be less than $400. This would certainly solve the problem, but it would not illustrate the use of inequalities. The original problem describes an inequality, and we should solve it as an inequality.

EXAMPLE 7   Tessie owns a piece of land on which she still owes $12,760 to the bank. Hard times have befallen Tessie, and she wants to sell this land and at least pay off the mortgage. The real estate agent gets 6% of the selling price, and her city has a $400 real estate transfer tax paid by the seller. What should the range of the selling price be in order for Tessie to get at least enough money to pay off her mortgage?

Solution   If x is the selling price, then the commission is $.06x$. We can write an inequality expressing the fact that the selling price minus the real estate commission minus the $400 tax must be at least $12,760.

$$x - .06x - 400 \geq 12,760$$

$$.94x - 400 \geq 12,760 \qquad 1 - .06 = .94$$

$$.94x \geq 13,160 \qquad \text{Add 400 to each side.}$$

$$x \geq \frac{13,160}{.94} \qquad \text{Divide each side by .94.}$$

$$x \geq 14,000$$

The selling price must be at least $14,000 for Tessie to pay off the mortgage.   ◄

Warm-ups

True or false?

1. $0 < 0$ **2.** $-300 > -2$ **3.** $-60 \leq -60$

4. The inequality $6 < x$ is equivalent to $x < 6$.

5. The inequality $-2x < 10$ is equivalent to $x < -5$.

6. The inequality $3x \geq -12$ is equivalent to $x \leq -4$.

7. The inequality $-x > 4$ is equivalent to $-4 > x$.

8. If x is no larger than 8, then $x \leq 8$.

9. If m is any real number, then exactly one of the following is true: $m < 0$, $m = 0$, or $m > 0$.

10. The number -2 is a member of the solution set to the inequality $3 - 4x \leq 11$.

2.5 EXERCISES

Determine whether each inequality is true or false. See Example 1.

1. $-3 < -9$

2. $-8 > -7$

3. $0 \leq 8$

4. $-6 \geq -8$

5. $(-3)20 > (-3)40$

6. $(-1)(-3) < (-1)(5)$

7. $9 - (-3) \leq 12$

8. $(-4)(-5) + 2 \geq 21$

Graph each inequality. See Example 2.

9. $x \leq -1$

10. $x \geq -7$

11. $x > 20$

12. $x < 30$

13. $3 < x$

14. $-2 > x$

15. $x > 2.3$

16. $x < 4.5$

Solve each inequality. Graph the solution set and write it in set notation. See Examples 3, 4, and 5.

17. $2x - 3 > 7$

18. $3x - 2 < 6$

19. $3 - 5x \leq 18$

20. $5 - 4x \geq 19$

21. $\dfrac{x - 3}{-5} < -2$

22. $\dfrac{2x - 3}{4} > 6$

23. $\dfrac{5 - 3x}{4} \leq 2$

24. $\dfrac{7 - 5x}{-2} \geq -1$

25. $3x - 8 > 0$

26. $7x < 0$

27. $4 - 3x < 0$

28. $9 - 2x \leq 0$

29. $-\dfrac{5}{6}x \geq -10$

30. $-\dfrac{2}{3}x \geq -8$

31. $3 - \dfrac{1}{4}x \geq 2$

32. $5 - \dfrac{1}{3}x > 2$

33. $\dfrac{1}{4}x - \dfrac{1}{2} < \dfrac{1}{2}x - \dfrac{2}{3}$

34. $\dfrac{1}{3}x - \dfrac{1}{6} < \dfrac{1}{6}x - \dfrac{1}{2}$

For each of the following word problems, write an inequality and solve it. See Examples 6 and 7.

35. Jennifer is going shopping for a new car. In addition to the price of the car, there is an 8% sales tax and a $172 title and license fee. If Jennifer decides that she will spend less than $10,000 total, then what is the price range for the car?

36. Tak Fong is going to buy a sewing machine in a city with a 10% sales tax. He has at most $700 to spend. In what price range should he look?

37. Sophia is shopping for a new truck in a city with a 9% sales tax. There is also an $80 title and license fee to pay. She wants to get a good truck and plans to spend at least $10,000. What is the price range for the truck?

38. Larry, Curly, and Moe are going to buy their mother a color television set. Larry has a better job than Curly and agrees to contribute twice as much as Curly. Moe is unemployed and can only spare $50. If the kind of television

Mama wants costs at least $600, then what is the price range for Curly's contribution?

39. Professor Jorgenson gives only a midterm exam and a final exam. The semester average is computed by taking 1/3 of the midterm exam score plus 2/3 of the final exam score. To get a C or better, Stanley must have a semester average of at least 70. If Stanley scored only 56 on the midterm, then for what range of scores on the final exam would he get a C or better?

40. Professor Brown counts her midterm as 2/3 of the grade and her final as 1/3 of the grade. Wilbert scored only 56 on the midterm. What range of scores on the final exam would put Wilbert's average above 70?

41. A pair of ordinary jeans at A-Mart costs $50 less than a designer pair of jeans at Enrico's. In fact, you can buy 4 pairs of A-Mart jeans for less than 1 pair of Enrico's jeans. What is the price range for a pair of A-Mart jeans?

42. Al and Rita both drive parcel delivery trucks. Al averages 20 miles per hour less than Rita. In fact, Al is so slow that in 5 hours he covered fewer miles than Rita did in 3 hours. What are the possible values for Al's rate of speed?

Solve each inequality and graph the solution set.

43. $4.273 + 2.8x \le 10.985$

44. $1.064 < 5.94 - 3.2x$

45. $3.25x - 27.39 > 4.06 + 5.1x$

46. $4.86(3.2x - 1.7) > 5.19 - x$

*47. $-\dfrac{1}{2}(2x - 3) + \dfrac{1}{3}(4 - 6x) \ge \dfrac{1}{4}(7 - 2x) - 3$

*48. $\dfrac{3}{5}(x - 3) - \dfrac{1}{4}(7 - 5x) < \dfrac{2}{3}(3 - x) - 5$

*49. $\dfrac{1}{2}\left(\dfrac{3}{5}x - \dfrac{1}{8}\right) + 1 > \dfrac{1}{8}\left(\dfrac{3}{4}x + \dfrac{1}{10}\right)$

*50. $-\dfrac{2}{3}\left(5x - \dfrac{3}{4}\right) + \dfrac{1}{8}\left(\dfrac{1}{2}x - 2\right) > \dfrac{3}{8}$

Write each set in interval notation.

*51. $\{x \mid x > 1\}$

*52. $\{x \mid x < 3\}$

*53. $\{x \mid x \le -3\}$

*54. $\{x \mid x \ge -2\}$

*55. $\{x \mid x < 5\}$

*56. $\{x \mid x > -7\}$

*57. $\{x \mid x \ge -4\}$

*58. $\{x \mid x \le -9\}$

2.6 Compound Inequalities

IN THIS SECTION:

- Basics
- Graphing the Solution Set
- Applications

In this section we will use the ideas of union and intersection from Chapter 1, along with our knowledge of inequalities from the last section to work, with compound inequalities.

Basics

The inequalities that we studied in the last section are referred to as **simple inequalities.** If we join two simple inequalities with the connective "and" or the connective "or," we get a **compound inequality.** The statement

$$x > 2 \quad \text{and} \quad x < 5$$

is a compound inequality. A number is considered a solution to this inequality if the statement is true when the number is used in place of x. A compound inequality using the connective "and" is true only if *both* simple inequalities are true. If we let $x = 3$, then the compound inequality

$$3 > 2 \quad \text{and} \quad 3 < 5$$

is true, and 3 is a member of the solution set to the compound inequality $x > 2$ and $x < 5$. If we let $x = 6$, then the compound inequality

$$6 > 2 \quad \text{and} \quad 6 < 5$$

is false because the first simple inequality is true but the second is false.
    Now consider the compound inequality

$$x < 3 \quad \text{or} \quad x \geq 7.$$

A compound inequality using the connective "or" is true if one or the other or both of the simple inequalities are true. If we let $x = 2$, then the compound inequality

$$2 < 3 \quad \text{or} \quad 2 \geq 7$$

is true, and 2 is a member of the solution set to the compound inequality $x < 3$ or $x \geq 7$. A compound inequality using "or" is false only if both simple inequalities are false. If we let $x = 4$, we get

$$4 < 3 \quad \text{or} \quad 4 \geq 7.$$

This statement is false because both simple inequalities are false.

EXAMPLE 1   Determine whether each compound inequality is true or false.

a) $-3 < 5$ and $-3 > -1$ **b)** $-2 > -4$ and $0 \leq 0$
c) $4 \leq 7$ or $4 \geq 6$ **d)** $7 < 5$ or $7 < 6$

Solution

a) False, because $-3 > -1$ is false.
b) True, because both simple inequalities are true.
c) True, because $4 \leq 7$ is true.
d) False, because both simple inequalities are false.   ◀

EXAMPLE 2   Determine whether or not the number 5 is a member of the solution set to each compound inequality.

a) $x < 6$ and $x < 9$ **b)** $x \leq 7$ or $x \leq 3$

Solution

a) Because $5 < 6$ and $5 < 9$ are both true, 5 is a member of the solution set.

b) Because $5 \leq 7$ is true, it does not matter that $5 \leq 3$ is false. So 5 is a member of the solution set. ◄

Graphing the Solution Set

Consider the compound inequality

$$x > 2 \quad \text{and} \quad x < 5.$$

A number is in the solution set to this compound inequality only if the number is larger than 2 and smaller than 5. The number must be in the solution set to $x > 2$ and in the solution set to $x < 5$. Thus the solution set to this compound inequality is the intersection of the two solution sets to the simple inequalities. In symbols,

$$\{x \mid x > 2 \quad \text{and} \quad x < 5\} = \{x \mid x > 2\} \cap \{x \mid x < 5\}.$$

EXAMPLE 3 Graph the solution set to the compound inequality $x > 2$ and $x < 5$.

Solution The solution set is the intersection of the two solution sets to the simple inequalities. To see what this intersection is, we first sketch the graph of each simple inequality. The graph of $x > 2$ is shown in Fig. 2.12, and the graph of $x < 5$ is shown in Fig. 2.13. The intersection of these two solution sets is the portion of the number line that is shaded on both graphs—just the part between 2 and 5, not including the endpoints. The graph of $\{x \mid x > 2 \text{ and } x < 5\}$ is shown in Fig. 2.14. We write this set in interval notation as $(2, 5)$.

Figure 2.12

Figure 2.13

Figure 2.14 ◄

Now consider the statement

$$x > 4 \quad \text{or} \quad x < -1.$$

A number is in the solution set to this compound inequality if it satisfies one or the other or both of the simple inequalities. The number must be in the union of the solution sets to the simple inequalities. In symbols,

$$\{x \mid x > 4 \quad \text{or} \quad x < -1\} = \{x \mid x > 4\} \cup \{x \mid x < -1\}.$$

EXAMPLE 4   Graph the solution set to the compound inequality $x > 4$ or $x < -1$.

Solution   The solution set is the union of the solution sets to the simple inequalities. To see what the graph of this set looks like, we start by graphing each simple inequality. The graph of $x > 4$ is shown in Fig. 2.15, and the graph of $x < -1$ is shown in Fig. 2.16. We graph the union of these two sets by putting both shaded regions together on the same line. The graph of $\{x \mid x > 4 \text{ or } x < -1\}$ is shown in Fig. 2.17. This set is written in interval notation as $(-\infty, -1) \cup (4, +\infty)$.

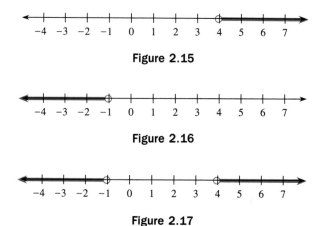

Figure 2.15

Figure 2.16

Figure 2.17

It is always easier to obtain the union of two solution sets than the intersection. To graph the union, we can just shade in each of the two solution sets on the same line. To graph the intersection, we must be careful not to shade too much. For intersection, shade only numbers that satisfy *both* inequalities. Numbers that satisfy one inequality but not the other are omitted.

It is not always necessary to graph the solution set to each simple inequality before graphing the solution set to the compound inequality. We can save time and work if we learn to think of the two preliminary graphs but draw only the final one.

EXAMPLE 5   Sketch the graph and write the solution set in set notation to each compound inequality.

a) $x < 3$ and $x < 5$ **b)** $x > 4$ or $x > 0$
c) $x < 2$ and $x > 6$ **d)** $x < 3$ or $x > 1$

Solution

a) Shade only the numbers that are both less than 3 and less than 5. Note that the numbers between 3 and 5 are not shaded in Fig. 2.18. The compound inequality

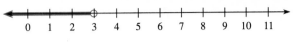

Figure 2.18

$x < 3$ and $x < 5$ is equivalent to the simple inequality $x < 3$. The solution set can be written as $\{x \mid x < 3\}$.

b) For union (or), we shade both regions on the same graph as shown in Fig. 2.19. The compound inequality $x > 4$ or $x > 0$ is equivalent to the simple inequality $x > 0$. The solution set is $\{x \mid x > 0\}$.

Figure 2.19

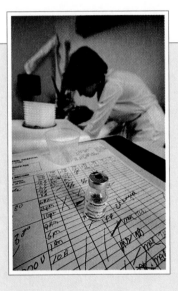

Math at Work

Herbologists in China, South America, and other parts of the world have long practiced the art of using plants to treat and prevent disease naturally. Modern technology enables doctors to prescribe chemical compounds that have been derived from plant sources. These synthetic drugs produce more consistent effects in the body.

For a drug to begin working, the concentration of drug particles within the bloodstream must exceed a certain value called the *minimum therapeutic level*. In a typical drug-dosage scheme, nurses administer a fixed amount (say 200 mg) of the drug at regular time intervals. The patient's body gradually eliminates this initial dosage. To keep the drug level above the minimum therapeutic level, a new dosage is given before the initial dosage has been completely eliminated.

The level of the drug in the body rises (above 200 mg) and falls again as each dosage is administered, but gradually the drug builds up in the patient. Doctors have found that the maximum build-up is a multiple of the quantity administered. For many drugs the maximum build-up level in the bloodstream may be considered unsafe. That level must be taken into account by doctors.

When prescribing medicine, a doctor must consider that a drug-dosage scheme involves two inequalities. The level of the drug in the patient's body must stay above the minimum therapeutic level so that the patient heals as quickly as possible, but at the same time the drug should not build up to a level that is considered unsafe. Suppose that a fixed amount of a drug is to be administered every 4 hours and under this condition the drug will build up to a level 4.32 times the dosage administered. If a build-up of over 1000 mg of the drug is considered unsafe, then what inequality must the dosage satisfy?

c) There are no numbers that are both less than 2 and greater than 6. The solution set is the empty set, $\emptyset$.

d) Graphing both inequalities on the same graph covers the entire line, as shown in Fig. 2.20. The solution set is the set of all real numbers, R.

Figure 2.20   ◄

If we start with a more complicated compound inequality, we first simplify each part of the compound inequality, and then find the union or intersection.

EXAMPLE 6   Solve $x + 2 > 3$ and $x - 6 < 7$. Graph the solution set.

Solution   First simplify each simple inequality.

$$x + 2 - 2 > 3 - 2 \quad \text{and} \quad x - 6 + 6 < 7 + 6$$
$$x > 1 \qquad\qquad \text{and} \qquad\qquad x < 13$$

The intersection of these two solution sets is the set of numbers between (but not including) 1 and 13. Its graph is shown in Fig. 2.21. The solution set is written in interval notation as (1, 13).

Figure 2.21   ◄

EXAMPLE 7   Graph the solution set to the inequality

$$5 - 7x \geq 12 \quad \text{or} \quad 3x - 2 < 7.$$

Solution   First solve each of the simple inequalities.

$$5 - 7x - 5 \geq 12 - 5 \quad \text{or} \quad 3x - 2 + 2 < 7 + 2$$
$$-7x \geq 7 \qquad\qquad \text{or} \qquad\qquad 3x < 9$$
$$x \leq -1 \qquad\qquad \text{or} \qquad\qquad x < 3$$

The union of the two solution sets is $\{x \mid x < 3\}$. The graph is shown in Fig. 2.22.

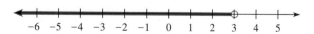

Figure 2.22   ◄

An inequality may be read from left to right or right to left. Consider the inequality $1 < x$. If we read it in the usual way, we say "1 is less than x." The meaning is clearer if we read the variable first. Reading from right to left, we say "x is greater than 1."

Another notation is commonly used for the compound inequality

$$x > 1 \quad \text{and} \quad x < 13.$$

This inequality can also be written as

$$1 < x < 13.$$

If we read this from left to right, it says "1 is less than x is less than 13." The meaning of this inequality is also clearer if we read the variable first and read the first inequality symbol from right to left. Reading the variable first, we say "x is greater than 1 and less than 13." This inequality says that x is between 1 and 13, and reading it this way makes that clearer. Note that this notation is used only to express the case in which x is between two numbers.

EXAMPLE 8 Solve the inequality and graph the solution set:

$$-2 \le 2x - 3 < 7.$$

Solution This inequality could be written as the compound inequality

$$2x - 3 \ge -2 \quad \text{and} \quad 2x - 3 < 7.$$

However, there is no need to write it this way, because we can solve it in its original form.

$$-2 + 3 \le 2x - 3 + 3 < 7 + 3 \qquad \textbf{Add 3 to each part.}$$

$$1 \le 2x < 10$$

$$\frac{1}{2} \le \frac{2x}{2} < \frac{10}{2} \qquad \textbf{Divide each part by 2.}$$

$$\frac{1}{2} \le x < 5$$

The solution set is $\left\{x \mid \dfrac{1}{2} \le x < 5\right\}$, and its graph is shown in Fig. 2.23. The interval notation for this set is $[1/2, 5)$.

Figure 2.23

EXAMPLE 9 Solve the inequality $-1 < 3 - 2x < 9$ and graph the solution set.

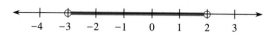

Figure 2.24

Solution

$$-1 - 3 < 3 - 2x - 3 < 9 - 3$$ Subtract 3 from each part of the inequality.

$$-4 < -2x < 6$$

$$2 > x > -3$$ Divide each part by -2 and reverse both inequality symbols.

$$-3 < x < 2$$ Rewrite the inequality with the smallest number on the left.

The solution set is $\{x \mid -3 < x < 2\}$, and its graph is shown in Figure 2.24.   ◄

Applications

EXAMPLE 10 Fiana made a score of 76 on her midterm exam. For her to get a B in the course, the average of her midterm exam and final exam must be between 80 and 89 inclusive. What possible scores on the final exam would give Fiana a B in the course?

Solution Let x represent her final exam score. Between 80 and 89 inclusive means that an average between 80 and 89 as well as an average of exactly 80 or 89 will get a B. So the average of the two scores must be greater than or equal to 80 and less than or equal to 89.

$$80 \le \frac{x + 76}{2} \le 89$$

$$160 \le x + 76 \le 178$$ Multiply by 2.

$$160 - 76 \le x \le 178 - 76$$ Subtract 76.

$$84 \le x \le 102$$

If Fiana scores between 84 and 102 inclusive, she will get a B in the course.   ◄

Warm-ups

True or false?

1. $3 < 5$ and $3 \le 10$

2. $3 < 5$ or $3 < 10$

3. $3 > 5$ and $3 < 10$

4. $3 \ge 5$ or $3 \le 10$

5. $4 < 8$ and $4 > 2$

6. $4 < 8$ or $4 > 2$

7. $-3 < 0 < -2$

8. $\{x \mid x > 3\} \cap \{x \mid x > 8\} = \{x \mid x > 8\}$

9. $\{x \mid x > 3\} \cup \{x \mid x \ge 8\} = \{x \mid x \ge 8\}$

10. $\{x \mid x > -2\} \cap \{x \mid x < 9\} = \{x \mid -2 < x < 9\}$

2.6 EXERCISES

Determine whether −4 *is a member of the solution set to each of the following compound inequalities. See Examples* 1 *and* 2.

1. $x < 5$ and $x > -3$

2. $x < 5$ or $x > -3$

3. $x \le -4$ and $x \le 0$

4. $x \ge -4$ or $x > 0$

5. $x < -3$ or $x < -9$

6. $x < 0$ and $x < 5$

Graph the solution set to each compound inequality. See Example 3, 4, *and* 5.

7. $x > -1$ and $x < 4$

8. $x \le 3$ and $x \le 0$

9. $x \ge 2$ or $x \ge 5$

10. $x < -1$ or $x < 3$

11. $x \le 6$ and $x > -2$

12. $x > -2$ and $x \le 4$

13. $x \le 6$ and $x > 9$

14. $x < 7$ or $x > 0$

15. $x \le 6$ or $x > 9$

16. $x \ge 4$ and $x \le -4$

17. $x \ge 6$ and $x \le 1$

18. $x > 3$ or $x < -3$

Solve each compound inequality. Write the solution set in set notation and graph it. See Examples 6 *and* 7.

19. $x - 3 > 7$ or $3 - x > 2$

20. $x - 5 > 6$ or $2 - x > 4$

21. $3 < x$ and $1 + x > 10$

22. $-.3x < 9$ and $.2x > 2$

23. $\frac{1}{2}x > 5$ or $-\frac{1}{3}x < 2$

24. $5 < x$ or $3 - \frac{1}{2}x < 7$

25. $2x - 3 \le 5$ and $x - 1 > 0$

26. $\frac{3}{4}x < 9$ and $-\frac{1}{3}x \le -15$

27. $.3x > 1$ or $4 - .2x > 3$

28. $\frac{1}{4}x - \frac{1}{3} > -\frac{1}{5}$ and $\frac{1}{2}x < 2$

29. $.5x < 2$ and $-.6x < -3$

30. $.3x < .6$ or $.05x > -4$

31. $\frac{2}{3}x > 8$ and $-\frac{1}{5}x > 3$

32. $3 - 2(x - 4) > 1$ and $2(x - 2) > x - 2$

33. $.2(x - 3) > .5$ or $.05x < 2$

34. $.1(x - 3) > .3$ or $.01x < 2$

Solve each compound inequality. Write the solution set in set notation and graph it. See Examples 8 *and* 9.

35. $5 < 2x - 3 < 11$

36. $-2 < 3x + 1 < 10$

37. $-1 < 5 - 3x \le 14$

38. $-1 \le 3 - 2x < 11$

39. $-3 < \dfrac{3m + 1}{2} \le 5$

40. $0 \le \dfrac{3 - 2x}{2} < 5$

41. $-2 < \dfrac{1 - 3x}{-2} < 7$

42. $-3 < \dfrac{2x - 1}{3} < 7$

43. $3 \le 3 - 5(x - 3) \le 8$

44. $2 \le 4 - \dfrac{1}{2}(x - 8) \le 10$

Solve each of the following word problems by using a compound inequality. See Example 10.

45. Abdul is shopping for a new truck in a city with an 8% sales tax. There is also an $84 title and license fee to pay. He wants to get a good truck, and he plans to spend at least $12,000 but not more than $15,000. What is the price range for the truck?

46. Professor Johnson gives only a midterm exam and a final exam. The semester average is computed by taking 1/3 of the midterm exam score plus 2/3 of the final exam score. To get a C, Beth must have a semester average between 70 and 79 inclusive. If Beth scored only 64 on the midterm, then for what range of scores on the final exam would Beth get a C?

47. Professor Davis counts his midterm as 2/3 of the grade, and his final as 1/3 of the grade. Jason scored only 64 on the midterm. What range of scores on the final exam would put Jason's average between 70 and 79 inclusive?

48. Renee wants to sell her car through a broker who charges a commission of 10% of the selling price. The book value of the car is $14,900, but Renee still owes $13,104 on it. While the car is in only fair condition and will not sell for more than the book value, Renee must get enough to at least pay off the loan. What is the range of the selling price?

49. In defending her smoking habit, Jane calculates that she smokes only 3 full cigarettes a day, 1 after each meal. The rest of the time she smokes on the run and smokes only half of the cigarette before forgetting where she put it. Figuring this in, she estimates that she smokes the equivalent of 5 to 12 cigarettes per day. How many times a day does she light up on the run?

50. The length of a rectangle is 20 meters longer than the width. The perimeter must be between 80 and 100 meters. What are the possible values for the width of the rectangle?

For Exercises 51–60, write either a simple or a compound inequality that has the given graph as its solution set.

51.

52.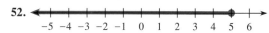

53.

54.

55.

56.

57.

58.

59.

60.

Solve each compound inequality and graph its solution set.

61. $2 < x < 7$ and $2x > 10$

62. $3 < 5 - x < 8$ or $-3x < 0$

*63. $-1 < 3x + 2 \le 5$ or $\dfrac{3}{2}x - 6 > 9$

*64. $0 < 5 - 2x \le 10$ and $-6 < 4 - x < 0$

Write each set in interval notation.

*65. $\{x \mid x > 3 \text{ or } x < 0\}$

*67. $\{x \mid x \leq 2 \text{ and } x \geq -9\}$

*69. $\{x \mid -3 < x < 8\}$

*71. $\{x \mid 8 \leq x < 9\}$

*66. $\{x \mid x < -1 \text{ or } x > 5\}$

*68. $\{x \mid x > 3 \text{ and } x < 10\}$

*70. $\{x \mid -5 \leq x < 9\}$

*72. $\{x \mid 3 \leq x \leq 6\}$

2.7 Absolute Value Equations and Inequalities

IN THIS SECTION:

- **Absolute Value Equations**
- **Absolute Value Inequalities**
- **Applications**

In Chapter 1 we learned that absolute value measures the distance of a number from 0 on the number line. In this section we will learn to solve equations and inequalities involving absolute value.

Absolute Value Equations

Solving equations involving absolute value requires some techniques that are different from those studied in previous sections. For example, the solution set to the equation

$$|x| = 5$$

is $\{5, -5\}$, because both 5 and -5 are five units from 0 on the number line (see Fig. 2.25). We can say that the equation $|x| = 5$ is equivalent to the compound equation

$$x = 5 \quad \text{or} \quad x = -5.$$

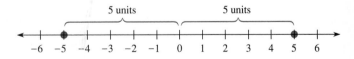

Figure 2.25

The equation $|x| = 0$ is equivalent to the equation $x = 0$, because 0 is the only number whose distance from 0 is zero. The solution set to $|x| = 0$ is $\{0\}$.

The equation $|x| = -7$ is inconsistent because absolute value measures distance, and distance is never negative. Thus the solution set is empty. These ideas are summarized below.

Basic Absolute Value Equations

Absolute Value Equation	Equivalent Equation	Solution Set
$\|x\| = k \ (k > 0)$	$x = k$ or $x = -k$	$\{k, -k\}$
$\|x\| = 0$	$x = 0$	$\{0\}$
$\|x\| = k \ (k < 0)$		$\varnothing$

We use these ideas in solving more complicated equations involving absolute value.

EXAMPLE 1   Solve the absolute value equations.

a) $|x - 7| = 2$ **b)** $|3x - 5| = 7$

Solution

a) $x - 7 = 2$ or $x - 7 = -2$ Equivalent equation
         $x = 9$ or $x = 5$
The solution set is $\{5, 9\}$. Note that 5 and 9 are both a distance of 2 units away from 7.

b) $3x - 5 = 7$ or $3x - 5 = -7$ Equivalent equation
       $3x = 12$ or $3x = -2$
       $x = 4$ or $x = -\dfrac{2}{3}$
The solution set is $\{-2/3, 4\}$.   ◀

EXAMPLE 2   Solve $|2(x - 6) + 7| = 0$.

Solution

$$2(x - 6) + 7 = 0 \qquad \text{Equivalent equation}$$
$$2x - 12 + 7 = 0$$
$$2x - 5 = 0$$
$$2x = 5$$
$$x = \frac{5}{2}$$

The solution set is $\{5/2\}$.   ◀

EXAMPLE 3   Solve $|3x - 6(5 - x)| + 4 = 2$.

Solution   First subtract 4 from each side to isolate the absolute value expression.

$$|3x - 6(5 - x)| = -2$$

The solution set is empty because there is no quantity that has a negative absolute value.   ◀

Absolute Value Inequalities

To solve absolute value inequalities, we again need to keep in mind that absolute value measures distance from 0 on the number line. The inequality

$$|x| > 5$$

should be thought of as saying that x is more than 5 units away from 0. This is true for any number to the right of 5 or any number to the left of -5. So this inequality is equivalent to the compound inequality

$$x > 5 \quad \text{or} \quad x < -5.$$

Remember that the solution set to this inequality is the union of the solution sets to the two simple inequalities. The solution set is $\{x \mid x > 5 \text{ or } x < -5\}$. The graph of $|x| > 5$ is shown in Fig. 2.26.

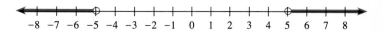

Figure 2.26

We should think of the inequality $|x| \le 3$ as saying that x is a number that lies less than or equal to 3 units away from 0. Any number between -3 and 3 inclusive satisfies that condition. An equivalent inequality is

$$x \ge -3 \quad \text{and} \quad x \le 3.$$

This inequality can also be written as

$$-3 \le x \le 3.$$

The graph of $|x| \le 3$ is shown in Fig. 2.27.

Figure 2.27

The four basic types of absolute value inequalities are summarized below.

Basic Absolute Value Inequalities

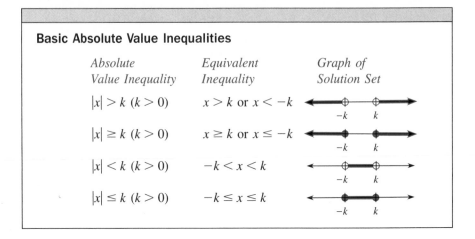

Absolute Value Inequality	Equivalent Inequality	Graph of Solution Set

We can solve more complicated inequalities in the same manner as simple ones.

EXAMPLE 4   Solve $|x - 9| < 2$ and graph the solution set.

Solution   Write the equivalent compound inequality from the above table.

$$-2 < x - 9 < 2$$
$$-2 + 9 < x - 9 + 9 < 2 + 9 \qquad \text{Add 9 to each part of the inequality.}$$
$$7 < x < 11$$

The graph of the solution set, $\{x \mid 7 < x < 11\}$, is shown in Fig. 2.28. Note that the graph consists of the numbers that are within 2 units of 9.

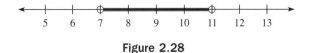

Figure 2.28   ◀

EXAMPLE 5   Solve $|3x + 5| > 2$ and graph the solution set.

Solution

$$3x + 5 > 2 \quad \text{or} \quad 3x + 5 < -2 \qquad \text{Equivalent inequality}$$
$$3x > -3 \quad \text{or} \qquad 3x < -7$$
$$x > -1 \quad \text{or} \qquad x < -\frac{7}{3}$$

The graph is shown in Fig. 2.29.

Figure 2.29   ◄

EXAMPLE 6   Solve $|5 - 3x| \le 6$ and graph the solution set.

Solution

$$-6 \le 5 - 3x \le 6 \qquad \text{Equivalent inequality}$$

$$-11 \le -3x \le 1 \qquad \text{Subtract 5 from each part.}$$

$$\frac{11}{3} \ge x \ge -\frac{1}{3} \qquad \begin{array}{l}\text{Divide by } -3 \text{ and reverse}\\ \text{each inequality symbol.}\end{array}$$

$$-\frac{1}{3} \le x \le \frac{11}{3} \qquad \begin{array}{l}\text{Write } -1/3 \text{ on the left, because}\\ \text{it is smaller than } 11/3.\end{array}$$

The graph is shown in Fig. 2.30.

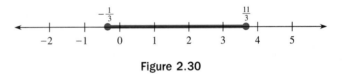

Figure 2.30   ◄

There are a few absolute value inequalities that do not fit the categories just discussed. They are easy to solve once we understand them.

EXAMPLE 7   Solve $3 + |7 - 2x| \ge 3$.

Solution   Subtract 3 from each side to isolate the absolute value expression.

$$|7 - 2x| \ge 0$$

Because the absolute value of any real number is greater than or equal to 0, the solution set is R, the set of all real numbers.   ◄

EXAMPLE 8   Solve $|5x - 12| < -2$.

Solution   We write an equivalent inequality only when the value of k is positive. With -2 on the right side, we do not write an equivalent inequality. We solve this by remembering that the absolute value of any quantity is greater than or equal to 0. No value for x can make this absolute value less than -2. The solution set is $\varnothing$, the empty set.   ◄

Applications

A simple example will show how absolute value inequalities can be used in applications.

EXAMPLE 9 The water temperature in a certain manufacturing process must be kept at 143°F. The computer is programmed to shut the process down if the water temperature is more than 7 degrees away from what it is supposed to be. For what temperature readings is the process shut down?

Solution If we let x represent the water temperature, then $x - 143$ represents the difference between the actual temperature and the desired temperature. The quantity $x - 143$ could be positive or negative. The process is shut down if the absolute value of $x - 143$ is greater than 7. This condition can be expressed with the absolute value inequality

$$|x - 143| > 7.$$

$$x - 143 > 7 \quad \text{or} \quad x - 143 < -7$$
$$x > 150 \quad \text{or} \quad x < 136$$

The process is shut down if the temperature is greater than 150°F or less than 136°F.   ◄

Warm-ups

True or false?

1. The equation $|x| = 2$ is equivalent to $x = 2$ or $x = -2$.
2. All absolute value equations have two solutions.
3. The equation $|2x - 3| = 7$ is equivalent to $2x - 3 = 7$ or $2x + 3 = 7$.
4. The inequality $|x| > 5$ is equivalent to $x > 5$ or $x < -5$.
5. The equation $|x| = -5$ is equivalent to $x = 5$ or $x = -5$.
6. There is only one solution to the equation $|3 - x| = 0$.
7. We should write the inequality $x > 3$ or $x < -3$ as $3 < x < -3$.
8. The inequality $|x| \leq 7$ is equivalent to $-7 \leq x \leq 7$.
9. The equation $|x| + 2 = 5$ is equivalent to $|x| = 3$.
10. The inequality $|x| < -2$ is equivalent to $x < 2$ and $x > -2$.

2.7 EXERCISES

Solve each absolute value equation. See Examples 1, 2, and 3.

1. $|a| = 5$

2. $|x| = 2$

3. $|x - 3| = 1$

4. $|x - 5| = 2$

5. $|3 - x| = 6$

6. $|7 - x| = 6$

7. $|3x - 4| = 12$

8. $|5x + 2| = -3$

9. $|5x - 9| = 0$

10. $|3 - 6x| = 12$

11. $|6 - x| = 0$

12. $|5 - x| = 0$

13. $|7x - 6| = -3$

14. $|2a + 3| = 15$

15. $|2(x - 4) + 3| = 5$

16. $|3(x - 2) + 7| = 6$

Write an absolute value inequality whose solution set is shown by the graph. See Examples 4–6.

17.

18.

19.

20.

21.

22.

23.

24.

Determine whether each absolute value inequality is equivalent to the inequality following it. See Examples 4–6.

25. $|x| < 3, \; x < 3$

26. $|x| > 3, \; x > 3$

27. $|x - 3| > 1, \; x - 3 > 1 \text{ or } x - 3 < -1$

28. $|x - 3| \le 1, \; -1 \le x - 3 \le 1$

29. $|x - 3| \ge 1, \; x - 3 \ge 1 \text{ or } x - 3 \le 1$

30. $|x - 3| > 0, \; x - 3 > 0$

31. $|4 - x| < 1, \; 4 - x < 1 \text{ and } -(4 - x) < 1$

32. $|4 - x| > 1, \; 4 - x > 1 \text{ or } -(4 - x) > 1$

33. $|4 - x| \le 1, \; -1 < 4 - x < 1$

34. $|4 - x| \ge 1, \; 4 - x > 1 \text{ or } 4 - x \le -1$

Solve each absolute value inequality and graph the solution set. See Examples 4–6.

35. $|x| > 6$

36. $|w| > 3$

37. $|a| < 3$

38. $|x| < 7$

39. $|x - 2| \ge 3$

40. $|x - 5| \ge 1$

41. $|2x - 4| < 5$

42. $|2x - 1| < 3$

43. $|5 - x| \le 7$

44. $|6 - x| \le 1$

45. $|3 - 2x| \ge 12$

46. $|5 - 2x| \ge 10$

47. $\left| \dfrac{2w + 7}{3} \right| > 1$

48. $\left| \dfrac{2y + 3}{2} \right| > 1$

Solve each absolute value inequality and graph the solution set. See Examples 7 and 8.

49. $|x - 2| > 0$

50. $|6 - x| \geq 0$

51. $|x - 5| \geq 0$

52. $|3x - 7| \geq -3$

53. $|3x - 7| < -3$

54. $|7x - 42| < -6$

55. $|2x + 3| > -6$

56. $|5 - x| > 0$

For each of the following word problems, write an absolute value equation or inequality and solve it. See Example 9.

57. Nadia and Yanick were born 3 years apart. If Nadia is 18, then what are the possibilities for Yanick's age?

58. Anthony and his sister Cleo have birthdays that are one week apart. If Anthony was born on October 15, then what are the possible dates for Cleo's birthday?

59. Research at a major university has shown that identical twins generally differ by less than 6 pounds in body weight. If Kim weighs 127 pounds, then in what range is the weight of her identical-twin sister Kathy?

60. Jude's IQ score is more than 15 points away from Sherry's. If Sherry scored 110, then in what range is Jude's score?

Photo for Exercise 59

Solve each absolute value equation.

61. $3 + |x| = 5$

62. $|x| - 10 = -3$

63. $2 - |x + 3| = -6$

64. $4 - 3|x - 2| = -8$

65. $|7.3x - 5.26| = 4.215$

66. $|5.74 - 2.17x| = 10.28$

*67. $|x - 5| = |2x + 1|$

*68. $|w - 6| = |3 - 2w|$

*69. $|5 - 2x| = |4 - x|$

*70. $|4x - 1| = |2x - 3|$

*71. $|x - 3| = |3 - x|$

*72. $|a - 6| = |6 - a|$

*73. $|w + 6| = |w - 6|$

*74. $|2m + 5| = |2m - 5|$

*75. $|t - 5| = t$

*76. $|w - 6| = w + 2$

*77. $|7 - w| = w + 5$

*78. $|w + 2| = -w$

*79. $x + |x| = 3$

*80. $x - |x| = 0$

Solve each absolute value inequality and graph the solution set.

81. $|5x| < 10$

82. $|3x| > 6$

83. $1 < |x + 2|$

84. $5 \geq |x - 4|$

85. $5 > |x| + 1$

86. $4 \leq |x| - 6$

87. $3 - 5|x| > -2$

88. $1 - 2|x| < -7$

89. $|5.67x - 3.124| < 1.68$

90. $|4.67 - 3.2x| \geq 1.43$

*91. $|2x - 1| < 3$ and $2x - 3 > -2$

*92. $|5 - 3x| \geq 3$ and $5 - 2x > 3$

*93. $|x - 2| < 3$ and $|x - 7| < 3$

*94. $|x - 5| < 4$ and $|x - 6| > 2$

*95. $|x| > x$

*96. $|x| > x + 6$

*97. $|x - 5| < 2x + 1$

*98. $|x - 2| > 1 - 3x$

Wrap-up

CHAPTER 2

SUMMARY

	Concepts	Examples
Solution set	The set of all numbers that satisfy an equation (or inequality)	$x + 2 = 6$ Solution set $\{4\}$
Equivalent equations	Equations with the same solution set	$2x + 1 = 5$ $2x = 4$ $x = 4$
Properties of equality	We may perform the same operation $(+, -, \cdot, \div)$ on each side of an equation without changing the solution set (excluding multiplication and division by 0).	$x + 1 = 5$ $x - 1 = 3$ $2x = 8$ $\dfrac{x}{2} = 2$
Identity	An equation that is satisfied by every number for which both sides are defined	$x + x = 2x$
Conditional equation	An equation whose solution set contains at least one real number but is not an identity	$5x - 10 = 0$
Inconsistent equation	An equation whose solution set is $\varnothing$	$x = x + 1$
Linear equation in one variable	An equation that can be written in the form $ax + b = 0$, with $a \neq 0$	$5x - 1 = 2x - 9$ $3x + 8 = 0$
Strategy for solving a linear equation	**1.** If fractions are present, multiply each side by the LCD to eliminate the fractions. **2.** Use the distributive property to remove parentheses. **3.** Combine any like terms. **4.** Use the addition and subtraction properties to get all variables on one side and numbers on the other. **5.** Use the multiplication or division property to get a single variable on one side. **6.** Check your work by replacing the variable in the original equation with your solution.	

Strategy for solving word problems	**1.** Read the problem. **2.** If possible, draw a diagram to illustrate the problem. **3.** Choose a variable and write down what it represents. **4.** Represent any other unknowns in terms of that variable. **5.** Write an equation that fits the situation. **6.** Solve the equation. **7.** Be sure that your solution answers the question posed in the original problem. **8.** Check your answer by using it to solve the original problem (not the equation).

Definition of linear inequality

Any inequality that can be written in the form $ax + b < 0$, with $a \neq 0$
In place of $<$ we can use $\leq$, $>$, or $\geq$.

$2x + 9 > 7$
$2x + 2 > 0$

Properties of inequality

We may perform the same operation $(+, -, \cdot, \div)$ on each side of an inequality just as we do in solving equations, with one exception: when multiplying or dividing by a negative number, the inequality symbol must be reversed.

$-3x > 6$
$x < -2$

Trichotomy property

For any two real numbers a and b, exactly one of the following is true:

$$a < b, \quad a = b, \quad \text{or} \quad a > b.$$

Compound inequality

Two simple inequalities connected with the word "and" or "or"

$x > 3$ or $x < 1$

    And corresponds to *intersection*.
    Or corresponds to *union*.

Basic absolute value equations

Absolute Value Equation	Equivalent Equation	Solution Set
$\lvert x \rvert = k \ (k > 0)$	$x = k$ or $x = -k$	$\{k, -k\}$
$\lvert x \rvert = 0$	$x = 0$	$\{0\}$
$\lvert x \rvert = k \ (k < 0)$		$\varnothing$

Basic absolute value inequalities

Absolute Value Inequality	Equivalent Inequality	Graph of Solution Set
$\lvert x \rvert > k \ (k > 0)$	$x > k$ or $x < -k$	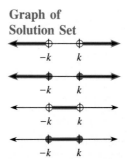
$\lvert x \rvert \geq k \ (k > 0)$	$x \geq k$ or $x \leq -k$	
$\lvert x \rvert < k \ (k > 0)$	$-k < x < k$	
$\lvert x \rvert \leq k \ (k > 0)$	$-k \leq x \leq k$	

REVIEW EXERCISES

2.1 *Solve each equation.*

1. $2x - 7 = 9$

2. $5x - 7 = 38$

3. $2(x - 3) - 5 = 5 - (3 - 2x)$

4. $2(x - 4) + 5 = -(3 - 2x)$

5. $5 - 4x = 11$

6. $7 - 3x = -8$

7. $x - x = 0$

8. $2x - x = 0$

9. $\frac{1}{4}x - \frac{1}{5} = \frac{1}{5}x + \frac{4}{5}$

10. $\frac{1}{2}x - 1 = \frac{1}{3}x$

11. $\frac{t}{2} - \frac{t - 2}{3} = \frac{3}{2}$

12. $\frac{y + 1}{4} - \frac{y - 1}{6} = y + 5$

13. $5 - 2(x - 4) + 3(x - 7) = -3$

14. $.04x - .06(x - 8) = .1x$

2.2 *Solve each equation mentally.*

15. $4x = 12$

16. $5x = 25$

17. $x - 6 = 5$

18. $x - 7 = 12$

19. $\frac{1}{2}x = 6$

20. $\frac{1}{3}x = \frac{1}{2}$

21. $-\frac{1}{3}x = 2$

22. $-\frac{1}{2}x = 5$

23. $2x - 5 = 0$

24. $5x - 3 = 0$

25. $-2x = 4$

26. $-3x = 9$

27. $2x - 3 = 1$

28. $5x - 4 = 1$

29. $.1x = 7$

30. $.1x = 6$

31. $-\frac{3}{5}x = 1$

32. $-\frac{2}{3}x = 1$

33. $\frac{3}{17}x = 0$

34. $-\frac{3}{8}x = 0$

35. $5x - 3x = 2x$

36. $.01x = 2$

37. $.01x = 12$

38. $2x + 3x = 5x$

39. $x + 3 = x$

40. $x + 1 = x$

41. $\frac{1}{x} = 3$

42. $\frac{1}{x} = \frac{2}{3}$

2.3 *Solve each equation for x.*

43. $ax + b = 0$

44. $mx + c = d$

45. $ax + 2 = cx$

46. $mx = 3 - x$

47. $mwx = P$

48. $xyz = 2$

49. $\frac{1}{x} + \frac{1}{2} = w$

50. $\frac{1}{x} + \frac{1}{a} = 2$

Solve each equation for y.

51. $3x - 2y = -6$

52. $4x - 3y + 9 = 0$

53. $y - 2 = -\frac{1}{3}(x - 6)$

54. $y + 6 = \frac{1}{2}(x - 4)$

55. $\frac{1}{2}x - \frac{1}{4}y = 5$

56. $-\frac{x}{3} + \frac{y}{2} = \frac{5}{8}$

2.4 *Solve each word problem.*

57. If the perimeter of a legal size note pad is 45 inches, and the pad is 5.5 inches longer than it is wide, then what are its length and width?

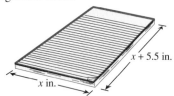

Figure for Exercise 57

58. The height of a trapezoid is 5 feet, and the upper base is 2 feet shorter than the lower base. If the area of the trapezoid is 45 square feet, then how long is the lower base?

59. Roy makes $8000 more per year than his wife makes. Roy saves 10% of his income for retirement, and his wife saves 8%. Together they save $5660 per year. How much does each make?

60. Duane makes $1000 less per year than his wife makes. Duane gives 5% of his income to charity, and his wife gives 10% of her income. Together they contribute $2500 to charity. How much does each make?

61. Sarah is buying a car for $7600. The dealer gave her a 20% discount off the list price. What was the list price?

Photo for Exercise 61

62. At 25% off, a gold chain sells for $465. What was the list price?

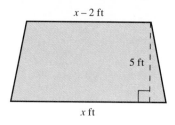

Figure for Exercise 58

63. Rebecca has 15 coins consisting of dimes and nickels. The total value of the coins is $.95. How many of each does she have?

64. Camille has 19 coins consisting of nickels, dimes, and quarters. The value of the coins is $1.60. If she has 6 times as many nickels as quarters, then how many of each does she have?

65. On a recent bicycle trip across the desert, Barbara rode for 5 hours. Her bicycle then developed mechanical difficulties and she walked the bicycle for 3 hours to the nearest town. Altogether, she covered 85 miles. If she rides 9 miles per hour faster than she walks, then how far did she walk?

66. Delmas flew to Detroit in 90 minutes and drove his new car back home in 6 hours. If he drove 150 miles per hour slower than he flew, then how fast did he fly?

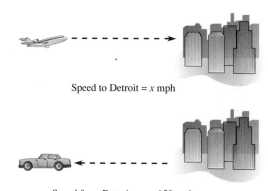

Speed to Detroit = x mph

Speed from Detroit = $x - 150$ mph

Figure for Exercise 66

2.5 *Solve each inequality and graph the solution set.*

67. $3 - 4x < 15$

68. $5 - 6x > 35$

69. $2(x - 3) > -6$

70. $4(5 - x) < 20$

71. $-\frac{3}{4}x \geq 6$

72. $-\frac{2}{3}x \leq 4$

73. $3(x + 2) > 5(x - 1)$

74. $4 - 2(x - 3) < 0$

75. $\frac{1}{2}x + 7 \le \frac{3}{4}x - 5$

76. $\frac{5}{6}x - 3 \ge \frac{2}{3}x + 7$

2.6 *Solve each compound inequality and graph the solution set.*

77. $x + 2 > 3$ or $x - 6 < -10$

78. $x - 2 > 5$ or $x - 2 < -1$

79. $x > 0$ and $x - 6 < 3$

80. $x \le 0$ and $x + 6 > 3$

81. $6 - x < 3$ or $-x < 0$

82. $-x > 0$ or $x + 2 < 7$

83. $2x < 8$ and $2(x - 3) < 6$

84. $\frac{1}{3}x > 2$ and $\frac{1}{4}x > 2$

85. $x - 6 > 2$ and $6 - x > 0$

86. $-\frac{1}{2}x < 6$ or $\frac{2}{3}x < 4$

87. $.5x > 10$ or $.1x < 3$

88. $.02x > 4$ and $.2x < 3$

89. $-2 \le \frac{2x - 3}{10} \le 1$

90. $-3 < \frac{4 - 3x}{5} < 2$

2.7 *Solve each absolute value equation or inequality and graph the solution set.*

91. $|2x| \ge 8$

92. $|5x - 1| \le 14$

93. $|x| = -5$

94. $|x| + 2 = 16$

95. $|5 - x| > 9$

96. $|6 - x| < 3$

97. $|x - 3| < -3$

98. $|x - 7| \le -4$

99. $|x| - 5 = -1$

100. $|x - 5| = -1$

101. $|x + 4| \ge -1$

102. $|6x - 1| \ge 0$

103. $2 - 3|x - 2| < -1$

104. $4 > 2|6 - x| - 3$

Miscellaneous

Solve each problem.

105. Chi Ching is going to open a video rental store. Industry statistics show that 45% of the rental price goes for overhead. If the maximum that anyone will pay to rent a tape is \$5 and Chi Ching wants a profit of at least \$1.65 per tape, then in what range should the rental price be?

106. Regina makes \$6.80 per hour working in the snack bar. To keep her grant, she may not earn more than \$51 per week. What is the range of the number of hours per week that she may work?

107. Dane's car was found abandoned at mile marker 86 on the interstate. If Dane was picked up by the police on the interstate exactly 5 miles away, then at what mile marker was he picked up?

108. Scott scored 72 points on the midterm and Katie's score was more than 16 points away from Scott's. What was Katie's score?

For each graph, write an equation or inequality that has the solution set shown by the graph. Use absolute value when possible.

109.

110.

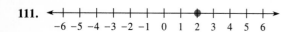

111.

112.

113.

114.

115.

116.

117.

118.

119.

120.

121.

122.

123.

124.

125.

126.

CHAPTER 2 TEST

Solve each equation.

1. $-10x - 5 + 4x = -4x + 3$

2. $\dfrac{y}{2} - \dfrac{y-3}{3} = \dfrac{y+6}{6}$

3. $|w| + 3 = 9$

4. $|3 - 2(5 - x)| = 3$

Solve for the indicated variable.

5. $2x - 5y = 20$ for y

6. $a = axt + S$ for a

Solve each inequality and graph the solution set.

7. $|w| > 4$

8. $|m - 6| \le 2$

9. $2 - 3(w - 1) < -2w$

10. $2 < \dfrac{5 - 2x}{3} < 7$

11. $3x - 2 < 7$ and $-3x \le 15$

12. $\dfrac{2}{3}y < 4$ or $y - 3 < 12$

Find the solution set to each of the following equations and inequalities.

13. $|2x - 7| = -3$

14. $x > 5$ or $x < 12$

15. $x < 0$ and $x > 7$

16. $|2x - 5| \leq 0$

17. $|x - 3| < 0$

18. $x + 3x = 4x$

19. $2(x + 7) = 2x + 9$

20. $|x - 6| > -6$

21. $5x - 7 = 0$

Write a complete solution to each of the following word problems.

22. The perimeter of a rectangle is 84 meters. If the width is 16 meters less than the length, then what is the width of the rectangle?

23. If the area of a triangle is 21 square inches and the base is 3 inches, then what is the height?

24. Joan bought a gold chain marked 30% off. If she paid $210, then what was the original price?

25. How many liters of an 11% alcohol solution should be mixed with 60 liters of a 5% alcohol solution to obtain a mixture that is 7% alcohol?

Tying It All Together

CHAPTERS 1–2

Simplify each expression.

1. $5x + 6x$

2. $5x \cdot 6x$

3. $\dfrac{6x + 2}{2}$

4. $5 - 4(2 - x)$

5. $(30 - 1)(30 + 1)$

6. $(30 + 1)^2$

7. $(30 - 1)^2$

8. $(2 + 3)^2$

9. $2^2 + 3^2$

10. $(8 - 3)(3 - 8)$

11. $(-1)(3 - 8)$

12. -2^2

13. $3x + 8 - 5(x - 1)$

14. $(-6)^2 - 4(-3)2$

15. $3^2 \cdot 2^3$

16. $4(-6) - (-5)(3)$

17. $-3x \cdot x \cdot x$

18. $(-1)(-1)(-1)(-1)(-1)(-1)$

Solve each equation.

19. $5x + 6x = 8x$

20. $5x + 6x = 11x$

21. $5x + 6x = 0$

22. $5x + 6 = 11x$

23. $3x + 1 = 0$

24. $5 - 4(2 - x) = 1$

25. $3x + 6 = 3(x + 2)$

26. $x - .01x = 990$

CHAPTER 3

Exponents and Polynomials

A circus clown is trying to perfect a new act. In this act he will toss a ball into the air, do a double back flip, and then catch the ball. Every time he tries this trick, the ball hits the ground before he completes his double back flip. If he can toss the ball straight up with a velocity of 64 feet per second, then how long does he have before the ball returns to the earth?

The distance D (in feet) that the ball is above the earth at any time t (in seconds) is given by the formula $D = -16t^2 + 64t$. The right side of this equation is a polynomial. At the moment the ball is tossed, t has a value of 0 and D has a value of 0. When the ball returns to the earth, D has a value of 0 again, but for a different value of t. This value of t will tell us how long the ball was in the air. To find the value of t, we must solve the polynomial equation $-16t^2 + 64t = 0$. In this chapter we will study polynomials and learn how to solve equations of this type. We will solve this problem as Exercise 39 of Section 3.9.

3.1 Integral Exponents

IN THIS SECTION

- Positive and Negative Exponents
- Product Rule
- Changing the Sign of an Exponent
- Quotient Rule

In Chapter 1 we defined positive integral exponents and learned to evaluate expressions involving exponents. In this section we will extend the definition of exponents to include all integers and learn some rules for working with integral exponents. In Chapter 5 we will learn that any rational number can be used as an exponent.

Positive and Negative Exponents

Positive integral exponents provide a convenient way to write repeated multiplication or very large numbers. For example,

$$2 \cdot 2 \cdot 2 = 2^3, \qquad y \cdot y \cdot y \cdot y = y^4, \qquad \text{and } 1{,}000{,}000{,}000 = 10^9.$$

We refer to 2^3 as "a power of 2," or "2 raised to the third power," or "2 cubed."

Recall that for any nonzero number a, the reciprocal of a is written as $1/a$. Negative exponents are used as a convenient way to represent multiplicative inverses or reciprocals. For example, we use 2^{-3} to represent the reciprocal of 2^3. Because $2^3 = 8$, $2^{-3} = 1/8$. Because 2^3 and 2^{-3} are reciprocals of each other, we can write

$$2^{-3} = \frac{1}{2^3} \qquad \text{In words, } 2^{-3} \text{ is the reciprocal of } 2^3.$$

and

$$2^3 = \frac{1}{2^{-3}}. \qquad \text{In words, } 2^3 \text{ is the reciprocal of } 2^{-3}.$$

Note that the cube of 2 is 8, and the reciprocal of 8 is 1/8. The reciprocal of 2 is 1/2, and 1/2 cubed is 1/8. In evaluating the expression 2^{-3}, we can evaluate the power and the reciprocal in either order.

For negative exponents we cannot allow the base to be zero, because zero does not have a reciprocal. The reason why negative exponents were chosen to represent reciprocals will be apparent when we study the product rule.

DEFINITION
Negative Integral Exponents

If a is a nonzero real number and n is a positive integer, then

$$a^{-n} = \frac{1}{a^n}. \qquad \text{If } n \text{ is positive, } -n \text{ is negative.}$$

We can evaluate the power and the reciprocal in either order:

$$a^{-n} = \left(\frac{1}{a}\right)^n$$

Note that if $n = 1$, $a^{-1} = \frac{1}{a}$, the reciprocal of a. For example,

$$2^{-1} = \frac{1}{2}, \quad \left(\frac{2}{3}\right)^{-1} = \frac{3}{2}, \quad \text{and} \quad \left(\frac{1}{4}\right)^{-1} = 4.$$

EXAMPLE 1   Evaluate each expression.

a) 3^{-2} **b)** $(-3)^{-2}$ **c)** $\left(\frac{3}{4}\right)^{-2}$ **d)** $2^5 \cdot 2^{-4}$ **e)** $\frac{1}{5^{-2}}$

Solution

a) $3^{-2} = \frac{1}{3^2}$ Because 3^{-2} is the reciprocal of 3^2

$\qquad = \frac{1}{9}$ The square of 3 is 9.

b) $(-3)^{-2} = \frac{1}{(-3)^2}$ Because $(-3)^{-2}$ is the reciprocal of $(-3)^2$

$\qquad\quad = \frac{1}{9}$ The square of -3 is 9.

We get the same result if we square -3 and then find the reciprocal:

$$(-3)^{-2} = 9^{-1} = \frac{1}{9}$$

c) $\left(\frac{3}{4}\right)^{-2} = \left(\frac{4}{3}\right)^2$ The reciprocal of 3/4 is 4/3.

$\qquad\quad = \frac{16}{9}$ The square of 4/3 is 16/9.

d) $2^5 \cdot 2^{-4} = 2^5 \cdot \frac{1}{2^4} = 32 \cdot \frac{1}{16} = 2$

e) $\frac{1}{5^{-2}} = 5^2 = 25$ The reciprocal of 5^{-2} is 5^2.

◀

Product Rule

In the following examples of multiplying exponential expressions we use facts we already know to simplify the expression. By looking closely at each example we can discover a rule that will simplify the expression quickly.

$$2^3 \cdot 2^5 = \overbrace{(2 \cdot 2 \cdot 2)}^{3} \cdot \overbrace{(2 \cdot 2 \cdot 2 \cdot 2 \cdot 2)}^{5} = 2^8 \qquad \textit{Note: } 3 + 5 = 8$$

$$\underbrace{}_{\text{eight 2s}}$$

$$2^{-4} \cdot 2^{-5} = \frac{1}{2^4} \cdot \frac{1}{2^5} = \frac{1}{2^9} = 2^{-9} \qquad \textit{Note: } -4 + (-5) = -9$$

$$\underbrace{\phantom{\frac{1}{2^4} \cdot \frac{1}{2^5}}}_{\text{nine 2s}}$$

$$2^{-4} \cdot 2^6 = \frac{2^6}{2^4} = \frac{\cancel{2} \cdot \cancel{2} \cdot \cancel{2} \cdot \cancel{2} \cdot 2 \cdot 2}{\cancel{2} \cdot \cancel{2} \cdot \cancel{2} \cdot \cancel{2}} = 2^2 \qquad \textit{Note: } -4 + 6 = 2$$

Note that in each case the exponent of the answer is the sum of the exponents in the two original exponential expressions. However, if we apply this idea to the product of 2^3 and 2^{-3}, we get

$$2^3 \cdot 2^{-3} = 2^0. \qquad 3 + (-3) = 0$$

Because $2^3 = 8$ and $2^{-3} = 1/8$, their product is 1. So to be consistent, we define 2^0 to be 1. We define the zero power of any nonzero number to be 1, so that we can make a consistent product rule for integral exponents. Because the base must be nonzero when we use negative exponents, we have no reason to define the zero power of zero.

DEFINITION
Zero Exponent

If a is any nonzero real number, then $a^0 = 1$.

Now that we have defined zero exponents, we can state the product rule and allow the exponents to be any integers.

Product Rule

If m and n are integers and $a \neq 0$, then
$$a^m \cdot a^n = a^{m+n}.$$

To apply the product rule, the bases must be identical. The product rule cannot be applied to $2^3 \cdot 3^2$. Note that when the bases are identical the product rule does not allow us to multiply the bases. For example, $2^5 \cdot 2^4 \neq 4^9$. By the product rule,

$$2^5 \cdot 2^4 = 2^9.$$

EXAMPLE 2 Simplify each expression. Write answers with positive exponents and assume that all variables represent nonzero real numbers.

a) $y^4 \cdot y^6$ b) $3^{50} \cdot 3^{-47}$ c) $3x^{-3} \cdot 5x$ d) $2a^4 \cdot b^{-6} \cdot a^2 \cdot 3b^2$

Solution

a) $y^4 \cdot y^6 = y^{10}$ Product rule

b) $3^{50} \cdot 3^{-47} = 3^3$ Product rule

              $= 27$

c) $3x^{-3} \cdot 5x = 3 \cdot 5 \cdot x^{-3} \cdot x$

             $= 15x^{-2}$ Because $x = x^1$, $x^{-3} \cdot x^1 = x^{-2}$.

             $= \dfrac{15}{x^2}$ Write the answer with positive exponents.

d) $2a^4 \cdot b^{-6} \cdot a^2 \cdot 3b^2 = 6a^4 \cdot a^2 \cdot b^{-6} \cdot b^2$

                       $= 6a^6 \cdot b^{-4}$ Product rule

                       $= \dfrac{6a^6}{b^4}$ Write the answers with positive exponents. ◀

Changing the Sign of an Exponent

Because 5^{-2} and 5^2 are reciprocals of each other, we write

$$5^{-2} = \frac{1}{5^2} \quad \text{and} \quad 5^2 = \frac{1}{5^{-2}}.$$

Note that the sign of the exponent changes as the exponential expression moves from numerator to denominator or vice versa. It is convenient to remember this fact when we simplify expressions involving exponents.

Sign-change Rule

If a is a nonzero real number and n is *any integer*, then

$$a^n = \frac{1}{a^{-n}}.$$

The sign-change rule looks like the definition of negative exponent, but notice that in the sign-change rule n can be any integer.

EXAMPLE 3 Write each expression without negative exponents and simplify. All variables represent nonzero real numbers.

a) $\dfrac{5^{-3}}{2^{-2}}$ b) $3x^{-7}$ c) $2^3 \cdot 3^{-2}$ d) $\dfrac{x^{-3}}{y^{-2}z^3}$

Solution

a) $\dfrac{5^{-3}}{2^{-2}} = 5^{-3} \cdot \dfrac{1}{2^{-2}} = \dfrac{1}{5^3} \cdot 2^2 = \dfrac{1}{125} \cdot 4 = \dfrac{4}{125}$

Note that the sign-change rule allows us to move an exponential expression from numerator to denominator (or vice versa) as long as we change the sign of the exponent.

b) $3x^{-7} = 3 \cdot \dfrac{1}{x^7} = \dfrac{3}{x^7}$ Do not move the 3. The negative exponent is only on the x.

c) $2^3 \cdot 3^{-2} = 2^3 \cdot \dfrac{1}{3^2} = \dfrac{2^3}{3^2} = \dfrac{8}{9}$ Note that the product rule does not apply because the bases are different.

d) $\dfrac{x^{-3}}{y^{-2}z^3} = x^{-3} \cdot \dfrac{1}{y^{-2}} \cdot \dfrac{1}{z^3} = \dfrac{1}{x^3} \cdot y^2 \cdot \dfrac{1}{z^3} = \dfrac{y^2}{x^3 z^3}$ ◀

We must be very careful with the sign-change rule. *It does not allow us to move exponential expressions from numerator to denominator (or vice versa) when addition or subtraction is involved.* For example,

$$\frac{2^{-1} + 1^{-1}}{1^{-1}} \neq \frac{1}{2 + 1}.$$

To see that these two expressions are not equal, we simplify each side of this statement. The right side is equal to $1/3$. Because $2^{-1} = 1/2$ and $1^{-1} = 1$, the left side is equal to $3/2$.

Quotient Rule

We can use arithmetic to simplify the quotient of two exponential expressions. For example,

$$\frac{2^5}{2^3} = \frac{2 \cdot 2 \cdot 2 \cdot 2 \cdot 2}{2 \cdot 2 \cdot 2} = 2^2.$$

There are five 2s in the numerator and three 2s in the denominator. After dividing, two 2s remain.

We can also simplify a quotient by using the rules of exponents. In the next example, we move the expression in the denominator to the numerator, changing the sign of the exponent, and then use the product rule to add the exponents:

$$\frac{2^{50}}{2^{30}} = 2^{50} \cdot 2^{-30} = 2^{50+(-30)} = 2^{20}$$

Note that changing the sign of the exponent in the denominator and adding it to the exponent in the numerator is exactly the same as subtracting the exponents. In the next example we simply subtract the exponents:

$$\frac{2^{-3}}{2^{-7}} = 2^{-3-(-7)} = 2^{-3+7} = 2^4$$

We showed more steps than are necessary in the above example. The purpose of the rules of exponents is to make it easier to simplify certain types of expressions. With practice we can learn to subtract the exponents mentally and write only the answer.

Quotient Rule

If m and n are any integers and $a \neq 0$, then

$$\frac{a^m}{a^n} = a^{m-n}.$$

Note that to apply the quotient rule the bases must be identical.

EXAMPLE 4 Use the rules of exponents to simplify the following. Write answers with positive exponents only. All variables represent nonzero real numbers.

a) $\dfrac{4^{13}}{4^6}$ b) $\dfrac{2x^{-7}}{x^9}$ c) $\dfrac{w(2w^{-4})}{3w^{-2}}$ d) $\dfrac{x^{-3}y^5}{x^{-2}y^2}$

Solution

a) $\dfrac{4^{13}}{4^6} = 4^{13-6} = 4^7$

b) $\dfrac{2x^{-7}}{x^9} = 2x^{-7-9} = 2x^{-16} = \dfrac{2}{x^{16}}$

c) $\dfrac{w(2w^{-4})}{3w^{-2}} = \dfrac{2w^{-3}}{3w^{-2}}$ By the product rule, $w \cdot w^{-4} = w^{-3}$.

     $= \dfrac{2w^{-3-(-2)}}{3}$ Quotient rule

     $= \dfrac{2w^{-1}}{3}$

     $= \dfrac{2}{3w}$ Sign-change rule

d) $\dfrac{x^{-3}y^5}{x^{-2}y^2} = x^{-3-(-2)}y^{5-2} = x^{-1}y^3 = \dfrac{y^3}{x}$ ◀

Warm-ups

True or false?

1. $3^5 3^4 = 3^9$ 2. $3 \cdot 3 \cdot 3^{-1} = 1/27$ 3. $3^{-2} = 1/9$

4. $1^{-1} = 1$ 5. $\dfrac{2^5}{2^{-2}} = 2^7$ 6. $2^3 \cdot 5^2 = 10^5$

7. $-2^{-2} = -1/4$ 8. $5 \cdot 5^{-1} = 1$ 9. $2 \cdot 3^{-5} = \dfrac{1}{2 \cdot 3^5}$

10. $\dfrac{3^{-2}}{x^{-2}} = \dfrac{x^2}{9}$ for $x \neq 0$.

3.1 EXERCISES

For all exercises in this section assume that the variables represent nonzero real numbers, and use only positive exponents in your answers.

Use the rules of exponents to simplify each expression. See Example 1.

1. 4^2

2. 3^3

3. 4^{-2}

4. 3^{-3}

5. $(-4)^{-2}$

6. -3^{-3}

7. 5^{-1}

8. $\left(\dfrac{3}{5}\right)^{-1}$

9. $\left(-\dfrac{3}{4}\right)^{-2}$

10. $(.5)^{-1}$

11. $10^2 \cdot 10^{-2}$

12. $6^2 \cdot 2^{-1}$

13. $2^{-3} \cdot 4^2$

14. $3 \cdot 10^{-1}$

15. $500 \cdot 10^{-2}$

16. $5 \cdot 5^{-1}$

17. $7^4 \cdot 7^{-4}$

18. -2^{-4}

19. -3^{-2}

20. $\dfrac{1}{2^{-4}}$

21. $\dfrac{1}{7^{-2}}$

22. $\dfrac{2}{2^{-3}}$

Simplify. See Example 2.

23. $2^5 \cdot 2^{12}$

24. $2x^2 \cdot 3x^3$

25. $2y^3 \cdot y^2 \cdot y^7$

26. $2 \cdot 10^6 \cdot 5 \cdot 10^7$

27. $5x^{-3} \cdot 6x^4$

28. $(x - y)^0$

29. $2w^{-3} \cdot w^7 \cdot w^{-4}$

30. $4a^{13}a^{-4}$

31. $(x^2 - y^2)^0$

32. $9y^8y^{-8}$

33. $x^2y^{-9} \cdot x^7y^{14}$

34. $5y^2z \cdot y^{-3}z^{-1}$

35. $3x^{-8} \cdot 2x^3$

36. $7m^{-2} \cdot 3m^{-9}$

37. $a^{-8}b^0a^{-9}$

38. $x^2y^2 \cdot x^{-2}y^{-2}$

Write each expression without negative exponents and simplify. See Example 3.

39. $\dfrac{2}{4^{-2}}$

40. $\dfrac{5}{10^{-3}}$

41. $\dfrac{3^{-1}}{10^{-2}}$

42. $\dfrac{2y^{-2}}{3^{-1}}$

43. $\dfrac{4x^{-3}}{5y^{-2}}$

44. $5^{-1}xy^{-3}$

45. $\dfrac{x^3x^{-6}}{x^{-3}x^2}$

46. $\dfrac{y^{-4}y^{-6}}{y^2y^{-7}}$

Use the quotient rule along with the other rules of exponents to simplify each expression. See Example 4.

47. $\dfrac{x^5}{x^3}$

48. $\dfrac{a^8}{a^3}$

49. $\dfrac{x^6}{x^{-2}}$

50. $\dfrac{w^2}{w^{-5}}$

51. $\dfrac{a^{-5}}{a^{-2}}$

52. $\dfrac{a^{-3}}{a^{-4}}$

53. $\dfrac{w^{-5}}{w^3}$

54. $\dfrac{x^{-6}}{x^2}$

55. $4x^{-3} \div x^2$

56. $2x^{-1} \div x^{-5}$

57. $x \div x^{-1}$

58. $a^{-1} \div a^{-1}$

59. $\dfrac{3^3w^2}{3^{-5}w^{-3}}$

60. $\dfrac{2^{-3}w^5}{2^5w^{-7}}$

61. $(x^2y)(x^{-3}y^3)$

62. $(5x^2y)(2x^{-3}y^3)$

63. $\dfrac{6x^{-5} \cdot x^{-4}}{2x^{-3} \cdot x^7}$

64. $\dfrac{10a^2a^{-5}}{2a^6a^{-3}}$

65. $\dfrac{3x^{-6}y^{-1}}{6x^{-5}y^{-2}}$

66. $\dfrac{2r^{-3}t^{-1}}{10r^5t^{-3}}$

Use the rules of exponents to simplify each expression.

67. $(-2)^3$

68. $(-3)^3$

69. $(-2)^4$

70. $(-5)^2$

71. $\left(\dfrac{1}{2}\right)^2$

72. $\left(\dfrac{1}{3}\right)^2$

73. $\left(\dfrac{1}{3}\right)^{-3}$

74. $\left(\dfrac{1}{4}\right)^{-3}$

75. -2^4

76. -3^4

77. -2^3

78. $(-3)^4$

79. $\left(1+\dfrac{1}{2}\right)^2$

80. $\left(2+\dfrac{1}{2}\right)^2$

81. $-(-2)^{-3}$

82. $-(-3)^{-1}$

83. $2^3 \cdot 3^2$

84. $2^3 \cdot 5^2$

85. $(2-3)^{-1}$

86. $(4-3)^{-2}$

87. $2^{-1}+3^{-1}$

88. $3^{-1}+4^{-1}$

89. $(2+3)^{-1}$

90. $(3-4)^{-1}$

91. $5a^2 \cdot 3a^2$

92. $2x^2 \cdot 5x^{-5}$

93. $5a^2 + 3a^2$

94. $2x^2 \cdot 5y^{-5}$

95. $\dfrac{3a^5}{6a^3}$

96. $\dfrac{6ab^{-2}}{-2a^2b^{-3}}$

97. $\dfrac{-3x^3y^{-3}}{-9x^2y}$

98. $\dfrac{-2x^{-5}y^6}{6x^{-6}y^2}$

99. $\dfrac{8x^8}{2x^2}$

100. $\dfrac{10y^{12}}{5y^3}$

101. $\dfrac{-12a^6}{-3a^2}$

102. $\dfrac{30b^{12}}{-5b^6}$

103. $(.036)^{-2} + (4.29)^3$

104. $3(4.71)^2 - 5(.471)^{-3}$

105. $\dfrac{(5.73)^{-1} + (4.29)^{-1}}{(3.762)^{-1}}$

106. $[5.29 + (.374)^{-1}]^3$

3.2 The Power Rules

IN THIS SECTION:

- Raising an Exponential Expression to a Power
- Raising a Product to a Power
- Raising a Quotient to a Power
- Summary of Rules
- Applications

In Section 3.1 we learned some of the basic rules for working with exponents. All of the rules of exponents are designed to make it easier to work with exponential expressions. In this section we will extend our list of rules to include three new rules.

Raising an Exponential Expression to a Power

Consider the following examples of raising an exponential expression to a power. In each case, we can use rules that we already know to obtain the answer, but we will discover a simple rule that can be done mentally. We will assume that all variables represent nonzero real numbers.

$$(x^3)^2 = x^3 \cdot x^3 \qquad \text{Exponent 2 indicates two factors of } x^3.$$
$$= x^6 \qquad \text{Product rule: } 3 + 3 = 6$$

Note that $2 \cdot 3 = 6$ also.

$$(5^{-3})^4 = 5^{-3} \cdot 5^{-3} \cdot 5^{-3} \cdot 5^{-3} \qquad \text{Four factors of } 5^{-3}$$
$$= 5^{-12} \qquad \text{Product rule: } -3 + (-3) + (-3) + (-3) = -12$$

Note that $-3 \cdot 4 = -12$ also.

$$(2^{-3})^{-4} = \frac{1}{(2^{-3})^4} = \frac{1}{2^{-3} \cdot 2^{-3} \cdot 2^{-3} \cdot 2^{-3}} = \frac{1}{2^{-12}} = 2^{12}$$

Note that in each example the exponent of the answer is the product of the two original exponents. We can state the general rule for raising a power to a power as follows.

Power of a Power Rule

If m and n are any integers and $a \neq 0$, then

$$(a^m)^n = a^{mn}.$$

EXAMPLE 1 Use the rules of exponents to simplify each expression. Write the answer with positive exponents only.

a) $(2^3)^5$ **b)** $(x^2)^{-6}$ **c)** $3(y^{-3})^{-2}y^{-5}$ **d)** $\dfrac{(x^2)^{-1}}{(x^{-3})^3}$

Solution

a) $(2^3)^5 = 2^{15}$ Power of a power rule

b) $(x^2)^{-6} = x^{-12}$ Power of a power rule

$$= \frac{1}{x^{12}} \qquad \text{Write the answer with positive exponents only.}$$

c) $3(y^{-3})^{-2}y^{-5} = 3y^6 y^{-5}$ Power of a power rule

$$= 3y \qquad \text{Product rule}$$

d) $\dfrac{(x^2)^{-1}}{(x^{-3})^3} = \dfrac{x^{-2}}{x^{-9}}$ Power of a power rule

$$= x^7 \qquad \text{Quotient rule}$$

◀

Raising a Product to a Power

In the next two examples we simplify a product raised to a positive power, and then a product raised to a negative power.

$$\overset{\text{3 factors of } 2x}{(2x)^3 = 2x \cdot 2x \cdot 2x} = 2^3 \cdot x^3 = 8x^3$$

$$(ay)^{-3} = \frac{1}{(ay)^3} = \frac{1}{(ay)(ay)(ay)} = \frac{1}{a^3y^3} = a^{-3}y^{-3}$$

Note that in each case we have rewritten the expression with the power distributed over each factor of the product. These examples illustrate the power of a product rule.

Power of a Product Rule

If a and b are nonzero real numbers and n is any integer, then

$$(ab)^n = a^n \cdot b^n.$$

EXAMPLE 2 Simplify. Assume that the variables represent nonzero real numbers. Write the answer with positive exponents only.

a) $(-3x)^4$ **b)** $(-2x^2)^3$ **c)** $(3x^{-2}y^3)^{-2}$

Solution

a) $(-3x)^4 = (-3)^4x^4 = 81x^4$

b) $(-2x^2)^3 = (-2)^3(x^2)^3 = -8x^6$

c) $(3x^{-2}y^3)^{-2} = (3)^{-2}(x^{-2})^{-2}(y^3)^{-2} = \frac{1}{9}x^4y^{-6} = \frac{x^4}{9y^6}$ ◄

Raising a Quotient to a Power

Now consider an example of applying rules that we already know to a power of a quotient.

$$\left(\frac{x}{5}\right)^3 = \frac{x}{5} \cdot \frac{x}{5} \cdot \frac{x}{5} = \frac{x^3}{5^3}$$

We get a similar result with a negative power:

$$\left(\frac{x}{5}\right)^{-3} = \left(\frac{5}{x}\right)^3 = \frac{5}{x} \cdot \frac{5}{x} \cdot \frac{5}{x} = \frac{5^3}{x^3} = \frac{x^{-3}}{5^{-3}}$$

Note that in the final result the original power is distributed over each term of the quotient. We can state the power of a quotient rule as follows.

Power of a Quotient Rule

If a and b are nonzero real numbers and n is any integer, then

$$\left(\frac{a}{b}\right)^n = \frac{a^n}{b^n}.$$

EXAMPLE 3   Use the rules of exponents to simplify each expression. Write your answers with positive exponents only. Assume that the variables are nonzero real numbers.

a) $\left(\dfrac{x}{2}\right)^3$ 　　　　 **b)** $\left(\dfrac{2x^3}{3y^2}\right)^3$ 　　　 **c)** $\left(\dfrac{x^{-2}}{2^3}\right)^{-1}$ 　　　 **d)** $\left(\dfrac{3}{4x^3}\right)^{-2}$

Math at Work

Without a doubt, the computer has revolutionized more areas of our society than any other modern invention. From space exploration to weather prediction, international business, and telecommunications, the world as we know it depends on the speed and the power of computers.

Computer designers measure the changing power of computers in "generations" of technological development. The first generation, built in the 1940s, was a single machine called ENIAC. Constructed out of thousands of vacuum tubes, ENIAC performed calculations at a rate of 10^3 every second—slow by today's standards, but faster than any contemporary device.

Subsequent generations of computers improved on ENIAC's speed by using transistors rather than vacuum tubes. By the late 1950s, computers could perform 10^4 calculations per second. Computers of the 1970s performed 10^6 operations, making them 1000 times faster than ENIAC.

In the 1980s the microchip replaced transistor circuits and again increased the power of the computer to make over 10^8 operations per second. Throughout the 1980s, computer designers struggled to reach the next barrier: a computer capable of 10^9 calculations per second.

The Y-MP, introduced by Cray Research in 1988, did just that. Researchers have already begun using this supercomputer not only to solve mathematical problems and store data, but also to electronically imitate real-world situations such as stock market behavior and weather patterns.

Competing companies have already announced intentions to build a computer capable of 22×10^9 calculations per second. How much faster will this computer be compared to the original ENIAC?

Solution

a) $\left(\dfrac{x}{2}\right)^3 = \dfrac{x^3}{2^3} = \dfrac{x^3}{8}$

b) $\left(\dfrac{2x^3}{3y^2}\right)^3 = \dfrac{2^3 x^9}{3^3 y^6}$ Because $(x^3)^3 = x^9$ and $(y^2)^3 = y^6$

$\qquad\quad = \dfrac{8x^9}{27y^6}$

c) $\left(\dfrac{x^{-2}}{2^3}\right)^{-1} = \dfrac{x^2}{2^{-3}} = 8x^2$

d) $\left(\dfrac{3}{4x^3}\right)^{-2} = \dfrac{3^{-2}}{4^{-2}x^{-6}} = \dfrac{4^2 x^6}{3^2} = \dfrac{16x^6}{9}$ ◀

When a quotient is raised to a negative power, it is often simpler to get the reciprocal first, then use the power of a quotient rule.

EXAMPLE 4   Simplify. Assume that the variables are nonzero real numbers and write the answer with positive exponents only.

a) $\left(\dfrac{3}{4}\right)^{-3}$
b) $\left(\dfrac{x^2}{3}\right)^{-2}$
c) $\left(\dfrac{2x^3}{3}\right)^{-2}$

Solution

a) $\left(\dfrac{3}{4}\right)^{-3} = \left(\dfrac{4}{3}\right)^3 = \dfrac{64}{27}$ Cube the reciprocal of 3/4.

b) $\left(\dfrac{x^2}{3}\right)^{-2} = \left(\dfrac{3}{x^2}\right)^2 = \dfrac{3^2}{(x^2)^2} = \dfrac{9}{x^4}$

c) $\left(\dfrac{2x^3}{3}\right)^{-2} = \left(\dfrac{3}{2x^3}\right)^2 = \dfrac{9}{4x^6}$ ◀

Summary of Rules

The definitions and rules that were introduced in the last two sections are summarized in the following box.

Rules for Integral Exponents

If a and b are nonzero real numbers and m and n are integers, then

1. $a^{-n} = \dfrac{1}{a^n} = \left(\dfrac{1}{a}\right)^n$ Negative-exponent definition (power and reciprocal in either order)

2. $a^{-1} = \dfrac{1}{a}$ *The reciprocal of a is a^{-1}.*

3. $a^0 = 1$ Zero-exponent definition

4. $a^m a^n = a^{m+n}$ Product rule

5. $a^n = \dfrac{1}{a^{-n}}$ Sign-change rule

6. $\dfrac{a^m}{a^n} = a^{m-n}$ Quotient rule

7. $(a^m)^n = a^{mn}$ Power of a power rule

8. $(ab)^n = a^n b^n$ Power of a product rule

9. $\left(\dfrac{a}{b}\right)^n = \dfrac{a^n}{b^n}$ Power of a quotient rule

Applications

Both positive and negative exponents are used in formulas for compound interest. Suppose P dollars (present value) is invested in a savings account with the annual interest rate r.

Amount Formula

The amount A of the account after interest is compounded annually with rate r for t years is given by the formula

$$A = P(1 + r)^t.$$

To find out how much must be deposited now (present value) to amount to A dollars after t years, we can solve the formula for P by multiplying each side by $(1 + r)^{-t}$.

$$A \cdot (1 + r)^{-t} = P(1 + r)^t \cdot (1 + r)^{-t}$$
$$A \cdot (1 + r)^{-t} = P(1 + r)^0$$
$$P = A(1 + r)^{-t}$$

Present-value Formula

The present value P that will amount to A dollars after t years with annual interest rate r is given by the formula

$$P = A(1 + r)^{-t}.$$

EXAMPLE 5 If $100 dollars is deposited in an account paying 5% compounded annually, then how much is in the account after 2 years?

Solution Use $t = 2$, $P = \$100$, and $r = .05$ in the amount formula.

$$A = P(1 + r)^t$$
$$A = \$100(1 + .05)^2$$
$$= \$100(1.05)^2$$
$$= \$100(1.1025)$$
$$= \$110.25$$

The amount in the account after 2 years is $\$110.25$. ◄

EXAMPLE 6 How much would a person have to deposit today in an account paying 10% compounded annually to have $\$1,000,000$ after 40 years?

Solution Use $r = .10$, $t = 40$, and $A = 1,000,000$ in the present-value formula.

$$P = A(1 + r)^{-t}$$
$$P = 1,000,000(1 + .10)^{-40}$$
$$P = 1,000,000(1.1)^{-40} \qquad \text{Use a calculator with an exponent key.}$$
$$P = 22,094.93$$

A deposit of $\$22,094.93$ today in an account paying 10% compounded annually would amount to $\$1,000,000$ after 40 years. ◄

Warm-ups

True or false? Assume that all variables are nonzero real numbers.

1. $(2^2)^3 = 2^5$

2. $(2^{-3})^{-1} = 8$

3. $(x^{-3})^3 = x^{-9}$

4. $(2^3)^3 = 2^{27}$

5. $(2x)^3 = 6x^3$

6. $(-3y^3)^2 = 9y^9$

7. $\left(\dfrac{2}{3}\right)^{-1} = \dfrac{3}{2}$

8. $\left(\dfrac{2}{3}\right)^3 = \dfrac{8}{27}$

9. $\left(\dfrac{x^2}{2}\right)^3 = \dfrac{x^6}{8}$

10. $\left(\dfrac{2}{x}\right)^{-2} = \dfrac{x^2}{4}$

3.2 EXERCISES

For all exercises in this section assume that the variables represent nonzero real numbers, and use only positive exponents in your answers.

Use the rules of exponents to simplify each expression. See Example 1.

1. $(y^2)^5$

2. $(x^6)^2$

3. $(2^2)^3$

4. $(3^2)^2$

5. $(x^{-2})^7$

6. $(m^{-3})^{-6}$

7. $(a^{-3})^{-3}$

8. $(x^2)^{-4}$

9. $(x^{-2})^3(x^{-3})^{-2}$

10. $(m^{-3})^{-1}(m^2)^{-4}$

11. $\dfrac{(x^3)^{-4}}{(x^2)^{-5}}$

12. $\dfrac{(a^2)^{-3}}{(a^{-2})^4}$

Simplify. See Example 2.

13. $(9y)^2$

14. $(2a)^3$

15. $(-5w)^2$

16. $(-2w^{-5})^3$

17. $(x^3y^{-2})^3$

18. $(a^2b^{-3})^2$

19. $(3ab)^2$

20. $(2xy^2)^3$

21. $\dfrac{2xy^{-2}}{(3xy)^{-1}}$

22. $\dfrac{3ab^{-1}}{(3ab)^{-1}}$

23. $\dfrac{(ab)^{-2}}{ab^2}$

24. $\dfrac{(xy)^{-3}}{xy^3}$

Simplify. See Example 3.

25. $\left(\dfrac{w}{2}\right)^3$

26. $\left(\dfrac{m}{5}\right)^2$

27. $\left(\dfrac{3}{4}\right)^3$

28. $\left(\dfrac{2}{3}\right)^4$

29. $\left(\dfrac{2x}{y}\right)^2$

30. $\left(\dfrac{-3x^3}{y}\right)^{-2}$

31. $\left(\dfrac{-2y^2}{x}\right)^{-3}$

32. $\left(\dfrac{2ab}{3}\right)^3$

Simplify. See Example 4.

33. $\left(\dfrac{2}{5}\right)^{-2}$

34. $\left(\dfrac{3}{4}\right)^{-2}$

35. $\left(-\dfrac{1}{2}\right)^{-3}$

36. $\left(-\dfrac{2}{3}\right)^{-2}$

37. $\left(-\dfrac{2x}{3}\right)^{-3}$

38. $\left(-\dfrac{ab}{c}\right)^{-1}$

39. $\left(\dfrac{2x^2}{3y}\right)^{-3}$

40. $\left(\dfrac{ab^{-3}}{a^2b}\right)^{-2}$

Solve each problem. See Examples 5 and 6.

41. Suppose $3000 is deposited in an account paying 9% compounded annually. What is the value of the account after 3 years?

42. Melissa borrowed $40,000 at 12% compounded annually and made no payments for 3 years. How much did she owe the bank at the end of the 3 years? (Use the compound-interest formula.)

43. Mr. Watkins wants to have $10,000 in a savings account when his little Wanda is ready for college. How much must he deposit today in an account paying 7% compounded annually to have $10,000 in 18 years?

44. If Domingo wants to have $2,000,000 when he retires in 45 years, then how much must he deposit now in an account paying 8% compounded annually?

Use the rules of exponents to simplify each expression. Do these mentally. Write down only the answer.

45. $3x^4 \cdot 2x^5$

46. $(3x^4)^2$

47. $(-2x^2)^3$

48. $3x^2 \cdot 2x^{-4}$

49. $\dfrac{3x^{-2}y^{-1}}{z^{-1}}$

50. $\dfrac{2^{-1}x^2}{y^{-2}}$

51. $\left(\dfrac{-2}{3}\right)^{-1}$

52. $\left(\dfrac{-1}{5}\right)^{-2}$

53. $\left(\dfrac{2x^3}{3}\right)^2$

54. $\left(\dfrac{-2y^4}{x}\right)^3$

55. $(-2x^{-2})^{-1}$

56. $(-3x^{-2})^3$

Use the rules of exponents to simplify each expression.

57. $\left(-\dfrac{x}{2}\right)^{-2}$

58. $\left(-\dfrac{x}{3}\right)^4$

59. $(-x^3)^{-2}$

60. $(w^{-5})^3$

61. $\left(\dfrac{x^2y}{xy^2}\right)^{-3}$

62. $\left(\dfrac{x^3y^2}{xy^3}\right)^{-1}$

63. $\dfrac{(a^{-1}b^2)^3}{(ab^{-2})^3}$

64. $\dfrac{(m^2n^{-3})^4}{mn^5}$

65. $\dfrac{(x^{-2}y)^{-3}}{(xy^{-1})^2}$

66. $\dfrac{(x^{-1}y^3)^{-2}}{(xy^{-1})^3}$

67. $\left(\dfrac{a^{-2}b^3}{c^4}\right)^{-2}(a^{-1}b^2)^3$

68. $(xz^2)^{-3}\left(\dfrac{xy^{-1}}{z}\right)^2$

*69. $\left(\dfrac{a^{-2}b^3}{a^{-3}b^5}\right)^4\left(\dfrac{a^{-5}b^4a^2}{b^{-6}a^5}\right)^{-3}$

*70. $\dfrac{(x^2y^3z^{-3})^5(2x^{-4}y^3z^{-1})^{-1}}{(-3x^2y^4)^{-2}(-xyz)^{-3}}$

*71. $\dfrac{(-2x^{-3}yz)^{-3}(-xy^{-1}z^2)^2(3^{-1}xyz^2)^2}{(2^{-1}x^2y)^3(3^{-2}xy^{-1})^{-1}(-xz^2)^{-3}}$

*72. $\left(\dfrac{2^{-1}ab^{-2}}{3a^2b^{-3}}\right)^{-2}\left(\dfrac{-5a^{-1}b^3}{4ab^{-3}}\right)^2\left(\dfrac{-2^{-1}ab}{3a^{-4}b}\right)^3$

*73. $\dfrac{2^{3m-1}(2^{5-m})^2}{(2^{1-m})^3(2^{m+1})^{-1}}$

*74. $\left(\dfrac{3^{n-1}}{2^{n+5}}\right)^2\cdot\left(\dfrac{2^{n+4}}{3^{2n-1}}\right)^{-3}$

3.3 Addition, Subtraction, and Multiplication of Polynomials

IN THIS SECTION:
- Polynomials
- Evaluating Polynomials
- Addition and Subtraction of Polynomials
- Multiplication of Polynomials

A polynomial is a particular type of algebraic expression that serves as a fundamental building block in algebra. We used polynomials in Chapters 1 and 2, but we did not identify them as polynomials. In this section we will learn to recognize polynomials and to perform three basic operations with them.

Polynomials

An expression such as

$$3x^3 - 15x^2 + 7x - 2$$

is called a **polynomial** in one variable. Because this expression could also be written as

$$3x^3 + (-15x^2) + 7x + (-2),$$

we say that this polynomial is a sum of four terms

$$3x^3, \quad -15x^2, \quad 7x, \quad \text{and} \quad -2.$$

A **term** of a polynomial is the product of a number and one or more variables raised to whole-number powers. The number preceding the variable in each term is called the **coefficient** of that variable. In the polynomial just shown, 3 is the coefficient of

x^3, -15 is the coefficient of x^2, and 7 is the coefficient of x. In algebra a number is frequently referred to as a **constant,** so the last term, -2, is called the **constant term.** It can be thought of as the product of a number and a variable once we realize that -2 is the same as $-2x^0$. A **polynomial** is defined as a single term or a finite sum of terms. As in the example shown, a sum is usually written as a difference when a term has a negative coefficient.

EXAMPLE 1   Determine whether each algebraic expression is a polynomial.

a) -3 　　　　　　　　　　　**b)** $3x + 2^{-1}$ 　　　　　　　　　**c)** $3x^{-2} + 4y^2$

d) $\dfrac{1}{x} + \dfrac{1}{x^2}$ 　　　　　　　　　　　　　　　　　**e)** $x^{49} - 8x^2 + 11x - 2$

Solution
a) This expression is a polynomial of one term, a constant term.
b) Only the variables must have whole-number exponents. Therefore, this expression is a polynomial of two terms.
c) This expression is not a polynomial because x has a negative exponent.
d) $\dfrac{1}{x} + \dfrac{1}{x^2} = x^{-1} + x^{-2}$
　　This expression is not a polynomial because the exponents of the variables are negative.
e) This expression is a polynomial.　　　　　　　　　　　　　　　　　◀

　　　For simplicity, we generally write polynomials in one variable with the exponents in decreasing order from left to right. Thus, we would write
$$3x^3 - 15x^2 + 7x - 2 \quad \text{rather than} \quad -15x^2 - 2 + 7x + 3x^3.$$
When a polynomial is written this way, the coefficient of the first term is called the **leading** coefficient.
　　　Certain polynomials have special names. A **monomial** is a polynomial that has one term, a **binomial** is a polynomial that has two terms, and a **trinomial** is a polynomial that has three terms. The **degree** of a polynomial in one variable is the highest power of the variable in the polynomial. The number 0 is considered to be a monomial without degree, because $0 = 0x^n$, where n could be any number.

EXAMPLE 2   For each polynomial, state the degree of the polynomial and the coefficient of x^2. Determine whether the polynomial is monomial, binomial, or trinomial.

a) $\dfrac{x^2}{3} - 5x^3 + 7$ 　　　　　　**b)** $x^{48} - x^2$ 　　　　　　　　**c)** 6

Solution
a) This is a third-degree trinomial; the coefficient of x^2 is $1/3$.
b) This is a binomial whose degree is 48; the coefficient of x^2 is -1.

c) Because $6 = 6x^0$, this is a monomial with degree 0. Because x^2 does not appear in this polynomial, the coefficient of x^2 is 0. ◄

Although we are mainly concerned here with polynomials in one variable, we will also deal with polynomials in more than one variable, such as

$$4x^2 - 5xy + 6y^2, \qquad x^2 + y^2 + z^2, \qquad \text{and} \qquad ab^2 - c^3.$$

Evaluating Polynomials

The formula $D = -16t^2 + 64t$ was given in the introduction to this chapter as a formula for the location of a ball tossed into the air. The formula expresses the ball's distance above the earth in terms of the time t. The value of the polynomial $-16t^2 + 64t$ depends on which number is used to replace the variable t. To emphasize that the value of D depends on the value of t, we often write $D(t)$ in place of D:

$$D(t) = -16t^2 + 64t$$

We read $D(t)$ as "D of t." To find the distance when $t = 2$, we replace t by 2 in the formula:

$$D(2) = -16 \cdot 2^2 + 64 \cdot 2$$
$$= -16 \cdot 4 + 128$$
$$= 64$$

The notation $D(2)$ stands for the value of the polynomial when $t = 2$. The statement $D(2) = 64$ means that the distance is 64 feet when the time is 2 seconds.

EXAMPLE 3   Let $Q(x) = 2x^3 - 3x^2 - 7x - 6$. Find $Q(3)$ and $Q(-1)$.

Solution

$$Q(3) = 2(3)^3 - 3(3)^2 - 7(3) - 6$$
$$= 54 - 27 - 21 - 6 = 0$$
$$Q(-1) = 2(-1)^3 - 3(-1)^2 - 7(-1) - 6$$
$$= -2 - 3 + 7 - 6 = -4$$ ◄

Addition and Subtraction of Polynomials

When we evaluate a polynomial, we get a real number. Thus the operations that we perform with real numbers can be performed with polynomials. Actually, we have been adding and subtracting polynomials since Chapter 1. To add two polynomials, add the like terms. To add the polynomials $x^3 + 3x^2 - 5x$ and $-3x^3 + 6x + 2$, we write

$$(x^3 + 3x^2 - 5x) + (-3x^3 + 6x + 2) = -2x^3 + 3x^2 + x + 2.$$

We may also write addition of polynomials vertically. When we do this, we must line up the like terms as shown:

$$x^3 + 3x^2 - 5x$$
$$\underline{-3x^3 \qquad\quad + 6x + 2} \qquad \text{Add.}$$
$$-2x^3 + 3x^2 + \ x + 2$$

When we subtract polynomials, we subtract like terms. Because $a - b = a + (-b)$, we can change the sign of b and add it to a. To subtract

$$x^2 - 3x + 5 \text{ from } 3x^2 + 5x - 7,$$

we write

$$(3x^2 + 5x - 7) - (x^2 - 3x + 5) = (3x^2 + 5x - 7) + (-x^2 + 3x - 5) \qquad \text{Change signs.}$$

$$= 2x^2 + 8x - 12. \qquad \text{Add.}$$

If we subtract polynomials vertically, we change the signs of the terms on the bottom and add. Subtract:

$$3x^2 + 5x - 7 \qquad\qquad\qquad\qquad 3x^2 + 5x - \ 7$$
$$\underline{x^2 - 3x + 5} \leftarrow \text{Change signs and add.} \rightarrow \underline{-x^2 + 3x - \ 5}$$
$$2x^2 + 8x - 12$$

EXAMPLE 4 Perform the indicated operation.

a) $(x^2 - 5x - 7) + (7x^2 - 4x + 10)$
b) $(3x^3 - 5x^2 - 7) + (4x^2 - 2x + 3)$
c) $(x^2 - 7x - 2) - (5x^2 + 6x - 4)$
d) $(6y^3z - 5yz + 7) - (4y^2z - 3yz - 9)$

Solution

a) $(x^2 - 5x - 7) + (7x^2 - 4x + 10) = x^2 + 7x^2 - 5x - 4x - 7 + 10$
$$= 8x^2 - 9x + 3 \qquad \text{Combine like terms.}$$

b) For illustration, we will write this addition vertically.

$$3x^3 - 5x^2 \qquad\ - 7$$
$$\underline{\qquad\ 4x^2 - 2x + 3} \qquad \text{Line up like terms.}$$
$$3x^3 - \ x^2 - 2x - 4$$

c) $(x^2 - 7x - 2) - (5x^2 + 6x - 4) = x^2 - 7x - 2 - 5x^2 - 6x + 4 \qquad \text{Change signs.}$

$$= -4x^2 - 13x + 2 \qquad \text{Combine like terms.}$$

d) For illustration, we will write this subtraction vertically. Subtract:

$$6y^3z \qquad\ - 5yz + 7 \qquad\qquad 6y^3z \qquad\ - 5yz + \ 7$$
$$\underline{\qquad 4y^2z - 3yz - 9} \leftarrow \text{Change signs} \rightarrow \underline{\qquad\ -4y^2z + 3yz + \ 9}$$
$$\qquad\qquad\qquad\qquad \text{and add.} \qquad\quad 6y^3z - 4y^2z - 2yz + 16 \qquad\qquad \blacktriangleleft$$

It is certainly not necessary to write out all of the steps shown in Example 4, but we must use the following rule.

Addition and Subtraction of Polynomials

> To add two polynomials, add the like terms.
>
> To subtract two polynomials, subtract the like terms.

The operations of addition and subtraction of polynomials can often be done mentally once we have had some practice.

Multiplication of Polynomials

We learned how to multiply monomials when we learned the product rule in Section 3.1. For example,

$$-2x^3 \cdot 4x^2 = -8x^5.$$

To multiply a monomial and a binomial, we apply the distributive property. For example,

$$3x(x^3 - 5) = 3x^4 - 15x.$$

To multiply the binomial $x + 2$ and the trinomial $x^2 + 3x - 5$, we apply the distributive property twice. First we multiply the binomial and each term of the trinomial:

$$(x + 2)(x^2 + 3x - 5) = (x + 2)x^2 + (x + 2)3x + (x + 2)(-5) \qquad \text{Distributive property}$$
$$= x^3 + 2x^2 + 3x^2 + 6x - 5x - 10 \qquad \text{Distributive property}$$
$$= x^3 + 5x^2 + x - 10 \qquad \text{Combine like terms.}$$

Multiplication of polynomials can often be done more easily by setting it up like multiplication of whole numbers. Multiply:

$$
\begin{array}{r}
x^2 + 3x - 5 \\
x + 2 \\
\hline
2x^2 + 6x - 10 \\
x^3 + 3x^2 - 5x \\
\hline
x^3 + 5x^2 + x - 10
\end{array}
$$

2 times -5, $3x$, and x^2
x times -5, $3x$, and x^2
Add.

Multiplication of Polynomials

> To multiply polynomials, multiply each term of the first polynomial by each term of the second polynomial and then combine like terms.

EXAMPLE 5 Multiply the polynomials.

a) $2ab^2 \cdot 3a^2b$ **b)** $(-1)(5 - x)$ **c)** $(x^3 - 5x + 2)(-3x)$
d) $(x - 3)(2x + 5)$ **e)** $(x + y)(z + 4)$

Solution

a) $2ab^2 \cdot 3a^2b = 6a^3b^3$
b) $(-1)(5 - x) = -5 + x = x - 5$
c) $(x^3 - 5x + 2)(-3x) = -3x^4 + 15x^2 - 6x$
d) $(x - 3)(2x + 5) = (x - 3)2x + (x - 3)5$ Distributive property
$\qquad\qquad\qquad\quad = 2x^2 - 6x + 5x - 15$ Distributive property
$\qquad\qquad\qquad\quad = 2x^2 - x - 15$

e)

$$
\begin{array}{r}
x + y \\
z + 4 \\
\hline
4x + 4y \\
zx + zy \\
\hline
zx + zy + 4x + 4y
\end{array}
$$
◄

Note what happened to the binomial in Example 5b when we multiplied it by -1. If we multiply any difference by -1 we get the same type of result:

$$-1(a - b) = -a + b = b - a.$$

Because multiplying by -1 is the same as taking the opposite, we can write this equation as

$$-(a - b) = b - a.$$

This equation says that $a - b$ and $b - a$ are opposites, or additive inverses, of each other. Note that the opposite of $a + b$ is $-a - b$, not $a - b$.

Warm-ups

True or false?

1. The expression $3x^{-2} - 5x + 2$ is a trinomial.
2. In the polynomial $3x^2 - 5x + 3$, the coefficient of x is 5.
3. The degree of the polynomial $x^2 + 3x - 5x^3 + 4$ is 2.
4. If $C(x) = x^2 - 3$, then $C(5) = 22$.
5. If $P(t) = 30t + 10$, then $P(0) = 40$.
6. $(2x^2 - 3x + 5) + (x^2 + 5x - 7) = 3x^2 + 2x - 2$ for any value of x.
7. $(x^2 - 5x) - (x^2 - 3x) = -8x$ for any values of x.
8. $-2x(3x - 4x^2) = 8x^3 - 6x^2$ for any value of x.
9. $-(x - 7) = 7 - x$ for any value of x.
10. The opposite of $y + 5$ is $y - 5$ for any value of y.

3.3 EXERCISES

Determine whether each algebraic expression is a polynomial. See Example 1.

1. $3x$

2. -9

3. $x^{-1} + 4$

4. $3x^{-3} + 4x - 1$

5. $x^2 - 3x + 1$

6. $\dfrac{x^3}{3} - \dfrac{3x^2}{2} + 4$

7. $\dfrac{1}{x} + y - 3$

8. $x^{50} - 9y^2$

Identify each polynomial as a monomial, binomial, or trinomial. State the degree of each polynomial and the coefficient of x^3. See Example 2.

9. $x^4 - 8x^3$

10. $15 - x^3$

11. $x^3 + 3x^4 - 5x^6$

12. $\dfrac{x^3}{2} + \dfrac{5x}{2} - 7$

13. $\dfrac{x^7}{15}$

14. $5x^4$

For each given polynomial find the indicated value of the polynomial. See Example 3.

15. $P(x) = x^4 - 1$, $P(3)$

16. $P(x) = x^2 - x - 2$, $P(-1)$

17. $M(x) = -3x^2 + 4x - 9$, $M(-2)$

18. $C(w) = 3w^2 - w$, $C(0)$

19. $R(x) = x^5 - x^4 + x^3 - x^2 + x - 1$, $R(1)$

20. $T(a) = a^7 + a^6$, $T(-1)$

21. Suppose that the cost in dollars of x cubic yards of gravel is given by the formula $C(x) = 20x + 15$. Find $C(3)$, the cost of 3 cubic yards of gravel.

22. Suppose Sheila's annual bonus in dollars for selling n life insurance policies is given by the formula $B(n) = .1n^2 + 3n + 50$. Find $B(20)$, her bonus for selling 20 policies.

Perform the indicated operations. See Example 4.

23. Add
$$2a - 3$$
$$a + 5$$

24. Add
$$2w - 6$$
$$w + 5$$

25. Subtract
$$7xy + 30$$
$$2xy + \ 5$$

26. Subtract
$$5ab + 7$$
$$3ab + 6$$

27. Add
$$x^3 + 3x^2 - 5x - 2$$
$$-x^3 + 8x^2 + 3x - 7$$

28. Add
$$x^2 - 3x + 7$$
$$-2x^2 - 5x + 2$$

29. Subtract
$$5x + 2$$
$$4x - 3$$

30. Subtract
$$4x + 3$$
$$2x - 6$$

31. Subtract
$$-x^2 + 3x - 5$$
$$5x^2 - 2x - 7$$

32. Subtract
$$-3x^2 + 5x - 2$$
$$x^2 - 5x - 6$$

33. Add
$$x - y$$
$$x + y$$

34. Add
$$-w + 4$$
$$2w - 3$$

Find the following products. See Example 5.

35. Multiply
$$2x - 3$$
$$-5x$$

36. Multiply
$$3a^3 - 5a^2 + 7$$
$$-2a$$

37. Multiply
$$x + 6$$
$$2x - 3$$

38. Multiply
$$3x^2 + 2$$
$$2x^2 - 5$$

39. Multiply
$$a + b$$
$$a - b$$

40. Multiply
$$x^2 + xy + y^2$$
$$x - y$$

41. Multiply
$$a^2 - ab + b^2$$
$$a + b$$

42. Multiply
$$x + 5$$
$$x + 5$$

Perform the indicated operations. See Examples 4 and 5.

43. $(x^2 - 3x) + (5x - 9)$

44. $(2y^2 - 3y - 8) + (y^2 + 4y - 1)$

45. $(x - 7) + (2x - 3) + (5 - x)$

46. $(5x - 3) + (x^3 + 3x - 2) + (-2x - 3)$

47. $-3x^2 \cdot 5x^4$

48. $(-ab^5)(-2a^2b)$

49. $(2xy - 3) - (xy - 5)$

50. $(2x - 5) - (x^2 - 3x + 2)$

51. $(x - 2)(x + 2)$

52. $(x - 1)(x + 1)$

53. $-1(3x - 2)$

54. $-1(-x^2 + 3x - 9)$

55. $(xt - 2y) - (xt + 2y) - (xt + 5y)$

56. $(ws + 2) - (ws + 5) - (-3ws + 4)$

57. $(a^2 - 5a + 3) + (3a^2 - 6a - 7)$

58. $(w^2 - 3w + 2) + (2w - 3 + w^2)$

59. $(w^2 - 7w - 2) - (w - 3w^2 + 5)$

60. $(a^3 - 3a) - (1 - a - 2a^2)$

61. $(x^2 + x + 2)(2x - 3)$

62. $(x^2 - 3x + 2)(x - 4)$

63. $(x - 2)(x^2 + 2x + 4)$

64. $(a - 3)(a^2 + 3a + 9)$

65. $5x^2(3x - 4)$

66. $3y(8y^2 - 3y + 2)$

67. $(x - w)(z + 2w)$

68. $(w - a)(t + 3)$

69. $(az + w)(az + w)$

70. $(mt + n)(mt + n)$

71. $(2x - 1)(3x + 5)$

72. $(3a - 4)(a + 8)$

Perform the indicated operations.

73. $\left(\frac{1}{2}x + 2\right) + \left(\frac{1}{4}x - \frac{1}{2}\right)$

74. $\left(\frac{1}{3}x + 1\right) + \left(\frac{1}{3}x - \frac{3}{2}\right)$

75. $\left(\frac{1}{2}x^2 + \frac{1}{3}x - \frac{1}{5}\right) - \left(x^2 - \frac{2}{3}x - \frac{1}{5}\right)$

76. $\left(\frac{2}{3}x^2 - \frac{1}{3}x + \frac{1}{6}\right) - \left(-\frac{1}{3}x^2 + x + 1\right)$

77. $\left(\frac{1}{2}x - 1\right)\left(\frac{1}{2}x + 3\right)$

78. $\left(\frac{1}{3}x - 1\right)\left(\frac{1}{3}x + 1\right)$

79. $(3.759x^2 - 4.71x + 2.85) + (11.61x^2 + 6.59x - 3.716)$

80. $(43.19x^3 - 3.7x^2 - 5.42x + 3.1) - (62.7x^3 - 7.36x - 12.3)$

81. $(2.31x - 5.4)(6.25x + 1.8)$

82. $(x - .28)(x^2 - 34.6x + 21.2)$

*83. $[x^2 - 3 - (x^2 + 5x - 4)] - [x - 3(x^2 - 5x)]$

*84. $[x^3 - 4x(x^2 - 3x + 2) - 5x] + [x^2 - 5(4 - x^2) + 3]$

*85. $[5x - 4(x - 3)][3x - 7(x + 2)]$

*86. $[x^2 - (5x - 2)][x^2 + (5x - 2)]$

*87. $[x^2 - (m + 2)][x^2 + (m + 2)]$

*88. $[3x^2 - (x - 2)][3x^2 + (x + 2)]$

*89. $2x(5x - 4) - 3x[5x^2 - 3x(x - 7)]$

*90. $-3x(x - 2) - 5[2x - 4(x + 6)]$

3.4 Multiplying Binomials

IN THIS SECTION:
- **The FOIL Method**
- **The Square of a Binomial**
- **Product of a Sum and Difference**

In Section 3.3 we learned to multiply polynomials. In this section, we will learn rules to make multiplication of binomials simpler.

The FOIL Method

Consider how we find the product of the two binomials $x + 3$ and $x + 5$.

$$(x + 3)(x + 5) = (x + 3)x + (x + 3)5 \qquad \text{Distributive property}$$
$$= x^2 + 3x + 5x + 15 \qquad \text{Distributive property}$$
$$= x^2 + 8x + 15 \qquad \text{Combine like terms.}$$

There are four terms in the product. The term x^2 is the product of the first term of each binomial. The term $5x$ is the product of the two outer terms, 5 and x. The term $3x$ is the product of the two inner terms, 3 and x. The term 15 is the product of the last two terms in each binomial, 3 and 5. It may be helpful to connect the terms multiplied by lines.

$$(x + 3)(x + 5)$$

F = *First* terms
O = *Outer* terms
I = *Inner* terms
L = *Last* terms

This is simply a way of organizing the multiplication of binomials. It is called the **FOIL method.** (The name should make it easier to remember.)

Now we apply FOIL to the product $(x - 3)(x + 4)$.

$$(x - 3)(x + 4) = x^2 + 4x - 3x - 12 = x^2 + x - 12$$

The outer and inner products are frequently like terms, and we can learn to add them mentally. We can then find the product of two binomials without writing anything but the answer.

EXAMPLE 1 Use FOIL to find the products of the binomials.

a) $(2x - 3)(3x + 4)$
b) $(m + w)(2m - w)$
c) $(2x^3 + 5)(2x^3 - 5)$
d) $(a + b)(a - 3)$

Solution

a) $(2x - 3)(3x + 4) = 6x^2 + 8x - 9x - 12 = 6x^2 - x - 12$
b) $(m + w)(2m - w) = 2m^2 + mw - w^2$
 We combine $2mw$ and $-mw$ mentally to get mw.

F   O   I   L

c) $(2x^3 + 5)(2x^3 - 5) = 4x^6 - 10x^3 + 10x^3 - 25 = 4x^6 - 25$

d) $(a + b)(a - 3) = a^2 - 3a + ab - 3b$ Note that no terms can be combined. ◄

The Square of a Binomial

To compute $(a + b)^2$, the square of a binomial, we can write it as $(a + b)(a + b)$ and use FOIL.

$$(a + b)(a + b) = a^2 + ab + ab + b^2 = a^2 + 2ab + b^2$$

The square of a binomial occurs so frequently that we do not have to use FOIL to find it. There are two rules that make squaring a binomial easier. To square a sum, we use the following rule.

Rule for the Square of a Sum

$$(a + b)^2 = a^2 + 2ab + b^2$$

Restated: To square $a + b$, we square the first term (a^2), add twice the product of the two terms ($2ab$), then add the square of the last term (b^2).

EXAMPLE 2 Square each sum, using the new rule.

a) $(x + 5)^2$ **b)** $(2w + 3)^2$ **c)** $(2y^4 + 3)^2$

Solution

a) $(x + 5)^2 = x^2 \;+\; 2(x)(5) \;+\; 5^2 = x^2 + 10x + 25$

               Square   Twice   Square
               of   the   of
               first   product   last

The equation $(x + 5)^2 = x^2 + 10x + 25$ is an identity. This equation is true for any value of x. Note that $(x + 5)^2 = x^2 + 25$ is *not* true for all real numbers.

b) $(2w + 3)^2 = (2w)^2 + 2(2w)(3) + 3^2$
                 $= 4w^2 + 12w + 9$

c) $(2y^4 + 3)^2 = (2y^4)^2 + 2(2y^4)(3) + 3^2$
                 $= 4y^8 + 12y^4 + 9$ ◄

When we use FOIL to compute $(a - b)^2$, we see that

$$(a - b)(a - b) = a^2 - ab - ab + b^2 = a^2 - 2ab + b^2.$$

We can use instead the following rule for squaring a difference.

Rule for the Square of a Difference

$$(a - b)^2 = a^2 - 2ab + b^2$$

Restated: Square the first term, subtract twice the product of the two terms, and add the square of the last term.

EXAMPLE 3 Square each difference, using the new rule.

a) $(x - 6)^2$ **b)** $(3w - 5y)^2$

c) $(-4 - st)^2$ **d)** $(3 - 5a^3)^2$

Solution

a) $(x - 6)^2 = x^2 - 2(x)(6) + 6^2$
$$= x^2 - 12x + 36$$

b) $(3w - 5y)^2 = (3w)^2 - 2(3w)(5y) + (5y)^2$
$$= 9w^2 - 30wy + 25y^2$$

c) $(-4 - st)^2 = (-4)^2 - 2(-4)(st) + (st)^2$
$$= 16 + 8st + s^2t^2$$

d) $(3 - 5a^3)^2 = 3^2 - 2(3)(5a^3) + (5a^3)^2$
$$= 9 - 30a^3 + 25a^6$$ ◀

Product of a Sum and a Difference

If we multiply the sum $a + b$ and the difference $a - b$ by using FOIL, we get

$$(a + b)(a - b) = a^2 - ab + ab - b^2 = a^2 - b^2.$$

The inner and outer products add up to zero, canceling each other out. So *the product of a sum and a difference is the difference of two squares,* as shown in the following rule.

Rule for the Product of a Sum and a Difference

$$(a + b)(a - b) = a^2 - b^2$$

EXAMPLE 4 Find the products.

a) $(x + 3)(x - 3)$ **b)** $(a^3 + 8)(a^3 - 8)$

c) $(3x^2 - y^3)(3x^2 + y^3)$

Solution

a) $(x + 3)(x - 3) = x^2 - 9$

b) $(a^3 + 8)(a^3 - 8) = a^6 - 64$

c) $(3x^2 - y^3)(3x^2 + y^3) = 9x^4 - y^6$ ◀

The square of a sum, the square of a difference, and the product of a sum and a difference are referred to as **special products.** Although the special products can be found using FOIL, they occur so frequently in algebra that it is essential to learn how to find them using the new rules.

Warm-ups

True or false?

1. $(x + 2)(x + 5) = x^2 + 7x + 10$ for any value of x.
2. $(2x - 3)(3x + 5) = 6x^2 + x - 15$ for any value of x.
3. $(2 + 3)^2 = 2^2 + 3^2$
4. $(x + 7)^2 = x^2 + 14x + 49$ for any value of x.
5. $(8 - 3)^2 = 64 - 9$
6. The product of a sum and a difference of the same two terms is equal to the difference of two squares.
7. $(60 - 1)(60 + 1) = 3600 - 1$
8. $(x - y)^2 = x^2 - 2xy + y^2$ for any values of x and y.
9. $(x - 3)^2 = x^2 - 3x + 9$ for any value of x.
10. The expression $3x \cdot 5x$ is a product of two binomials.

3.4 EXERCISES

Find the products mentally. Write down only the answer. See Example 1.

1. $(x - 2)(x + 4)$
2. $(x - 3)(x + 5)$
3. $(a + 3)(a + 2)$
4. $(b + 1)(b + 2)$
5. $(2x - 1)(x + 3)$
6. $(2y - 3)(y + 2)$
7. $(2a - 3)(a + 5)$
8. $(3x - 5)(x + 6)$
9. $(w + 6z)(w - z)$
10. $(w + 4y)(w - 2y)$
11. $(2m - 3)(4m + 3)$
12. $(2x - 7)(x + 1)$
13. $(2 + t)(1 + t)$
14. $(3 + z)(5 + z)$
15. $(3x^2 - 4)(x^2 + 1)$
16. $(2a^2 - 3)(3a^2 + 5)$
17. $(x - 3)(y + w)$
18. $(z - 1)(y + 2)$
19. $(s^3 + t)(s^3 - 6t)$
20. $(x^2 + 3y)(5x^2 + y)$
21. $(a + b)(c + d)$
22. $(m - p)(n + q)$

Find the square of each sum or difference. Where possible write down only the answer. See Examples 2 *and* 3.

23. $(m + 3)^2$
24. $(a + 2)^2$
25. $(a - 4)^2$
26. $(b - 3)^2$
27. $(2w + 1)^2$
28. $(2m - 1)^2$
29. $(3t - 5u)^2$
30. $(3w + 2x)^2$
31. $(x - 1)^2$
32. $(d - 5)^2$
33. $(a - 3y)^2$
34. $(3m + 5n)^2$

35. $(3x - 1)^2$

36. $\left(x + \dfrac{1}{2}\right)^2$

37. $\left(a - \dfrac{1}{3}\right)^2$

38. $\left(2x + \dfrac{1}{2}\right)^2$

39. $(2 + gt)^2$

40. $(1 + 3rs)^2$

41. $(8y^2 - 3)^2$

42. $(3x^2 + 9)^2$

43. $(4 - x)^2$

44. $(3 + x)^2$

45. $(xy - 2)^2$

46. $(x^2y + 4)^2$

Find the products mentally. See Example 4.

47. $(w - 9)(w + 9)$

48. $(m - 4)(m + 4)$

49. $(w + y)(w - y)$

50. $(a - x)(a + x)$

51. $(2x - 7)(2x + 7)$

52. $(5x + 3)(5x - 3)$

53. $(3x^2 - 2)(3x^2 + 2)$

54. $(4y^2 + 1)(4y^2 - 1)$

55. $(a^2 - 1)(a^2 + 1)$

56. $(x^3 - 1)(x^3 + 1)$

Perform the operations and simplify.

57. $(x - 6)(x + 9)$

58. $(2x^2 - 3)(3x^2 + 4)$

59. $(5 - x)(5 + x)$

60. $(4 - ab)(4 + ab)$

61. $(3x - 4a)(2x - 5a)$

62. $(x^5 + 2)(x^5 - 2)$

63. $(2t - 3)(t + w)$

64. $(5x - 9)(ax + b)$

65. $(3x^2 + 2y^3)^2$

66. $(5a^4 - 2b)^2$

67. $(2y + 2)(3y - 5)$

68. $(3b - 3)(2b + 3)$

69. $(50 + 2)(50 - 2)$

70. $(100 - 1)(100 + 1)$

71. $(3 + 7x)^2$

72. $(1 - pq)^2$

73. $(5y - 3)^2$

74. $(3a + 2)^2$

75. $\left(x + \dfrac{1}{2}\right)\left(x + \dfrac{1}{3}\right)$

76. $\left(x + \dfrac{1}{2}\right)\left(x - \dfrac{1}{4}\right)$

77. $\left(3y + \dfrac{1}{2}\right)^2$

78. $\left(2y - \dfrac{1}{3}\right)^2$

Solve each problem.

79. Suppose that the length of a rectangular room is $x + 3$ meters and the width is $x + 1$ meters. Find a trinomial that can be used to represent the area of the room.

80. Barbie and Ken have plans for a square house with area x^2 square feet. They want to lengthen one side of the square by 20 feet and shorten the other side by 6 feet. Find a trinomial that can be used to represent the area of the larger house.

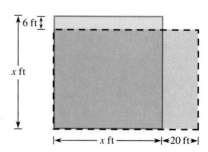

Figure for Exercise 80

Find the products. Assume that all variables are nonzero and that variables used as exponents represent integers.

*81. $(x^m + 2)(x^{2m} + 3)$

*82. $(a^n - b)(a^n + b)$

*83. $a^{n+1}(a^{2n} + a^n - 3)$

*84. $x^{3b}(x^{-3b} + 3x^{-b} + 5)$

*85. $(a^m + a^n)^2$

*87. $(5y^m + 8z^k)(3y^{2m} + 4z^{3-k})$

*86. $(x^w - x^t)^2$

*88. $(-4x^{a-1} + 3y^{b+5})(x^{2a-3} - 2y^{4-b})$

Use a calculator to help you to perform the following operations.

89. $(3.2x - 4.5)(5.1x + 3.9)$

91. $(3.6y + 4.4)^2$

90. $(5.3x - 9.2)^2$

92. $(3.3a - 7.9b)(3.3a + 7.9b)$

3.5 Factoring Polynomials

IN THIS SECTION:

- Factoring Out the Greatest Common Factor
- Factoring Out a Negative Factor
- Linear and Quadratic Polynomials
- Factoring the Difference of Two Squares
- Factoring Perfect-square Trinomials
- Factoring a Polynomial Completely

In Sections 3.3 and 3.4 we learned how to multiply polynomials. In this section we learn to reverse certain of those multiplications. We start with a polynomial and write it as a product of simpler polynomials. This technique will be used for solving equations and problems later in this chapter.

Factoring Out the Greatest Common Factor

A natural number larger than 1 that has no factors other than itself and 1 is called a **prime number.** The numbers

$$2, \ 3, \ 5, \ 7, \ 11, \ 13, \ 17, \ 19, \ 23$$

are the first nine prime numbers.

To factor a natural number **completely** means to write it as a product of prime numbers. If we want to factor 12, we might write $12 = 4 \cdot 3$. However, 12 is not factored completely as $4 \cdot 3$ because 4 is not a prime. The complete factorization of 12 is $12 = 2 \cdot 2 \cdot 3$ (or $2^2 \cdot 3$).

We use the distributive property to multiply a monomial and a binomial:

$$6x(2x - 1) = 12x^2 - 6x$$

If we start with $12x^2 - 6x$ we can use the distributive property to get

$$12x^2 - 6x = 6x(2x - 1).$$

We have **factored out** $6x$, which is a common factor of $12x^2$ and $-6x$. When we write a polynomial as a product of two or more polynomials, we call it **factoring.** We could have factored out just 3 to get

$$12x^2 - 6x = 3(4x^2 - 2x),$$

but this would not be factoring out the greatest common factor. The **greatest common factor** is a monomial that includes every number or variable that is a factor of all terms of the polynomial.

We can use the following strategy for finding the greatest common factor of a group of terms.

STRATEGY

Finding the Greatest Common Factor

1. Factor each term completely.
2. Write a product using each factor that is common to all of the terms.
3. On each of these factors, use an exponent equal to the smallest exponent that appears on that factor in any of the terms.

EXAMPLE 1   Find the greatest common factor for each group of terms.

a) $8x^2y$, $20xy^3$ b) $30a^2$, $45a^3b^2$, $75a^4b$

Solution

a) $8x^2y = 2^3x^2y$ and $20xy^3 = 2^2 \cdot 5xy^3$

The factors common to both terms are 2, x, and y. In the greatest common factor we use the smallest exponent that appears on each factor in either of the terms. The greatest common factor is $2^2xy = 4xy$.

b) $30a^2 = 2 \cdot 3 \cdot 5a^2$ $45a^3b^2 = 3^2 \cdot 5a^3b^2$ $75a^4b = 3 \cdot 5^2a^4b$

The greatest common factor is $3 \cdot 5 \cdot a^2 = 15a^2$.   ◀

To factor out the greatest common factor from the polynomial $60x^5 + 24x^3 + 36x^2$, we first factor each coefficient completely:

$$60x^5 = 2^2 \cdot 3 \cdot 5x^5, \qquad 24x^3 = 2^3 \cdot 3x^3, \qquad 36x^2 = 2^2 \cdot 3^2x^2.$$

The greatest common factor of $60x^5$, $24x^3$, and $36x^2$ is

$$2^2 \cdot 3 \cdot x^2 = 12x^2.$$

So $60x^5 + 24x^3 + 36x^2 = 12x^2(5x^3 + 2x + 3)$. Check by multiplying.

EXAMPLE 2   Factor the following polynomials by factoring out the greatest common factor.

a) $5x^4 - 10x^3 + 15x^2$
b) $8xy^2 + 20x^2y$
c) $(x + 3)w + (x + 3)a$

Solution

a) $5x^4 - 10x^3 + 15x^2 = 5x^2(x^2 - 2x + 3)$

b) $8xy^2 + 20x^2y = 4xy(2y + 5x)$

c) Although $x + 3$ is not a monomial, it can be treated like one for factoring.
$(x + 3)w + (x + 3)a = (x + 3)(w + a)$ ◀

Factoring Out a Negative Factor

Generally, we factor out a common factor with no sign in front of it, but we could factor out a common factor with a negative sign preceding it. It will be necessary to do this in Example 6 of this section.

EXAMPLE 3 Factor out the greatest common factor, first using a factor with no sign and then a factor with a negative sign.

a) $5x - 5y$ b) $-x^2 - 3$ c) $-x^3 + 3x^2 - 5x$

Solution

a) $5x - 5y = 5(x - y)$
$= -5(-x + y)$ Note: $-5(-x + y) = -5(y - x)$

b) $-x^2 - 3 = 1(-x^2 - 3)$ The greatest common factor is 1.
$= -1(x^2 + 3)$

c) $-x^3 + 3x^2 - 5x = x(-x^2 + 3x - 5)$
$= -x(x^2 - 3x + 5)$ ◀

Linear and Quadratic Polynomials

A first-degree polynomial in one variable, such as $3x - 5$, is called a linear polynomial. (The equation $3x - 5 = 0$ is a linear equation.) In general, we have the following definition.

DEFINITION
Linear Polynomial

If a and b are real numbers with $a \neq 0$, then $ax + b$ is called a **linear polynomial.**

A second-degree polynomial, such as $x^2 + 5x - 6$ or $3x^2 - 7$, is called a quadratic polynomial. In general, we have the following definition.

DEFINITION
Quadratic Polynomial

If a, b, and c are real numbers with $a \neq 0$, then $ax^2 + bx + c$ is called a **quadratic polynomial.**

Factoring the Difference of Two Squares

To factor a polynomial means to write it as a product of simpler polynomials. Consider the quadratic polynomial $x^2 - 25$. We recognize that $x^2 - 25$ is a difference of two squares, $x^2 - 5^2$. We say that x is the **square root** of x^2 and 5 is the **square root** of 25. (We will discuss square roots extensively in Chapter 5.) Now recall the rule for the product of a sum and a difference.

Factoring the Difference of Two Squares

$$a^2 - b^2 = (a + b)(a - b)$$

The difference of two squares factors as the product of the sum and difference of the square roots. We factor $x^2 - 25$ by letting $a = x$ and $b = 5$ in the formula for the difference of two squares:

$$x^2 - 25 = (x + 5)(x - 5)$$

The above equation expresses a quadratic polynomial as a product of two linear factors.

EXAMPLE 4   Factor each polynomial.

a) $y^2 - 36$ **b)** $9x^2 - 1$ **c)** $4x^2 - y^2$

Solution   Each of these binomials is a difference of two squares. Each binomial factors into a product of a sum and a difference of the square roots of the two terms.

a) $y^2 - 36 = (y + 6)(y - 6)$ **b)** $9x^2 - 1 = (3x + 1)(3x - 1)$
c) $4x^2 - y^2 = (2x + y)(2x - y)$ ◀

Factoring Perfect-square Trinomials

The trinomial that results from squaring a binomial is called a **perfect-square trinomial.** We can rewrite the rules from Section 3.4 for the square of a sum or difference as rules for factoring.

Factoring Perfect-square Trinomials

$$a^2 + 2ab + b^2 = (a + b)^2$$
$$a^2 - 2ab + b^2 = (a - b)^2$$

Consider the polynomial $x^2 + 6x + 9$. If we recognize that

$$x^2 + 6x + 9 = x^2 + 2 \cdot x \cdot 3 + 3^2,$$

then we can see that it is a perfect-square trinomial. It fits the rule if $a = x$ and $b = 3$:

$$x^2 + 6x + 9 = (x + 3)^2$$

Perfect-square trinomials are easy to identify using the following strategy.

STRATEGY
Identifying Perfect-square
Trinomials

A trinomial is a perfect-square trinomial if
1. the first and last terms are perfect squares, and
2. the middle term is 2 or -2 times the product of the square roots of the first and last terms.

Consider the trinomial

$$x^2 - 8x + 16.$$

The first term, x^2, and the last term, 16, are perfect squares. The middle term, $-8x$, is -2 times the product of x and 4 (the square roots of the first and last terms). This trinomial is a perfect-square trinomial. Because of the negative middle term, it is the square of a difference:

$$x^2 - 8x + 16 = (x - 4)^2$$

The middle term in a perfect-square trinomial may have a positive or a negative coefficient, but the first and last terms must be positive.

EXAMPLE 5   Factor the quadratic polynomials.

a) $x^2 - 4x + 4$ **b)** $a^2 + 14a + 49$ **c)** $4x^2 + 12x + 9$

Solution   Each of these trinomials is a perfect-square trinomial. We can write each trinomial as a square of a binomial.

a) $x^2 - 4x + 4 = x^2 - 2(2x) + 2^2 = (x - 2)^2$

                Twice   Square root of first
                        Square root of last

b) $a^2 + 14a + 49 = (a + 7)^2$
c) $4x^2 + 12x + 9 = (2x)^2 + 2(3)(2x) + 3^2 = (2x + 3)^2$

                Twice   Square root of first
                        Square root of last   ◀

Factoring a Polynomial Completely

A polynomial is factored when it is written as a product of simpler polynomials. Polynomials that cannot be factored are called **prime polynomials**. Examples are $x + 5$, $a - 6$, and $3x + 1$. Any monomial is also considered a prime polynomial. A polynomial is **factored completely** when it is written as a product of prime polynomials.

EXAMPLE 6   Factor each polynomial completely.

a) $5x^2 - 20$ b) $3a^3 - 30a^2 + 75a$ c) $-2b^3 - 24b^2 - 72b$

Solution

a) $5x^2 - 20 = 5(x^2 - 4)$ Factor out the common factor.

   $\qquad\qquad = 5(x - 2)(x + 2)$ Factor the difference of two squares.

b) $3a^3 - 30a^2 + 75a = 3a(a^2 - 10a + 25)$ Factor out the common factor.

   $\qquad\qquad\qquad = 3a(a - 5)^2$ Factor the perfect-square trinomial.

c) $-2b^3 - 24b^2 - 72b = -2b(b^2 + 12b + 36)$ Factor out $-2b$ to make
   the next step easier.

   $\qquad\qquad\qquad = -2b(b + 6)^2$ Factor the perfect-square trinomial.   ◄

Warm-ups

True or false?

1. For the polynomial $3x^2y - 6xy^2$ we can factor out either $3xy$ or $-3xy$.
2. The greatest common factor for the polynomial $8a^3 - 15b^2$ is 1.
3. $x - 3 = 3 - x$ of any value of x.
4. $2x - 4 = -2(2 - x)$ for any value of x.
5. The difference of two squares can be factored as a product of a sum and a difference.
6. $x^2 - 16 = (x - 4)(x + 4)$ for any value of x.
7. The polynomial $x^2 + 6x + 36$ is a perfect-square trinomial.
8. The polynomial $y^2 + 16$ is a perfect-square trinomial.
9. $9x^2 + 21x + 49 = (3x + 7)^2$ for any value of x.
10. $4a^2 - 12a + 9 = (2a - 3)^2$ for any value of a.

3.5 EXERCISES

Find the greatest common factor for each group of terms. See Example 1.

1. $48, 36x$
2. $42a, 28a^2$
3. $9wx, 21wy, 15xy$
4. $70x^2, 84x, 42x^3$
5. $24x^2y, 42xy^2$
6. $15a^2, 14b^2$

Factor out the greatest common factor in each expression. See Example 2.

7. $x^3 - 5x$
8. $10x^2 - 20y^3$
9. $48wx + 36wy$
10. $42wz + 28wa$
11. $(x - 6)a + (x - 6)b$
12. $(y - 4)3 + (y - 4)b$
13. $(y - 1)^2y + (y - 1)^2z$
14. $(w - 2)^2 \cdot w + (w - 2)^2 \cdot 3$
15. $2x^3 - 4x^2 + 6x$
16. $6x^3 - 12x^2 + 18x$

Factor each of these binomials, first using no sign on the greatest common factor and then using a negative sign on the greatest common factor. See Example 3.

17. $2x - 2y$

18. $-3x + 6$

19. $-3x + 6x^2$

20. $5x + 10x^2$

21. $-w^3 + 3w^2$

22. $-2w^4 + 6w^3$

23. $a^2 + a$

24. $-2a^2 - 4a$

25. $2x - 7$

26. $w - 5$

Factor each polynomial. See Example 4.

27. $x^2 - 100$

28. $81 - y^2$

29. $4y^2 - 49$

30. $16b^2 - 1$

31. $9x^2 - 25a^2$

32. $121a^2 - b^2$

33. $144w^2z^2 - 1$

34. $x^2y^2 - 9c^2$

Factor each polynomial. See Example 5.

35. $x^2 - 20x + 100$

36. $y^2 + 10y + 25$

37. $w^2 - 2wt + t^2$

38. $4m^2 - 4m + 1$

39. $9t^2 + 30t + 25$

40. $4r^2 + 20rt + 25t^2$

Factor each polynomial completely. See Example 6.

41. $2x^2 - 8$

42. $3x^3 - 27x$

43. $x^3 + 10x^2 + 25x$

44. $5a^4m - 45a^2m$

45. $4x^2 + 4x + 1$

46. $ax^2 - 8ax + 16a$

47. $(x + 3)x + (x + 3)7$

48. $(x - 2)x - (x - 2)5$

49. $6y^2 + 3y$

50. $4y^2 - y$

51. $4x^2 - 20x + 25$

52. $a^3x^3 - 6a^2x^2 + 9ax$

53. $(2x - 3)x - (2x - 3)2$

54. $(2x + 1)x + (2x + 1)3$

55. $9a^3 - aw^2$

56. $2bn^2 - 4b^2n + 2b^3$

57. $-5a^2 + 30a - 45$

58. $-2x^2 + 50$

59. $-3y^3 - 18y^2 - 27y$

60. $-2m^2n - 8mn - 8n$

61. $-7a^2b^2 + 7$

62. $-17a^2 - 17a$

63. $(x + 2)x^2 - (x + 2)9$

64. $(a - 1)a^2 - (a - 1)$

3.6 Factoring $ax^2 + bx + c$

IN THIS SECTION:

- **Factoring Trinomials with Leading Coefficient 1**
- **Factoring Trinomials with Leading Coefficient Other Than 1**
- **Another Method**

In Section 3.5 we learned to factor certain special polynomials. In this section we will learn to factor general quadratic polynomials. We first factor $ax^2 + bx + c$ with $a = 1$, and then we consider the case $a \neq 1$.

Factoring Trinomials with Leading Coefficient 1

Let's look closely at an example of finding the product of two binomials using the distributive property.

$$(x + 3)(x + 4) = (x + 3)x + (x + 3)4 \qquad \text{Distributive property}$$
$$= x^2 + 3x + 4x + 12 \qquad \text{Distributive property}$$
$$= x^2 + 7x + 12$$

To factor $x^2 + 7x + 12$, we need to reverse these steps. First observe that the coefficient 7 is the sum of two numbers that have a product of 12. The only numbers that have a product of 12 and a sum of 7 are 3 and 4. So write $7x$ as $3x + 4x$:

$$x^2 + \underline{7x} + 12 = x^2 + \underline{3x + 4x} + 12$$

Now factor the common factor x out of the first two terms and the common factor 4 out of the last two terms. This is called **factoring by grouping.**

Factor out x. Factor out 4.

$$x^2 + 7x + 12 = \overbrace{x^2 + 3x} + \overbrace{4x + 12} \qquad \text{Rewrite } 7x \text{ as } 3x + 4x.$$
$$= (x + 3)x + (x + 3)4 \qquad \text{Factor out common factors.}$$
$$= (x + 3)(x + 4) \qquad \text{Factor out the common factor } x + 3.$$

To factor $x^2 + 9x + 18$ we must find two numbers that have a product of 18 and a sum of 9. The pairs of numbers with a product of 18 are 1 and 18, 2 and 9, and 3 and 6. Only 3 and 6 have a sum of 9. So write $9x$ as $3x + 6x$ and factor by grouping:

$$x^2 + 9x + 18 = x^2 + 3x + 6x + 18$$
$$= (x + 3)x + (x + 3)6$$
$$= (x + 3)(x + 6) \qquad \text{Check by using FOIL.}$$

We can actually skip most of the steps just shown. For example, to factor $x^2 - x - 6$ we must find a pair of numbers with a product of -6 and a sum of -1. Because the numbers are -3 and 2, we can write

$$x^2 - x - 6 = (x - 3)(x + 2). \qquad \text{Check by using FOIL.}$$

EXAMPLE 1 Factor each quadratic polynomial.

a) $x^2 + 4x + 3$ **b)** $x^2 + 3x - 10$ **c)** $a^2 - 5a + 6$

Solution

a) $x^2 + 4x + 3 = (x + 1)(x + 3)$ To get a product of 3 and a sum of 4, use 1 and 3.

b) $x^2 + 3x - 10 = (x + 5)(x - 2)$ To get a product of -10 and a sum of 3, use 5 and -2.

c) $a^2 - 5a + 6 = (a - 3)(a - 2)$ To get a product of 6 and a sum of -5, use -3 and -2.   ◄

Factoring Trinomials with Leading Coefficient Other Than 1

If the leading coefficient of a quadratic trinomial is not 1, we can again use grouping to factor the trinomial. However, the procedure is slightly different.

Consider the trinomial $2x^2 + 11x + 12$. First find the product of the leading coefficient and the constant term. In this case it is $2 \cdot 12 = 24$. Now find two numbers with a product of 24 and a sum of 11. The pairs of numbers with a product of 24 are 1 and 24, 2 and 12, 3 and 8, and 4 and 6. Only 3 and 8 have a product of 24 and a sum of 11. Now replace $11x$ by $3x + 8x$ and factor by grouping.

$$2x^2 + 11x + 12 = 2x^2 + 3x + 8x + 12$$
$$= (2x + 3)x + (2x + 3)4$$
$$= (2x + 3)(x + 4)$$

The strategy for factoring a quadratic trinomial is summarized below. The steps listed here work whether or not the leading coefficient is 1.

STRATEGY

Factoring $ax^2 + bx + c$ by the ac Method

To factor the trinomial $ax^2 + bx + c$,

1. find two numbers that have a product equal to ac and a sum equal to b,
2. replace bx by two terms using the two new numbers as coefficients,
3. factor the resulting four-term polynomial by grouping.

EXAMPLE 2 Factor each trinomial.

a) $2x^2 + 9x + 4$

b) $2x^2 + 5x - 12$

Solution

a) Because $2 \cdot 4 = 8$, we need two numbers with a product of 8 and a sum of 9. The numbers are 1 and 8. Replace $9x$ by $x + 8x$ and factor by grouping.

$$2x^2 + 9x + 4 = 2x^2 + x + 8x + 4$$
$$= (2x + 1)x + (2x + 1)4$$
$$= (2x + 1)(x + 4) \qquad \text{Check by FOIL.}$$

b) Because $2(-12) = -24$, we need two numbers with a product of -24 and a sum of 5. The pairs of numbers with a product of 24 are 1 and 24, 2 and 12, 3 and 8, and 4 and 6. To get a product of -24, one of the numbers must be negative and the other positive. To get a sum of positive 5, we need -3 and 8.

$$2x^2 + 5x - 12 = 2x^2 - 3x + 8x - 12$$
$$= (2x - 3)x + (2x - 3)4$$
$$= (2x - 3)(x + 4) \qquad \text{Check by FOIL.} \qquad \blacktriangleleft$$

Another Method

After we have gained some experience factoring by grouping we can often find the factors without going through the steps of grouping. Consider

$$6w^2 + 23w - 18.$$

The factors of $6w^2$ might be $6w$ and w, or they might be $2w$ and $3w$. The last term, -18, also has several factors, one of which must be positive and the other negative. We can list all of the possibilities that give the correct first and last terms, without putting in the signs.

$$(2w \quad 1)(3w \quad 18) \qquad (6w \quad 1)(w \quad 18)$$
$$(2w \quad 18)(3w \quad 1) \qquad (6w \quad 18)(w \quad 1)$$
$$(2w \quad 2)(3w \quad 9) \qquad (6w \quad 2)(w \quad 9)$$
$$(2w \quad 9)(3w \quad 2) \qquad (6w \quad 9)(w \quad 2)$$
$$(2w \quad 3)(3w \quad 6) \qquad (6w \quad 3)(w \quad 6)$$
$$(2w \quad 6)(3w \quad 3) \qquad (6w \quad 6)(w \quad 3)$$

We then examine each possibility using FOIL, considering one + and one − sign, to see which one gives $23w$ (the middle term) as the sum of the outer and inner products. For example, for the possibility $(6w \quad 1)(w \quad 18)$, the outer and inner products are $108w$ and $1w$. These will not total $23w$ for any choice of signs. In fact, we could realize that $6 \cdot 18$ is too large even before multiplying. We can see at a glance that some of the possibilities will not work. For example, the second binomial of $(2w \quad 1)(3w \quad 18)$ has a common factor of 3. If this was the correct factorization, there would be a common factor of 3 in all three terms of the original trinomial, but there is not. *If the original trinomial has no common factor, there can be no common factor in either binomial of the correct factorization.* The correct factorization is

$$6w^2 + 23w - 18 = (2w + 9)(3w - 2).$$

Even though there are a lot of possibilities in some factoring problems, we will find that we can often discover the correct factors without writing down every possibility. We can use a bit of guesswork in factoring trinomials. *Try* whichever possibility you think might work. *Check it by multiplying.* If it is not right, then *try again.* This method is called **trial-and-error.**

EXAMPLE 3   Factor each quadratic trinomial using trial and error.

a) $2x^2 + 5x - 3$ **b)** $3x^2 - 11x + 6$

Solution

a) Because $2x^2$ factors only as $2x \cdot x$ and 3 factors only as $1 \cdot 3$, there are only two possible ways to factor this trinomial to get the correct first and last terms:

$$(2x \quad 1)(x \quad 3) \quad \text{and} \quad (2x \quad 3)(x \quad 1)$$

Because the last term of the trinomial is negative, one of the missing signs must be $+$, and the other must be $-$. The correct factorization is

$$2x^2 + 5x - 3 = (2x - 1)(x + 3). \qquad \text{Check by using FOIL.}$$

b) There are four possible factorizations:

$$(3x \quad 1)(x \quad 6) \qquad\qquad (3x \quad 2)(x \quad 3)$$
$$(3x \quad 6)(x \quad 1) \qquad\qquad (3x \quad 3)(x \quad 2)$$

Because the last term is positive and the middle term is negative, both signs must be negative. The correct factorization is

$$3x^2 - 11x + 6 = (3x - 2)(x - 3). \qquad \text{Check by using FOIL.} \qquad \blacktriangleleft$$

If the terms of a polynomial have a common factor other than 1, then factor out the greatest common factor *before* using the *ac* method or trial and error.

EXAMPLE 4   Factor each polynomial completely.

a) $2w^3 + 12w^2 - 80w$ \qquad\qquad\qquad b) $12x^2 - 24x + 9$

Solution

a) $2w^3 + 12w^2 - 80w = 2w(w^2 + 6w - 40)$
$$= 2w(w + 10)(w - 4)$$

b) $12x^2 - 24x + 9 = 3(4x^2 - 8x + 3)$
$$= 3(2x - 3)(2x - 1) \qquad \blacktriangleleft$$

Warm-ups

True or false? Answer true if the correct factorization is given and false if the factorization is incorrect.

1. $x^2 + 9x + 18 = (x + 3)(x + 6)$
2. $y^2 + 2y - 35 = (y + 5)(y - 7)$
3. $x^2 + 4 = (x + 2)(x + 2)$
4. $x^2 - 5x - 6 = (x - 3)(x - 2)$
5. $x^2 - 4x - 12 = (x - 6)(x + 2)$
6. $x^2 + 15x + 36 = (x + 4)(x + 9)$
7. $3x^2 + 4x - 15 = (3x + 5)(x - 3)$
8. $4x^2 + 4x - 3 = (4x - 1)(x + 3)$
9. $4x^2 - 4x - 3 = (2x + 1)(2x - 3)$
10. $4x^2 + 8x + 3 = (2x + 1)(2x + 3)$

3.6 EXERCISES

Factor each polynomial. See Example 1.

1. $x^2 + 4x + 3$

2. $y^2 + 5y + 6$

3. $a^2 + 15a + 50$

4. $t^2 + 11t + 24$

5. $x^2 - 6x + 8$

6. $y^2 - 13y + 30$

7. $y^2 - 5y - 14$

8. $a^2 - 12a + 27$

9. $x^2 - 3x - 18$

10. $x^2 - x - 30$

11. $a^2 + 7a - 30$

12. $w^2 - 29w - 30$

Factor each polynomial. See Examples 2 and 3.

13. $2x^2 - 5x - 3$

14. $2a^2 + 3a - 2$

15. $6w^2 + 5w + 1$

16. $6x^2 - 5x + 1$

17. $6m^2 - m - 12$

18. $6y^2 + 17y + 12$

19. $4x^2 + 11x + 6$

20. $4x^2 + 16x + 15$

21. $12y^2 + y - 1$

22. $12x^2 + 5x - 2$

23. $6a^2 + a - 5$

24. $30b^2 - b - 3$

Factor each polynomial completely. See Example 4.

25. $2x^2 + 20x + 50$

26. $3a^2 + 6a + 3$

27. $a^3 - 36a$

28. $x^3 - 5x^2 - 6x$

29. $5a^2 + 25a - 30$

30. $2a^2 - 2a - 84$

31. $2x^2 - 128y^2$

32. $a^3 - 6a^2 + 9a$

33. $-3x^2 + 3x + 36$

34. $xy^2 - 3xy - 70x$

35. $m^5 + 20m^4 + 100m^3$

36. $4a^2 - 16a + 16$

37. $6x^2 + 23x + 20$

38. $2y^2 - 13y + 6$

39. $y^2 - 12y + 36$

40. $m^2 - 2m + 1$

41. $9m^2 - 25n^2$

42. $m^2n^2 - 2mn^3 + n^4$

43. $5a^2 + 20a - 60$

44. $-3y^2 + 9y + 30$

45. $-2w^2 + 18w + 20$

46. $x^2z + 2xyz + y^2z$

47. $w^2x^2 - 100x^2$

48. $9x^2 + 30x + 25$

49. $9x^2 - 1$

50. $6w^2 - 19w - 36$

51. $8x^2 - 2x - 15$

52. $4w^2 + 12w + 9$

53. $4x^2 - 20x + 25$

54. $9m^2 - 121$

55. $9a^2 + 60a + 100$

56. $4m^2 + 20m + 25$

57. $90y^2 - 1000$

58. $16m^2 + 24m + 9$

59. $25x^2 - 10x + 1$

60. $20x^2 - 7x - 3$

61. $10x^2 + 7x - 12$

62. $10w^2 + 20w - 80$

63. $4a^2 + 24a + 32$

64. $20x^2 - 23x + 6$

Factor each of the following polynomials. Assume that the variables used as exponents represent positive integers.

*65. $x^{2a} + 2x^a - 15$

*66. $y^{2b} + y^b - 20$

*67. $x^{2a} - y^{2b}$

*68. $w^{4m} - a^2$

*69. $x^{a+2} - x^a$

*70. $y^{2a+1} - y$

*71. $x^{2a} + 6x^a + 9$

*72. $x^{2a} - 2x^a y^b + y^{2b}$

3.7 Division of Polynomials

IN THIS SECTION:

- Dividing a Polynomial by a Monomial
- Dividing a Polynomial by a Binomial
- Using Division in Factoring

We began our study of polynomials in Section 3.3 by learning how to add, subtract, and multiply polynomials. In this section we will learn how to divide polynomials, and we will use division to discover two new rules for factoring polynomials.

Dividing a Polynomial by a Monomial

We learned how to divide monomials in Section 3.1. For example,

$$6x^3 \div 3x = \frac{6x^3}{3x} = 2x^2.$$

We check by multiplying. Because $2x^2 \cdot 3x = 6x^3$, this answer is correct.

We know that

$$3x(2x^2 + 5x - 4) = 6x^3 + 15x^2 - 12x.$$

If we divide $6x^3 + 15x^2 - 12x$ by the monomial $3x$, we get $2x^2 + 5x - 4$. We can perform this division by dividing $3x$ into each term of $6x^3 + 15x^2 - 12x$.

$$\frac{6x^3 + 15x^2 - 12x}{3x} = \frac{6x^3}{3x} + \frac{15x^2}{3x} - \frac{12x}{3x} = 2x^2 + 5x - 4$$

Note that because $3x$ is a common factor of $6x^3 + 15x^2 - 12x$, $3x$ divides evenly into each term with 0 remainder. We call $3x$ the **divisor,** $6x^3 + 15x^2 - 12x$ the **dividend,** and $2x^2 + 5x - 4$ the **quotient.**

The remainder is not always 0. For example, if we divide 17 by 5 we get

$$
\begin{array}{r}
3 \leftarrow \text{Quotient} \\
\text{Divisor} \rightarrow 5\overline{)17} \leftarrow \text{Dividend} \\
\underline{15} \\
2 \leftarrow \text{Remainder}
\end{array}
$$

Note that $5 \cdot 3 + 2 = 17$. It is always true that

$$(\text{divisor})(\text{quotient}) + (\text{remainder}) = \text{dividend}.$$

EXAMPLE 1   Find the quotient for each indicated division.

a) $-12x^5 \div 2x^3$ **b)** $(-20x^6 + 8x^4 - 4x^2) \div 4x^2$

Solution

a) $-12x^5 \div 2x^3 = -6x^2$

     The quotient is $-6x^2$. Check: $-6x^2 \cdot 2x^3 = -12x^5$.

b) $\dfrac{-20x^6 + 8x^4 - 4x^2}{4x^2} = \dfrac{-20x^6}{4x^2} + \dfrac{8x^4}{4x^2} - \dfrac{4x^2}{4x^2} = -5x^4 + 2x^2 - 1$

     The quotient is $-5x^4 + 2x^2 - 1$. Check:

$$4x^2(-5x^4 + 2x^2 - 1) = -20x^6 + 8x^4 - 4x^2. \qquad \blacktriangleleft$$

Dividing a Polynomial by a Binomial

We know that

$$(x - 2)(x + 5) = x^2 + 3x - 10.$$

So if we divide $x^2 + 3x - 10$ by the factor $x - 2$, we should get the other factor, $x + 5$. This division is not done like division by a monomial, it is done like long division of whole numbers. We get the first term of the quotient by dividing the first term of $x - 2$ into the first term of $x^2 + 3x - 10$. Divide x^2 by x to get x.

Multiply.
$$
\begin{array}{r}
x \\
x - 2\overline{)x^2 + 3x - 10} \\
x^2 - 2x
\end{array}
\qquad x^2 \div x = x
$$

Subtract. $5x$ $3x - (-2x) = 5x$

$x \cdot (x - 2) = x^2 - 2x$

Now bring down -10. We get the second term of the quotient (below) by dividing the first term of $x - 2$ into the first term of $5x - 10$. Divide $5x$ by x to get 5.

Multiply.
$$
\begin{array}{r}
x + 5 \\
x - 2\overline{)x^2 + 3x - 10} \\
x^2 - 2x \\
5x - 10 \\
5x - 10
\end{array}
\qquad 5x \div x = 5
$$

$5 \cdot (x - 2) = 5x - 10$

Subtract. 0 $-10 - (-10) = 0$

It is usually best to write the terms of the divisor and the dividend in descending order of the exponents. If any terms are missing, as in the next example, we insert terms with a coefficient of 0 for them.

EXAMPLE 2 Divide $2x^3 - 4 - 16x$ by $2x - 4$ and identify the quotient and the remainder.

Solution We first rearrange $2x^3 - 4 - 16x$ as $2x^3 - 16x - 4$ and insert the term $0x^2$. We start by dividing $2x^3$ by $2x$ to get x^2.

$$
\begin{array}{r}
x^2 + 2x - 4 \\
2x - 4 \overline{)\,2x^3 + 0x^2 - 16x - 4} \\
\underline{2x^3 - 4x^2} \\
4x^2 - 16x \\
\underline{4x^2 - 8x} \\
-8x - 4 \\
\underline{-8x + 16} \\
-20
\end{array}
$$

$0x^2 - (-4x^2) = 4x^2$

$-16x - (-8x) = -8x$

$-4 - 16 = -20$

The quotient is $x^2 + 2x - 4$, and the remainder is -20. The degree of the remainder is 0, and the degree of the divisor is 1. When dividing polynomials, we do not stop until the degree of the remainder is smaller than the degree of the divisor. To check the answer we must verify that

$$(2x - 4)(x^2 + 2x - 4) - 20 = 2x^3 - 16x - 4. \qquad \blacktriangleleft$$

EXAMPLE 3 Divide $3x^4 - 2 - 5x$ by $x^2 - 3x$ and identify the quotient and remainder.

Solution Rearrange $3x^4 - 2 - 5x$ as $3x^4 - 5x - 2$ and insert the terms $0x^3$ and $0x^2$.

$$
\begin{array}{r}
3x^2 + 9x + 27 \\
x^2 - 3x \overline{)\,3x^4 + 0x^3 + 0x^2 - 5x - 2} \\
\underline{3x^4 - 9x^3} \\
9x^3 + 0x^2 \\
\underline{9x^3 - 27x^2} \\
27x^2 - 5x \\
\underline{27x^2 - 81x} \\
76x - 2
\end{array}
$$

$0x^3 - (-9x^3) = 9x^3$

The quotient is $3x^2 + 9x + 27$, and the remainder is $76x - 2$. Note that the degree of the remainder is 1, and the degree of the divisor is 2. We stop the division when the degree of the remainder is smaller than the degree of the divisor. To check, verify that

$$(x^2 - 3x)(3x^2 + 9x + 27) + 76x - 2 = 3x^4 - 5x - 2. \qquad \blacktriangleleft$$

To divide 17 by 5, we can write $17 \div 5$ or $\dfrac{17}{5}$. We can express the result by saying that the quotient is 3 and the remainder is 2, or we can write

$$\frac{17}{5} = 3 + \frac{2}{5}.$$

In general, we have

$$\frac{\text{dividend}}{\text{divisor}} = \text{quotient} + \frac{\text{remainder}}{\text{divisor}}.$$

EXAMPLE 4   Express $\dfrac{-4x}{x-3}$ in the form quotient $+ \dfrac{\text{remainder}}{\text{divisor}}$.

Solution   Use long division to get the required form.

$$
\begin{array}{r}
-4 \\
x-3\overline{)-4x} \\
\underline{-4x + 12} \\
-12
\end{array}
$$

The quotient is -4, and the remainder is -12.

$$\frac{-4x}{x-3} = -4 + \frac{-12}{x-3}$$

To check, verify that $(-4)(x-3) - 12 = -4x$.   ◀

Using Division in Factoring

We always have

$$\text{dividend} = (\text{divisor})(\text{quotient}) + \text{remainder}.$$

If the remainder is 0, then

$$\text{dividend} = (\text{divisor})(\text{quotient}).$$

Therefore, *the dividend factors as the divisor times the quotient if and only if the remainder is 0.* We can use division to help us discover factors of polynomials. To use this idea, however, we must know a factor or a possible factor to use as the divisor.

EXAMPLE 5   Is $2x + 1$ a factor of $6x^3 - 5x^2 + 4x + 4$?

Solution

$$
\begin{array}{r}
3x^2 - 4x \;+ 4 \\
2x + 1\overline{)6x^3 - 5x^2 + 4x + 4} \\
\underline{6x^3 + 3x^2} \\
-8x^2 + 4x \\
\underline{-8x^2 - 4x} \\
8x + 4 \\
\underline{8x + 4} \\
0
\end{array}
$$

Because the remainder is 0, $2x + 1$ is a factor, and

$$6x^3 - 5x^2 + 4x + 4 = (2x + 1)(3x^2 - 4x + 4).$$ ◄

We can use division to discover two new factoring rules. Consider the division of a difference of two cubes $a^3 - b^3$ by $a - b$ and the division of a sum of two cubes $a^3 + b^3$ by $a + b$.

$$
\begin{array}{r}
a^2 + \; ab + \; b^2 \\
a - b\overline{)a^3 + \;\;\; 0 + \;\;\; 0 - b^3} \\
\underline{a^3 - a^2b} \\
a^2b + \;\;\; 0 \\
\underline{a^2b - ab^2} \\
ab^2 - b^3 \\
\underline{ab^2 - b^3} \\
0
\end{array}
\qquad
\begin{array}{r}
a^2 - \; ab + \; b^2 \\
a + b\overline{)a^3 + \;\;\; 0 + \;\;\; 0 + b^3} \\
\underline{a^3 + a^2b} \\
-a^2b + \;\;\; 0 \\
\underline{-a^2b - ab^2} \\
ab^2 + b^3 \\
\underline{ab^2 + b^3} \\
0
\end{array}
$$

These divisions show us that the difference of two cubes and the sum of two cubes can be factored as follows.

Factoring the Difference and Sum of Two Cubes

$$a^3 - b^3 = (a - b)(a^2 + ab + b^2)$$
$$a^3 + b^3 = (a + b)(a^2 - ab + b^2)$$

EXAMPLE 6   Factor each polynomial.

a) $x^3 - 8$ **b)** $y^3 + 1$ **c)** $8z^3 - 27$

Solution   Each of these binomials is either a difference or a sum of two cubes.

a) $x^3 - 8 = x^3 - 2^3$ Express $x^3 - 8$ as a difference of two cubes.

$\qquad\quad = (x - 2)(x^2 + 2x + 4)$ Replace a by x and b by 2 in the formula for the difference of two cubes.

b) $y^3 + 1 = y^3 + 1^3$ Sum of two cubes
$$= (y + 1)(y^2 - y + 1)$$ Let $a = y$ and $b = 1$ in the formula for the sum of two cubes.

c) $8z^3 - 27 = (2z)^3 - 3^3$ Difference of two cubes
$$= (2z - 3)(4z^2 + 6z + 9)$$ Let $a = 2z$ and $b = 3$ in the formula for the difference of two cubes.

◄

Warm-ups

True or false?

1. The quotient times the dividend plus the remainder equals the divisor.
2. $(x + 2)(x + 3) + 1 = x^2 + 5x + 7$ is true for any value of x.
3. The quotient of $(x^2 + 5x + 7) \div (x + 3)$ is $x + 2$.
4. If $x^2 + 5x + 7$ is divided by $x + 2$, the remainder is 1.
5. If the remainder is 0, then the divisor is a factor of the dividend.
6. If $x^3 - 8$ is divided by $x - 2$, the remainder is 0.
7. The polynomial $x + 1$ is a factor of $x^3 + 1$.
8. The polynomial $x + 2$ is a factor of $x^3 - 8$.
9. $x^3 - 27 = (x - 3)(x^2 + 6x + 9)$ for any value of x.
10. $x^3 - 8 = (x - 2)^3$ for any value of x.

3.7 EXERCISES

Find the quotient for each division. See Example 1.

1. $36x^7 \div 3x^3$
2. $-30x^3 \div (-5x)$
3. $16x^2 \div (-8x^2)$
4. $-22a^3 \div 11a^2$
5. $(6b - 9) \div 3$
6. $(8x^2 - 6x) \div 2x$
7. $(3x^2 + 6x) \div (3x)$
8. $(5x^3 - 10x^2 + 20x) \div (5x)$
9. $(10x^4 - 8x^3 + 6x^2) \div (-2x^2)$
10. $(-9x^3 + 6x^2 - 12x) \div (-3x)$
*11. $(7x^3 - 4x^2) \div (2x)$
*12. $(6x^3 - 5x^2) \div (4x^2)$

Find the quotient and remainder for each division. Check by using the formula: dividend = (divisor)(quotient) + remainder. See Examples 2 and 3.

13. $(x^2 + 8x + 13) \div (x + 3)$
14. $(x^2 + 5x + 7) \div (x + 3)$
15. $(x^2 - 2x) \div (x + 2)$
16. $(2x) \div (x + 5)$
17. $(3x) \div (x - 1)$
18. $(x^2) \div (x + 1)$
19. $(x^3 + 8) \div (x + 2)$
20. $(y^3 - 1) \div (y - 1)$
21. $(a^3 + 4a - 5) \div (a - 2)$
22. $(w^3 + w^2 - 3) \div (w - 2)$
23. $(x^3 - x^2 + x - 3) \div (x + 1)$
24. $(a^2 - 4) \div (a + 2)$
25. $(x^4 - x + x^3 - 1) \div (x - 2)$
26. $(3x^4 + 6 - x^2 + 3x) \div (x + 2)$
27. $(5x^2 - 3x^4 + x - 2) \div (x^2 - 2)$

28. $(x^4 - 2 + x^3) \div (x^2 + 3)$ ***29.** $(x^2 - 4x + 2) \div (2x - 3)$ ***30.** $(x^2 - 5x + 1) \div (3x + 6)$

*31. $(2x^2 - x + 6) \div (3x - 2)$ ***32.** $(3x^2 + 4x - 1) \div (2x + 1)$

Divide each of the following. Express each answer in the form: $\text{quotient} + \dfrac{\text{remainder}}{\text{divisor}}$. *See Example 4.*

33. $\dfrac{2x}{x - 5}$ **34.** $\dfrac{x}{x - 1}$ **35.** $\dfrac{x^2}{x + 1}$

36. $\dfrac{x^2 + 9}{x + 3}$ **37.** $\dfrac{x^2 + 1}{x - 1}$ **38.** $\dfrac{-x}{x + 4}$

39. $\dfrac{-3x}{x + 2}$ **40.** $\dfrac{x^2}{x - 1}$ **41.** $\dfrac{x^3}{x + 2}$

42. $\dfrac{x^3 - 1}{x - 2}$ **43.** $\dfrac{x^3 + 2x}{x^2}$ **44.** $\dfrac{2x^2 + 3}{2x}$

*45. $\dfrac{x^2}{2x + 3}$ ***46.** $\dfrac{x}{3x - 1}$ ***47.** $\dfrac{x^3 - 5x}{2x - 1}$

*48. $\dfrac{x^2 - 4}{3x + 1}$

For each pair of polynomials, determine whether the first polynomial is a factor of the second. See Example 5.

49. $x^2 - 2, \quad x^4 + 3x^3 - 6x - 4$ **50.** $x^2 - 3, \quad x^4 + 2x^3 - 4x^2 - 6x + 3$

51. $x + 4, \quad x^3 + x^2 - 11x + 8$ **52.** $x + 4, \quad x^3 + x^2 + x + 48$

53. $x - 4, \quad x^3 - 13x - 12$ **54.** $x - 1, \quad x^3 + 3x^2 - 5x$

55. $2x - 3, \quad 2x^3 - 3x^2 - 4x + 6$ **56.** $3x - 5, \quad 6x^2 - 7x - 6$

57. $3x - 1, \quad 3x^3 + 2x^2 - 4x + 2$ **58.** $x + 5, \quad x^3 - 6x^2 + 30$

59. $x - 3, \quad x^3 - 8x - 3$ **60.** $x - 2, \quad x^6 - 64$

Factor. See Example 6.

61. $a^3 - 1$ **62.** $w^3 + 1$ **63.** $w^3 + 27$

64. $x^3 - 64$ **65.** $8x^3 - 1$ **66.** $27x^3 + 1$

67. $a^3 + 8$ **68.** $m^3 - 8$ **69.** $m^3 - n^3$

70. $x^3 - y^3$ **71.** $8 - 27x^3$ **72.** $27 - 64y^3$

*73. $w^3x^6 - 125y^9$ ***74.** $216m^{12} - 8a^3b^6c^9$

*75. $125a^9 + 8c^3b^6$ ***76.** $343w^{18} + 1000a^{15}$

*77. $.001x^3 - .008y^3$ ***78.** $.125a^6 + 8w^3$

Find the quotient and remainder for each indicated division.

79. $-81y^{12} \div 27y^3$ **80.** $(x^2 - 1) \div (x - 1)$

81. $(z^2 + 6z + 9) \div (z + 3)$ **82.** $(z^2 + 6z + 9) \div z$

83. $(m^3 - 27) \div (m^2 + 3m + 9)$ **84.** $(w^3 + 8) \div (w + 2)$

85. $(w^3 + 8) \div (w - 2)$ **86.** $(6y^7 - 15y^4 + 18y^2) \div (-3y^2)$

87. $(x^2 + 9) \div (x + 3)$ **88.** $(x^2 + 36) \div (x - 6)$

3.8 Factoring Strategy

IN THIS SECTION:

- Prime Polynomials
- Factoring by Substitution
- Factoring by Grouping
- Summary

In previous sections we established the general idea of factoring and some special cases. In this section we will study two more types of factoring and then a general strategy for factoring polynomials.

Prime Polynomials

Monomials are considered prime polynomials. Binomials with no common factors, such as $2x + 1$ and $a - 3$, are also prime polynomials. To determine if a polynomial such as $x^2 + 1$ is a prime polynomial, we must try all possibilities for factoring it. If $x^2 + 1$ could be factored as a product of two binomials, the only possibilities that would give a first term of x^2 and a last term of 1 are $(x + 1)^2$, $(x - 1)^2$, and $(x + 1)(x - 1)$. However, we know that

$$(x + 1)^2 = x^2 + 2x + 1,$$
$$(x - 1)^2 = x^2 - 2x + 1,$$

and

$$(x + 1)(x - 1) = x^2 - 1.$$

Because none of these possibilities results in $x^2 + 1$, the polynomial $x^2 + 1$ is a prime polynomial. Note that $x^2 + 1$ is a sum of two squares. A sum of two squares (that have no common factor) is always a prime polynomial.

EXAMPLE 1 Determine whether the polynomial $x^2 + 3x + 4$ is a prime polynomial.

Solution To factor $x^2 + 3x + 4$, we must find two integers with a product of 4 and a sum of 3. The only pairs of positive integers with a product of 4 are 1 and 4, and 2 and 2. Because the product is positive 4, both numbers must be negative or both positive. Under these conditions it is impossible to get a sum of positive 3. The polynomial is prime.   ◀

Factoring by Substitution

So far, we have factored only polynomials of degree 2 or 3. When we need to factor polynomials of high degree, such as $x^4 + x^2 - 2$, we can use a method called **substitution** to get a polynomial of degree 2 or 3 that we can factor easily. If we let $a = x^2$, then $a^2 = x^4$. We substitute a and a^2 into the original polynomial and factor the resulting polynomial.

$$x^4 + x^2 - 2 = a^2 + a - 2 \qquad \text{Substitute } a \text{ for } x^2 \text{ and } a^2 \text{ for } x^4.$$
$$= (a + 2)(a - 1) \qquad \text{Factor.}$$
$$= (x^2 + 2)(x^2 - 1) \qquad \text{Substitute back } x^2 \text{ for } a.$$
$$= (x^2 + 2)(x - 1)(x + 1) \qquad \text{Factor completely.}$$

The idea in using substitution is to convert the polynomial into a second-degree or third-degree polynomial that we know how to factor.

EXAMPLE 2 Use substitution to factor each polynomial completely.

a) $3x^6 - 3$

b) $(a + 3)^2 + 5(a + 3) + 6$

Solution

a) To factor this polynomial, we must first factor out the common factor 3 and then recognize that x^6 is a perfect square: $x^6 = (x^3)^2$.

$$3x^6 - 3 = 3(x^6 - 1) \qquad \text{Factor out the common factor.}$$
$$= 3((x^3)^2 - 1) \qquad \text{Recognize the difference of two squares.}$$
$$= 3(a^2 - 1) \qquad \text{Substitute } a \text{ for } x^3.$$
$$= 3(a - 1)(a + 1) \qquad \text{Factor the difference of two squares.}$$
$$= 3(x^3 - 1)(x^3 + 1) \qquad \text{Substitute } x^3 \text{ for } a.$$
$$= 3(x - 1)(x^2 + x + 1)(x + 1)(x^2 - x + 1) \qquad \text{Factor completely.}$$

b)
$$(a + 3)^2 + 5(a + 3) + 6 = m^2 + 5m + 6 \qquad \text{Let } m = a + 3.$$
$$= (m + 2)(m + 3) \qquad \text{Factor.}$$
$$= (a + 3 + 2)(a + 3 + 3) \qquad \text{Substitute } a + 3 \text{ for } m.$$
$$= (a + 5)(a + 6) \qquad \text{Combine like terms.} \qquad \blacktriangleleft$$

In Example 2a we recognized $x^6 - 1$ as a difference of two squares. However, $x^6 - 1$ is also a difference of two cubes, and we can factor it using the rule for the difference of two cubes:

$$x^6 - 1 = (x^2)^3 - 1 = (x^2 - 1)(x^4 + x^2 + 1)$$

To complete this factorization, we can factor $x^2 - 1$, but it is difficult to see how to factor $x^4 + x^2 + 1$. (It is not prime.) Although x^6 can be thought of as a perfect square or a perfect cube, in this case it is better to think of it as a perfect square.

You will find that you can often factor polynomials of higher degree without actually doing the substitution step. For example, $x^8 + 5x^4 + 6$ can be factored by substitution, but it is not too difficult to factor it by trial and error:

$$x^8 + 5x^4 + 6 = (x^4 + 3)(x^4 + 2)$$

Factoring by Grouping

In Section 3.6 we wrote a trinomial as a polynomial with four terms, and then used factoring by grouping. Factoring by grouping can also be used on other types of polynomials with four terms. For example, to factor the polynomial

$$x^3 + x^2 + 4x + 4,$$

note that the first two terms have a common factor of x^2, and the last two terms have a common factor of 4:

$$x^3 + x^2 + 4x + 4 = x^2(x + 1) + 4(x + 1) \qquad \text{Factor by grouping.}$$
$$= (x^2 + 4)(x + 1) \qquad \text{Factor out } x + 1.$$

Note that the factor $x^2 + 4$ is prime.

EXAMPLE 3   Use grouping to factor each polynomial completely.

a) $3x^3 - x^2 - 27x + 9$ **b)** $am + aw + m^2 + mw$

Solution

a) We can factor x^2 out of the first two terms, and 9 or -9 from the last two terms. We choose -9 to get the factor $3x - 1$ in each case.

$$3x^3 - x^2 - 27x + 9 = x^2(3x - 1) - 9(3x - 1) \qquad \text{Factor by grouping.}$$
$$= (x^2 - 9)(3x - 1) \qquad \text{Factor out } 3x - 1.$$
$$= (x - 3)(x + 3)(3x - 1) \qquad \text{Factor completely.}$$

b) $am + aw + m^2 + mw = a(m + w) + m(m + w) \qquad \text{Factor by grouping.}$
$$= (a + m)(m + w) \qquad \text{Factor out } m + w. \qquad \blacktriangleleft$$

Summary

A strategy for factoring polynomials is summarized in the following box.

STRATEGY
Factoring Polynomials

1. If there are any common factors, factor them out first.
2. When factoring a binomial, look for the special cases: difference of two squares, difference of two cubes, and sum of two cubes. Remember that a sum of two squares is prime.
3. When factoring a trinomial, check to see whether it is a perfect-square trinomial.

4. When factoring a trinomial that is not a perfect square, use grouping or trial and error.

5. When factoring a polynomial of high degree, use substitution to get a polynomial of degree 2 or 3, or use trial and error.

6. If the polynomial has four terms, try factoring by grouping.

EXAMPLE 4 Factor the following polynomials completely.

a) $3w^3 - 3w^2 - 18w$ b) $10x^2 + 160$ c) $16a^2 - 80a + 100$

Solution

a) $3w^3 - 3w^2 - 18w = 3w(w^2 - w - 6)$ Factor out the common factor $3w$.

$\qquad\qquad\qquad\quad = 3w(w - 3)(w + 2)$ Factor completely.

b) $10x^2 + 160 = 10(x^2 + 16)$ The factor $x^2 + 16$ is prime.

c) $16a^2 - 80a + 100 = 4(4a^2 - 20a + 25)$

$\qquad\qquad\qquad\qquad = 4(2a - 5)^2$ ◀

Warm-ups

True or false?

1. $x^2 - 9 = (x - 3)^2$ for any value of x.

2. The polynomial $4x^2 + 12x + 9$ is a perfect-square trinomial.

3. A sum of two squares (with no common factor) is prime.

4. The polynomial $x^4 - 16$ is factored completely as $(x^2 - 4)(x^2 + 4)$.

5. $y^3 - 27 = (y + 3)(y^2 + 3y - 9)$ for any value of y.

6. The polynomial $y^6 - 1$ is the difference of two squares.

7. $x^2 + 36 = (x + 6)^2$ for any value of x.

8. The polynomial $x^2 - 4x - 4$ is a prime polynomial.

9. The polynomial $a^6 - 1$ is the difference of two cubes.

10. The polynomial $x^2 + 3x - ax + 3a$ can be factored by grouping.

3.8 EXERCISES

Determine whether each polynomial is a prime polynomial. See Example 1.

1. $y^2 + 100$

2. $3x^2 + 27$

3. $x^2 - 2x - 3$

4. $x^2 - 2x + 3$

5. $x^2 + 2x + 3$

6. $x^2 + 4x + 3$

7. $x^2 - 4x - 3$

8. $x^2 + 4x - 3$

9. $6x^2 + 3x - 4$

10. $4x^2 - 5x - 3$

Use substitution to factor each polynomial completely. See Example 2.

11. $a^4 - 10a^2 + 25$ **12.** $9y^4 + 12y^2 + 4$ **13.** $x^4 - 6x^2 + 8$

14. $x^6 + 2x^3 - 3$ **15.** $(3x - 5)^2 - 1$ **16.** $2y^6 - 128$

17. $(m + 2)^2 + 2(m + 2) - 3$ **18.** $6 - 6y^6$ **19.** $(2w - 3)^2 - 2(2w - 3) - 15$

20. $(2x + 1)^2 - 4$ **21.** $32a^4 - 18$ **22.** $2a^4 - 32$

23. $x^4 - (x - 6)^2$ **24.** $y^4 - (2y + 1)^2$

25. $3(y - 1)^2 + 11(y - 1) - 20$ **26.** $2(w + 2)^2 + 5(w + 2) - 3$

Use grouping to factor each polynomial completely. See Example 3.

27. $ax + ay + bx + by$ **28.** $7x + 7z + kx + kz$

29. $x^3 + x^2 - 9x - 9$ **30.** $x^3 + x^2 - 25x - 25$

31. $aw - bw - 3a + 3b$ **32.** $wx - wy + 2x - 2y$

33. $a^4 + 3a^3 + 27a + 81$ **34.** $ac - bc - 5a + 5b$

35. $y^4 - 5y^3 + 8y - 40$ **36.** $x^3 + ax - 3a - 3x^2$

37. $ady + d - w - awy$ **38.** $xy + by + ax + ab$

39. $x^2y - y + ax^2 - a$ **40.** $a^2c - b^2c + a^2d - b^2d$

41. $y^4 + y + by^3 + b$ **42.** $ab + bm + aw^2 + mw^2$

Use the factoring strategy to factor each polynomial completely.

43. $9x^2 - 24x + 16$ **44.** $-3x^2 + 18x + 48$ **45.** $12x^2 - 13x + 3$

46. $2x^2 - 3x - 6$ **47.** $3a^4 + 81a$ **48.** $a^3 - 25a$

49. $32 + 2x^2$ **50.** $x^3 + 4x^2 + 4x$ **51.** $6x^2 - 5x + 12$

52. $x^4 + 2x^3 - x - 2$ **53.** $(x + y)^2 - 1$ **54.** $x^3 + 9x$

55. $a^3b - ab^3$ **56.** $2m^3 - 250$ **57.** $x^4 + 2x^3 - 8x - 16$

58. $(x + 5)^2 - 4$ **59.** $m^2n + 2mn^2 + n^3$ **60.** $a^2b - 6ab + 9b$

61. $2m + 2n + wm + wn$ **62.** $aw + bw - 5a - 5b$ **63.** $4w^2 + 4w - 3$

64. $4w^2 + 8w - 63$ **65.** $t^4 + 4t^2 - 21$ **66.** $m^4 + 5m^2 + 4$

67. $a^3 + 7a^2 - 30a$ **68.** $2y^4 + 3y^3 - 20y^2$ **69.** $(y + 5)^2 - 2(y + 5) - 3$

70. $(2m - 1)^2 + 7(2m - 1) + 10$ **71.** $2w^4 - 1250$ **72.** $5a^5 - 5a$

73. $8a^3 + 8a$ **74.** $awx + ax$ **75.** $(w + 5)^2 - 9$

76. $(a - 6)^2 - 1$ **77.** $4aw^2 - 12aw + 9a$ **78.** $9an^3 + 15an^2 - 14an$

79. $x^2 - 6x + 9$ **80.** $x^3 + 12x^2 + 36x$ **81.** $3x^4 - 75x^2$

82. $3x^2 + 9x + 12$ **83.** $m^3n - n$ **84.** $m^4 + 16m^2$

85. $12x^2 + 2x - 30$ **86.** $90x^2 + 3x - 60$

87. $2a^3 - 32$ **88.** $12x^2 - 28x + 15$

Factor completely. Assume that variables used as exponents represent positive integers.

*89. $a^{3m} - 1$ ***90.** $x^{6a} + 8$ ***91.** $a^{3w} - b^{6n}$

*92. $x^{2n} - 9$ ***93.** $t^{8n} - 16$ ***94.** $a^{3n+2} + a^2$

*95. $a^{2n+1} - 2a^{n+1} - 15a$ ***96.** $x^{3m} + x^{2m} - 6x^m$

*97. $a^{2n} - 3a^n + a^nb - 3b$ ***98.** $x^mz + 5z + x^{m+1} + 5x$

3.9 Solving Equations by Factoring

IN THIS SECTION:

- **The Zero Factor Property**
- **Applications**

The techniques of factoring can be used to solve equations involving polynomials that cannot be solved by the other methods that we have learned. After we learn to solve equations by factoring, we will use this technique to solve some new applied problems in this section and in Chapters 4 and 6.

The Zero Factor Property

Consider the equation

$$x^2 + x - 12 = 0.$$

We can factor the left side and write $(x + 4)(x - 3) = 0$. This means that the product of $x + 4$ and $x - 3$ is zero. The product of two numbers is zero only when one or the other of the numbers is zero. Therefore, the equation

$$(x + 4)(x - 3) = 0$$

is equivalent to the compound equation

$$x + 4 = 0 \quad \text{or} \quad x - 3 = 0.$$

To solve the original equation, we proceed as follows:

$$x^2 + x - 12 = 0$$
$$(x + 4)(x - 3) = 0 \qquad \text{Factor the left side.}$$
$$x + 4 = 0 \quad \text{or} \quad x - 3 = 0 \qquad \text{Write the equivalent compound equation.}$$
$$x = -4 \quad \text{or} \quad x = 3 \qquad \text{Solve the compound equation.}$$

The solution set is $\{-4, 3\}$. Check these solutions in the original equation $x^2 + x - 12 = 0$. Letting $x = -4$ and $x = 3$ we get

$$(-4)^2 + (-4) - 12 = 16 - 4 - 12 = 0$$

and

$$(3)^2 + 3 - 12 = 9 + 3 - 12 = 0.$$

The idea used here is called the **zero factor property,** and it is stated as follows.

Zero Factor Property

> The equation $a \cdot b = 0$ is equivalent to the compound equation
> $$a = 0 \qquad \text{or} \qquad b = 0.$$

The zero factor property is used only in solving polynomial equations that have zero on one side and a polynomial that can be factored on the other side. The polynomials that we factored most often were the quadratic polynomials. The equations that we will solve most often using the zero factor property will be **quadratic equations.**

DEFINITION
Quadratic Equation

> If a, b, and c are real numbers, with $a \neq 0$, then the equation
> $$ax^2 + bx + c = 0$$
> is called a **quadratic equation.**

In Chapter 6 we will learn to solve quadratic equations that cannot be solved by factoring. Keep the following strategy in mind when solving equations by factoring.

STRATEGY
Solving Equations by Factoring

> 1. Write the equation with 0 on the right side.
> 2. Factor the left side.
> 3. Use the zero factor property to get simpler equations. (Set each factor equal to 0.)
> 4. Solve the simpler equations.
> 5. Check the answers in the original equation.

EXAMPLE 1   Solve $10x^2 = 5x$.

Solution

$$10x^2 - 5x = 0 \qquad \text{Rewrite the equation with zero on the right side.}$$
$$5x(2x - 1) = 0 \qquad \text{Factor the left side.}$$
$$5x = 0 \quad \text{or} \quad 2x - 1 = 0 \qquad \text{Zero factor property}$$
$$x = 0 \quad \text{or} \qquad x = \frac{1}{2}$$

The solution set is $\{0, \frac{1}{2}\}$. Check each solution in the original equation.   ◀

If there are more than two factors, we can write an equivalent equation by setting each factor equal to zero.

EXAMPLE 2 Solve $2x^3 - 3x^2 - 8x + 12 = 0$.

Solution

$$x^2(2x - 3) - 4(2x - 3) = 0 \qquad \text{Factor by grouping.}$$
$$(x^2 - 4)(2x - 3) = 0 \qquad \text{Factor out } 2x - 3.$$
$$(x - 2)(x + 2)(2x - 3) = 0 \qquad \text{Factor completely.}$$
$$x - 2 = 0, \text{ or } x + 2 = 0, \text{ or } 2x - 3 = 0 \qquad \text{Zero factor property}$$
$$x = 2, \text{ or } \qquad x = -2, \text{ or } \qquad x = \frac{3}{2}$$

The solution set is $\{-2, \frac{3}{2}, 2\}$. Check each solution in the original equation. ◀

The equation in the next example involves absolute value.

EXAMPLE 3 Solve $|x^2 - 2x - 16| = 8$.

Solution $|x^2 - 2x - 16| = 8$ is equivalent to the compound equation

$$\begin{array}{lll} x^2 - 2x - 16 = 8 & \text{or} & x^2 - 2x - 16 = -8 \\ x^2 - 2x - 24 = 0 & \text{or} & x^2 - 2x - 8 = 0 \\ (x - 6)(x + 4) = 0 & \text{or} & (x - 4)(x + 2) = 0 \\ x - 6 = 0 \text{ or } x + 4 = 0 & \text{or} & x - 4 = 0 \text{ or } x + 2 = 0 \\ x = 6 \text{ or } \quad x = -4 & \text{or} & x = 4 \text{ or } \quad x = -2 \end{array}$$

The solution set is $\{-2, -4, 4, 6\}$. Check each solution. ◀

Applications

Many applied problems can be solved by using equations like those we have been solving.

EXAMPLE 4 Ronald's living room is 2 feet longer than it is wide, and its area is 168 square feet. What are the dimensions of the room?

Solution Let x be the width and $x + 2$ be the length (see Fig. 3.1). Because the area of a rectangle is the length times the width, we can write the equation

$$x(x + 2) = 168.$$

We solve the equation by factoring.

$$x^2 + 2x - 168 = 0$$
$$(x - 12)(x + 14) = 0$$
$$x - 12 = 0 \quad \text{or } x + 14 = 0$$
$$x = 12 \text{ or } \qquad x = -14$$

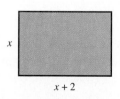

x

$x + 2$

Figure 3.1

Because the width of a room is a positive number, we disregard the solution $x = -14$. We use $x = 12$ and get a width of 12 feet and a length of 14 feet. Check this by multiplying 12 and 14 to get 168. ◀

Applications involving quadratic equations often require a theorem from geometry called the **Pythagorean Theorem.** This theorem states that in any right triangle, the sum of the squares of the lengths of the legs is equal to the hypotenuse squared.

The Pythagorean Theorem

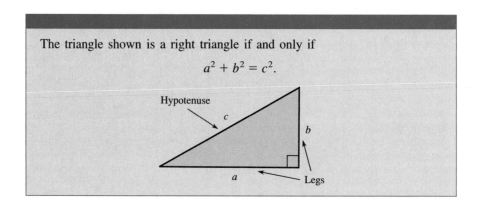

The triangle shown is a right triangle if and only if

$$a^2 + b^2 = c^2.$$

Hypotenuse

c

b

a

Legs

We will use the Pythagorean Theorem to solve the following geometric problem.

EXAMPLE 5   Shirley used 14 meters of fencing to enclose a rectangular area, and then 5 meters more to fence across the rectangle on the diagonal. What are the length and width of the rectangle?

Solution   The perimeter of a rectangle is twice the length plus twice the width, $P = 2L + 2W$. Because the perimeter is 14 meters, the sum of one length and one width is 7 meters. If we let x represent the width, then $7 - x$ is the length. We use the Pythagorean Theorem to get a relationship between the length, width, and diagonal. See Fig. 3.2.

$$x^2 + (7 - x)^2 = 5^2 \qquad \text{By the Pythagorean Theorem}$$
$$x^2 + 49 - 14x + x^2 = 25 \qquad \text{Simplify.}$$
$$2x^2 - 14x + 24 = 0 \qquad \text{Simplify.}$$
$$x^2 - 7x + 12 = 0 \qquad \text{Divide each side by 2.}$$
$$(x - 3)(x - 4) = 0 \qquad \text{Factor the left side.}$$
$$x - 3 = 0 \text{ or } x - 4 = 0 \qquad \text{Zero factor property}$$
$$x = 3 \text{ or } \qquad x = 4$$
$$7 - x = 4 \text{ or } 7 - x = 3$$

Solving the equation gives two possible rectangles: a 3 by 4 rectangle, or a 4 by 3 rectangle—however, those are identical rectangles. The rectangle is 3 meters by 4 meters. ◀

5

x

$7 - x$

Figure 3.2

Warm-ups

True or false?

1. The equation $(x - 1)(x + 3) = 12$ is equivalent to $x - 1 = 3$ or $x + 3 = 4$.

2. Equations solved by factoring may have two solutions.

3. The equation $c \cdot d = 0$ is equivalent to $c = 0$ or $d = 0$.

4. The equation $|x^2 + 4| = 5$ is equivalent to the compound equation $x^2 + 4 = 5$ or $x^2 - 4 = 5$.

5. The solution set to the equation $(2x - 1)(3x + 4) = 0$ is $\{1/2, -4/3\}$.

6. The Pythagorean Theorem states that the sum of the squares of any two sides of any triangle is equal to the square of the third side.

7. If the perimeter of a rectangular room is 38 feet, then the sum of the length and width is 19 feet.

8. If two numbers have a sum of 8, then they can be represented by x and $8 - x$.

9. The solution set to the equation $x(x - 1)(x - 2) = 0$ is $\{1, 2\}$.

10. The solution set to the equation $3(x + 2)(x - 5) = 0$ is $\{3, -2, 5\}$.

3.9 EXERCISES

Solve each equation. See Examples 1 and 2.

1. $(x - 5)(x + 4) = 0$
2. $(a - 6)(a + 5) = 0$
3. $(2x - 5)(3x + 4) = 0$
4. $(3k + 8)(4k - 3) = 0$
5. $w^2 + 5w - 14 = 0$
6. $t^2 - 6t - 27 = 0$
7. $m^2 - 7m = 0$
8. $h^2 - 5h = 0$
9. $a^2 - a = 20$
10. $p^2 - p = 42$
11. $3x^2 - 3x - 36 = 0$
12. $-2x^2 - 16x - 24 = 0$
13. $z^2 + \dfrac{3}{2}z = 10$
14. $m^2 + \dfrac{11}{3}m = -2$
15. $x^3 - 4x = 0$
16. $16x - x^3 = 0$
17. $w^3 + 4w^2 - 25w - 100 = 0$
18. $a^3 + 2a^2 - 16a - 32 = 0$
19. $n^3 - 2n^2 - n + 2 = 0$
20. $w^3 - w^2 - 25w + 25 = 0$

Solve each equation. See Example 3.

21. $|x^2 - 5| = 4$
22. $|x^2 - 17| = 8$
23. $|x^2 + 2x - 36| = 12$
24. $|x^2 + 2x - 19| = 16$
25. $|x^2 + 4x + 2| = 2$
26. $|x^2 + 8x + 8| = 8$
27. $|x^2 + 6x + 1| = 8$
28. $|x^2 - x - 21| = 9$

Solve each problem. See Examples 4 and 5.

29. The length of a new "super-size" color print is 2 inches more than the width. If the area is 24 square inches, what are the length and width?

30. The rows in Felicia's garden are each 3 feet wide. If the rows run north and south, she can have 2 more rows than if they run east and west. If the area of Felicia's garden is 135 square feet, then what are the length and width?

31. The sum of two numbers is 13, and their product is 36. Find the numbers.

32. The sum of two numbers is 6.5, and their product is 9. Find the numbers.

33. The length of Yolanda's closet is 2 feet longer than twice its width. If the diagonal measures 13 feet, then what are the length and width?

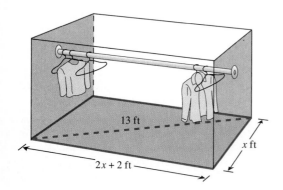

13 ft

x ft

$2x + 2$ ft

Figure for Exercise 33

34. The base of a ski ramp forms a right triangle. One leg of the triangle is 2 meters longer than the other. If the hypotenuse is 10 meters, then what are the lengths of the legs?

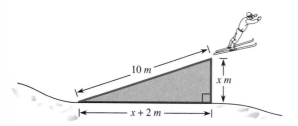

10 m

x m

$x + 2$ m

Figure for Exercise 34

35. The perimeter of a rectangle is 34 feet, and the diagonal is 13 feet. What are the length and width of the rectangle?

36. The perimeter of a rectangle is 28 inches, and the diagonal measures 10 inches. What are the length and width of the rectangle?

37. The sum of the squares of two consecutive integers is 25. Find the integers.

38. In the plans for their dream house, the Baileys have a master bedroom 240 square feet in size. If they increase the width by 3 feet, they must decrease the length by 4 feet in order to keep the original area. What are the original dimensions of the bedroom?

39. A circus clown tosses a ball into the air at 64 feet per second. The distance D (in feet) that the ball is above the earth at any time t (in seconds) is given by the formula $D = -16t^2 + 64t$. Determine the length of time that the ball is in the air.

40. If an object is dropped from a height of s_0 feet, then its altitude after t seconds is given by the formula $S = -16t^2 + s_0$. If a bomb is dropped from an airplane at a height of 1600 feet, then how long does it take for it to reach the ground?

41. Mary Gold has a rectangular flower bed that measures 4 feet by 6 feet. If she wants to increase the length and width by the same amount in order to have a flower bed of 48 square feet, then what will the new dimensions be?

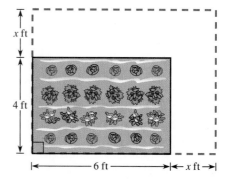

x ft

4 ft

6 ft

x ft

Figure for Exercise 41

42. Mr. Converse has 112 students in his algebra class with an equal number in each row. If he arranges the desks so that he has 1 fewer rows, he will have 2 more students in each row. How many rows did he have originally?

Solve each equation.

43. $2x^2 - x = 6$

44. $3x^2 + 14x = 5$

45. $|x^2 + 5x| = 6$

46. $|x^2 + 6x - 4| = 12$

47. $x^2 + 5x = 6$

48. $(x + 2)(x + 1) = 12$

49. $x + 5x = 6$

50. $(x + 2)(x + 3) = 20$

51. $y^3 + 9y^2 + 20y = 0$

52. $m^3 - 2m^2 - 3m = 0$

53. $5a^3 = 45a$

54. $5x^3 = 125x$

Solve each equation for y. Assume a and b are positive numbers.

55. $y^2 + by = 0$

56. $y^2 + ay + by + ab = 0$

57. $a^2y^2 - b^2 = 0$

58. $9y^2 + 6ay + a^2 = 0$

59. $4y^2 + 4by + b^2 = 0$

60. $y^2 - b^2 = 0$

61. $ay^2 + 3y - ay = 3$

62. $a^2y^2 + 2aby + b^2 = 0$

Wrap-up

CHAPTER 3

SUMMARY

	Concepts	Examples
Definition of negative integral exponents	If a is a nonzero real number and n is a positive integer, then $$a^{-n} = \frac{1}{a^n} = \left(\frac{1}{a}\right)^n$$ Find the power and reciprocal in either order. If $n = -1$, $a^{-1} = \frac{1}{a}$, the reciprocal of a.	$2^{-3} = \frac{1}{2^3} = \frac{1}{8}$ $\left(\frac{2}{3}\right)^{-2} = \left(\frac{3}{2}\right)^2$ $5^{-1} = \frac{1}{5}, \left(\frac{1}{3}\right)^{-1} = 3$
Definition of zero exponent	If a is any nonzero real number, then $a^0 = 1$. The expression 0^0 is undefined.	$3^0 = 1$

Rules of Exponents		Examples
	If a and b are nonzero real numbers and m and n are integers, then the following rules hold.	
Product rule	$a^m \cdot a^n = a^{m+n}$	$3^5 \cdot 3^7 = 3^{12}$, $2^{-3} \cdot 2^{10} = 2^7$
Sign-change rule	$a^n = \dfrac{1}{a^{-n}}$	$5^2 = \dfrac{1}{5^{-2}}$, $6^{-3} = \dfrac{1}{6^3}$
Quotient rule	$\dfrac{a^m}{a^n} = a^{m-n}$	$\dfrac{x^8}{x^5} = x^3$, $\dfrac{5^4}{5^{-7}} = 5^{11}$
Power of a power rule	$(a^m)^n = a^{mn}$	$(5^2)^3 = 5^6$ $(2x)^3 = 8x^3$
Power of a product rule	$(ab)^n = a^n b^n$	$(2x^3)^4 = 16x^{12}$
Power of a quotient rule	$\left(\dfrac{a}{b}\right)^n = \dfrac{a^n}{b^n}$	$\left(\dfrac{x}{3}\right)^2 = \dfrac{x^2}{9}$

Polynomials		Examples
Term of a polynomial	The product of a number (coefficient) and one or more variables raised to whole-number powers	$3x^4$, $-2xy^2$, 5
Polynomial	A single term or a finite sum of terms	$x^5 - 3x^2 + 7$
Adding or subtracting polynomials	Add or subtract the like terms.	$(x + 3) + (x - 7) = 2x - 4$ $(x^2 - 2x) - (3x^2 - x) = -2x^2 - x$
Multiplying two polynomials	Multiply each term of the first polynomial by each term of the second polynomial, then combine like terms.	$\begin{array}{r} x^2 + 2x + 3 \\ x - 1 \\ \hline -x^2 - 2x - 3 \\ x^3 + 2x^2 + 3x \\ \hline x^3 + x^2 + x - 3 \end{array}$

Shortcuts for Multiplying Two Binomials		Examples
FOIL	The product of two binomials can be found quickly by multiplying their First, Outer, Inner, and Last terms.	$(x + 2)(x + 3) = x^2 + 5x + 6$
Square of a sum	$(a + b)^2 = a^2 + 2ab + b^2$	$(x + 5)^2 = x^2 + 10x + 25$
Square of a difference	$(a - b)^2 = a^2 - 2ab + b^2$	$(m - 3)^2 = m^2 - 6m + 9$
Product of a sum and a difference	$(a + b)(a - b) = a^2 - b^2$	$(x + 3)(x - 3) = x^2 - 9$

Other Important Facts		**Examples**

Dividing polynomials	Long division	$$\begin{array}{r} x - 7 \\ x + 2{\overline{\smash{\big)}\,x^2 - 5x + 3}} \\ \underline{x^2 + 2x} \\ -7x + 3 \\ \underline{-7x - 14} \\ 17 \end{array}$$
Factoring a polynomial	Write a polynomial as a product of two or more polynomials. A polynomial is factored completely if it is a product of prime polynomials.	$$\begin{aligned} 3x^2 - 3 &= 3(x^2 - 1) \\ &= 3(x + 1)(x - 1) \end{aligned}$$

Types of Factoring		**Examples**

Common factors	Factor out the greatest common factor.	$2x^3 - 6x = 2x(x^2 - 3)$
Difference of two squares	$a^2 - b^2 = (a + b)(a - b)$ (Sum of two squares is prime.)	$m^2 - 25 = (m - 5)(m + 5)$
Difference of two cubes	$a^3 - b^3 = (a - b)(a^2 + ab + b^2)$	$x^3 - 8 = (x - 2)(x^2 + 2x + 4)$
Sum of two cubes	$a^3 + b^3 = (a + b)(a^2 - ab + b^2)$	$x^3 + 27 = (x + 3)(x^2 - 3x + 9)$
Grouping	Factor out common factors from groups of terms.	$$\begin{aligned} &3x + 3w + bx + bw \\ &= 3(x + w) + b(x + w) \\ &= (3 + b)(x + w) \end{aligned}$$
Perfect-square trinomials	$a^2 + 2ab + b^2 = (a + b)^2$ $a^2 - 2ab + b^2 = (a - b)^2$	$x^2 + 10x + 25 = (x + 5)^2$ $x^2 - 6x + 9 = (x - 3)^2$
Factoring $ax^2 + bx + c$	By grouping: 1. Find two numbers that have a product equal to ac and a sum equal to b. 2. Replace bx by two terms using the two new numbers as coefficients. 3. Factor the resulting four-term polynomial by grouping.	$$\begin{aligned} &2x^2 + 7x + 3 \\ &= 2x^2 + x + 6x + 3 \\ &= (2x + 1)x + (2x + 1)3 \\ \\ &= (2x + 1)(x + 3) \end{aligned}$$
	By trial and error: Try possibilities by considering factors of the first term and factors of the last term. Check them by FOIL.	$$\begin{aligned} &12x^2 + 19x - 18 \\ &= (3x - 2)(4x + 9) \end{aligned}$$
Substitution	Use substitution on higher-degree polynomials to reduce the degree to 2 or 3.	$x^4 - 3x^2 - 18$ Let $a = x^2$. $a^2 - 3a - 18$

	Solving Equations by Factoring	Examples

Strategy
1. Write the equation with 0 on the right side.
2. Factor the left side.
3. Set each factor equal to zero.
4. Solve the simpler equations.
5. Check the answers in the original equation.

$x^2 - 3x - 18 = 0$
$(x - 6)(x + 3) = 0$
$x - 6 = 0$ or $x + 3 = 0$
   $x = 6$ or $x = -3$
$6^2 - 3(6) - 18 = 0$

REVIEW EXERCISES

3.1 *Simplify each expression. Assume that all variables represent nonzero real numbers. Write your answers with positive exponents.*

1. $2 \cdot 2 \cdot 2^{-1}$
2. $5^{-1} \cdot 5$
3. $2^2 \cdot 3^2$
4. $3^2 \cdot 5^2$
5. $(-3)^{-3}$
6. $(-2)^{-2}$
7. $-(-1)^{-3}$
8. $3^4 \cdot 3^7$
9. $2x^3 \cdot 4x^{-6}$
10. $-3a^{-3} \cdot 4a^{-4}$
11. $\dfrac{y^{-5}}{y^{-3}}$
12. $\dfrac{w^3}{w^{-3}}$
13. $\dfrac{a^5 \cdot a^{-2}}{a^{-4}}$
14. $\dfrac{2m^3 \cdot m^6}{2m^{-2}}$
15. $\dfrac{6x^{-2}}{3x^2}$
16. $\dfrac{-5y^2x^{-3}}{5y^{-2}x^7}$

3.2 *Simplify each expression. Assume that all variables represent nonzero real numbers. Write your answers with positive exponents.*

17. $\left(\dfrac{2}{3}\right)^2$
18. $\left(\dfrac{1}{2} + \dfrac{1}{3}\right)^2$
19. $\left(\dfrac{3}{4}\right)^{-1}$
20. $\left(\dfrac{1}{3}\right)^{-2}$
21. $\left(\dfrac{1}{2} - \dfrac{1}{3}\right)^{-2}$
22. $\left(-\dfrac{4}{5}\right)^{-1}$
23. $\left(-\dfrac{1}{2}\right)^{-2}$
24. $(2x^3)^3 \cdot (3x^2)^2$
25. $(x^2)^4 \cdot x^{-3}$
26. $(a^{-3})^{-2} \cdot a^{-7}$
27. $(-3x^{-2}y)^{-4}$
28. $(a^{-1}b)^{-1}$
29. $\left(-\dfrac{3x}{2y^3}\right)^{-2}$
30. $\left(-\dfrac{x}{y}\right)^{-3}$
31. $(m^2n^3)^{-2}(m^{-3})^4$
32. $(w^{-3}xy)^{-1}(wx^{-3}y)^2$

3.3 *Perform the indicated operations.*

33. $(2w - 3) + (6w + 5)$
34. $(3a - 2xy) + (5xy - 7a)$
35. $(x^2 - 3x - 4) - (x^2 + 3x - 7)$
36. $(7 - 2x - x^2) - (x^2 - 5x + 6)$
37. $(x^2 - 2x + 4)(x - 2)$
38. $(x + 5)(x^2 - 2x + 10)$
39. $xy + 7z - 5(xy - 3z)$
40. $7 - 4(x - 3)$
41. $m^2(5m^3 - m + 2)$
42. $(a + 2)^3$
43. $5x + 3x(x^2 - 5x + 4)$
44. $2 + 3x^2(x - 5)$

3.4 *Perform the following computations mentally. Write down only the answers.*

45. $(x - 3)(x + 7)$
46. $(k - 5)(k + 4)$
47. $(z - 5y)(z + 5y)$
48. $(m - 3)(m + 3)$
49. $(m + 8)^2$
50. $(b + 2a)^2$
51. $(w - 6x)(w - 4x)$
52. $(k - 3)^2$
53. $(n - 5)^2$

54. $(m^2 - 5)(m^2 + 5)$ **55.** $(2w - 3)(w + 6)$ **56.** $(3k - 5t)(2k + 6t)$

57. $(2m + 1)^2$ **58.** $(2 - 3z)^2$

59. $(8 - a)^2$ **60.** $(t + 7)(t + 5)$

3.5 *Complete the factoring by filling in the parentheses.*

61. $3x - 6 = 3(\quad)$ **62.** $7x^2 - x = x(\quad)$ **63.** $4a - 20 = -4(\quad)$

64. $w^2 - w = -w(\quad)$ **65.** $3w - w^2 = -w(\quad)$ **66.** $3x - 6 = (\quad)(2 - x)$

67. $5m - 10 = (\quad)(m - 2)$ **68.** $3x - 9 = (\quad)(x - 3)$ **69.** $a - b = -1(\quad)$

70. $w - 8 = -1(\quad)$ **71.** $a + b = -(\quad)$ **72.** $x - 3 = (\quad)(3 - x)$

Factor each polynomial.

73. $y^2 - 81$ **74.** $r^2 t^2 - 9v^2$ **75.** $4x^2 + 28x + 49$

76. $y^2 - 20y + 100$ **77.** $t^2 - 18t + 81$ **78.** $4w^2 + 4ws + s^2$

3.6 *Factor each polynomial.*

79. $x^2 - 7x - 30$ **80.** $y^2 + 4y - 32$ **81.** $w^2 - 3w - 28$

82. $6t^2 - 5t + 1$ **83.** $2m^2 + 5m - 7$ **84.** $12x^2 - 17x + 6$

3.7 *Find the quotient and remainder for each division.*

85. $(x^3 + x^2 - 11x + 10) \div (x - 2)$ **86.** $(a^9 - 8) \div (a^3 - 2)$

87. $(m^4 - 1) \div (m + 1)$ **88.** $(a^2 - b^2) \div (a - b)$

89. $(3m^3 + 6m^2 - 18m) \div (3m)$ **90.** $(m - 1) \div (1 - m)$

91. $(w - 3) \div (3 - w)$ **92.** $(x^4 - 1) \div (x - 1)$

93. $(12x^3 - 8x^2 + 6x) \div (2x)$ **94.** $(2x^3 + 5x^2 + 9) \div (x + 3)$

Express each expression in the form: quotient $+ \dfrac{remainder}{divisor}$.

95. $\dfrac{x^2 - 5}{x - 1}$ **96.** $\dfrac{x^2 + 3x + 2}{x + 3}$ **97.** $\dfrac{3x}{x - 2}$

98. $\dfrac{4x}{x - 5}$ **99.** $\dfrac{x}{1 - x}$ **100.** $\dfrac{y}{y + 1}$

101. $\dfrac{x^3}{x^2 + 1}$ **102.** $\dfrac{5x}{5 - x}$ **103.** $\dfrac{-2x}{x + 5}$

104. $\dfrac{-3x}{x + 3}$

Use division to determine whether the first polynomial is a factor of the second.

105. $x + 2,\quad x^3 - 2x^2 + 3x + 22$ **106.** $x - 2,\quad x^3 + x - 10$

107. $x - 5,\quad x^3 - x - 120$ **108.** $x + 3,\quad x^3 + 2x + 15$

109. $x - 1,\quad x^3 + x^2 - 3$ **110.** $x - 1,\quad x^3 + 1$

111. $x^2 + 2,\quad x^4 + x^3 + 5x^2 + 2x + 6$ **112.** $x^2 + 1,\quad x^4 - 1$

3.8 *Factor each polynomial completely.*

113. $5x^3 + 40$

114. $w^3 - 6w^2 + 9w$

115. $9x^2 + 9x + 2$

116. $ax^3 + a$

117. $x^3 + x^2 - x - 1$

118. $16x^2 - 4x - 2$

119. $x^2y - 16y$

120. $-5m^2 + 5$

121. $a^3b^2 - 2a^2b^2 + ab^2$

122. $-2w^2 - 16w - 32$

123. $x^3 - x^2 + 9x - 9$

124. $w^4 + 2w^2 - 3$

125. $x^4 - x^2 - 12$

126. $8x^3 - 1$

127. $a^6 - a^3$

128. $a^2 - ab + 2a - 2b$

129. $8m^2 + 24m + 18$

130. $-3x^2 - 9x + 30$

131. $(2x - 3)^2 - 16$

132. $(m - 6)^2 - (m - 6) - 12$

3.9 *Solve each equation.*

133. $x^3 - 5x^2 = 0$

134. $2m^2 + 10m + 12 = 0$

135. $(a - 2)(a - 3) = 6$

136. $(w - 2)(w + 3) = 50$

137. $2m^2 - 9m - 5 = 0$

138. $m^3 + 4m^2 - 9m - 36 = 0$

139. $w^3 + 5w^2 - w - 5 = 0$

140. $12x^2 + 5x - 3 = 0$

141. $|x^2 - 5| = 4$

142. $|x^2 - 3x - 7| = 3$

Solve each problem.

143. According to the manufacturer, a 10-foot ladder is the safest when the distance from the ground to the top of the ladder, measured along the wall, is 2 feet longer than the distance of the bottom of the ladder to the wall. How far from the wall should you place the bottom of this ladder?

144. Find three consecutive integers such that the sum of their squares is 50.

CHAPTER 3 TEST

Simplify each expression. Assume that all variables represent nonzero real numbers. Exponents in your answers should be positive exponents.

1. 3^{-2}

2. $\dfrac{1}{6^{-2}}$

3. $\left(\dfrac{1}{2}\right)^{-3}$

4. $3x^4 \cdot 4x^3$

5. $\dfrac{8y^9}{2y^{-3}}$

6. $(4a^2b)^3$

7. $\left(\dfrac{x^2}{3}\right)^{-3}$

8. $\dfrac{(2^{-1}a^2b)^{-3}}{4a^{-9}}$

Perform the indicated operations.

9. $(3x^3 - x^2 + 6) + (4x^2 - 2x - 3)$

10. $(x^2 - 6x - 7) - (3x^2 + 2x - 4)$

11. $(x^2 - 3x + 7)(x - 2)$

12. $(x^3 + 7x^2 + 7x - 15) \div (x + 3)$

13. $(x - 2)^3$

14. $(x - 3) \div (3 - x)$

Find the products.

15. $(x - 7)(x + 3)$

16. $(x - 6)^2$

17. $(y - 5)(y + 5)$

18. $(2x + 5)^2$

Express in the form: $\text{quotient} + \dfrac{\text{remainder}}{\text{divisor}}$.

19. $\dfrac{5x}{x + 3}$

20. $\dfrac{x^2 + 3x - 6}{x - 2}$

Factor completely.

21. $a^2 - 2a - 24$

22. $4x^2 + 28x + 49$

23. $3m^3 - 24$

24. $2x^2y - 32y$

25. $2xa + 3a - 10x - 15$

26. $x^4 + 3x^2 - 4$

Solve each equation.

27. $2m^2 + 7m - 15 = 0$

28. $x^3 - 4x = 0$

29. $|x^2 + x - 9| = 3$

Write a complete solution to the word problem.

30. A portable television is advertised as having a 10-inch diagonal measure screen. If the width of the screen is 2 inches more than the height, then what are the dimensions of the screen?

Tying It All Together

CHAPTERS 1–3

Simplify each of the following.

1. 4^2

2. $4(-2)$

3. 4^{-2}

4. $2^3 \cdot 4^{-1}$

5. $2^{-1} + 2^{-1}$

6. $2^{-1} \cdot 3^{-1}$

7. $3^{-1} - 2^{-2}$

8. $3^2 - 4(5)(-2)$

9. $2^7 - 2^6$

10. $.08(32) + .08(68)$

11. $3 - 2|5 - 7 \cdot 3|$

12. $5^{-1} + 6^{-1}$

Solve each equation.

13. $.05x - .04(x - 50) = 4$

14. $15x - 27 = 0$

15. $2x^2 + 15x - 27 = 0$

16. $2x^2 + 15x = 0$

17. $|15x - 27| = 3$

18. $|15x - 27| = 0$

19. $|15x - 27| = -78$

20. $|x^2 + x - 4| = 2$

21. $(2x - 1)(x + 5) = 0$

22. $|3x - 1| + 6 = 9$

23. $x - \dfrac{2}{3} = \dfrac{4}{5}$

24. $-\dfrac{2}{3}x = \dfrac{4}{5}$

Rational Expressions

Charles's mother has always known that Charles can eat cookies faster than she can bake them, so one day she decided to get a head start on her son. She made enough cookies to have the cookie jar full when Charles came home from school and continued baking and stocking the cookie jar as he was working on emptying it. If it takes his mother 2 hours to bake enough cookies to fill the cookie jar, and it takes Charles 1 1/2 hours to empty it, then how long will it take him to empty it with his mother continuously restocking it?

This type of problem is known as a work problem. Each hour, Charles empties a fraction of the cookie jar, but his mother fills a fraction of the cookie jar. Since Charles is working against his mother, the cookie jar will eventually get emptied. This problem can be solved by using an equation involving rational expressions. In this chapter we will study rational expressions and equations involving rational expressions. This particular problem appears as Exercise 45 of Section 4.6.

4.1 Properties of Rational Expressions

IN THIS SECTION:

- Definition of Rational Expressions
- Domain
- Reducing to Lowest Terms
- Building Up the Denominator

Rational expressions are ratios of polynomials. Rational expressions are as fundamental to algebra as rational numbers are to arithmetic. We often use the basic properties of rational numbers that we learned in arithmetic without being fully aware of why they work. In this section we will look carefully at some of the properties of rational numbers and see how they extend to rational expressions.

Definition of Rational Expressions

A rational number is the ratio of two integers, with the denominator not equal to 0. For example,

$$\frac{2}{7}, \qquad -\frac{3}{5}, \qquad \frac{4}{1}, \qquad \text{and} \qquad 0$$

are rational numbers. A **rational expression** is the ratio of two polynomials with the denominator not equal to 0. For example,

$$\frac{2}{3}, \qquad 3a + 5, \qquad \frac{x - 3}{2x^2 - 2}, \qquad \frac{y + 2}{5y}, \qquad \text{and} \qquad \frac{x - 2}{x + 1}$$

are rational expressions. Because an integer is a monomial, a rational number is also a rational expression. If the denominator of a rational expression is 1, it is usually omitted, as in the expression $3a + 5$.

Domain

The **domain** of a rational expression is the set of all real numbers that can be used in place of the variable. Because the denominator of a rational expression cannot be 0, the domain of a rational expression consists of the set of real numbers except those that cause the denominator to be zero. The domain of

$$\frac{x}{x + 5}$$

is the set of all real numbers excluding -5. In set-builder notation, this set is written

$$\{x \mid x \neq -5\}.$$

EXAMPLE 1 Find the domain of each rational expression.

a) $\dfrac{x-2}{x+9}$
b) $\dfrac{y+2}{5y}$
c) $\dfrac{x-3}{2x^2-2}$

Solution

a) The denominator will be zero if $x + 9 = 0$, or if $x = -9$. The domain is $\{x \mid x \neq -9\}$.

b) The denominator will be zero if $5y = 0$; that is, if $y = 0$. The domain is $\{y \mid y \neq 0\}$.

c) The denominator will be zero if $2x^2 - 2 = 0$. Solving this equation gives the values for x that are *not* part of the domain.

$$2x^2 - 2 = 0$$
$$2(x^2 - 1) = 0 \qquad \text{Factor out 2.}$$
$$2(x+1)(x-1) = 0 \qquad \text{Factor completely.}$$
$$x + 1 = 0 \quad \text{or} \quad x - 1 = 0 \qquad \text{Zero factor property}$$
$$x = -1 \quad \text{or} \quad x = 1$$

The domain is the set of all real numbers except 1 and -1. This set is written as $\{x \mid x \neq 1 \text{ and } x \neq -1\}$. ◄

Reducing to Lowest Terms

Each rational number can be written in infinitely many equivalent forms. For example,

$$\frac{2}{3} = \frac{4}{6} = \frac{6}{9} = \frac{8}{12} = \frac{10}{15} = \cdots$$

Each equivalent form of 2/3 is obtained from 2/3 by multiplying both numerator and denominator by the same nonzero number. For example,

$$\frac{2}{3} = \frac{2}{3} \cdot 1 = \frac{2}{3} \cdot \frac{2}{2} = \frac{4}{6} \qquad \text{and} \qquad \frac{2}{3} = \frac{2}{3} \cdot \frac{3}{3} = \frac{6}{9}.$$

Note that we are actually multiplying 2/3 by equivalent forms of 1, the multiplicative identity.

If we start with 4/6 and convert it into 2/3, we are **reducing** 4/6 to **lowest terms.** This process is written as

$$\frac{4}{6} = \frac{\cancel{2} \cdot 2}{\cancel{2} \cdot 3} = \frac{2}{3}.$$

A rational number is expressed in lowest terms when the numerator and denominator have no common factors other than 1. In reducing 4/6, we divide the numerator and denominator by the common factor 2. We can multiply or divide both numerator and denominator of a rational number by the same nonzero number without changing the value of the rational number. This idea is called the **basic principle of rational numbers** and can be written in symbols as follows.

Basic Principle of Rational Numbers

If a/b is a rational number and c is a nonzero real number, then

$$\frac{a}{b} = \frac{ac}{bc} \quad \text{and} \quad \frac{a}{b} = \frac{a \div c}{b \div c}.$$

Although it is true that

$$\frac{5}{6} = \frac{2+3}{2+4},$$

we cannot divide out the 2's in this expression because the 2's are not factors. We can divide out only common factors when reducing fractions.

Each rational expression can also be expressed in infinitely many equivalent forms. To reduce rational expressions to lowest terms, we follow exactly the same procedure as we do for rational numbers: Factor the numerator and denominator completely, then divide out all common factors.

EXAMPLE 2   Reduce each rational expression to lowest terms.

a) $\dfrac{18}{42}$
b) $\dfrac{2x^2 - 8}{x^2 + x - 6}$
c) $\dfrac{a^7 b}{a^2 b^3}$
d) $\dfrac{2a - 2b}{3b - 3a}$

Solution

a) $\dfrac{18}{42} = \dfrac{2 \cdot 3^2}{2 \cdot 3 \cdot 7} = \dfrac{3}{7}$ $\begin{array}{l} 18 = 2 \cdot 9 = 2 \cdot 3^2 \\ 42 = 2 \cdot 21 = 2 \cdot 3 \cdot 7 \end{array}$

b) $\dfrac{2x^2 - 8}{x^2 + x - 6} = \dfrac{2(x^2 - 4)}{(x - 2)(x + 3)} = \dfrac{2(x - 2)(x + 2)}{(x - 2)(x + 3)} = \dfrac{2x + 4}{x + 3}$

c) $\dfrac{a^7 b}{a^2 b^3} = \dfrac{a^5}{b^2}$ We used the quotient rule to reduce.

d) $\dfrac{2a - 2b}{3b - 3a} = \dfrac{2(a - b)}{3(b - a)} = \dfrac{-2(b - a)}{3(b - a)} = \dfrac{-2}{3}$ Factoring out -2 gives the common factor $b - a$.   ◀

To say that the two rational expressions, such as those of Example 2b, are equivalent means that they have the same numerical value for any replacement of the variables, provided that the replacement is in the domain of both expressions. In other words, the equation of Example 2b,

$$\frac{2x^2 - 8}{x^2 + x - 6} = \frac{2x + 4}{x + 3},$$

is an identity.

The main points to remember for reducing rational expressions are summarized in the following strategy for reducing rational expressions.

STRATEGY
Reducing Rational
Expressions

1. All reducing is done by dividing out common factors.
2. Factor the numerator and denominator completely to see the common factors.
3. Use the quotient rule to reduce a ratio of two monomials involving exponents.
4. We may have to factor out a common factor with a negative sign to get identical factors in the numerator and denominator.

Building Up the Denominator

In Section 4.3 we will find that only rational expressions with identical denominators can be added or subtracted. Fractions without identical denominators can be converted to equivalent fractions with a common denominator by reversing the procedure for reducing fractions to lowest terms. This procedure is called **building up the denominator.**

Consider converting the fraction $1/3$ into an equivalent fraction with a denominator of 51. Any fraction equivalent to $1/3$ can be obtained by multiplying the numerator and denominator of $1/3$ by the same nonzero number. Because $51 = 3 \cdot 17$, we multiply the numerator and denominator of $1/3$ by 17 to get an equivalent fraction with a denominator of 51:

$$\frac{1}{3} = \frac{1}{3} \cdot 1 = \frac{1}{3} \cdot \frac{17}{17} = \frac{17}{51}.$$

We use the same procedure for rational expressions. To convert the rational expression

$$\frac{2}{x - 3}$$

into an equivalent rational expression with denominator $2x^2 - 5x - 3$, we first factor $2x^2 - 5x - 3$:

$$2x^2 - 5x - 3 = (2x + 1)(x - 3).$$

Now we multiply the numerator and denominator by $2x + 1$, the missing factor:

$$\frac{2}{x - 3} = \frac{2(2x + 1)}{(x - 3)(2x + 1)} = \frac{4x + 2}{2x^2 - 5x - 3}.$$

EXAMPLE 3 Convert each rational expression into an equivalent rational expression that has the indicated denominator.

a) $\dfrac{2}{7}, \dfrac{?}{42}$

b) $\dfrac{5}{a}, \dfrac{?}{ab}$

c) $\dfrac{5}{2a - 2b}, \dfrac{?}{6b - 6a}$

d) $\dfrac{x + 2}{x + 3}, \dfrac{?}{x^2 + 7x + 12}$

Solution

a) Factor 42 as $42 = 2 \cdot 3 \cdot 7$, then multiply numerator and denominator by the missing factors, 2 and 3.

$$\frac{2}{7} = \frac{2 \cdot 2 \cdot 3}{7 \cdot 2 \cdot 3} = \frac{12}{42}$$

b) Multiply numerator and denominator by b:

$$\frac{5}{a} = \frac{5b}{ab}$$

c) Factor both $2a - 2b$ and $6b - 6a$ to see what factor is missing in $2a - 2b$. Note that we factor out -6 from $6b - 6a$ to get the factor $a - b$:

$$2a - 2b = 2(a - b)$$
$$6b - 6a = -6(a - b) = -3 \cdot 2(a - b)$$

Now multiply the numerator and denominator by the missing factor, -3.

$$\frac{5}{2a - 2b} = \frac{5(-3)}{(2a - 2b)(-3)} = \frac{-15}{6b - 6a}$$

d) Because $x^2 + 7x + 12 = (x + 3)(x + 4)$, multiply numerator and denominator by $x + 4$.

$$\frac{x + 2}{x + 3} = \frac{(x + 2)(x + 4)}{(x + 3)(x + 4)} = \frac{x^2 + 6x + 8}{x^2 + 7x + 12} \qquad \blacktriangleleft$$

Warm-ups

True or false?

1. A rational number is a rational expression.

2. The expression $\dfrac{2 + x}{x - 1}$ is a rational expression.

3. The domain of the rational expression $\dfrac{3}{x - 2}$ is $\{2\}$.

4. The domain of the expression $\dfrac{2x + 5}{(x - 9)(2x + 1)}$ is $\{x \mid x \neq 9 \text{ and } x \neq -1/2\}$.

5. The domain of the rational expression $\dfrac{x - 1}{x + 2}$ is $\{x \mid x \neq -2 \text{ and } x \neq 1\}$.

6. The rational expression $\dfrac{5x + 2}{15}$ reduces to $\dfrac{x + 2}{3}$.

7. To convert the rational expression $\dfrac{x}{x - 1}$ into a rational expression with denominator $x^2 - 1$, we multiply numerator and denominator by x.

8. The expression $\dfrac{2}{3 - x}$ is equivalent to $\dfrac{-2}{x - 3}$.

9. The equation $\dfrac{4x^3}{6x} = \dfrac{2x^2}{3}$ is an identity.

10. The expression $\dfrac{x^2 - y^2}{x - y}$ reduced to lowest terms is $x - y$.

4.1 EXERCISES

Find the domain of each rational expression. See Example 1.

1. $\dfrac{3x}{x - 1}$

2. $\dfrac{x}{x + 5}$

3. $\dfrac{5x - 1}{x^2 - 4}$

4. $\dfrac{2x - 1}{x^2 - 9}$

5. $\dfrac{7}{2x - 5}$

6. $\dfrac{12}{5 - 4x}$

7. $\dfrac{2x - 3}{x^2 + 5x + 6}$

8. $\dfrac{3x + 1}{2x^2 - 7x - 4}$

9. $\dfrac{x - 1}{x^2 + 4x}$

10. $\dfrac{2x}{3x^2 + 9x}$

11. $\dfrac{x + 1}{x^2}$

12. $\dfrac{x^2 - 3x - 4}{x}$

Reduce each rational expression to lowest terms. See Example 2.

13. $\dfrac{6}{57}$

14. $\dfrac{14}{91}$

15. $\dfrac{42}{210}$

16. $\dfrac{242}{154}$

17. $\dfrac{2x + 2}{4}$

18. $\dfrac{3a + 3}{3}$

19. $\dfrac{3x - 6y}{10y - 5x}$

20. $\dfrac{5b - 10a}{5b - 5a}$

21. $\dfrac{ab^2}{a^3b}$

22. $\dfrac{a^3b^2}{a^3 + a^4}$

23. $\dfrac{a - b}{2b - 2a}$

24. $\dfrac{2m - 2n}{4n - 4m}$

25. $\dfrac{3x + 6}{3x}$

26. $\dfrac{a^3 - b^3}{a - b}$

27. $\dfrac{x^2 - 4}{x^2 + 4x + 4}$

28. $\dfrac{2a^2 - 2b^2}{2a^2 + 2b^2}$

29. $\dfrac{2x^2 + 2x - 12}{4x^2 - 36}$

30. $\dfrac{2x^2 + 10x + 12}{2x^2 - 8}$

31. $\dfrac{w^2 x^3 y}{w x^5 y^2}$

32. $\dfrac{x^3 + 7x^2 - 4x}{x^3 - 16x}$

33. $\dfrac{27x^3 + y^3}{6x + 2y}$

34. $\dfrac{ab + 3a - by - 3y}{a^2 - y^2}$

35. $\dfrac{2x^4 - 32}{4x - 8}$

36. $\dfrac{2x^2 - 5x - 3}{2x^2 + 11x + 5}$

Convert each rational expression into an equivalent rational expression that has the indicated denominator. See Example 3.

37. $\dfrac{1}{5}, \dfrac{?}{50}$

38. $\dfrac{2}{3}, \dfrac{?}{9}$

39. $\dfrac{1}{x}, \dfrac{?}{3x^2}$

40. $\dfrac{3}{ab^2}, \dfrac{?}{a^3 b^5}$

41. $\dfrac{5}{x - 1}, \dfrac{?}{x^2 - 2x + 1}$

42. $\dfrac{x}{x - 3}, \dfrac{?}{x^2 - 9}$

43. $\dfrac{1}{2x + 2}, \dfrac{?}{-6x - 6}$

44. $\dfrac{1}{a - b}, \dfrac{?}{2b - 2a}$

45. $\dfrac{x + 2}{x + 3}, \dfrac{?}{x^2 + 2x - 3}$

46. $\dfrac{x}{x - 5}, \dfrac{?}{x^2 - x - 20}$

47. $\dfrac{7}{x - 1}, \dfrac{?}{1 - x}$

48. $\dfrac{3}{x + 2}, \dfrac{?}{x^3 + 8}$

49. $5, \dfrac{?}{a}$

50. $3, \dfrac{?}{a + 1}$

51. $\dfrac{3}{x + 1}, \dfrac{?}{x^2 - 1}$

52. $\dfrac{x}{x + 1}, \dfrac{?}{x^2 + x}$

53. $x, \dfrac{?}{7 - x}$

54. $\dfrac{c}{a}, \dfrac{?}{4a^2}$

In place of the question mark, put an expression that will make these rational expressions equivalent. Try to do these mentally.

55. $\dfrac{1}{3} = \dfrac{?}{21}$

56. $4 = \dfrac{?}{3}$

57. $5 = \dfrac{10}{?}$

58. $\dfrac{3}{4} = \dfrac{12}{?}$

59. $\dfrac{-3}{6} = \dfrac{3}{?}$

60. $\dfrac{10}{20} = \dfrac{?}{-10}$

61. $\dfrac{-5}{x} = \dfrac{?}{-2x}$

62. $\dfrac{-1}{2x} = \dfrac{2}{?}$

63. $\dfrac{3}{a} = \dfrac{?}{a^2}$

64. $\dfrac{5}{y} = \dfrac{10}{?}$

65. $\dfrac{2}{a - b} = \dfrac{?}{b - a}$

66. $\dfrac{3}{x - 4} = \dfrac{?}{4 - x}$

67. $\dfrac{2}{x - 1} = \dfrac{?}{x^2 - 1}$

68. $\dfrac{5}{x + 3} = \dfrac{?}{x^2 - 9}$

69. $\dfrac{2}{w - 3} = \dfrac{-2}{?}$

70. $\dfrac{-2}{5 - x} = \dfrac{2}{?}$

71. $\dfrac{2x + 4}{6} = \dfrac{?}{3}$

72. $\dfrac{2x - 3}{4x - 6} = \dfrac{1}{?}$

73. $\dfrac{x + 4}{x^2 - 16} = \dfrac{1}{?}$

74. $\dfrac{2x + 2}{2x} = \dfrac{x + 1}{?}$

75. $\dfrac{3a + 3}{3a} = \dfrac{?}{a}$

76. $\dfrac{x - 3}{x^2 - 9} = \dfrac{1}{?}$

4.2 Multiplication and Division

IN THIS SECTION:

- Multiplying Rational Numbers
- Multiplying Rational Expressions
- Dividing $a - b$ by $b - a$
- Dividing Rational Numbers
- Dividing Rational Expressions

In Chapter 3 we learned to add, subtract, multiply, and divide polynomials. In this chapter we will learn to perform the same operations with rational expressions. We will begin in this section with multiplication and division.

Multiplying Rational Numbers

We multiply two rational numbers by multiplying their numerators and multiplying their denominators. For example,

$$\frac{6}{7} \cdot \frac{14}{15} = \frac{84}{105} = \frac{21 \cdot 4}{21 \cdot 5} = \frac{4}{5}.$$

Instead of reducing the rational number after multiplying, it is often easier to reduce before multiplying. We first factor all terms, then divide out the common factors, then multiply:

$$\frac{6}{7} \cdot \frac{14}{15} = \frac{2 \cdot \cancel{3}}{\cancel{7}} \cdot \frac{2 \cdot \cancel{7}}{\cancel{3} \cdot 5} = \frac{4}{5}$$

The definition of multiplication of rational numbers follows.

DEFINITION
Multiplication of Rational Numbers

If a/b and c/d are rational numbers, then

$$\frac{a}{b} \cdot \frac{c}{d} = \frac{ac}{bd}.$$

Multiplying Rational Expressions

We multiply rational expressions in the same way that we multiply rational numbers.

EXAMPLE 1 Multiply the following rational expressions.

a) $\dfrac{3x - 3y}{6} \cdot \dfrac{10x}{x^2 - y^2}$

b) $\dfrac{x^2 + 7x + 12}{x^2 + 3x} \cdot \dfrac{x^2}{x^2 - 16}$

c) $(a^2 - 1) \cdot \dfrac{6}{2a^2 + 4a + 2}$

d) $\dfrac{a^3 - b^3}{b - a} \cdot \dfrac{6}{2a^2 + 2ab + 2b^2}$

Solution

a) $\dfrac{3x - 3y}{6} \cdot \dfrac{10x}{x^2 - y^2} = \dfrac{3(x - y)}{2 \cdot 3} \cdot \dfrac{2 \cdot 5x}{(x - y)(x + y)}$ Factor.

$\qquad = \dfrac{5x}{x + y}$ Divide out common factors.

b) $\dfrac{x^2 + 7x + 12}{x^2 + 3x} \cdot \dfrac{x^2}{x^2 - 16} = \dfrac{(x + 3)(x + 4)}{x(x + 3)} \cdot \dfrac{x^2}{(x - 4)(x + 4)}$

$\qquad = \dfrac{x}{x - 4}$

Because x is not a factor of the denominator, this expression cannot be further reduced.

c) $(a^2 - 1) \cdot \dfrac{6}{2a^2 + 4a + 2} = \dfrac{(a + 1)(a - 1)}{1} \cdot \dfrac{2 \cdot 3}{2(a + 1)^2}$ $a^2 - 1 = \dfrac{a^2 - 1}{1}$

$\qquad = \dfrac{3(a - 1)}{a + 1}$

$\qquad = \dfrac{3a - 3}{a + 1}$ We usually remove parentheses in the numerator.

d) Note that $a - b$ is a factor of $a^3 - b^3$, and $b - a$ occurs in the denominator. We can factor $b - a$ as $-1(a - b)$ to get a common factor.

$\dfrac{a^3 - b^3}{b - a} \cdot \dfrac{6}{2a^2 + 2ab + 2b^2} = \dfrac{(a - b)(a^2 + ab + b^2)}{-1(a - b)} \cdot \dfrac{2 \cdot 3}{2(a^2 + ab + b^2)}$

$\qquad = \dfrac{3}{-1}$

$\qquad = -3$ ◀

Dividing $a - b$ by $b - a$

Because $a - b = -1(b - a)$, we can write

$$\dfrac{a - b}{b - a} = -1.$$

In Example 1d we factored -1 out of $b - a$ to get a common factor of $a - b$. It is easier to use the fact that $(a - b) \div (b - a) = -1$. Note how we use this fact in the following product:

$$\frac{m-4}{3} \cdot \frac{6}{4-m} = \frac{\overset{-1}{\cancel{m-4}}}{\cancel{3}} \cdot \frac{\overset{2}{\cancel{6}}}{\cancel{4-m}} = -2.$$

We divided 6 by 3 to get 2, and we divided $m - 4$ by $4 - m$ to get -1.

Dividing Rational Numbers

We divide rational numbers by multiplying by the reciprocal or multiplicative inverse of the divisor. For example,

$$\frac{3}{4} \div \frac{15}{2} = \frac{3}{4} \cdot \frac{2}{15} = \frac{\cancel{3}}{2 \cdot \cancel{2}} \cdot \frac{\cancel{2} \cdot 1}{\cancel{3} \cdot 5} = \frac{1}{10}.$$

The definition of division of rational numbers follows.

DEFINITION
Division of Rational Numbers

If a/b and c/d are rational numbers with $c/d \neq 0$, then

$$\frac{a}{b} \div \frac{c}{d} = \frac{a}{b} \cdot \frac{d}{c}.$$

Dividing Rational Expressions

We divide rational expressions in the same way that we divide rational numbers.

EXAMPLE 2   Perform the indicated operations.

a) $\dfrac{10}{3x} \div \dfrac{6}{5x}$
    b) $\dfrac{25 - x^2}{x^2 + x} \div \dfrac{x - 5}{x^2 - 1}$
    c) $\dfrac{5x^2}{2} \div 4x$

Solution

a) $\dfrac{10}{3x} \div \dfrac{6}{5x} = \dfrac{10}{3x} \cdot \dfrac{5x}{6} = \dfrac{\cancel{2} \cdot 5}{3\cancel{x}} \cdot \dfrac{5\cancel{x}}{\cancel{2} \cdot 3} = \dfrac{25}{9}$

b) $\dfrac{25 - x^2}{x^2 + x} \div \dfrac{x - 5}{x^2 - 1} = \dfrac{25 - x^2}{x^2 + x} \cdot \dfrac{x^2 - 1}{x - 5}$ Invert and multiply.

$$= \frac{\overset{-1}{\cancel{(5 - x)}}(5 + x)}{x\cancel{(x + 1)}} \cdot \frac{\cancel{(x + 1)}(x - 1)}{\cancel{x - 5}} \qquad \frac{5 - x}{x - 5} = -1$$

$$= \frac{-1(5 + x)(x - 1)}{x} \qquad \text{Divide out the common factors.}$$

$$= \frac{-x^2 - 4x + 5}{x} \qquad \text{Remove parentheses.}$$

c) $\dfrac{5x^2}{2} \div 4x = \dfrac{5x^2}{2} \cdot \dfrac{1}{4x} = \dfrac{5x}{8}$ $\quad x^2 \div x = x$ ◀

When division of rational expressions is indicated by a fraction bar, we still invert and multiply.

EXAMPLE 3 Perform the operations indicated.

a) $\dfrac{\dfrac{a+b}{3}}{\dfrac{1}{2}}$
b) $\dfrac{\dfrac{x^2-4}{2}}{\dfrac{x-2}{3}}$
c) $\dfrac{\dfrac{m^2+1}{5}}{3}$

Solution

a) $\dfrac{\dfrac{a+b}{3}}{\dfrac{1}{2}} = \dfrac{a+b}{3} \cdot \dfrac{2}{1} = \dfrac{2a+2b}{3}$ **Invert and multiply.**

b) $\dfrac{\dfrac{x^2-4}{2}}{\dfrac{x-2}{3}} = \dfrac{x^2-4}{2} \cdot \dfrac{3}{x-2}$

$= \dfrac{(x-2)(x+2)}{2} \cdot \dfrac{3}{x-2}$

$= \dfrac{3x+6}{2}$

c) $\dfrac{\dfrac{m^2+1}{5}}{3} = \dfrac{m^2+1}{5} \cdot \dfrac{1}{3} = \dfrac{m^2+1}{15}$ **Multiply by 1/3, the inverse of 3.** ◀

Warm-ups

True or false?

1. We can multiply only fractions that have identical denominators.

2. $\dfrac{2}{7} \cdot \dfrac{3}{7} = \dfrac{6}{7}$

3. To divide rational expressions, we invert the divisor and multiply.

4. $a \div b = \dfrac{1}{a} \cdot b$ for any nonzero a and b.

5. $\dfrac{1}{2x} \cdot 8x^2 = 4x$ for any nonzero value of x.

6. One-half of one-third is one-sixth.

7. One-third divided by one-half is two-thirds.

8. The quotient of $w - z$ divided by $z - w$ is -1, provided $z - w \neq 0$.

9. $\dfrac{x}{3} \div 2 = \dfrac{x}{6}$ for any value of x.

10. $\dfrac{a}{b} \div \dfrac{b}{a} = 1$ for any nonzero values of a and b.

4.2 EXERCISES

Perform the indicated operations. See Example 1.

1. $\dfrac{12}{42} \cdot \dfrac{35}{22}$

2. $\dfrac{3}{8} \cdot \dfrac{20}{21}$

3. $\dfrac{3a}{10b} \cdot \dfrac{5b^2}{6}$

4. $\dfrac{3x}{7y} \cdot \dfrac{14y^2}{9x}$

5. $\dfrac{10x + 5}{5x^2 + 5} \cdot \dfrac{2x^2 + x - 1}{4x^2 - 1}$

6. $\dfrac{x^3 + x}{5} \cdot \dfrac{5}{x^2 + x}$

7. $\dfrac{ax + aw + bx + bw}{x^2 - w^2} \cdot \dfrac{x - w}{a^2 - b^2}$

8. $(a^2 - 4) \cdot \dfrac{7}{a - 2}$

9. $\dfrac{a^2 - 2a + 4}{a^3 + 8} \cdot \dfrac{(a + 2)^3}{2a + 4}$

10. $\dfrac{w^3 - 1}{(w - 1)^2} \cdot \dfrac{w^2 - 1}{w^2 + w + 1}$

Perform the indicated operations. See Example 2.

11. $\dfrac{15}{17} \div \dfrac{10}{17}$

12. $\dfrac{3}{4} \div \dfrac{1}{8}$

13. $\dfrac{x^2 + 6x + 9}{18} \div \dfrac{(x + 3)^3}{36}$

14. $(w + 1) \div \dfrac{w^2 - 1}{w}$

15. $\dfrac{a - 3}{2} \div \dfrac{4}{3 - a}$

16. $\dfrac{x - y}{5} \div \dfrac{y^2 - x^2}{2}$

Perform the indicated operations. See Example 3.

17. $\dfrac{\dfrac{x^2 - 25}{3}}{\dfrac{x - 5}{6}}$

18. $\dfrac{\dfrac{3x^2 + 3}{5}}{\dfrac{3x + 3}{5}}$

19. $\dfrac{\dfrac{a - b}{2}}{3}$

20. $\dfrac{\dfrac{10}{a + b}}{5}$

21. $\dfrac{a^2 - b^2}{\dfrac{a + b}{3}}$

22. $\dfrac{x^2 + 5x + 6}{\dfrac{x + 2}{x + 3}}$

Perform the indicated operations mentally. Write down only the answer.

23. $\dfrac{5x}{2} \div 3$

24. $\dfrac{x}{a} \div 2$

25. $\dfrac{3}{4} \div \dfrac{1}{4}$

26. $\dfrac{1}{4} \div \dfrac{1}{2}$

27. $7 \div \dfrac{1}{2}$

28. $\dfrac{1}{2} \div \dfrac{1}{4}$

29. $5 \div 3$

30. $6 \div \dfrac{1}{3}$

31. $\dfrac{1}{3} \div 6$

32. $\dfrac{1}{5} \div 2$

33. One-half of 6

34. One-half of 5

35. One-half of 1/6

36. One-half of 1/4

37. One-half of b/a

38. One-third of x/y

39. One-half of $4x/3$

40. One-third of $6x/y$

41. $(a - b) \div (b - a)$

42. $(a - b) \div (-1)$

43. $\dfrac{x - y}{3} \cdot \dfrac{6}{y - x}$

44. $\dfrac{5x - 5y}{x} \cdot \dfrac{1}{x - y}$

45. $\dfrac{2a + 2b}{a} \cdot \dfrac{1}{2}$

46. $\dfrac{x - y}{y - x} \cdot \dfrac{1}{2}$

47. $5xy \div \dfrac{1}{x}$

48. $3x \div \dfrac{1}{2}$

49. $\dfrac{a + b}{\dfrac{1}{2}}$

50. $\dfrac{x + 3}{\dfrac{1}{3}}$

51. $\dfrac{\dfrac{3x}{5}}{y}$

52. $\dfrac{\dfrac{b^2 - 4a}{2}}{a}$

53. $\dfrac{\dfrac{3a}{5b}}{2}$

54. $\dfrac{\dfrac{6x}{a}}{x}$

55. $\dfrac{\dfrac{b}{a}}{2}$

56. $\dfrac{\dfrac{3w}{5m}}{\dfrac{3}{m}}$

57. $\dfrac{x^3}{2} \cdot \dfrac{x^2}{3}$

58. $\dfrac{2}{x} \cdot (x^3 + x^2)$

59. $\dfrac{a - 3}{7} \cdot \dfrac{14}{3 - a}$

60. $\dfrac{a^3 - 7}{3} \cdot \dfrac{12}{7 - a^3}$

61. $\dfrac{x + 2}{3} \cdot \dfrac{6}{x - 2}$

62. $\dfrac{a - 3}{5} \cdot \dfrac{15}{3 + a}$

Perform the indicated operations.

63. $\dfrac{3x^2 + 13x - 10}{x} \cdot \dfrac{x^3}{9x^2 - 4} \cdot \dfrac{7x - 35}{x^2 - 25}$

64. $\dfrac{x^2 + 5x + 6}{x} \cdot \dfrac{x^2}{3x + 6} \cdot \dfrac{9}{x^2 - 4}$

65. $\dfrac{(a^2b^3c)^2}{(-2ab^2c)^3} \cdot \dfrac{(a^3b^2c)^3}{(abc)^4}$

66. $\dfrac{(-wy^2)^3}{3w^2y} \cdot \dfrac{(2wy)^2}{4wy^3}$

67. $\dfrac{(2mn)^3}{6mn^2} \div \dfrac{2m^2n^3}{(m^2n)^4}$

68. $\dfrac{(rt)^3}{rt^4} \div \dfrac{(rt^2)^3}{r^2t^3}$

69. $\dfrac{2x^2 + 7x - 15}{4x^2 - 100} \cdot \dfrac{2x^2 - 9x - 5}{4x^2 - 1}$

70. $\dfrac{x^3 + 1}{x^2 - 1} \cdot \dfrac{3x - 3}{x^3 - x^2 + x}$

71. $\dfrac{k^2 + 2km + m^2}{k^2 - 2km + m^2} \cdot \dfrac{m^2 + 3m - mk - 3k}{m^2 + mk + 3m + 3k}$

72. $\dfrac{a^2 + 2ab + b^2}{ac + bc - ad - bd} \div \dfrac{ac + ad - bc - bd}{c^2 - d^2}$

*73. $\dfrac{15w^2 - wt - 2t^2}{27w^3 + t^3} \cdot \dfrac{27w^2 - 9wt + 3t^2}{25w^2 - 4t^2} \div \dfrac{10w + 4t}{25w^2 + 20wt + 4t^2}$

*74. $\dfrac{6k^2 - k - 2}{2k^2 + 3kp - 2p^2} \cdot \dfrac{2k^2 + kp - p^2}{3k^2 - 10k + 3} \div \dfrac{2k^2 + k + p + 2kp}{k^2 + 2kp - 3k - 6p}$

4.3   Addition and Subtraction

IN THIS SECTION:

- Least Common Denominator
- Adding and Subtracting Rational Numbers
- Adding and Subtracting Rational Expressions
- Shortcuts

We can multiply or divide any rational expressions or fractions, but we add or subtract only rational expressions or fractions with identical denominators. So when the denominators are not the same, we must find equivalent forms of the expressions that have identical denominators. In this section we will review the idea of least common denominator and learn to use it for addition and subtraction of rational expressions.

Least Common Denominator

Suppose 4 and 6 are the denominators of two rational numbers. Because $12 = 3 \cdot 4$ and $12 = 2 \cdot 6$, denominators of 4 and 6 can both be converted to a denominator of 12 by multiplying the rational numbers by 3 and 2, respectively. The denominator 12 is a multiple of both 4 and 6, and it is the smallest number that is a multiple of both 4 and 6. The **least common denominator** (LCD) is the smallest number that is a multiple of all denominators involved.

To find the LCD for larger denominators such as 24 and 126, first factor the numbers completely.

$$24 = 2^3 \cdot 3$$
$$126 = 2 \cdot 3^2 \cdot 7$$

Any number that is a multiple of both 24 and 126 must have all of the factors of 24 and all of the factors of 126 in its factorization. Thus, in the LCD we use the factors 2, 3, and 7, and for each factor we use the highest power that appears on that factor in either of the factorizations. The highest power of 2 that appears in these two numbers is 3, the highest power of 3 that appears is 2, and the highest power of 7 is 1. So the LCD of 24 and 126 is $2^3 \cdot 3^2 \cdot 7$. If we write this product without exponents, we can see clearly that it is a multiple of both 24 and 126.

$$\underbrace{2 \cdot 2 \cdot \overbrace{2 \cdot 3}^{126} \cdot 3 \cdot 7}_{24} = 504. \qquad \begin{array}{l} 504 = 126 \cdot 4 \\ 504 = 24 \cdot 21 \end{array}$$

If we omitted any one of the factors in $2 \cdot 2 \cdot 2 \cdot 3 \cdot 3 \cdot 7$, we would not have a multiple of both 24 and 126. That is what makes 504 the *least* common denominator.

To find the LCD for the two polynomial denominators $4x^2 + 24x + 36$ and $6x^2 - 54$, we proceed in the same manner. First, we factor each polynomial completely.

$$4x^2 + 24x + 36 = 4(x^2 + 6x + 9) = 2^2(x + 3)^2$$
$$6x^2 - 54 = 6(x^2 - 9) = 2 \cdot 3(x - 3)(x + 3)$$

In the LCD we use each factor the maximum number of times it appears in either factorization. The LCD is

$$2^2 \cdot 3(x - 3)(x + 3)^2.$$

The strategy for finding the LCD for polynomial denominators can be stated as follows.

STRATEGY
Finding the LCD for
Polynomial Denominators

1. Factor each denominator completely. Use exponent notation for repeated factors.
2. Write the product of all of the different factors that appear in the denominators.
3. On each factor use the highest power that appears on that factor in any of the denominators.

EXAMPLE 1 Find the least common denominator for each group of denominators.

a) 18, 30

b) $x^2 + 5x + 6$, $x^2 + 6x + 9$

c) a^2bc, ab^3c^2, a^3bc

Solution

a) $18 = 2 \cdot 3^2$, $30 = 2 \cdot 3 \cdot 5$
   In the LCD we use the factors 2, 3, and 5 the maximum number of times they appear in either factorization. The LCD is $2 \cdot 3^2 \cdot 5 = 90$.

b) $x^2 + 5x + 6 = (x + 2)(x + 3)$, $x^2 + 6x + 9 = (x + 3)^2$
   The LCD is $(x + 2)(x + 3)^2$.

c) The expressions a^2bc, ab^3c^2, and a^3bc are already factored. In the LCD we use a, b, and c the maximum number of times that each appears in any of the expressions. The LCD is $a^3b^3c^2$. ◀

Adding and Subtracting Rational Numbers

We can add or subtract only rational numbers that have identical denominators. For example,

$$\frac{1}{7} + \frac{3}{7} = \frac{4}{7} \quad \text{and} \quad \frac{3}{5} - \frac{2}{5} = \frac{1}{5}.$$

In general, we have the following definition.

DEFINITION
Addition and Subtraction of
Rational Numbers

If $b \neq 0$, then

$$\frac{a}{b} + \frac{c}{b} = \frac{a+c}{b} \qquad \text{and} \qquad \frac{a}{b} - \frac{c}{b} = \frac{a-c}{b}.$$

If the rational numbers have different denominators, we must convert them to equivalent forms that have identical denominators and then add or subtract. Of course, it is most efficient to use the least common denominator as in the following example.

EXAMPLE 2   Perform the addition: $\dfrac{1}{6} + \dfrac{1}{8}$.

Solution

$$\frac{1}{6} + \frac{1}{8} = \frac{1}{2 \cdot 3} + \frac{1}{2^3} \qquad \text{Factor each denominator. The LCD is } 2^3 \cdot 3 = 24.$$

$$= \frac{1(2 \cdot 2)}{2 \cdot 3(2 \cdot 2)} + \frac{1(3)}{2 \cdot 2 \cdot 2(3)} \qquad \begin{array}{l}\text{Multiply by the missing factors to} \\ \text{get } 2^3 \cdot 3 \text{ in each denominator.}\end{array}$$

$$= \frac{4}{24} + \frac{3}{24}$$

$$= \frac{7}{24} \qquad\qquad\qquad\qquad\qquad\qquad\qquad\qquad\qquad\qquad \blacktriangleleft$$

Adding and Subtracting Rational Expressions

Rational expressions are added or subtracted just like rational numbers. We can add or subtract only rational expressions that have identical denominators.

EXAMPLE 3   Perform the indicated operations.

a) $\dfrac{5x - 3}{x - 1} + \dfrac{5 - 7x}{x - 1}$

b) $\dfrac{x^2 + 4x + 7}{x^2 - 1} - \dfrac{x^2 - 2x + 1}{x^2 - 1}$

Solution

a) $\dfrac{5x - 3}{x - 1} + \dfrac{5 - 7x}{x - 1} = \dfrac{5x - 3 + 5 - 7x}{x - 1}$ Add the numerators.

$$= \frac{-2x + 2}{x - 1} \qquad \text{Combine like terms.}$$

$$= \frac{-2(x - 1)}{x - 1} \qquad \text{Factor.}$$

$$= -2 \qquad \text{Reduce to lowest terms.}$$

b) The polynomials in the numerators are treated as if they were in parentheses.

$$\frac{x^2 + 4x + 7}{x^2 - 1} - \frac{x^2 - 2x + 1}{x^2 - 1} = \frac{x^2 + 4x + 7 - (x^2 - 2x + 1)}{x^2 - 1}$$

$$= \frac{6x + 6}{x^2 - 1} \qquad \begin{array}{l} x^2 - x^2 = 0 \\ 4x - (-2x) = 6x \end{array}$$

$$= \frac{6(x + 1)}{(x + 1)(x - 1)} \qquad \text{Factor.}$$

$$= \frac{6}{x - 1} \qquad \text{Reduce to lowest terms.} \quad \blacktriangleleft$$

To add or subtract rational expressions with denominators that are not identical, we follow the same steps as for rational numbers: first find the LCD, then convert each expression into an equivalent one with that denominator.

EXAMPLE 4 Perform the indicated operations.

a) $\dfrac{3}{a^2 b} + \dfrac{5}{ab^3}$

b) $\dfrac{x + 1}{6} - \dfrac{2x - 3}{4}$

Solution

a) The LCD for $a^2 b$ and ab^3 is $a^2 b^3$. To get this denominator, multiply the first expression by b^2 in its numerator and denominator, and multiply the second expression by a:

$$\frac{3}{a^2 b} + \frac{5}{ab^3} = \frac{3(b^2)}{a^2 b(b^2)} + \frac{5(a)}{ab^3(a)}$$

$$= \frac{3b^2}{a^2 b^3} + \frac{5a}{a^2 b^3}$$

$$= \frac{3b^2 + 5a}{a^2 b^3}$$

b) $\dfrac{x + 1}{6} - \dfrac{2x - 3}{4} = \dfrac{(x + 1)(2)}{6(2)} - \dfrac{(2x - 3)(3)}{4(3)}$ 　Convert each denominator to the LCD 12.

$$= \frac{2x + 2}{12} - \frac{6x - 9}{12}$$

$$= \frac{2x + 2 - (6x - 9)}{12} \qquad \begin{array}{l} \text{Subtract the numerators. Note} \\ 6x - 9 \text{ is put in parentheses.} \end{array}$$

$$= \frac{-4x + 11}{12} \qquad \begin{array}{l} 2x - 6x = -4x \\ 2 - (-9) = 11 \end{array} \qquad \blacktriangleleft$$

EXAMPLE 5 Perform the indicated operations.

a) $\dfrac{1}{x^2 - 1} + \dfrac{2}{x^2 + x}$

b) $\dfrac{5}{a - 2} - \dfrac{3}{2 - a}$

Solution

a) $\dfrac{1}{x^2 - 1} + \dfrac{2}{x^2 + x} = \underbrace{\dfrac{1}{(x - 1)(x + 1)}}_{\text{Missing } x} + \underbrace{\dfrac{2}{x(x + 1)}}_{\text{Missing } x - 1}$ The LCD is
$x(x - 1)(x + 1)$.

$= \dfrac{1(x)}{(x - 1)(x + 1)(x)} + \dfrac{2(x - 1)}{x(x + 1)(x - 1)}$ Make the denominators the same.

$= \dfrac{x}{x(x - 1)(x + 1)} + \dfrac{2x - 2}{x(x - 1)(x + 1)}$

$= \dfrac{3x - 2}{x(x - 1)(x + 1)}$

For this type of answer, we usually leave the denominator in factored form. That way, if we need to work with the expression further, we do not have to factor the denominator again.

b) If we recognize that $-1(2 - a) = a - 2$, we need only convert the second expression to the same denominator as the first.

$\dfrac{5}{a - 2} - \dfrac{3}{2 - a} = \dfrac{5}{a - 2} - \dfrac{3(-1)}{(2 - a)(-1)}$

$= \dfrac{5}{a - 2} - \dfrac{-3}{a - 2}$

$= \dfrac{8}{a - 2}$ $5 - (-3) = 8$

Note that if we had changed the denominator of the first expression to $2 - a$, we would have obtained the answer

$$\dfrac{-8}{2 - a}.$$

This rational expression is equivalent to $\dfrac{8}{a - 2}$.   ◄

Shortcuts

Consider the following addition.

$$\dfrac{a}{b} + \dfrac{c}{d} = \dfrac{a(d)}{b(d)} + \dfrac{c(b)}{d(b)} = \dfrac{ad + cb}{bd} \qquad \text{The LCD is } bd.$$

We can use this result as a rule for adding simple fractions in which the LCD is the product of the denominators.

RULE
Adding Simple Fractions

If $b \neq 0$ and $d \neq 0$, then

$$\frac{a}{b} + \frac{c}{d} = \frac{ad + bc}{bd}.$$

Consider how easily the following sum is obtained by using this rule.

$$\frac{3}{8} + \frac{1}{3} = \frac{3}{8} + \frac{1}{3} = \frac{3 \cdot 3 + 8 \cdot 1}{8 \cdot 3} = \frac{17}{24}$$

If we learn to compute the numerator mentally, then we can easily perform this addition mentally.

A similar rule works for subtraction.

RULE
Subtracting Simple Fractions

If $b \neq 0$ and $d \neq 0$, then

$$\frac{a}{b} - \frac{c}{d} = \frac{ad - bc}{bd}.$$

With practice, we can use these rules to do simple addition and subtraction problems mentally. These rules can be applied to any rational expression, but they work best when the LCD is the product of the two denominators. When using these rules, always make sure the answer is in lowest terms. If the product of the two denominators is too large, these rules are not helpful because then reducing can be complicated.

EXAMPLE 6 Perform the computations mentally.

a) $\dfrac{1}{2} + \dfrac{1}{3}$

b) $\dfrac{1}{a} - \dfrac{1}{x}$

c) $\dfrac{1}{6} - \dfrac{1}{5}$

d) $\dfrac{a}{5} + \dfrac{a}{3}$

e) $x + \dfrac{2}{3}$

f) $\dfrac{1}{4} + \dfrac{1}{6}$

Solution

a) For the numerator, compute $ad + bc = 1 \cdot 3 + 2 \cdot 1 = 5$. Use $2 \cdot 3 = 6$ for the denominator.

$$\frac{1}{2} + \frac{1}{3} = \frac{5}{6}$$

b) $\dfrac{1}{a} - \dfrac{1}{x} = \dfrac{x-a}{ax}$

c) $\dfrac{1}{6} - \dfrac{1}{5} = \dfrac{-1}{30}$

d) $\dfrac{8a}{15}$

e) $x + \dfrac{2}{3} = \dfrac{x}{1} + \dfrac{2}{3} = \dfrac{3x+2}{3}$

f) $\dfrac{1}{4} + \dfrac{1}{6} = \dfrac{10}{24} = \dfrac{5}{12}$ ◀

Warm-ups

True or false?

1. The least common denominator for the denominators 6 and 10 is 60.
2. The least common denominator for $6a^2b$ and $8ab^3$ is $24ab$.
3. The LCD for the denominators $x^2 - 1$ and $x - 1$ is $x^2 - 1$.
4. The LCD for the rational expressions $\dfrac{5}{x}$ and $\dfrac{x-3}{x+1}$ is $x+1$.
5. $\dfrac{1}{2} + \dfrac{2}{3} = \dfrac{3}{5}$
6. $5 + \dfrac{1}{x} = \dfrac{6}{x}$ for any nonzero value of x.
7. $\dfrac{7}{a} + 3 = \dfrac{7+3a}{a}$ for any $a \ne 0$.
8. $\dfrac{c}{3} - \dfrac{d}{5} = \dfrac{5c-3d}{15}$ for any values of c and d.
9. $\dfrac{2}{3} + \dfrac{3}{4} = \dfrac{17}{12}$
10. $\dfrac{3}{m} - \dfrac{5}{q} = \dfrac{5m-3q}{mq}$ for any nonzero values of m and q.

4.3 EXERCISES

Find the least common denominator for each group of denominators. See Example 1.

1. 24, 20
2. 12, 18, 22
3. $a^3b,\ ab^4$
4. $x^2yz,\ xy^2z^3$

5. $x^2 - 1, x^2 + 2x + 1$

6. $x^2 - 9, x^3 + 27$

7. $x, x + 2, x - 2$

8. $y, y - 5, y + 2$

9. $x^2 - 4x, x^2 - 16, x^2 + 6x + 8$

10. $y, y - 3, 3y$

Perform the indicated operations. Reduce answers to lowest terms. See Example 2.

11. $\dfrac{1}{28} + \dfrac{3}{35}$

12. $\dfrac{7}{48} - \dfrac{5}{36}$

13. $\dfrac{7}{24} - \dfrac{4}{15}$

14. $\dfrac{7}{52} + \dfrac{3}{40}$

Perform the indicated operations. Reduce answers to lowest terms. See Example 3.

15. $\dfrac{3x - 4}{2x - 4} + \dfrac{2x - 6}{2x - 4}$

16. $\dfrac{a^3}{a + b} + \dfrac{b^3}{a + b}$

17. $\dfrac{x^2 + 4x - 6}{x^2 - 9} - \dfrac{x^2 + 2x - 12}{x^2 - 9}$

18. $\dfrac{3x - 3}{x - 4} - \dfrac{4x - 7}{x - 4}$

Perform the indicated operations. Reduce answers to lowest terms. See Examples 4 and 5.

19. $\dfrac{x}{2a} + \dfrac{3x}{5a}$

20. $\dfrac{x}{6y} - \dfrac{3x}{8y}$

21. $2x - \dfrac{1}{3}$

22. $\dfrac{5}{a + 2} - \dfrac{7}{a}$

23. $\dfrac{1}{4} + a$

24. $\dfrac{1}{a - b} + \dfrac{2}{a + b}$

25. $\dfrac{2}{x + 1} - \dfrac{3}{x}$

26. $\dfrac{3}{wz^2} + \dfrac{5}{w^2z}$

27. $\dfrac{5}{x + 2} + \dfrac{3}{x - 2}$

28. $\dfrac{9}{4y} - x$

29. $\dfrac{2}{a^2b} - \dfrac{3}{ab^2}$

30. $\dfrac{b^2}{4a} - c$

31. $\dfrac{1}{a - b} + \dfrac{1}{b - a}$

32. $\dfrac{3}{x - 5} + \dfrac{7}{5 - x}$

33. $\dfrac{5}{2x - 4} - \dfrac{3}{2 - x}$

34. $\dfrac{4}{3x - 9} - \dfrac{7}{3 - x}$

35. $\dfrac{x}{x^2 - 9} + \dfrac{3}{x - 3}$

36. $\dfrac{x}{x^2 - 25} + \dfrac{5}{x - 5}$

37. $\dfrac{5}{x^2 + x - 2} + \dfrac{6}{x^2 + 2x - 3}$

38. $\dfrac{2}{x^2 - 4} - \dfrac{5}{x^2 - 3x - 10}$

39. $\dfrac{x}{x^2 - 9} + \dfrac{6}{x^2 + 4x + 3}$

40. $\dfrac{2x - 1}{x^2 - x - 12} + \dfrac{x + 5}{x^2 + 5x + 6}$

41. $\dfrac{1}{x} + \dfrac{2}{x - 1} + \dfrac{3}{x + 2}$

42. $\dfrac{2}{a} - \dfrac{3}{a + 1} + \dfrac{5}{a - 1}$

Perform the following operations mentally. Write down only the answer. See Example 6.

43. $\dfrac{1}{2} + \dfrac{1}{3}$

44. $\dfrac{3}{5} + \dfrac{1}{4}$

45. $\dfrac{1}{8} - \dfrac{3}{5}$

46. $\dfrac{a}{2} + \dfrac{5}{3}$

47. $\dfrac{x}{3} + \dfrac{1}{2}$

48. $\dfrac{1}{4} - \dfrac{2}{3}$

49. $\dfrac{a}{b} - \dfrac{2}{3}$

50. $m + \dfrac{1}{9}$

51. $a + \dfrac{2}{3}$

52. $\dfrac{m}{3} + 1$

53. $\dfrac{3}{a} + 1$

54. $\dfrac{1}{x} + 1$

55. $\dfrac{5}{x} - 2$

56. $\dfrac{1}{x + 2} + \dfrac{3}{x - 2}$

57. $\dfrac{3 + x}{x} - 1$

58. $\dfrac{a+2}{a} + 3$

59. $\dfrac{m}{n} - \dfrac{r}{s}$

60. $\dfrac{1}{a-b} - \dfrac{2}{a}$

61. $\dfrac{2}{3} + \dfrac{1}{4x}$

62. $\dfrac{2}{a+3} + \dfrac{3}{a+2}$

63. $\dfrac{1}{x+2} + \dfrac{2}{x+3}$

64. $\dfrac{1}{5} + \dfrac{1}{5x}$

65. $\dfrac{1}{a-1} + \dfrac{1}{a+1}$

66. $\dfrac{1}{x^3-1} + \dfrac{1}{x^3+1}$

Perform the indicated operations.

67. $\dfrac{x^2-3x}{x^3-1} + \dfrac{4}{x-1}$

68. $\dfrac{a-3}{a^3+8} - \dfrac{2}{a+2} - \dfrac{a-3}{a^2-2a+4}$

69. $\dfrac{w^2-3}{3w^3+81} - \dfrac{2}{6w+18} - \dfrac{w-4}{w^2-3w+9}$

70. $\dfrac{1}{a^3-1} - \dfrac{1}{a^3+1}$

*71. $\dfrac{w^2+3}{w^3-8} - \dfrac{2w}{w^2-4}$

*72. $\dfrac{x+5}{x^3+27} - \dfrac{x-1}{x^2-9}$

*73. $\dfrac{1}{x^3-1} - \dfrac{1}{x^2-1} + \dfrac{1}{x-1}$

*74. $\dfrac{x-4}{x^3-1} + \dfrac{x-2}{x^2-1}$

4.4 Complex Fractions

IN THIS SECTION:

- An Example
- Simplifying Complex Fractions
- Simplifying Expressions with Negative Exponents

In this section we will use the techniques of Section 4.3 to simplify complex fractions. As their name suggests, complex fractions are rather messy looking expressions. To illustrate how complex fractions might arise, we will first consider a problem that can be solved by using a complex fraction.

An Example

Eastside Elementary has the same number of students as Westside Elementary. One-half of the students at Eastside ride buses to school, and two-thirds of the students at Westside ride buses to school. One-sixth of the students at Eastside are female, and one-third of the students at Westside are female. If all of the female students ride the buses, then what percentage of the students who ride the buses are female?

To find the required percentage, we must divide the number of females who ride the buses by the total number of students who ride the buses. Let

x = the number of students at Eastside.

Because the number of students at Westside is also x, we have

$$\frac{1}{2}x + \frac{2}{3}x = \text{the total number of students who ride the buses}$$

and

$$\frac{1}{6}x + \frac{1}{3}x = \text{the total number of female students.}$$

Because all of the female students ride the buses, we can express the percentage of bus riders who are female by the rational expression:

$$\frac{\dfrac{1}{6}x + \dfrac{1}{3}x}{\dfrac{1}{2}x + \dfrac{2}{3}x}$$

This rational expression is called a **complex fraction.** We will simplify it following Example 1.

Simplifying Complex Fractions

A complex fraction is a fraction having rational expressions in the numerator, the denominator, or both. For example,

$$\frac{\dfrac{1}{2} + \dfrac{1}{3}}{\dfrac{1}{4} + \dfrac{1}{5}}, \qquad \frac{3 - \dfrac{2}{x}}{\dfrac{1}{x^2} - \dfrac{1}{4}}, \qquad \text{and} \qquad \frac{\dfrac{x + 2}{x^2 + 6x + 9}}{\dfrac{x}{x^2 - 9} + \dfrac{4}{x + 3}}$$

are complex fractions. We will illustrate two methods for simplifying complex fractions.

EXAMPLE 1 Simplify $\dfrac{\dfrac{1}{2} + \dfrac{1}{3}}{\dfrac{1}{4} + \dfrac{1}{5}}$.

Solution (**Method A**) For this method we perform the computations of the numerator and denominator separately and then divide. Because $\dfrac{1}{2} + \dfrac{1}{3} = \dfrac{5}{6}$ and $\dfrac{1}{4} + \dfrac{1}{5} = \dfrac{9}{20}$, we can write the following.

$$\frac{\dfrac{1}{2} + \dfrac{1}{3}}{\dfrac{1}{4} + \dfrac{1}{5}} = \frac{\dfrac{5}{6}}{\dfrac{9}{20}} = \frac{5}{6} \cdot \frac{20}{9} = \frac{5 \cdot 2 \cdot 10}{2 \cdot 3 \cdot 9} = \frac{50}{27}$$

(Method B) The LCD of the denominators 2, 3, 4, and 5 is 60. We can multiply the numerator and denominator of the complex fraction by 60 to eliminate the fractions as follows:

$$\frac{\frac{1}{2}+\frac{1}{3}}{\frac{1}{4}+\frac{1}{5}} = \frac{\left(\frac{1}{2}+\frac{1}{3}\right)60}{\left(\frac{1}{4}+\frac{1}{5}\right)60} = \frac{30+20}{15+12} = \frac{50}{27}$$

$(1/2)60 = 30$
$(1/3)60 = 20$
$(1/4)60 = 15$
$(1/5)60 = 12$

◄

In most cases method B is the faster method, and we will continue to use it. We will use method B to simplify the complex fraction that we obtained above to find what percentage of the bus riders are female. The LCD for the denominators 2, 3, and 6 is 6.

$$\frac{\left(\frac{1}{6}x+\frac{1}{3}x\right)6}{\left(\frac{1}{2}x+\frac{2}{3}x\right)6} = \frac{x+2x}{3x+4x} = \frac{3x}{7x} = \frac{3}{7} \approx .43, \text{ or } 43\%$$

Thus 43% of the students who ride the buses are female.

EXAMPLE 2 Simplify $\dfrac{3-\dfrac{2}{x}}{\dfrac{1}{x^2}-\dfrac{1}{4}}$.

Solution The LCD of x, x^2, and 4 is $4x^2$.

$$\frac{3-\dfrac{2}{x}}{\dfrac{1}{x^2}-\dfrac{1}{4}} = \frac{\left(3-\dfrac{2}{x}\right)(4x^2)}{\left(\dfrac{1}{x^2}-\dfrac{1}{4}\right)(4x^2)}$$ Multiply the numerator and denominator by $4x^2$.

$$= \frac{3(4x^2)-\dfrac{2}{x}(4x^2)}{\dfrac{1}{x^2}(4x^2)-\dfrac{1}{4}(4x^2)}$$

$$= \frac{12x^2-8x}{4-x^2}$$

◄

EXAMPLE 3 Simplify $\dfrac{\dfrac{x+2}{x^2+6x+9}}{\dfrac{x}{x^2-9}+\dfrac{4}{x+3}}$.

Solution

$$\frac{\dfrac{x+2}{(x+3)^2}}{\dfrac{x}{(x+3)(x-3)} + \dfrac{4}{x+3}}$$

Factor all denominators.
The LCD is $(x + 3)^2(x - 3)$.

$$= \frac{\dfrac{x+2}{(x+3)^2}(x+3)^2(x-3)}{\dfrac{x}{(x+3)(x-3)}(x+3)^2(x-3) + \dfrac{4}{x+3}(x+3)^2(x-3)}$$

Multiply the numerator and denominator by the LCD.

$$= \frac{(x+2)(x-3)}{x(x+3) + 4(x+3)(x-3)}$$

Simplify.

$$= \frac{(x+2)(x-3)}{(x+3)[x + 4(x-3)]}$$

$$= \frac{(x+2)(x-3)}{(x+3)(5x-12)}$$

◀

Simplifying Expressions with Negative Exponents

Sometimes complex fractions are written with negative exponents. For example,

$$\frac{\dfrac{1}{a} - \dfrac{1}{2}}{1 - \dfrac{1}{b}} = \frac{a^{-1} - 2^{-1}}{1 - b^{-1}}.$$

We can simplify this expression by multiplying the numerator and denominator by $2ab$, the LCD of the fractions. Remember that $a^{-1} \cdot a = a^0 = 1$.

$$\frac{a^{-1} - 2^{-1}}{1 - b^{-1}} = \frac{(a^{-1} - 2^{-1})2ab}{(1 - b^{-1})2ab} = \frac{a^{-1}(2ab) - 2^{-1}(2ab)}{1(2ab) - b^{-1}(2ab)} = \frac{2b - ab}{2ab - 2a}$$

EXAMPLE 4 Eliminate negative exponents and simplify $\dfrac{a^{-1} + b^{-2}}{ab^{-2} + ba^{-3}}$.

Solution The variables with negative exponents could be placed back in denominators with positive exponents as just shown. If we did that, the denominators would be a, b^2, b^2, and a^3, and the LCD would be a^3b^2. We can eliminate the negative exponents and simplify, if we multiply the numerator and denominator by a^3b^2.

$$\frac{a^{-1} + b^{-2}}{ab^{-2} + ba^{-3}} = \frac{(a^{-1} + b^{-2})a^3b^2}{(ab^{-2} + ba^{-3})a^3b^2}$$

$$= \frac{a^{-1}(a^3b^2) + b^{-2}(a^3b^2)}{ab^{-2}(a^3b^2) + ba^{-3}(a^3b^2)}$$

$$= \frac{a^2b^2 + a^3}{a^4 + b^3} \qquad \begin{array}{l} b^{-2}b^2 = b^0 = 1 \\ a^{-3}a^3 = a^0 = 1 \end{array}$$

Note that the positive exponents of a^3b^2 are just large enough to cancel out all of the negative exponents when we multiply. ◀

The next example is not exactly a complex fraction, but we can use the same technique as in the previous example.

EXAMPLE 5 Eliminate negative exponents and simplify $p + p^{-1}q^{-2}$.

Solution We can perform the addition and eliminate the negative exponents at the same time if we multiply the numerator and denominator by pq^2:

$$p + p^{-1}q^{-2} = \frac{(p + p^{-1}q^{-2})}{1} \cdot \frac{pq^2}{pq^2}$$

$$= \frac{p^2q^2 + 1}{pq^2} \qquad \begin{array}{l} p \cdot pq^2 = p^2q^2 \\ p^{-1}q^{-2} \cdot pq^2 = 1 \end{array} \qquad ◀$$

Warm-ups

True or false?

1. The LCD for the denominators 2, x, 6, and x^2 is $6x^3$.

2. The LCD for the denominators $a - b$, $2b - 2a$, and 6 is $6a - 6b$.

3. To simplify the complex fraction

$$\frac{\dfrac{1}{a} + \dfrac{2}{b}}{\dfrac{2}{a} + \dfrac{3}{b}}$$

by using method B, we multiply numerator and denominator by $\dfrac{ab}{ab}$.

4. $\dfrac{\dfrac{1}{2} + \dfrac{1}{3}}{1 + \dfrac{1}{2}} = \dfrac{5}{6} \div \dfrac{3}{2}$

5. $2^{-1} + 3^{-1} = (2 + 3)^{-1}$

6. $(2^{-1} + 3^{-1})^{-1} = 2 + 3$

7. $2 + 3^{-1} = 5^{-1}$

8. $x + 2^{-1} = \dfrac{x}{2}$ for any value of x.

9. To simplify $\dfrac{a^{-1} - b^{-1}}{a - b}$, we multiply numerator and denominator by ab.

10. To simplify $\dfrac{ab^{-2} + a^{-5}b^2}{a^{-3}b - a^5b^{-1}}$, we multiply numerator and denominator by a^5b^2.

4.4 EXERCISES

Simplify the following complex fractions. Use either method A or method B. See Example 1.

1. $\dfrac{\dfrac{1}{2} - \dfrac{1}{3}}{\dfrac{1}{4} - \dfrac{1}{5}}$

2. $\dfrac{\dfrac{1}{3} + \dfrac{1}{4}}{\dfrac{1}{5} + \dfrac{1}{6}}$

3. $\dfrac{\dfrac{2}{3} + \dfrac{5}{6} - \dfrac{1}{2}}{\dfrac{1}{8} - \dfrac{1}{3} + \dfrac{1}{12}}$

4. $\dfrac{\dfrac{2}{5} - \dfrac{x}{9} - \dfrac{1}{3}}{\dfrac{1}{3} + \dfrac{x}{5} + \dfrac{2}{15}}$

Simplify the complex fractions. Use method B. See Example 2.

5. $\dfrac{\dfrac{a + b}{b}}{\dfrac{a - b}{ab}}$

6. $\dfrac{\dfrac{m - n}{m^2}}{\dfrac{m - 3}{mn^3}}$

7. $\dfrac{a + \dfrac{3}{b}}{\dfrac{b}{a} + \dfrac{1}{b}}$

8. $\dfrac{m - \dfrac{2}{n}}{\dfrac{1}{m} - \dfrac{3}{n}}$

9. $\dfrac{\dfrac{x - 3y}{xy}}{\dfrac{1}{x} + \dfrac{1}{y}}$

10. $\dfrac{\dfrac{2}{w} + \dfrac{3}{t}}{\dfrac{w - t}{4wt}}$

11. $\dfrac{3 - \dfrac{m - 2}{6}}{\dfrac{4}{9} + \dfrac{2}{m}}$

12. $\dfrac{6 - \dfrac{2 - z}{z}}{\dfrac{1}{3z} - \dfrac{1}{6}}$

13. $\dfrac{\dfrac{a^2 - b^2}{a^2b^3}}{\dfrac{a + b}{a^3b}}$

14. $\dfrac{\dfrac{4x^2 - 1}{x^2y}}{\dfrac{4x - 2}{xy^2}}$

Simplify the following. See Examples 1, 2, and 3.

15. $\dfrac{x + \dfrac{4}{x + 4}}{x - \dfrac{4x + 4}{x + 4}}$

16. $\dfrac{x - \dfrac{x + 6}{x + 2}}{x - \dfrac{4x + 15}{x + 2}}$

17. $\dfrac{1 - \dfrac{1}{y - 1}}{3 + \dfrac{1}{y + 1}}$

18. $\dfrac{2 - \dfrac{3}{a-2}}{4 - \dfrac{1}{a+2}}$

19. $\dfrac{\dfrac{2}{3-x} - 4}{\dfrac{1}{x-3} - 1}$

20. $\dfrac{\dfrac{x}{x-5} - 2}{\dfrac{2x}{5-x} - 1}$

21. $\dfrac{\dfrac{w+2}{w-1} - \dfrac{w-3}{w}}{\dfrac{w+4}{w} + \dfrac{w-2}{w-1}}$

22. $\dfrac{\dfrac{x-1}{x+2} - \dfrac{x-2}{x+3}}{\dfrac{x-3}{x+3} + \dfrac{x+1}{x+2}}$

23. $\dfrac{\dfrac{1}{a-b} - \dfrac{3}{a+b}}{\dfrac{2}{b-a} + \dfrac{4}{b+a}}$

24. $\dfrac{\dfrac{3}{2+x} - \dfrac{4}{2-x}}{\dfrac{1}{x+2} - \dfrac{3}{x-2}}$

25. $\dfrac{3 - \dfrac{4}{a-1}}{5 - \dfrac{3}{1-a}}$

26. $\dfrac{\dfrac{x}{3} - \dfrac{x-1}{9-x}}{\dfrac{x}{6} - \dfrac{2-x}{x-9}}$

27. $\dfrac{\dfrac{2}{m-3} + \dfrac{4}{m}}{\dfrac{3}{m-2} + \dfrac{1}{m}}$

28. $\dfrac{\dfrac{1}{y+2} - \dfrac{4}{3y}}{\dfrac{3}{y} - \dfrac{2}{y+3}}$

29. $\dfrac{\dfrac{3}{x^2-1} - \dfrac{x-2}{x^3-1}}{\dfrac{3}{x^2+x+1} + \dfrac{x-3}{x^3-1}}$

30. $\dfrac{\dfrac{2}{a^3+8} - \dfrac{3}{a^2-2a+4}}{\dfrac{4}{a^2-4} + \dfrac{a-3}{a^3+8}}$

Simplify. See Examples 4 and 5.

31. $\dfrac{1 - x^{-1}}{1 - x^{-2}}$

32. $\dfrac{4 - a^{-2}}{2 - a^{-1}}$

33. $\dfrac{a^{-2} + b^{-2}}{a^{-1}b}$

34. $\dfrac{m^{-3} + n^{-3}}{mn^{-2}}$

35. $\dfrac{w^{-1} + y^{-1}}{z^{-1} + y^{-1}}$

36. $\dfrac{a^{-1} - b^{-1}}{a^{-1} + b^{-1}}$

37. $\dfrac{1 - 8x^{-3}}{x^{-1} + 2x^{-2} + 4x^{-3}}$

38. $\dfrac{a + 27a^{-2}}{1 - 3a^{-1} + 9a^{-2}}$

39. $\dfrac{x^{-1} + x^{-2}}{x + x^{-2}}$

40. $\dfrac{x - x^{-2}}{1 - x^{-2}}$

41. $\dfrac{2m^{-1} - 3m^{-2}}{m^{-2}}$

42. $\dfrac{4x^{-3} - 6x^{-5}}{2x^{-5}}$

43. $1 - a^{-1}$

44. $m^{-1} - a^{-1}$

45. $(x^{-1} + y^{-1})^{-1}$

46. $(a^{-1} - b^{-1})^{-2}$

47. $\dfrac{(a-b)^2}{a^{-2} - b^{-2}}$

48. $\dfrac{x^3 - y^3}{x^{-3} - y^{-3}}$

49. $\dfrac{a^{-1} - b^{-1}}{a - b}$

50. $\dfrac{a^2 - b^2}{a^{-2} - b^{-2}}$

Solve each problem.

51. Clarksville has three elementary schools. Northside has one-half as many students as Central, and Southside has two-thirds as many students as Central. One-third of the students at Northside are black, three-fourths of the stu-

dents at Central are black, and one-sixth of the students at Southside are black. What percentage of the city's elementary students are black?

52. Ramona drove from Clarksville to Leesville at 45 miles per hour. At Leesville she discovered she had forgotten her

purse and immediately returned to Clarksville at 55 miles per hour. What was her average speed for the trip? (The answer is *not* 50 miles per hour.)

4.5 Solving Equations Involving Rational Expressions

IN THIS SECTION:

- Multiplying by the LCD
- Extraneous Roots
- Proportions

Many problems in algebra require us to solve equations involving rational expressions. In this section we will learn how to solve equations that involve rational expressions, and in Section 4.6 we will solve problems using these equations.

Multiplying by the LCD

To solve equations involving rational expressions, we multiply each side of the equation by the LCD of the rational expressions.

EXAMPLE 1   Solve $\dfrac{1}{x} + \dfrac{1}{4} = \dfrac{1}{6}$.

Solution   The LCD for the denominators 4, 6, and x is $12x$.

$$12x\left(\frac{1}{x} + \frac{1}{4}\right) = 12x\left(\frac{1}{6}\right) \qquad \text{Multiply each side by } 12x.$$

$$12x\left(\frac{1}{x}\right) + \overset{3}{12}x\left(\frac{1}{\underset{1}{4}}\right) = \overset{2}{12}x\left(\frac{1}{\underset{}{6}}\right) \qquad \text{Distributive property}$$

$$12 + 3x = 2x \qquad \text{All denominators are eliminated.}$$

$$12 + x = 0$$

$$x = -12$$

The solution set is $\{-12\}$. Check this solution in the original equation.   ◀

EXAMPLE 2   Solve the equation $\dfrac{1}{x} + \dfrac{2}{3x} = \dfrac{1}{5}$.

Solution

$$15x\left(\frac{1}{x} + \frac{2}{3x}\right) = 15x\left(\frac{1}{5}\right) \qquad \text{Multiply each side by the LCD, } 15x.$$

$$15x \cdot \frac{1}{x} + 15x \cdot \frac{2}{3x} = 3x$$

$$15 + 10 = 3x$$

$$25 = 3x$$

$$\frac{25}{3} = x$$

The solution set is $\{25/3\}$. Check. ◄

EXAMPLE 3 Solve the equation $\dfrac{200}{x} + \dfrac{300}{x + 20} = 10$.

Solution

$$x(x + 20)\left(\frac{200}{x} + \frac{300}{x + 20}\right) = x(x + 20)10 \qquad \begin{array}{l}\text{Multiply each side}\\ \text{by } x(x + 20).\end{array}$$

$$\cancel{x}(x + 20)\frac{200}{\cancel{x}} + x\cancel{(x + 20)}\frac{300}{\cancel{x + 20}} = x(x + 20)10 \qquad \text{Distributive property}$$

$$(x + 20)200 + x(300) = (x^2 + 20x)10 \qquad \text{Simplify.}$$

$$200x + 4000 + 300x = 10x^2 + 200x$$

$$4000 + 300x = 10x^2 \qquad \text{Combine like terms.}$$

$$400 + 30x = x^2 \qquad \text{Divide each side by 10.}$$

$$0 = x^2 - 30x - 400$$

$$0 = (x - 40)(x + 10) \qquad \text{Factor.}$$

$$x - 40 = 0 \quad \text{or } x + 10 = 0 \qquad \text{Set each factor equal to 0.}$$

$$x = 40 \text{ or} \qquad x = -10$$

The solution set is $\{-10, 40\}$. Check these roots (or solutions) in the original equation. ◄

Extraneous Roots

A rational expression is not necessarily defined for all real numbers. Any number that causes the denominator to be zero is not in the domain of the rational expression. We must check every solution to see whether it causes a denominator to be zero. If it does, it is not a solution to the equation. Roots that do not check are called **extraneous roots.**

EXAMPLE 4 Solve the equation $\dfrac{3}{x} + \dfrac{6}{x - 2} = \dfrac{12}{x^2 - 2x}$.

Solution   Because $x^2 - 2x = x(x - 2)$, the LCD for x, $x - 2$, and $x^2 - 2x$ is $x(x - 2)$.

$$x(x - 2)\frac{3}{x} + x(x - 2)\frac{6}{x - 2} = x(x - 2)\frac{12}{x(x - 2)}$$ Multiply each side by $x(x - 2)$.

$$3(x - 2) + 6x = 12$$
$$3x - 6 + 6x = 12$$
$$9x - 6 = 12$$
$$9x = 18$$
$$x = 2$$

The domain of the rational expression $\dfrac{6}{x - 2}$ is the set of all real numbers except 2.

Math at Work

Need a light? Flip a switch. Want to watch TV? Turn the dial. Want to reheat dinner? Turn on the microwave.

Every day electricity performs hundreds of tasks for us dependably and cheaply. We take it for granted without questioning how it works or whether it will be there when we need it. Yet the way electricity works to bring us light, heat, and energy does not need to be a mystery.

Home lighting—like other electrically-powered appliances—results when an electrical circuit connects an energy source, such as the local power plant, with an energy receiver, such as a light bulb. When electricity flows through the circuit, the receiver resists some of the electrical current from passing through. If all of the receivers in a house were strung together like Christmas-tree lights, turning on one appliance would decrease the energy being received by other appliances.

To avoid this problem, electricians connect each receiver in parallel circuits. The electricity starts and ends at the same source but follows a separate path to each receiver. In a parallel circuit, the total resistance of the circuit is given by the formula

$$\frac{1}{R} = \frac{1}{R_1} + \frac{1}{R_2} + \cdots$$

Assume that an electrician is wiring a room that will have two receivers connected by a parallel circuit. If the resistance of the first is 5 ohms, and the resistance of the second is 10 ohms, find the total resistance.

One year later, the homeowner decides to add a third receiver in the room. If the resistance of the third receiver is 6 ohms, then what is the total resistance of the circuit?

So 2 is an extraneous root. Substituting 2 for x in the original equation would give us 0 in a denominator. The solution set is the empty set, $\varnothing$.   ◄

EXAMPLE 5   Solve the equation $x + 2 + \dfrac{x}{x - 2} = \dfrac{2}{x - 2}$.

Solution $(x - 2)(x + 2) + (x - 2)\dfrac{x}{x - 2} = (x - 2)\dfrac{2}{x - 2}$ Multiply each side by $x - 2$.

$$x^2 - 4 + x = 2$$
$$x^2 + x - 6 = 0$$
$$(x + 3)(x - 2) = 0$$
$$x + 3 = 0 \quad \text{or } x - 2 = 0$$
$$x = -3 \text{ or} \qquad x = 2$$

Check that -3 satisfies the original equation and 2 is an extraneous root. The solution set is $\{-3\}$.   ◄

Proportions

An equation that expresses the equality of two rational expressions is called a **proportion.** The equation

$$\frac{a}{b} = \frac{c}{d}$$

is a proportion. The terms in the position of b and c are called the **means.** The terms in the position of a and d are called the **extremes.** If we multiply this proportion by the LCD, bd, we get

$$bd \cdot \frac{a}{b} = bd \cdot \frac{c}{d},$$

or

$$ad = bc.$$

The equation $ad = bc$ says that *the product of the extremes is equal to the product of the means.* When solving a proportion, we can omit multiplication by the LCD and just remember the result, $ad = bc,$ as the extremes-means property.

Extremes-means Property

If $\dfrac{a}{b} = \dfrac{c}{d},$ then $ad = bc.$

The extremes-means property makes it easier to solve proportions.

EXAMPLE 6   Solve $\dfrac{20}{x} = \dfrac{30}{x+20}$.

Solution

$$20(x+20) = 30x \qquad \text{The extremes-means property}$$
$$20x + 400 = 30x$$
$$400 = 10x$$
$$40 = x$$

The solution set is {40}. Check in the original equation.   ◄

EXAMPLE 7   Solve $\dfrac{2}{x} = \dfrac{x+3}{5}$.

Solution

$$x^2 + 3x = 10 \qquad \text{The extremes-means property}$$
$$x^2 + 3x - 10 = 0$$
$$(x+5)(x-2) = 0$$
$$x + 5 = 0 \quad \text{or} \quad x - 2 = 0$$
$$x = -5 \text{ or} \qquad x = 2$$

The solution set is {−5, 2}. Check in the original equation.   ◄

Warm-ups

True or false?

1. The first step in solving an equation involving rational expressions is to multiply each side by the LCD for all of the denominators.

2. To solve $\dfrac{1}{x} + \dfrac{1}{2x} = \dfrac{1}{3}$, we first change each rational expression to an equivalent rational expression with a denominator of $6x$.

3. Extraneous roots are not real numbers.

4. To solve $\dfrac{1}{x-2} + 3 = \dfrac{1}{x+2}$, we multiply each side by $x^2 - 4$.

5. The solution set to $\dfrac{x}{3x+4} - \dfrac{6}{2x+1} = \dfrac{7}{5}$ is {7/8, −1/2}.

6. The solution set to $\dfrac{3}{x} = \dfrac{2}{5}$ is {15/2}.

7. We should use the extremes-means property on the equation
$$\frac{x-2}{x+3} + 1 = \frac{5}{x-4}.$$

8. The equation $x^2 = 5x$ is equivalent to the equation $x = 5$.

9. The solution set to $(2x - 3)(3x + 4) = 0$ is $\{3/2, 4/3\}$.

10. The equation $\dfrac{2}{x+1} = \dfrac{x-1}{4}$ is equivalent to $x^2 - 1 = 8$.

4.5 EXERCISES

Find the solution set to each equation. See Examples 1–5.

1. $\dfrac{1}{x} + \dfrac{1}{4} = \dfrac{1}{6}$

2. $\dfrac{3}{x} + \dfrac{1}{5} = \dfrac{1}{2}$

3. $\dfrac{2}{3x} + \dfrac{1}{15x} = \dfrac{1}{2}$

4. $\dfrac{5}{6x} - \dfrac{1}{8x} = \dfrac{17}{24}$

5. $\dfrac{3}{x-2} + \dfrac{5}{x} = \dfrac{10}{x}$

6. $\dfrac{5}{x-1} + \dfrac{1}{2x} = \dfrac{1}{x}$

7. $\dfrac{x}{x-2} + \dfrac{3}{x} = 2$

8. $\dfrac{x}{x-5} + \dfrac{5}{x} = \dfrac{11}{6}$

9. $\dfrac{3x-5}{x-1} = 2 - \dfrac{2x}{x-1}$

10. $\dfrac{x-3}{x+2} = 3 - \dfrac{5-2x}{x+2}$

11. $\dfrac{100}{x} = \dfrac{150}{x+5} - 1$

12. $\dfrac{30}{x} = \dfrac{50}{x+10} + \dfrac{1}{2}$

Find the solution set to each equation. See Examples 6 and 7.

13. $\dfrac{2}{x} = \dfrac{3}{4}$

14. $\dfrac{5}{x} = \dfrac{7}{9}$

15. $\dfrac{a}{3} = \dfrac{-1}{4}$

16. $\dfrac{b}{5} = \dfrac{-3}{7}$

17. $-\dfrac{5}{7} = \dfrac{2}{x}$

18. $-\dfrac{3}{8} = \dfrac{5}{x}$

19. $\dfrac{10}{x} = \dfrac{20}{x+20}$

20. $\dfrac{x}{5} = \dfrac{x+2}{3}$

21. $\dfrac{x}{x-3} = \dfrac{x+2}{x}$

22. $\dfrac{x+1}{x-5} = \dfrac{x+2}{x-4}$

23. $\dfrac{x-2}{x-3} = \dfrac{x+5}{x+2}$

24. $\dfrac{x}{x+5} = \dfrac{x}{x-2}$

25. $\dfrac{2}{x+1} = \dfrac{x-1}{4}$

26. $\dfrac{3}{x-2} = \dfrac{x+2}{7}$

27. $\dfrac{x}{6} = \dfrac{5}{x-1}$

28. $\dfrac{x+5}{2} = \dfrac{3}{x}$

Solve each equation.

29. $\dfrac{a}{9} = \dfrac{4}{a}$

30. $\dfrac{y}{3} = \dfrac{27}{y}$

31. $\dfrac{1}{2x-4} + \dfrac{1}{x-2} = \dfrac{1}{4}$

32. $\dfrac{7}{3x-9} - \dfrac{1}{x-3} = \dfrac{4}{9}$

33. $\dfrac{x-2}{4} = \dfrac{x-2}{x}$

34. $\dfrac{y+5}{2} = \dfrac{y+5}{y}$

35. $\dfrac{5}{2x+4} - \dfrac{1}{x-1} = \dfrac{3}{x+2}$

36. $\dfrac{5}{2w+6} - \dfrac{1}{w-1} = \dfrac{1}{w+3}$

37. $\dfrac{5}{x-3} = \dfrac{x}{x-3}$

38. $\dfrac{6}{a+2} = \dfrac{a}{a+2}$

39. $\dfrac{w}{6} = \dfrac{3}{2w}$

40. $\dfrac{2m}{5} = \dfrac{10}{m}$

41. $\dfrac{5}{4x-2} - \dfrac{1}{1-2x} = \dfrac{7}{3x+6}$

42. $\dfrac{5}{x+1} - \dfrac{1}{1-x} = \dfrac{1}{x^2-1}$

43. $\dfrac{5}{x} = \dfrac{2}{5}$

44. $\dfrac{-3}{2x} = \dfrac{1}{-5}$

45. $\dfrac{5}{x^2-9} + \dfrac{2}{x+3} = \dfrac{1}{x-3}$

46. $\dfrac{1}{x-2} - \dfrac{2}{x+3} = \dfrac{11}{x^2+x-6}$

*47. $\dfrac{9}{x^3-1} - \dfrac{1}{x-1} = \dfrac{2}{x^2+x+1}$

*48. $\dfrac{x+4}{x^3+8} + \dfrac{x+2}{x^2-2x+4} = \dfrac{11}{2x+4}$

4.6 Applications

IN THIS SECTION:

- **Formulas**
- **Uniform-motion Problems**
- **Work Problems**
- **Miscellaneous Problems**

In this section we will use the techniques of Section 4.5 to rewrite formulas involving rational expressions and to solve some word problems.

Formulas

Rewriting formulas involving rational expressions is similar to solving equations involving rational expressions. Generally, the first step is to multiply each side by the LCD for the rational expressions.

EXAMPLE 1   The formula

$$\frac{y-y_1}{x-x_1} = m$$

is the point-slope form of the equation of a line. (We will study lines in detail in Chapter 7.) Solve the formula for y.

Solution

$$(x-x_1)\frac{y-y_1}{x-x_1} = (x-x_1)m \qquad \text{Multiply each side by the denominator } x-x_1.$$

$$y - y_1 = (x - x_1)m \qquad \text{Reduce.}$$

$$y = (x - x_1)m + y_1$$

◄

In the next example we are solving for a variable that occurs on both sides of the equation. Remember that when a formula is solved for a certain variable, that variable appears only once in the formula.

EXAMPLE 2 The formula

$$\frac{P}{P_w} = \frac{2L}{2L + d}$$

is used in physics to find the relative density of a substance. Solve it for L.

Solution

$$P(2L + d) = P_w(2L) \quad \text{The extremes-means property}$$

$$2PL + Pd = 2LP_w \quad \text{Simplify.}$$

$$Pd = 2LP_w - 2PL \quad \text{Get all terms involving } L \text{ onto the same side.}$$

$$Pd = (2P_w - 2P)L \quad \text{Factor out } L.$$

$$\frac{Pd}{2P_w - 2P} = L \qquad \blacktriangleleft$$

In the next example we find the value of one variable when given the values of the remaining variables.

EXAMPLE 3 In the formula

$$\frac{y - y_1}{x - x_1} = m,$$

let $x_1 = 2$, $y_1 = -3$, $y = -1$, and $m = 1/2$. Find x.

Solution Substitute all of the values into the formula and solve for x.

$$\frac{-1 - (-3)}{x - 2} = \frac{1}{2}$$

$$\frac{2}{x - 2} = \frac{1}{2}$$

$$x - 2 = 4 \quad \text{The extremes-means property}$$

$$x = 6 \quad \text{Check in the original formula.} \qquad \blacktriangleleft$$

Uniform-motion Problems

The uniform-motion problems that we will solve here are similar to those of Chapter 2. In this chapter, however, the problems are solved by using equations with rational expressions.

EXAMPLE 4 Michelle drove her empty rig 300 miles to Salina to pick up a load of cattle. When her rig was fully loaded, she averaged 10 miles per hour less than

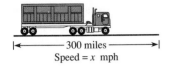

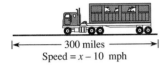

Figure 4.1

when the rig was empty. If the return trip took her 1 hour longer, then what was her average speed with the rig empty? (See Fig. 4.1).

Solution Let x be her average speed empty and $x - 10$ be her average speed full. Because the time can be determined from the distance and the rate, $T = D/R$, we can make the following table.

	Distance	Rate	Time
Empty	300	x	$\dfrac{300}{x}$
Full	300	$x - 10$	$\dfrac{300}{x - 10}$

We now write an equation expressing the fact that

the time empty = the time full − one hour.

$$\frac{300}{x} = \frac{300}{x - 10} - 1$$

$$x(x - 10)\frac{300}{x} = x(x - 10)\frac{300}{x - 10} - x(x - 10)1 \qquad \text{Multiply each side by } x(x - 10).$$

$$300x - 3000 = 300x - x^2 + 10x \qquad \text{Reduce.}$$

$$-3000 = -x^2 + 10x$$

$$x^2 - 10x - 3000 = 0 \qquad \text{Get 0 on one side.}$$

$$(x + 50)(x - 60) = 0 \qquad \text{Factor.}$$

$$x + 50 = 0 \qquad \text{or} \qquad x - 60 = 0 \qquad \text{The zero factor property}$$

$$x = -50 \qquad \text{or} \qquad x = 60$$

The answer $x = -50$ is a solution to the equation, but because it is negative, it does not indicate the speed of the truck. Michelle's average speed empty was 60 mph. Checking this answer, we find that if she traveled 300 miles at 60 miles per hour, it would take her 5 hours. If she traveled 300 miles at 50 miles per hour with the loaded rig, it would take her 6 hours. Because Michelle's time with the empty rig was 1 hour less than her time with the loaded rig, 60 miles per hour is the correct answer.   ◀

Work Problems

Problems involving different rates for completing a task are referred to as work problems. Work problems were not introduced before now because they are solved only with equations involving rational expressions.

EXAMPLE 5 Linda can mow the lawn with her riding lawn mower in 4 hours. When Linda uses the riding mower and Rebecca operates the push mower, it takes them 3 hours to mow the lawn. How long would it take Rebecca to mow the lawn by herself using the push mower?

Solution Let x be the number of hours it would take Rebecca working alone. Because Linda can mow the entire lawn in 4 hours, Linda mows 1/4 of the lawn per hour, whether she works alone or with Rebecca. Rebecca mows $1/x$ of the lawn per hour, whether working alone or with Linda. Together they mow 1/3 of the lawn per hour. We can classify all of the necessary information in the following table.

	Linda	Rebecca	Together
Time	4	x	3
Portion done per hour	$\dfrac{1}{4}$	$\dfrac{1}{x}$	$\dfrac{1}{3}$

Assuming that they do not interfere with one another, the portions they each do in one hour must total the portion they do together in one hour.

$$\frac{1}{4} + \frac{1}{x} = \frac{1}{3}$$

$$12x\frac{1}{4} + 12x\frac{1}{x} = 12x\frac{1}{3}$$

$$3x + 12 = 4x$$

$$12 = x$$

If it takes Rebecca 12 hours to mow the lawn by herself, then in the 3 hours that they work together, Rebecca mows 3/12, or 1/4, of the lawn. Because Linda can mow the lawn by herself in 4 hours, in the 3 hours that they work together she mows 3/4 of the lawn. Because 1/4 + 3/4 = 1, the lawn is completed in 3 hours. Thus we can be certain that it would take Rebecca 12 hours to mow the lawn by herself using the push mower.   ◀

Miscellaneous Problems

EXAMPLE 6 Patrick bought 50 pounds of meat consisting of hamburger and steak. Steak costs twice as much per pound as hamburger. If Patrick got $30 worth of hamburger and $90 worth of steak, then how many pounds of each did he buy?

Solution Let x be the number of pounds of hamburger and $50 - x$ be the number of pounds of steak. We can classify all of the given information in a table.

	Number of pounds	Total price	Price per pound
Hamburger	x	30	$\dfrac{30}{x}$
Steak	$50 - x$	90	$\dfrac{90}{50 - x}$

The following equation says that the price per pound of steak is twice that of hamburger.

$$2\left(\frac{30}{x}\right) = \frac{90}{50 - x}$$

$$\frac{60}{x} = \frac{90}{50 - x}$$

$$90x = 3000 - 60x \qquad \textbf{The extremes-means property}$$

$$150x = 3000$$

$$x = 20$$

$$50 - x = 30$$

Patrick purchased 20 pounds of hamburger and 30 pounds of steak. Check this answer.   ◄

Warm-ups

True or false?

1. The formula $w = \dfrac{1 - t}{t}$, solved for t, is $t = \dfrac{1 - t}{w}$.

2. To solve $\dfrac{1}{p} + \dfrac{1}{q} = \dfrac{1}{s}$ for s, we multiply each side by pqs.

3. If 50 pounds of steak costs x dollars, then the price is $50/x$ dollars per pound.

4. If Claudia drives x miles in 3 hours, then her rate is $x/3$ miles per hour.

5. If Takenori mows his entire lawn in $x + 2$ hours, then he mows $1/(x + 2)$ of the lawn per hour.

6. If Kareem drives 200 nails in 12 hours, then he is driving $200/12$ nails per hour.

7. If $200/x$ hours is 1 hour less than $200/(x - 3)$ hours, then this relationship is described by the equation $\dfrac{200}{x} - 1 = \dfrac{200}{x - 3}$.

8. If $A = \dfrac{mv^2}{B}$ and m and B are nonzero, then $v^2 = \dfrac{AB}{m}$.

9. If a and y are nonzero, and $a = \dfrac{x}{y}$, then $y = ax$.

10. If $500/x$ hours is 3 hours more than $500/(x + 20)$ hours, then this relationship is described by the equation $\dfrac{500}{x} + 3 = \dfrac{500}{x + 20}$.

4.6 EXERCISES

Solve each equation for y. See Example 1.

1. $\dfrac{y-3}{x-2} = 5$

2. $\dfrac{y-4}{x-7} = -6$

3. $\dfrac{y+1}{x-6} = -\dfrac{1}{3}$

4. $\dfrac{y+7}{x-2} = \dfrac{-2}{3}$

5. $\dfrac{y-a}{x-b} = m$

6. $\dfrac{y-h}{x-k} = a$

7. $\dfrac{y-2}{x+5} = -\dfrac{7}{3}$

8. $\dfrac{y-3}{x+1} = -\dfrac{9}{4}$

Solve each formula for the indicated variable. See Example 2.

9. $M = \dfrac{F}{f}$ for f

10. $P = \dfrac{A}{1+rt}$ for A

11. $A = \dfrac{\pi}{4} \cdot D^2$ for D^2

12. $V = \pi r^2 h$ for r^2

13. $F = k\dfrac{m_1 m_2}{r^2}$ for r^2

14. $F = \dfrac{mv^2}{r}$ for v^2

15. $\dfrac{1}{p} + \dfrac{1}{q} = \dfrac{1}{f}$ for q

16. $\dfrac{1}{R} = \dfrac{1}{R_1} + \dfrac{1}{R_2}$ for R_1

17. $e^2 = 1 - \dfrac{b^2}{a^2}$ for a^2

18. $e^2 = 1 - \dfrac{b^2}{a^2}$ for b^2

19. $\dfrac{P_1 V_1}{T_1} = \dfrac{P_2 V_2}{T_2}$ for T_1

20. $\dfrac{P_1 V_1}{T_1} = \dfrac{P_2 V_2}{T_2}$ for P_2

21. $V = \dfrac{4}{3}\pi r^2 h$ for h

22. $h = \dfrac{S - 2\pi r^2}{2\pi r}$ for S

Find the value of the indicated variable. See Example 3. For calculator problems, round answers to three decimal places.

23. In the formula of Exercise 9, if $M = 10$ and $F = 5$, find f.

24. In the formula of Exercise 10, if $A = 550$, $P = 500$, and $t = 2$, find r.

25. In the formula of Exercise 11, if $A = 6\pi$, find D^2.

26. In the formula of Exercise 12, if $V = 12\pi$ and $r = 3$, find h.

27. In the formula of Exercise 13, if $F = 32$, $r = 4$, $m_1 = 6$, and $m_2 = 8$, find k.

28. In the formula of Exercise 14, if $F = 10$, $m = 8$, and $v = 6$, find r.

29. In the formula of Exercise 15, if $f = 2.3$ and $q = 1.7$, find p.

30. In the formula of Exercise 16, if $R = 1.29$ and $R_1 = .045$, find R_2.

31. In the formula of Exercise 17, if $e = .62$ and $b = 3.5$, find a^2.

32. In the formula of Exercise 18, if $a = 3.61$ and $e = 2.4$, find b^2.

33. In the formula of Exercise 21, if $V = 25.6$ and $h = 3.2$, find r^2.

34. In the formula of Exercise 22, if $h = 3.6$ and $r = 2.45$, find S.

Solve the word problems. See Examples 5–7.

35. Karen can ride her bike from home to school in the same amount of time as she can walk from home to the post office. She rides 10 mph faster than she walks. The distance from her home to school is 7 miles, and the distance from her home to the post office is 2 miles. How fast does Karen walk?

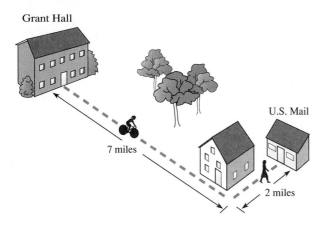

Grant Hall

U.S. Mail

7 miles

2 miles

Figure for Exercise 35

36. Beverly can drive 600 miles in the same time as it takes Susan to drive 500 miles. If Beverly drives 10 mph faster than Susan, then how fast does Beverly drive?

37. Patrick drives 40 miles to work, and Guy drives 60 miles to work. Guy claims that he drives at the same speed as Patrick, but it takes him only 12 minutes longer to get to work. If this is true, then how long does it take each of them to get to work? What are their speeds?

38. David and Keith are route drivers for a fast-photo company. David's route is 80 miles and Keith's is 100 miles. Keith averages 10 mph more than David and finishes his route 10 minutes before David. What is David's speed?

39. Every morning Yong Yi runs 5 miles, then walks 1 mile. He runs 6 mph faster than he walks. If his total time yesterday was 45 minutes, then how fast did he run?

40. Norma can row her boat 12 miles in the same time it takes Marietta to cover 36 miles in her motorboat. If Marietta's boat travels 15 mph faster than Norma's boat, then how fast is Norma rowing her boat?

41. A large pump can drain an 80,000-gallon pool in 3 hours. With a smaller pump also operating, the job takes only 2 hours. How long would it take the smaller pump to drain the pool by itself?

42. Lourdes can trim the hedges around her property in 8 hours by using an electric hedge trimmer. Rafael can do the same job in 15 hours by using a manual trimmer. How long would it take them to trim the hedges working together?

43. It takes 10 minutes to fill Alisha's bathtub and 12 minutes to drain the water out. How long would it take to fill it with the drain accidentally left open?

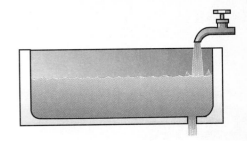

Figure for Exercise 43

44. It takes Gina 90 minutes to file the monthly invoices. If Hilda files twice as fast as Gina does, how long will it take them working together?

45. Charles can empty the cookie jar in 1 1/2 hours. It takes his mother 2 hours to bake enough cookies to fill it. If the cookie jar is full when Charles comes home from school, and his mother continues baking and restocking the cookie jar, then how long will it take him to empty the cookie jar?

46. Julie can paint a fence by herself in 12 hours. With Betsy's help, it takes only 5 hours. How long would it take Betsy by herself?

47. Madeline bought $5.28 worth of oranges and $8.80 worth of apples. She bought 2 more pounds of oranges than apples. If apples cost twice as much per pound as oranges, then how many pounds of each did she buy?

48. Luke raises rabbits and raccoons to sell for meat. The price of raccoon meat is three times the price of rabbit meat. One day Luke sold 160 pounds of meat, $72 worth of each type. What is the price per pound of each type of meat?

49. Muffy can eat a 25-pound bag of dog food in 28 days, while Missy eats a 25-pound bag in 23 days. How many days would it take them together to finish a 50-pound bag of dog food?

50. A pest-control specialist has found that 6 rats can eat an entire box of sugar-coated breakfast cereal in 13.6 minutes, and it takes a dozen mice 34.7 minutes to devour the same size box of cereal. How long would it take all 18 rodents, in a cooperative manner, to finish off a box of cereal?

Wrap-up

CHAPTER 4

SUMMARY

	Concepts	Examples
Rational expression	The ratio of two polynomials with the denominator not equal to 0	$\dfrac{x^2 - 1}{2x - 3}$
Domain of a rational expression	The set of all possible numbers that can be used as replacements for the variable	$D = \{x \mid x \neq 3/2\}$
Basic principle of rational numbers	If a/b is a rational number and c is a nonzero real number, then, $$\frac{a}{b} = \frac{ac}{bc} \quad \text{and} \quad \frac{a}{b} = \frac{a \div c}{b \div c}.$$	Used for reducing: $\dfrac{14}{16} = \dfrac{2 \cdot 7}{2 \cdot 8} = \dfrac{7}{8}$ Used for building: $\dfrac{2}{x} = \dfrac{2 \cdot 3}{x \cdot 3} = \dfrac{6}{3x}$
Multiplication of rational numbers	If a/b and c/d are rational numbers, then $$\frac{a}{b} \cdot \frac{c}{d} = \frac{ac}{bd}.$$	$\dfrac{3}{x} \cdot \dfrac{6}{x^2} = \dfrac{18}{x^3}$
Division of rational numbers	If a/b and c/d are rational numbers with $c/d \neq 0$, then $$\frac{a}{b} \div \frac{c}{d} = \frac{a}{b} \cdot \frac{d}{c}. \quad \text{(Invert and multiply.)}$$	$\dfrac{a}{x} \div \dfrac{5}{4x} = \dfrac{a}{x} \cdot \dfrac{4x}{5}$ $= \dfrac{4a}{5}$
Least common denominator	The LCD of a group of denominators is the smallest number that is a multiple of all of them.	$4a^3b, 6ab^2$ $\text{LCD} = 12a^3b^2$
Addition and subtraction	If $b \neq 0$, then $$\frac{a}{b} + \frac{c}{b} = \frac{a + c}{b} \quad \text{and} \quad \frac{a}{b} - \frac{c}{b} = \frac{a - c}{b}.$$	$\dfrac{2x}{x - 3} + \dfrac{7x}{x - 3} = \dfrac{9x}{x - 3}$

If the denominators are not identical, we must change each fraction to an equivalent fraction with the LCD as denominator.	$\dfrac{2}{x} + \dfrac{1}{3x} = \dfrac{6}{3x} + \dfrac{1}{3x} = \dfrac{7}{3x}$ $\text{LCD} = 3x$

Rules for adding and subtracting simple fractions

If $b \neq 0$ and $d \neq 0$, then

$$\frac{a}{b} + \frac{c}{d} = \frac{ad + bc}{bd},$$

and

$$\frac{a}{b} - \frac{c}{d} = \frac{ad - bc}{bd}.$$

$\dfrac{1}{2} + \dfrac{1}{3} = \dfrac{5}{6}$

$\dfrac{2}{5} - \dfrac{3}{7} = \dfrac{-1}{35}$

Simplifying complex fractions

Multiply the numerator and denominator by the LCD.

$$\frac{\left(\dfrac{1}{2} + \dfrac{1}{x}\right)6x}{\left(\dfrac{1}{x} - \dfrac{1}{3}\right)6x} = \frac{3x + 6}{6 - 2x}$$

Solving equations with rational expressions

Multiply each side by the LCD to eliminate all denominators.

$$\frac{1}{x} - \frac{1}{3} = \frac{1}{2x} - \frac{1}{6}$$

$$6x\left(\frac{1}{x} - \frac{1}{3}\right) = 6x\left(\frac{1}{2x} - \frac{1}{6}\right)$$

$$6 - 2x = 3 - x$$

Solving proportions by the extremes-means property

If $\dfrac{a}{b} = \dfrac{c}{d}$, then $ad = bc$.

$$\frac{2}{x - 3} = \frac{5}{6}$$

$$12 = 5x - 15$$

REVIEW EXERCISES

4.1 *Reduce each rational expression to lowest terms.*

1. $\dfrac{a^3bc^3}{a^5b^2c}$

2. $\dfrac{x^4 - 1}{3x^2 - 3}$

3. $\dfrac{68x^3}{51xy}$

4. $\dfrac{5x^2 - 15x + 10}{5x - 10}$

4.2 *Perform the indicated operations.*

5. $\dfrac{a^3b^2}{b^3a} \cdot \dfrac{ab - b^2}{ab - a^2}$

6. $\dfrac{x^3 - 1}{3x} \cdot \dfrac{6x^2}{x - 1}$

7. $\dfrac{w - 4}{3w} \div \dfrac{2w - 8}{9w}$

8. $\dfrac{x^3 - xy^2}{y} \div \dfrac{x^3 + 2x^2y + xy^2}{3y}$

4.3 *Find the least common denominator for each group of denominators.*

9. $6x,\ 3x - 6,\ x^2 - 2x$

10. $x^3 - 8,\ x^2 - 4,\ 2x + 4$

11. $6ab^3,\ 4a^5b^2$

12. $4x^2 - 9,\ 4x^2 + 12x + 9$

Perform the indicated operations.

13. $\dfrac{3}{2x-6}+\dfrac{1}{x^2-9}$

14. $\dfrac{3}{x-3}-\dfrac{5}{x+4}$

15. $\dfrac{w}{ab^2}-\dfrac{5}{a^2b}$

16. $\dfrac{x}{x-1}+\dfrac{3x}{x^2-1}$

4.4 *Simplify the complex fractions.*

17. $\dfrac{\dfrac{3}{2x}-\dfrac{5}{4x}}{\dfrac{1}{3}-\dfrac{2}{x}}$

18. $\dfrac{\dfrac{5}{x-2}-\dfrac{4}{4-x^2}}{\dfrac{3}{x+2}-\dfrac{1}{2-x}}$

19. $\dfrac{\dfrac{1}{y-2}-3}{\dfrac{5}{y-2}+4}$

20. $\dfrac{\dfrac{a}{b^2}-\dfrac{b}{a^3}}{\dfrac{a}{b}+\dfrac{b}{a^2}}$

21. $\dfrac{a^{-2}-b^{-3}}{a^{-1}b^{-2}}$

22. $p^{-1}+pq^{-2}$

4.5 *Solve each equation.*

23. $\dfrac{-3}{8}=\dfrac{2}{x}$

24. $\dfrac{2}{x}+\dfrac{5}{2x}=1$

25. $\dfrac{15}{a^2-25}+\dfrac{1}{a-5}=\dfrac{6}{a+5}$

26. $2+\dfrac{3}{x-5}=\dfrac{x-1}{x-5}$

4.6 *Solve each formula for the indicated variable.*

27. $\dfrac{y-b}{m}=x$ for y

28. $\dfrac{2A}{h}=b_1+b_2$ for A

29. $F=\dfrac{mv^2}{r}$ for m

30. $P=\dfrac{A}{1+rt}$ for r

31. $A=\dfrac{2}{3}\pi rh$ for r

32. $\dfrac{a}{w^2}=\dfrac{2}{b}$ for b

33. $\dfrac{y+3}{x-7}=2$ for y

34. $\dfrac{y-5}{x+4}=\dfrac{-1}{2}$ for y

Solve each word problem.

35. Nikita drove the 310 miles east of Louisville in the same time as it took him to drive the 360 miles west of Louisville. If his average speed west of Louisville was 10 mph more than east of Louisville, then how much time did this journey take?

36. A tug can push a barge 144 miles down the Mississippi River in the same time that it takes to push the barge 84 miles in the Gulf of Mexico. If the tug's speed is 5 mph greater going down the river, then what is its speed in the Gulf of Mexico?

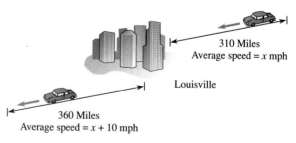

Figure for Exercise 35

37. Debbie can make a hand-sewn quilt in 2000 hours, and Rosalina, can make an identical quilt in 1000 hours. If Cheryl works just as fast as Rosalina, then how long will it take all three of them working together to make one quilt?

38. A small pump can pump all of the blood out of an average turnip in 30 minutes. A larger pump can pump all of the blood from the same turnip in 20 minutes. If both pumps are hooked to the turnip, then how long would it take to get all of the blood out?

Miscellaneous

For each of the following expressions, either perform the indicated operation or solve the equation, whichever is appropriate.

39. $\dfrac{1}{x} + \dfrac{1}{3x}$

40. $\dfrac{1}{y} + \dfrac{3}{2y} = 5$

41. $\dfrac{5}{3xy} + \dfrac{7}{6x}$

42. $\dfrac{2}{x-2} - \dfrac{3}{x}$

43. $\dfrac{5}{a-5} - \dfrac{3}{-a-5}$

44. $\dfrac{2}{x-2} - \dfrac{3}{x} = \dfrac{-1}{5x}$

45. $\dfrac{1}{x-2} - \dfrac{1}{x+2} = \dfrac{1}{15}$

46. $\dfrac{2}{x-3} \cdot \dfrac{6x-18}{30}$

47. $\dfrac{-3}{x+2} \cdot \dfrac{5x+10}{10}$

48. $\dfrac{x}{10} = \dfrac{10}{x}$

49. $\dfrac{x}{-3} = \dfrac{-27}{x}$

50. $\dfrac{x^2-4}{x} \div \dfrac{x^3-8}{x}$

51. $\dfrac{wx+wm+3x+3m}{w^2-9} \div \dfrac{x^2-m^2}{w-3}$

52. $\dfrac{-5}{7} = \dfrac{3}{x}$

53. $\dfrac{5}{a^2-25} + \dfrac{3}{a^2-4a-5}$

54. $\dfrac{3}{w^2-1} + \dfrac{2}{2w+2}$

55. $\dfrac{-7}{2a^2-18} - \dfrac{4}{a^2+5a+6}$

56. $\dfrac{-5}{3a^2-12} - \dfrac{1}{a^2-3a+2}$

57. $\dfrac{7}{a^2-1} + \dfrac{2}{1-a} = \dfrac{1}{a+1}$

58. $2 + \dfrac{4}{x-1} = \dfrac{3x+1}{x-1}$

59. $\dfrac{2x}{x-3} + \dfrac{3}{x-2} = \dfrac{6}{(x-2)(x-3)}$

60. $\dfrac{a-3}{a+3} \div \dfrac{9-a^2}{3}$

61. $\dfrac{x-2}{6} \div \dfrac{2-x}{2}$

62. $\dfrac{x}{x+4} - \dfrac{2}{x+1} = \dfrac{-2}{(x+1)(x+4)}$

63. $\dfrac{x-3}{x^2+3x+2} \cdot \dfrac{x^2-4}{3x-9}$

64. $\dfrac{x^2-1}{x^2+2x+1} \cdot \dfrac{x^3+1}{2x-2}$

65. $\dfrac{a+4}{a^3-8} - \dfrac{3}{2-a}$

66. $\dfrac{x+2}{5} = \dfrac{3}{x}$

67. $\dfrac{x^3-9x}{1-x^2} \div \dfrac{x^3+6x^2+9x}{x-1}$

68. $\dfrac{x+3}{2x+3} = \dfrac{x-3}{x-1}$

69. $\dfrac{a^2 + 3a + 3w + aw}{a^2 + 6a + 8} \cdot \dfrac{a^2 - aw - 2w + 2a}{a^2 + 3a - 3w - aw}$

70. $\dfrac{3}{4 - 2y} + \dfrac{6}{y^2 - 4} + \dfrac{3}{2 + y}$

71. $\dfrac{5}{x} - \dfrac{4}{x + 2} = \dfrac{1}{5} + \dfrac{1}{5x}$

72. $\dfrac{1}{x} + \dfrac{1}{x - 5} = \dfrac{2x + 1}{x^2 - 25} + \dfrac{9}{x^2 + 5x}$

In place of the question mark, put an expression that makes each of the following true. Do these mentally.

73. $\dfrac{6}{x} = \dfrac{?}{3x}$

74. $\dfrac{?}{a} = \dfrac{8}{4a}$

75. $\dfrac{3}{a - b} = \dfrac{?}{b - a}$

76. $\dfrac{-2}{a - x} = \dfrac{2}{?}$

77. $4 = \dfrac{?}{x}$

78. $5a = \dfrac{?}{b}$

79. $5x \div \dfrac{1}{2} = ?$

80. $3a \div \dfrac{1}{a} = ?$

81. $4a \div ? = 12a$

82. $14x \div ? = 28x^2$

83. $\dfrac{a - 3}{a^2 - 9} = \dfrac{1}{?}$

84. $\dfrac{?}{x^2 - 4} = \dfrac{1}{x - 2}$

85. $\dfrac{1}{2} - \dfrac{1}{5} = ?$

86. $\dfrac{1}{4} - \dfrac{1}{5} = ?$

87. $\dfrac{a}{3} + \dfrac{a}{2} = ?$

88. $\dfrac{x}{5} + \dfrac{x}{3} = ?$

89. $\dfrac{1}{a} - \dfrac{1}{b} = ?$

90. $\dfrac{3}{w} - \dfrac{2}{b} = ?$

91. $\dfrac{a}{3} - 1 = ?$

92. $\dfrac{x}{y} - 1 = ?$

93. $2 + \dfrac{1}{a} = ?$

94. $3 - \dfrac{1}{x} = ?$

95. $\dfrac{a}{5} - ? = \dfrac{a - 5}{5}$

96. $? - \dfrac{1}{y} = \dfrac{y^2 - 1}{y}$

97. $(x - 1) \div (-1) = ?$

98. $(a - 3) \div (3 - a) = ?$

99. $(m - 2) \div (2 - m) = ?$

100. $(x - 2) \div (2x - 4) = ?$

101. $\dfrac{\frac{b}{3a}}{2} = ?$

102. $\dfrac{a - 3}{\frac{1}{2}} = ?$

103. $\dfrac{a - 6}{5} \cdot \dfrac{10}{6 - a} = ?$

104. $\dfrac{3}{w - 2} + \dfrac{2}{2 - w} = ?$

105. $\dfrac{1}{3 - x} = \dfrac{?}{2x - 6}$

106. $\dfrac{18}{6} = \dfrac{3}{?}$

CHAPTER 4 TEST

State the domain of each rational expression.

1. $\dfrac{2x - 1}{x^2 - 9}$

2. $\dfrac{5}{4 - 3x}$

3. $\dfrac{5a}{a^2 + 4a}$

Perform the indicated operations. Write answers in lowest terms.

4. $\dfrac{5}{12} - \dfrac{4}{9}$

5. $\dfrac{3}{y} + 7$

6. $\dfrac{4}{a - 9} - \dfrac{1}{9 - a}$

7. $\dfrac{2}{x^2 - 4} - \dfrac{6}{x^2 - 3x - 10}$

8. $\dfrac{m^3 - 1}{(m - 1)^2} \cdot \dfrac{m^2 - 1}{3m^2 + 3m + 3}$

9. $\dfrac{a - b}{7} \div \dfrac{b^2 - a^2}{21}$

10. $\dfrac{3a^3 b}{20ab} \cdot \dfrac{2a^2 b}{9ab^3}$

Find the solution set to each equation.

11. $\dfrac{3}{x} = \dfrac{7}{4}$

12. $\dfrac{x}{x - 2} - \dfrac{5}{x} = \dfrac{3}{4}$

13. $\dfrac{3m}{2} = \dfrac{6}{m}$

Solve each formula for the indicated variable.

14. $W = \dfrac{a^2}{t}$ for t

15. $\dfrac{1}{a} + \dfrac{1}{b} = \dfrac{1}{2}$ for b

Solve each word problem.

16. When Geraldine's wading pool was new, it could be filled in 6 minutes with water from the hose. Now that the pool has several leaks, Geraldine can play only 8 minutes before the water in the pool is all gone. How long will it take to fill the leaky pool?

17. Milton and Bonnie are hiking the Appalachian Trail together. Milton averages 4 miles per hour, and Bonnie averages 3 miles per hour. If they start out together in the morning, but Milton gets to camp 2 hours and 30 minutes ahead of Bonnie, then how many miles did they hike that day?

Simplify.

18. $\dfrac{\dfrac{1}{x} + \dfrac{1}{3x}}{\dfrac{3}{4x} - \dfrac{1}{2}}$

19. $\dfrac{m^{-2} - w^{-2}}{w^{-1}m^{-2} + m^{-1}w^{-2}}$

20. $a^{-2} + a^{-1}$

21. $\dfrac{\dfrac{a^2 b^3}{4a}}{\dfrac{ab^3}{6a^2}}$

Tying It All Together

CHAPTERS 1–4

Find the solution set to each equation.

1. $\dfrac{3}{x} = \dfrac{4}{5}$

2. $\dfrac{2}{x} = \dfrac{x}{8}$

3. $\dfrac{x}{3} = \dfrac{4}{5}$

4. $\dfrac{3}{x} = \dfrac{x + 3}{6}$

5. $\dfrac{1}{x} = 4$

6. $\dfrac{2}{3}x = 4$

7. $2x + 3 = 4$

8. $2x + 3 = 4x$

9. $\dfrac{2a}{3} = \dfrac{6}{a}$ **10.** $\dfrac{12}{x} - \dfrac{14}{x+1} = \dfrac{1}{2}$ **11.** $|6x - 3| = 1$

12. $\dfrac{x}{2x+9} = \dfrac{3}{x}$ **13.** $4(6x-3)(2x+9) = 0$ **14.** $\dfrac{x-1}{x+2} - \dfrac{1}{5(x+2)} = 1$

Solve each equation for y. Assume A, B, and C are constants for which all expressions are defined.

15. $Ax + By = C$ **16.** $\dfrac{y-3}{x+5} = -\dfrac{1}{3}$ **17.** $Ay = By + C$

18. $\dfrac{A}{y} = \dfrac{y}{A}$ **19.** $\dfrac{A}{y} - \dfrac{1}{2} = \dfrac{B}{y}$ **20.** $\dfrac{A}{y} - \dfrac{1}{2} = \dfrac{B}{C}$

21. $3x - 4y = 6$ **22.** $y^2 - 2y - Ay + 2A = 0$

23. $A = \dfrac{1}{2}B(C + y)$ **24.** $y^2 + Cy = BC + By$

Simplify each expression.

25. $3x^5 \cdot 4x^8$ **26.** $3x^2(x^3 + 5x^6)$ **27.** $(5x^6)^2$

28. $(3a^3b^2)^3$ **29.** $\dfrac{12a^9b^4}{-3a^3b^{-2}}$ **30.** $\left(\dfrac{x^{-2}}{2}\right)^5$

31. $\left(\dfrac{2x^{-4}}{3y^5}\right)^{-3}$ **32.** $(-2a^{-1}b^3c)^{-2}$

CHAPTER **5**

Rational Exponents and Radicals

When a rectangular shape is used in construction it often includes the diagonal. We see rectangles with diagonals in such objects as bridges, fences, and shelves. The diagonal of a rectangle in construction adds strength to the rectangular shape.

How does the designer of a bridge figure out the length of the diagonal of a 30- by 40-foot rectangle? The builders certainly cannot put the steel beams in place and then cut them off to fit! The length of the diagonal is found by using the Pythagorean Theorem. In this case we must solve the equation $30^2 + 40^2 = x^2$. We have solved similar equations in Chapters 3 and 4 by factoring, but in this chapter we will learn to solve equations that cannot be factored. This particular diagonal will be found in Exercise 75 of Section 5.5.

5.1 Rational Exponents

IN THIS SECTION:

- ● **Roots**
- ● **Rational Exponents**
- ● **Using the Rules of Exponents**
- ● **Simplifying Expressions Involving Variables**

In Sections 3.1 and 3.2 we studied integral exponents. In this section we will extend the idea of exponent to include rational numbers as exponents.

Roots

If a square has a side of length 7 feet, then it has an area of $7^2 = 49$ square feet. If a square has an area of 36 square feet, then the length of its side is 6 feet. If we know the length of a side, then we square it to find the area. If we know the area, then we must reverse the process of squaring to find the length of a side. Reversing the process of squaring is called **taking the square root.** In general, reversing an nth power is referred to as **taking an nth root.**

Because we know that

$$3^2 = 9, \qquad (-4)^2 = 16, \qquad \text{and} \qquad 2^3 = 8,$$

we say that 3 is a square root of 9, -4 is a square root of 16, and 2 is the cube root of 8.

DEFINITION
***n*th Roots**

We say that a is an ***n*th root** of b, if

$$a^n = b.$$

Both 3 and -3 are square roots of 9, since $3^2 = 9$ and $(-3)^2 = 9$. The positive square root of 9 is called the **principal square root of 9.** Because $2^4 = 16$ and $(-2)^4 = 16$, there are two real fourth roots of 16, 2 and -2. The principal fourth root of 16 is 2. There are two even real roots of any positive number. When n is even, the exponent $1/n$ is used to indicate the principal nth root. The principal nth root can also be indicated by the radical symbol $\sqrt[n]{}$, which will be discussed in Section 5.2.

DEFINITION
Exponent 1/n, When n Is Even

If n is a positive even integer and a is positive, then $a^{1/n}$ denotes the **positive real nth root of a** and is called the **principal nth root of a**.

Note that an exponent of n indicates the nth power, an exponent of $-n$ indicates the reciprocal of the nth power, and an exponent of $1/n$ indicates nth root. We will see that choosing $1/n$ to indicate nth root fits in nicely with the rules of exponents that we have already studied.

EXAMPLE 1 Evaluate each of the following roots.

a) $4^{1/2}$ **b)** $16^{1/4}$ **c)** $-81^{1/4}$

Solution

a) Because $2^2 = 4$, $4^{1/2} = 2$.   Note that $4^{1/2} \neq -2$.

b) Because $2^4 = 16$, $16^{1/4} = 2$.   Note that $16^{1/4} \neq -2$.

c) Since there are no parentheses, we follow the order of operations. We find the root first and then take the opposite of it. Because $3^4 = 81$, $81^{1/4} = 3$. So $-81^{1/4} = -3$.   ◄

Note that $2^3 = 8$ but $(-2)^3 = -8$. The cube root of 8 is 2, and the cube root of -8 is -2. There is only one real odd root of a number whether the number is positive or negative.

DEFINITION
Exponent 1/n, When n Is Odd

If n is a positive odd integer and a is any real number, then $a^{1/n}$ denotes the real nth root of a.

EXAMPLE 2 Evaluate each of the following roots.

a) $8^{1/3}$ **b)** $(-27)^{1/3}$ **c)** $-32^{1/5}$

Solution

a) Because $2^3 = 8$, $8^{1/3} = 2$.

b) Because $(-3)^3 = -27$, $(-27)^{1/3} = -3$.

c) Because $2^5 = 32$, $-32^{1/5} = -2$.   ◄

We do not allow 0 as the base when we use negative exponents because division by zero is undefined. However, positive powers of zero are defined, and so are roots of zero: for example, $0^4 = 0$ and so $0^{1/4} = 0$.

DEFINITION
nth Root of Zero

> If n is a positive integer, then $0^{1/n} = 0$.

Note that the definition of roots did not include an even root of a negative number. For this reason we say that an expression such as $(-9)^{1/2}$ is *undefined*. The square root of a negative number is not defined now because there is no real number whose square is -9, and so there is no real square root of -9. However, in Chapter 6 we will define this type of expression when we study complex numbers.

The expression $3^{1/2}$ represents the unique positive real number whose square is 3. There is no rational number that has a square equal to 3. The number $3^{1/2}$ is an irrational number. If we use the square-root table of Appendix A or a calculator, we find that $3^{1/2}$ is approximately equal to the rational number 1.732. Since the square root of 3 is not a rational number, the simplest representation for the exact value of the square root of 3 is $3^{1/2}$.

Rational Exponents

When a rational number is used as an exponent, the denominator indicates "root" and the numerator indicates "power." For example, $8^{2/3}$ means $(8^{1/3})^2$. We take the cube root of 8 to get 2, then square 2 to get 4. Thus,

$$8^{2/3} = 4.$$

We now define rational exponents in terms of roots and powers.

DEFINITION
Rational Exponents

> If m and n are positive integers, then
> $$a^{m/n} = (a^{1/n})^m$$
> provided that $a^{1/n}$ is defined.

EXAMPLE 3 Evaluate the following expressions.

a) $16^{3/4}$ b) $27^{4/3}$ c) $(-8)^{2/3}$ d) $-4^{3/2}$

Solution

a) $16^{3/4} = (16^{1/4})^3 = 2^3 = 8$ Take the 4th root of 16, then cube it.
b) $27^{4/3} = (27^{1/3})^4 = 3^4 = 81$ Take the cube root of 27, then raise it to the 4th power.

c) $(-8)^{2/3} = [(-8)^{1/3}]^2 = (-2)^2 = 4$
d) $-4^{3/2} = -(4^{1/2})^3 = -2^3 = -8$

Note that $(-4)^{3/2}$ is not a real number because the square root of -4 is not a real number.   ◄

Recall that 2^{-3} was defined to be the reciprocal of 2^3. Negative rational exponents are also defined as reciprocals. For example, $8^{-2/3}$ is defined to be the reciprocal of $8^{2/3}$. We can express this relationship as

$$8^{-2/3} = \frac{1}{8^{2/3}} = \frac{1}{4}.$$

DEFINITION
Negative Rational Exponents

If m and n are positive integers, then

$$a^{-m/n} = \frac{1}{a^{m/n}},$$

provided that $a^{1/n}$ is defined and nonzero.

Note that the expression $8^{-2/3}$ involves three operations:

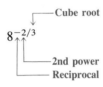

In general, we use the following strategy to evaluate expressions involving negative rational exponents:

STRATEGY
Evaluating $a^{-m/n}$

To evaluate the expression $a^{-m/n}$ there are three operations to perform:
1. Take the nth root.
2. Raise it to the mth power.
3. Find the reciprocal.

It is usually easiest to perform the operations in the above order. We will see below that we may perform the operations in any order.

EXAMPLE 4 Evaluate each expression.

a) $4^{-3/2}$ **b)** $(-27)^{-1/3}$ **c)** $(-16)^{-3/4}$

Solution

a) The square root of 4 is 2. The cube of 2 is 8. The reciprocal of 8 is 1/8. Thus,
$$4^{-3/2} = \frac{1}{8}.$$

b) The cube root of -27 is -3. The first power of -3 is -3. The reciprocal of -3 is $-1/3$. Thus, $(-27)^{-1/3} = -\dfrac{1}{3}$.

c) The expression $(-16)^{-3/4}$ is not a real number because it involves the fourth root of a negative number.   ◄

Using the Rules of Exponents

Consider the expressions

$$(8^{1/3})^2 \quad \text{and} \quad (8^2)^{1/3}.$$

If we evaluate the expression $(8^{1/3})^2$, we take the cube root first to get 2 and then we square to get 4. If we evaluate the expression $(8^2)^{1/3}$, we square first to get 64 and then take the cube root to get 4. The power of a power rule helps us to see why we get the same result. If we apply the power of a power rule, we would multiply the exponents and get $8^{2/3}$ for each expression. This example illustrates the fact that we could have defined the exponent m/n to be the nth root of the mth power.

ALTERNATIVE DEFINITION
Rational Exponents

If m and n are positive integers, then
$$a^{m/n} = (a^{1/n})^m = (a^m)^{1/n},$$
provided $a^{1/n}$ is defined.

The power of a power rule is not the only rule that can be applied to expressions with rational exponents. All of the rules for exponents hold for rational exponents as well as integral exponents. Of course we cannot apply the rules of exponents to expressions that are undefined.

Rules for
Rational Exponents

If a and b are nonzero real numbers and m and n are rational numbers for which the expressions are defined, then

1. $a^m a^n = a^{m+n}$ Product rule

2. $a^n = \dfrac{1}{a^{-n}}$ Sign-change rule

3. $\dfrac{a^m}{a^n} = a^{m-n}$ Quotient rule

4. $(a^m)^n = a^{mn}$ Power of a power rule

5. $(ab)^n = a^n b^n$ Power of a product rule

6. $\left(\dfrac{a}{b}\right)^n = \dfrac{a^n}{b^n}$ Power of a quotient rule

We can use the product rule to add rational exponents. For example,

$$16^{1/4} \cdot 16^{1/4} = 16^{2/4}.$$

The fourth root of 16 is 2, and 2 squared is 4. So $16^{2/4} = 4$. Because we also have $16^{1/2} = 4$, we see that rational exponents can be reduced to lowest terms. If an exponent can be reduced, it is usually simpler to reduce it before we evaluate the expression. We can simplify the above product as follows:

$$16^{1/4} \cdot 16^{1/4} = 16^{2/4} = 16^{1/2} = 4$$

EXAMPLE 5 Use the rules of exponents to simplify each expression.

a) $(3^{10})^{1/2}$ b) $2^{1/4} \cdot 2^{1/2}$ c) $\dfrac{5}{5^{1/2}}$

d) $\left(\dfrac{2^6}{3^9}\right)^{-1/3}$ e) $3^{1/2} \cdot 12^{1/2}$

Solution

a) $(3^{10})^{1/2} = 3^5$ Power of a power rule $10 \cdot \dfrac{1}{2} = 5$
    $= 243$

b) $2^{1/4} \cdot 2^{1/2} = 2^{1/4+1/2}$ Product rule
    $= 2^{3/4}$

Because the fourth root of 2 is not a whole number, we cannot simplify $2^{3/4}$ any further. We could find an approximate decimal value with a calculator.

c) $\dfrac{5}{5^{1/2}} = 5^{1-1/2} = 5^{1/2}$ We used the quotient rule to subtract the exponents.

d) $\left(\dfrac{2^6}{3^9}\right)^{-1/3} = \dfrac{2^{-2}}{3^{-3}} = \dfrac{3^3}{2^2} = \dfrac{27}{4}$ We used the power of a quotient, power of a power, and sign-change rules.

e) $3^{1/2} \cdot 12^{1/2} = (3 \cdot 12)^{1/2} = 36^{1/2} = 6$

Note that because the bases are different we could not use the product rule to add the exponents. We used the power of a product rule to remove the 1/2 power from both 3 and 12 and place it outside the parentheses.   ◀

Simplifying Expressions Involving Variables

In simplifying expressions involving rational exponents and variables, we must take special care. Remember that there are two square roots for every positive number, and the exponent 1/2 indicates the positive square root. Thus a statement such as

$$(x^2)^{1/2} = x$$

looks like we are correctly applying the power of a power rule, but this statement is false if x is negative. The 1/2 power on the left side indicates a positive square root. If x represents a negative number, we would have a positive number equal to a negative. For example,

$$[(-3)^2]^{1/2} \neq -3$$

because $(-3)^2 = 9$ and the positive square root of 9 is 3. To make a correct statement concerning the principal square root of x^2, we can use absolute value as follows.

Square Root of x^2

$$(x^2)^{1/2} = |x| \text{ for any real number } x.$$

It is also necessary to use absolute value when finding other even roots of expressions involving variables.

Remember also that there are no even roots of negative numbers. This means that expressions such as

$$a^{1/2}, \qquad x^{-3/4}, \qquad \text{and} \qquad y^{1/6}$$

are undefined if the variable has a negative value. We usually assume that the variables represent only positive numbers when we are working with expressions involving rational exponents and variables. That way we do not have to be concerned with undefined expressions and absolute value.

EXAMPLE 6   Use the rules of exponents to simplify the following. Write your answers with positive exponents. Assume that all variables represent positive real numbers.

a) $x^{2/3}x^{4/3}$

b) $\dfrac{a^{1/2}}{a^{1/4}}$

c) $(x^{1/2}y^{-3})^{1/2}$

d) $\left(\dfrac{x^2}{y^{1/3}}\right)^{-1/2}$

Solution

a) $x^{2/3}x^{4/3} = x^{6/3}$ **Product rule to add the exponents**
$\qquad\qquad\;\; = x^2$

b) $\dfrac{a^{1/2}}{a^{1/4}} = a^{1/2-1/4} = a^{1/4}$ **Quotient rule to subtract the exponents**

c) $(x^{1/2}y^{-3})^{1/2} = (x^{1/2})^{1/2}(y^{-3})^{1/2}$ **Power of a product rule**
$\qquad\qquad\;\; = x^{1/4}y^{-3/2}$ **Power of a power rule**
$\qquad\qquad\;\; = \dfrac{x^{1/4}}{y^{3/2}}$ **Sign-change rule**

d) Because this is a negative power of a quotient, we can first find the reciprocal of the quotient, then apply the power of a power rule.

$$\left(\frac{x^2}{y^{1/3}}\right)^{-1/2} = \left(\frac{y^{1/3}}{x^2}\right)^{1/2} = \frac{y^{1/6}}{x} \qquad \frac{1}{3}\cdot\frac{1}{2} = \frac{1}{6} \qquad \blacktriangleleft$$

Warm-ups

True or false?

1. $4^{-1/2} = \dfrac{1}{2}$　　　　　　**2.** $16^{1/2} = 8$

3. $(3^{2/3})^3 = 9$　　　　　　**4.** $8^{-2/3} = -4$

5. $2^{1/2} \cdot 2^{1/2} = 2$　　　　**6.** $\left(\dfrac{1}{4}\right)^{1/2} = \dfrac{1}{2}$

7. $\dfrac{3}{3^{1/2}} = 3^{1/2}$　　　　**8.** $(2^9)^{1/2} = 2^3$

9. $3^{1/3} \cdot 6^{1/3} = 18^{2/3}$　　**10.** $2^{3/4} \cdot 2^{1/4} = 4$

5.1 EXERCISES

Simplify. Some of these are undefined expressions. See Examples 1 and 2.

1. $100^{1/2}$　　**2.** $27^{1/3}$　　**3.** $-9^{1/2}$　　**4.** $-4^{1/2}$
5. $(-25)^{1/2}$　　**6.** $(-16)^{1/4}$　　**7.** $1000^{1/3}$　　**8.** $169^{1/2}$
9. $(-64)^{1/3}$　　**10.** $32^{1/5}$　　**11.** $-128^{1/7}$　　**12.** $(-64)^{1/6}$

Evaluate each expression. Some of these are undefined expressions. See Examples 3 and 4.

13. $32^{3/5}$　　**14.** $25^{3/2}$　　**15.** $(-32)^{3/5}$　　**16.** $-25^{3/2}$
17. $32^{-3/5}$　　**18.** $(-25)^{3/2}$　　**19.** $(-32)^{-3/5}$　　**20.** $-25^{-3/2}$
21. $(-27)^{4/3}$　　**22.** $(-27)^{2/3}$　　**23.** $64^{5/6}$　　**24.** $25^{3/2}$
25. $4^{-1/2}$　　**26.** $9^{-1/2}$　　**27.** $16^{-3/4}$　　**28.** $27^{-2/3}$
29. $8^{-4/3}$　　**30.** $4^{-3/2}$　　**31.** $(-9)^{-1/2}$　　**32.** $(-4)^{-1/2}$

Use the rules of exponents to simplify each expression. See Example 5.

33. $(2^6)^{1/3}$　　**34.** $(3^{10})^{1/5}$　　**35.** $(3^8)^{1/2}$　　**36.** $(3^{-6})^{1/3}$
37. $(2^{-4})^{1/2}$　　**38.** $(5^4)^{1/2}$　　**39.** $5^{1/4} \cdot 5^{-1/4}$　　**40.** $3^{1/3} \cdot 3^{1/4}$
41. $2^{1/2} \cdot 2^{1/3}$　　**42.** $3^{1/3} \cdot 3^{-1/3}$　　**43.** $18^{1/2} \cdot 2^{1/2}$　　**44.** $8^{1/2} \cdot 2^{1/2}$
45. $2^{1/2} \cdot 2^{-1/4}$　　**46.** $3^{1/4} \cdot 27^{1/4}$　　**47.** $3^{.26}3^{.74}$　　**48.** $(9^2)^{1/2}$
49. $3^{2/3} \cdot 9^{2/3}$　　**50.** $2^{1.5}2^{.5}$　　**51.** $(4^{16})^{1/2}$　　**52.** $9^{-1}9^{1/2}$
53. $4^{3/4} \div 4^{1/4}$　　**54.** $9^{1/4} \div 9^{3/4}$　　**55.** $\dfrac{8^{1/3}}{8^{2/3}}$　　**56.** $\dfrac{27^{-2/3}}{27^{-1/3}}$
57. $\left(\dfrac{1}{4}\right)^{1/2}$　　**58.** $\left(\dfrac{4}{9}\right)^{1/2}$　　**59.** $\left(\dfrac{1}{8}\right)^{-1/3}$　　**60.** $\left(\dfrac{1}{27}\right)^{-1/3}$

61. $\left(-\dfrac{8}{27}\right)^{2/3}$ **62.** $\left(-\dfrac{8}{27}\right)^{-1/3}$ **63.** $\left(-\dfrac{1}{16}\right)^{-3/4}$ **64.** $\left(\dfrac{9}{16}\right)^{-1/2}$

Simplify. Assume that all variables represent positive numbers. Write answers with positive exponents only. See Example 6.

65. $x^{1/2}x^{1/4}$ **66.** $y^{1/3}y^{1/3}$ **67.** $\dfrac{x^{1/2}y}{x}$ **68.** $\dfrac{x^{1/3}y^{-1/2}}{xy^{-1}}$

69. $(x^{1/2}y)(x^{-3/4}y^{1/2})$ **70.** $(a^{1/2}b^{-1/3})(ab)$ **71.** $x(x^2y^{-1})^{-1/2}$ **72.** $\dfrac{w^{1/3}}{w^3}$

73. $\dfrac{a^{1/2}}{a^2}$ **74.** $\left(\dfrac{a^{-1/2}}{b^{-1/4}}\right)^{-4}$ **75.** $\left(\dfrac{a^{1/2}}{b^{1/3}}\right)^6$ **76.** $(x^9)^{1/2}$

77. $(x^{16})^{1/2}$ **78.** $(a^8)^{1/3}$ **79.** $\dfrac{a^{1/2}b^{-2}}{ab^{1/3}}$ **80.** $(x^9)^{1/3}$

81. $(x^{-1/2}yz^{1/2})^{-1/2}$ **82.** $(x^8y^{-10}z^{12})^{1/2}$ **83.** $x^{-1/3}\cdot x^{1/2}\cdot x^{-1/6}$ **84.** $(x^{1/2})^{1/2}$

85. $(a^{1/2}b)^{1/2}(ab^{1/2})$ **86.** $(m^{1/4}n^{1/2})^2(m^2n^3)^{1/2}$ **87.** $(km^{1/2})^3(k^3m^5)^{1/2}$ **88.** $(tv^{1/3})^2(t^2v^{-3})^{-1/2}$

89. $(t^{2/3})^3$ **90.** $(w^{3/4})^4$ **91.** $(a^{-2/3})^{-3}$ **92.** $(x^{-1/2})^{-2}$

93. $(x^{-2/5})^{-5}$ **94.** $(x^{4/5})^5$ **95.** $\left(\dfrac{x^{10}}{w^{12}}\right)^{-1/2}$ **96.** $\left(\dfrac{a^6b^9}{c^{-12}}\right)^{-2/3}$

*97. $\left(\dfrac{w^{-3/4}a^{-2/3}}{w^2a^{-1/2}}\right)^6\left(\dfrac{w^{1/2}a^{-1/2}}{w^{-1/2}a^3}\right)^{-1/2}$

*98. $\left(\dfrac{-2m^{-3/5}p^{-1/4}}{2^{-2}m^{-1/2}p^{1/3}}\right)^2\left(\dfrac{3m^{-2}p^4}{9^{-1}mp^{-2}}\right)^{1/3}$

Solve each problem.

99. The length of the diagonal of a box can be found from the formula $D = (L^2 + W^2 + H^2)^{1/2}$, where L, W, and H represent the length, width, and height of the box. If the box is 12 inches long, 4 inches wide, and 3 inches high, then what is the length of the diagonal?

100. The radius of a sphere is given in terms of its volume by the formula $r = (.75V/\pi)^{1/3}$. Find the radius of a ball that has a volume of $32\pi/3$ cubic meters.

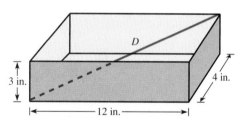

Figure for Exercise 99

Use a scientific calculator with a power key (x^y) to find the decimal value of each of the following. Round answers to four decimal places.

101. $2^{1/3}$ **102.** $5^{1/2}$ **103.** $-2^{1/2}$

104. $(-3)^{1/3}$ **105.** $1024^{1/10}$ **106.** $7776^{.2}$

107. $8^{.33}$ **108.** $289^{.5}$ **109.** $\left(\dfrac{64}{15625}\right)^{-1/6}$

110. $\left(\dfrac{32}{243}\right)^{-3/5}$ **111.** $\left(\dfrac{5^{-2/3}}{6^{1/4}}\right)^{-1/2}$ **112.** $\left(\dfrac{5^{1/3}\cdot 3^{1/2}}{(-6)^{1/3}}\right)^{1/3}$

5.2 Radicals

IN THIS SECTION:

- Radical Notation
- Product Rule for Radicals
- Quotient Rule for Radicals
- Rationalizing the Denominator
- Simplifying Radicals
- Simplifying Radicals Involving Variables

In Section 5.1 we used the exponent $1/n$ to signify the nth root. The symbol $\sqrt[n]{}$ has exactly the same meaning. In this section we will learn to use the rules of exponents with this new notation.

Radical Notation

Any expression involving an nth root can be written using radical notation. The symbol $\sqrt{}$ is called the **radical symbol.**

DEFINITION
Radicals

> If n is a positive integer and a is a number for which $a^{1/n}$ is defined, then the expression $\sqrt[n]{a}$ is called a **radical** and
> $$\sqrt[n]{a} = a^{1/n}.$$
> If $n = 2$, omit the 2 and write $\sqrt[2]{a}$ as $\sqrt{a}$.

The number a is called the **radicand.** The number n is called the **index** of the radical. The expressions

$$\sqrt{-4}, \qquad \sqrt[4]{-16}, \qquad \text{and} \qquad \sqrt[6]{-2}$$

are undefined, because no even roots of negative numbers are real numbers.

EXAMPLE 1 Write each exponential expression using radical notation and each radical expression using exponential notation. Assume that the variables represent positive numbers. Do not simplify.

a) $\sqrt{36}$ b) $\sqrt[3]{-8}$ c) $x^{3/4}$ d) $\sqrt[3]{y^6}$ e) $x^{-5/2}$

Solution

a) $\sqrt{36} = 36^{1/2}$

b) $\sqrt[3]{-8} = (-8)^{1/3}$

c) $x^{3/4} = \sqrt[4]{x^3} = (\sqrt[4]{x})^3$ Note that we can take the root or the power first.

d) $\sqrt[3]{y^6} = y^{6/3}$

e) $x^{-5/2} = \sqrt{x^{-5}} = (\sqrt{x})^{-5}$ ◀

EXAMPLE 2   Simplify each radical expression.

a) $\sqrt{49}$ b) $\sqrt[3]{-125}$ c) $\sqrt[4]{81}$ d) $\sqrt{x^{20}}$ e) $\sqrt[3]{2^{21}}$

Solution

a) $\sqrt{49} = 7$, because $7^2 = 49$.

b) $\sqrt[3]{-125} = -5$, because $(-5)^3 = -125$.

c) $\sqrt[4]{81} = 3$, because $3^4 = 81$.

d) $\sqrt{x^{20}} = x^{20/2} = x^{10}$

e) $\sqrt[3]{2^{21}} = 2^{21/3} = 2^7 = 128$ ◀

Product Rule for Radicals

Recall that the power of a product rule is valid for rational exponents as well as integers. For example, the power of a product rule allows us to write

$$(4y)^{1/2} = 4^{1/2} \cdot y^{1/2} \quad \text{and} \quad (8 \cdot 7)^{1/3} = 8^{1/3} \cdot 7^{1/3}.$$

The above examples can be written using radical notation as

$$\sqrt{4y} = \sqrt{4} \cdot \sqrt{y} \quad \text{and} \quad \sqrt[3]{8 \cdot 7} = \sqrt[3]{8} \cdot \sqrt[3]{7}.$$

The power of a product rule (for the power $1/n$) can be stated using radical notation. In this form the rule is called the **product rule for radicals.**

Product Rule for Radicals

> The nth root of a product is equal to the product of the nth roots. In symbols,
>
> $$\sqrt[n]{ab} = \sqrt[n]{a} \cdot \sqrt[n]{b},$$
>
> provided that all of the roots are defined.

The numbers 1, 4, 9, 16, 25, 36, 49, and so on, are called **perfect squares,** because they are the squares of the positive integers. If the radicand of a square root has a perfect square (other than 1) as a factor, the product rule can be used to simplify the radical expression. For example, the radicand of $\sqrt{50}$ has 25 as a factor, so we can use the product rule to factor $\sqrt{50}$ into a product of two square roots:

$$\sqrt{50} = \sqrt{25} \cdot \sqrt{2} = 5\sqrt{2}$$

When simplifying a cube root, we check the radicand for factors that are **perfect cubes:** 8, 27, 64, 125, and so on. In general, when simplifying an nth root we look for a perfect nth power as a factor of the radicand.

EXAMPLE 3 Use the product rule for radicals to simplify each expression. Assume that all variables represent positive numbers.

a) $\sqrt{4y}$ b) $\sqrt{18}$ c) $\sqrt[3]{56}$ d) $\sqrt[4]{32}$

Solution

a) $\sqrt{4y} = \sqrt{4} \cdot \sqrt{y} = 2\sqrt{y}$

b) The radicand 18 has a factor of 9. Thus,

$$\sqrt{18} = \sqrt{9} \cdot \sqrt{2} = 3\sqrt{2}.$$

c) This is a cube root, and 56 has a factor of 8. Thus,

$$\sqrt[3]{56} = \sqrt[3]{8} \cdot \sqrt[3]{7} = 2\sqrt[3]{7}.$$

d) This is a fourth root, and the perfect fourth power 16 is a factor of 32. Thus,

$$\sqrt[4]{32} = \sqrt[4]{16} \cdot \sqrt[4]{2} = 2\sqrt[4]{2}. \qquad \blacktriangleleft$$

Quotient Rule for Radicals

The power of a quotient rule is also valid for integral and rational exponents. This rule allows us to write

$$\left(\frac{y}{9}\right)^{1/2} = \frac{y^{1/2}}{9^{1/2}} = \frac{y^{1/2}}{3} \quad \text{and} \quad \left(\frac{3}{32}\right)^{1/5} = \frac{3^{1/5}}{32^{1/5}} = \frac{3^{1/5}}{2}.$$

The above examples can be written using radical notation as

$$\sqrt{\frac{y}{9}} = \frac{\sqrt{y}}{\sqrt{9}} = \frac{\sqrt{y}}{3} \quad \text{and} \quad \sqrt[5]{\frac{3}{32}} = \frac{\sqrt[5]{3}}{\sqrt[5]{32}} = \frac{\sqrt[5]{3}}{2}.$$

The power of a quotient rule (for the power $1/n$) can be stated using radical notation. When written with radicals, it is called the **quotient rule for radicals.**

Quotient Rule for Radicals

> The nth root of a quotient is equal to the quotient of the nth roots. In symbols,
>
> $$\sqrt[n]{\frac{a}{b}} = \frac{\sqrt[n]{a}}{\sqrt[n]{b}},$$
>
> provided that all of the roots are defined.

The quotient rule is used to change the root of a quotient into the quotient of the roots. For example,

$$\sqrt{\frac{3}{4}} = \frac{\sqrt{3}}{\sqrt{4}} = \frac{\sqrt{3}}{2}.$$

EXAMPLE 4   Use the quotient rule and product rule to simplify each expression.

a) $\sqrt{\dfrac{8}{9}}$

b) $\sqrt[3]{\dfrac{40y}{27}}$

Solution

a) $\sqrt{\dfrac{8}{9}} = \dfrac{\sqrt{8}}{\sqrt{9}} = \dfrac{\sqrt{4} \cdot \sqrt{2}}{3} = \dfrac{2\sqrt{2}}{3}$

b) $\sqrt[3]{\dfrac{40y}{27}} = \dfrac{\sqrt[3]{40y}}{\sqrt[3]{27}} = \dfrac{\sqrt[3]{8} \cdot \sqrt[3]{5y}}{3} = \dfrac{2\sqrt[3]{5y}}{3}$ ◀

Math at Work

Is a cold morning in Chicago any colder than a cold morning somewhere else? Surprisingly, sometimes the answer is yes.

Temperature alone does not give a clear indication of how cold the air will feel when you step outdoors. The speed of the wind influences our perception of the weather. Higher wind speeds make the temperature seem colder because the winds cause greater heat loss from the skin. Thus a 20-degree morning in the Windy City feels much colder than a 20-degree morning in a less windy part of the country.

Meteorologists use the term "wind-chill temperature" to take both temperature and wind speed into account. The wind-chill temperature is what the air temperature would have to be with no wind to give the same chilling effect on the skin. The wind-chill temperature indicates how cold the air temperature *feels*, regardless of the actual temperature.

Through experimentation in Antarctica, Paul Siple developed a formula in 1945 that gives the wind chill temperature, W, from the velocity of the wind, v, and the Fahrenheit temperature, t:

$$W = 9.14 - \frac{(10.5 + 6.7\sqrt{v} - .45v)(457 - 5t)}{110}$$

Using this formula, explain who will feel colder: a person in Minneapolis, where the temperature is $-10°F$ and the wind is 35 mph, or a person in Chicago, where the temperature is $5°F$ and the wind is 55 mph?

Rationalizing the Denominator

Numbers like $\sqrt{2}$, $\sqrt{3}$, and $\sqrt{5}$ are irrational numbers. If any irrational numbers appear in the denominator of a fraction, we can change the denominator into a rational number, or **rationalize** it. We rationalize a denominator by multiplying both numerator and denominator by another radical that makes the denominator rational. For example,

$$\frac{\sqrt{3}}{\sqrt{5}} = \frac{\sqrt{3}}{\sqrt{5}} \cdot \frac{\sqrt{5}}{\sqrt{5}} = \frac{\sqrt{15}}{5}.$$ Multiply both numerator and denominator by $\sqrt{5}$ to rationalize the denominator.

Note that the product rule allows us to write $\sqrt{3} \cdot \sqrt{5}$ as $\sqrt{15}$.

Simplifying Radicals

When we simplify any expression we try to make it look "simpler." When we simplify a radical we have three specific rules to follow.

Rules for Simplifying Radicals of Index _n_

A simplified radical of index n has
1. *no* perfect nth powers as factors of the radicand,
2. *no* fractions inside the radical, and
3. *no* radicals in the denominator.

The radical expressions in the next example do not satisfy the three rules given above. We use the product rule, quotient rule, and rationalizing the denominator to get a simplified form for each expression.

EXAMPLE 5   Simplify.

a) $\dfrac{\sqrt{10}}{\sqrt{6}}$ b) $\sqrt[3]{\dfrac{5}{2}}$ c) $\sqrt[4]{\dfrac{1}{3}}$

Solution

a) $\dfrac{\sqrt{10}}{\sqrt{6}} = \dfrac{\sqrt{10}}{\sqrt{6}} \cdot \dfrac{\sqrt{6}}{\sqrt{6}} = \dfrac{\sqrt{60}}{6} = \dfrac{\sqrt{4}\,\sqrt{15}}{6} = \dfrac{2\sqrt{15}}{6} = \dfrac{\sqrt{15}}{3}$

Note that $\sqrt{15} \div 3 \neq \sqrt{5}$.

b) Observe that $\sqrt[3]{2} \cdot \sqrt[3]{2} = \sqrt[3]{4}$, which is irrational. Therefore, we must multiply $\sqrt[3]{2}$ by $\sqrt[3]{4}$ to get $\sqrt[3]{8} = 2$, which is rational.

$$\sqrt[3]{\frac{5}{2}} = \frac{\sqrt[3]{5}}{\sqrt[3]{2}} = \frac{\sqrt[3]{5}}{\sqrt[3]{2}} \cdot \frac{\sqrt[3]{4}}{\sqrt[3]{4}} = \frac{\sqrt[3]{20}}{\sqrt[3]{8}} = \frac{\sqrt[3]{20}}{2}$$

c) $\sqrt[4]{\dfrac{1}{3}} = \dfrac{\sqrt[4]{1}}{\sqrt[4]{3}} = \dfrac{1}{\sqrt[4]{3}} \cdot \dfrac{\sqrt[4]{27}}{\sqrt[4]{27}} = \dfrac{\sqrt[4]{27}}{\sqrt[4]{81}} = \dfrac{\sqrt[4]{27}}{3}$ Do not omit the index of the radical in any step.   ◀

Simplifying Radicals Involving Variables

To simplify radicals involving variables, we must learn to recognize exponential expressions that are perfect squares, perfect cubes, and so on. Any even power of a variable is called a **perfect square.** For example,

$$x^2, \qquad w^4, \qquad y^8, \qquad z^{14}, \qquad \text{and} \qquad x^{50}$$

are perfect squares. They are called perfect squares because they are squares of variables with integral powers. If we assume that the variables represent positive numbers, we can write

$$\sqrt{x^2} = x, \qquad \sqrt{w^4} = w^2, \qquad \sqrt{y^8} = y^4, \qquad \sqrt{z^{14}} = z^7, \qquad \text{and} \qquad \sqrt{x^{50}} = x^{25}.$$

Note that to get the square root, we take one-half of the exponent.

EXAMPLE 6   Simplify each expression. Assume that all variables represent positive real numbers.

a) $\sqrt{12x^6}$ **b)** $\sqrt{98x^5y^9}$ **c)** $\sqrt{\dfrac{a}{b}}$ **d)** $\sqrt{\dfrac{x^3}{y^5}}$

Solution

a) Use the product rule to place all perfect squares under the first radical symbol and the remaining factors under the second.

$$\sqrt{12x^6} = \sqrt{4x^6 \cdot 3} = \sqrt{4x^6} \cdot \sqrt{3} = 2x^3\sqrt{3}$$

b) $\sqrt{98x^5y^9} = \sqrt{49x^4y^8}\,\sqrt{2xy} = 7x^2y^4\sqrt{2xy}$

c) $\sqrt{\dfrac{a}{b}} = \dfrac{\sqrt{a}}{\sqrt{b}} = \dfrac{\sqrt{a}\,\sqrt{b}}{\sqrt{b}\,\sqrt{b}} = \dfrac{\sqrt{ab}}{b}$

d) $\sqrt{\dfrac{x^3}{y^5}} = \dfrac{\sqrt{x^3}}{\sqrt{y^5}} = \dfrac{\sqrt{x^2}\,\sqrt{x}}{\sqrt{y^4}\,\sqrt{y}} = \dfrac{x\sqrt{x}}{y^2\sqrt{y}} = \dfrac{x\sqrt{x}}{y^2\sqrt{y}}\,\dfrac{\sqrt{y}}{\sqrt{y}} = \dfrac{x\sqrt{xy}}{y^2\,y} = \dfrac{x\sqrt{xy}}{y^3}$ ◀

Any variable with an exponent that is a multiple of 3 is called a **perfect cube.** For example,

$$a^3, \; b^6, \; c^{15}, \text{ and } w^{39}$$

are perfect cubes. Each of these expressions is the cube of a variable with an integral exponent. For any values of the variables, we can write

$$\sqrt[3]{a^3} = a, \qquad \sqrt[3]{b^6} = b^2, \qquad \sqrt[3]{c^{15}} = c^5, \qquad \text{and} \qquad \sqrt[3]{w^{39}} = w^{13}.$$

Note that to get the cube root we take one-third of the exponent. If the exponent is a multiple of 4, we have a perfect 4th power, if the exponent is a multiple of 5, we have a perfect 5th power, and so on. In the next example, we simplify radicals with index higher than 2.

EXAMPLE 7 Simplify. Assume that the variables represent positive numbers.

a) $\sqrt[3]{40x^8}$ b) $\sqrt[4]{x^{12}y^5}$ c) $\sqrt[3]{\dfrac{x}{y}}$

Solution

a) Use the product rule to place the largest perfect cube factors under the first radical and the remaining factors under the second.

$$\sqrt[3]{40x^8} = \sqrt[3]{8x^6}\,\sqrt[3]{5x^2} = 2x^2\sqrt[3]{5x^2}$$

b) $\sqrt[4]{x^{12}y^5} = \sqrt[4]{x^{12}y^4}\,\sqrt[4]{y} = x^3y\,\sqrt[4]{y}$

c) $\sqrt[3]{\dfrac{x}{y}} = \dfrac{\sqrt[3]{x}}{\sqrt[3]{y}} = \dfrac{\sqrt[3]{x}}{\sqrt[3]{y}}\dfrac{\sqrt[3]{y^2}}{\sqrt[3]{y^2}} = \dfrac{\sqrt[3]{xy^2}}{\sqrt[3]{y^3}} = \dfrac{\sqrt[3]{xy^2}}{y}$ ◄

Warm-ups

True or false?

1. $2^{1/2} = \sqrt{2}$ 2. $3^{1/3} = \sqrt{3}$ 3. $2^{2/3} = \sqrt[3]{4}$

4. $\sqrt{81} = \sqrt{9}$ 5. $\sqrt{19^2} = 19$ 6. $\sqrt[3]{a^{27}} = a^3$

7. $\dfrac{\sqrt{2}}{2} = \dfrac{1}{\sqrt{2}}$ 8. $\dfrac{\sqrt{10}}{2} = \sqrt{5}$ 9. $\sqrt{2^{-4}} = \dfrac{1}{4}$

10. $\dfrac{\sqrt{6}}{\sqrt{3}} = \sqrt{2}$

5.2 EXERCISES

All variables in the following exercises represent positive numbers.

Write the radical expressions in exponential notation and the exponential expressions in radical notation. Do not simplify. See Example 1.

1. $\sqrt{x^5}$ 2. $\sqrt{a^3}$ 3. $\sqrt[3]{a^{-12}}$ 4. $\sqrt[3]{w^{-27}}$

5. $3^{1/2}$ 6. $6^{1/3}$ 7. $-5^{1/3}$ 8. $-7^{1/2}$

9. $2^{2/5}$ 10. $3^{2/3}$ 11. $x^{-2/3}$ 12. $x^{-2/5}$

Simplify each radical expression. See Example 2.

13. $\sqrt{121}$ 14. $\sqrt{64}$ 15. $\sqrt[3]{-1000}$ 16. $\sqrt[5]{-1}$

17. $\sqrt[4]{-81}$ 18. $\sqrt{2^8}$ 19. $\sqrt[4]{3^{12}}$ 20. $\sqrt[3]{2^{-9}}$

21. $\sqrt{10^{-2}}$ 22. $\sqrt{-10^{-4}}$

Write each of the following in exponential notation, then simplify. See Examples 1 and 2.

23. $\sqrt{x^6}$

24. $\sqrt{x^{16}}$

25. $\sqrt[3]{x^{15}}$

26. $\sqrt[3]{a^{18}}$

27. $\sqrt{9^2}$

28. $\sqrt[4]{a^4}$

29. $\sqrt[4]{a^{16}}$

30. $\sqrt{9^{16}}$

31. $\sqrt{4^{16}}$

32. $\sqrt[5]{w^{30}}$

33. $\sqrt[5]{a^{20}}$

34. $\sqrt{w^4}$

Simplify each expression. See Example 3.

35. $\sqrt{20}$

36. $\sqrt{50}$

37. $\sqrt{45w}$

38. $\sqrt{48t}$

39. $\sqrt{288}$

40. $\sqrt{242}$

41. $\sqrt[3]{54}$

42. $\sqrt[3]{-48}$

43. $\sqrt[4]{32a}$

44. $\sqrt[4]{80xy}$

Simplify each expression. See Example 4.

45. $\sqrt{\dfrac{50}{9}}$

46. $\sqrt{\dfrac{18}{25}}$

47. $\sqrt[3]{\dfrac{16x}{27}}$

48. $\sqrt[3]{\dfrac{-81a}{1000}}$

49. $\sqrt{\dfrac{8x}{49}}$

50. $\sqrt{\dfrac{12b}{121}}$

51. $\sqrt[4]{\dfrac{32a}{81}}$

52. $\sqrt[4]{\dfrac{162y}{625}}$

Simplify. See Example 5.

53. $\sqrt{\dfrac{1}{2}}$

54. $\sqrt{\dfrac{1}{3}}$

55. $\sqrt{\dfrac{3}{8}}$

56. $\sqrt{\dfrac{2}{27}}$

57. $\dfrac{2}{\sqrt{5}}$

58. $\dfrac{5}{\sqrt{3}}$

59. $\dfrac{\sqrt{5}}{\sqrt{12}}$

60. $\dfrac{\sqrt{7}}{\sqrt{18}}$

61. $\dfrac{\sqrt{3}}{\sqrt{12}}$

62. $\dfrac{\sqrt{2}}{\sqrt{18}}$

63. $\sqrt[3]{\dfrac{1}{4}}$

64. $\dfrac{\sqrt[3]{5}}{\sqrt[3]{2}}$

65. $\dfrac{\sqrt[3]{6}}{\sqrt[3]{5}}$

66. $\sqrt[4]{\dfrac{2}{27}}$

Simplify. See Examples 6 and 7.

67. $\sqrt{12x^8}$

68. $\sqrt{72x^{10}}$

69. $\sqrt{60a^9}$

70. $\sqrt{63w^{15}}$

71. $\sqrt{x/y}$

72. $\sqrt{\dfrac{x^2}{a}}$

73. $\dfrac{\sqrt{a^3}}{\sqrt{b^7}}$

74. $\dfrac{\sqrt{w^5}}{\sqrt{y^8}}$

75. $\dfrac{\sqrt{ab^3}}{\sqrt{a^3b^2}}$

76. $\dfrac{\sqrt{m^3n^5}}{\sqrt{m^5n}}$

77. $\sqrt[3]{x^{13}}$

78. $\sqrt[3]{x^{17}}$

79. $\sqrt[4]{x^9y^6}$

80. $\sqrt[3]{x^{12}y^5z^3}$

81. $\sqrt[5]{64x^{22}}$

82. $\sqrt[4]{w^6y^9}$

83. $\sqrt[3]{\dfrac{a}{b}}$

84. $\sqrt[3]{\dfrac{a}{w^2}}$

*85. $\dfrac{\sqrt[3]{a^2b}}{\sqrt[3]{4ab^2}\ \sqrt[3]{3ab^5}}$

*86. $\dfrac{\sqrt[3]{5xy^2}}{\sqrt[3]{18x^2y}}$

*87. $\dfrac{6m \cdot \sqrt[5]{5mn^2}}{\sqrt[5]{12m^3n}}$

*88. $\dfrac{3ab \cdot \sqrt[5]{2b}}{\sqrt[5]{24a^2}}$

Use a calculator to find a decimal approximation to the following radical expressions. Round to three decimal places.

89. $\dfrac{5}{\sqrt{3}}$

90. $\sqrt{\dfrac{2}{27}}$

91. $\sqrt[3]{\dfrac{1}{3}}$

92. $\sqrt[3]{56}$

93. $\dfrac{\sqrt[3]{9}}{\sqrt[3]{4}}$

94. $\dfrac{\sqrt[4]{25}}{\sqrt{5}}$

95. $\dfrac{\sqrt[6]{16}}{\sqrt[3]{4}}$

96. $\sqrt[5]{2.48832}$

5.3   Operations with Radicals

IN THIS SECTION:

- **Adding and Subtracting Radicals**
- **Multiplying Radicals**
- **Conjugates**

In this section we will learn to add, subtract, and multiply radical expressions. The skills of simplifying radicals that we learned in Section 5.2 will be useful here.

Adding and Subtracting Radicals

If we need to know the sum of $\sqrt{2}$ and $\sqrt{3}$, we can look in the square-root table of Appendix A or use a calculator to find that $\sqrt{2} \approx 1.414$ and $\sqrt{3} \approx 1.732$. (The symbol $\approx$ means "is approximately equal to.") We can then add the decimal numbers and get

$$\sqrt{2} + \sqrt{3} \approx 1.414 + 1.732 = 3.146.$$

We cannot write an exact decimal form for $\sqrt{2} + \sqrt{3}$; the number 3.146 is an approximation of $\sqrt{2} + \sqrt{3}$. To represent the exact value of $\sqrt{2} + \sqrt{3}$, we use the form $\sqrt{2} + \sqrt{3}$. This form cannot be simplified any further. However, a sum of like radicals can be simplified. **Like radicals** are radicals that have the same index and same radicand.

To simplify the sum $3\sqrt{2} + 5\sqrt{2}$, we can use the fact that $3x + 5x = 8x$ is true for any value of x. By substituting $\sqrt{2}$ for x, we have $3\sqrt{2} + 5\sqrt{2} = 8\sqrt{2}$. Thus like radicals can be combined just as like terms are combined.

EXAMPLE 1   Simplify the following expressions. Assume that the variables represent positive numbers.

a) $3\sqrt{5} + 4\sqrt{5}$

b) $\sqrt[4]{w} - 6\sqrt[4]{w}$

c) $\sqrt{3} + \sqrt{5} - 4\sqrt{3} + 6\sqrt{5}$

d) $3\sqrt[3]{6x} + 2\sqrt[3]{x} + \sqrt[3]{6x} + \sqrt[3]{x}$

Solution

a) $3\sqrt{5} + 4\sqrt{5} = 7\sqrt{5}$

b) $\sqrt[4]{w} - 6\sqrt[4]{w} = -5\sqrt[4]{w}$

c) $\sqrt{3} + \sqrt{5} - 4\sqrt{3} + 6\sqrt{5} = -3\sqrt{3} + 7\sqrt{5}$ Note that only like radicals are combined.

d) $3\sqrt[3]{6x} + 2\sqrt[3]{x} + \sqrt[3]{6x} + \sqrt[3]{x} = 4\sqrt[3]{6x} + 3\sqrt[3]{x}$ ◀

We may have to simplify the radicals to determine which ones can be combined.

EXAMPLE 2   Simplify. Assume that the variables represent positive numbers.

a $\sqrt{8} + \sqrt{18}$

b) $\sqrt{\dfrac{1}{5}} + \sqrt{20}$

c) $\sqrt{2x^3} - \sqrt{4x^2} + \sqrt{18x^3}$

d) $\sqrt[3]{16} - \sqrt[3]{2}$

Solution

a) $\sqrt{8} + \sqrt{18} = \sqrt{4} \cdot \sqrt{2} + \sqrt{9} \cdot \sqrt{2}$

$\qquad\qquad\qquad = 2\sqrt{2} + 3\sqrt{2}$ Simplify each radical.

$\qquad\qquad\qquad = 5\sqrt{2}$ Note that $\sqrt{8} + \sqrt{18} \neq \sqrt{26}$.

b) $\sqrt{\dfrac{1}{5}} + \sqrt{20} = \dfrac{\sqrt{5}}{5} + 2\sqrt{5}$ Because $\sqrt{\dfrac{1}{5}} = \dfrac{1}{\sqrt{5}}\,\dfrac{\sqrt{5}}{\sqrt{5}} = \dfrac{\sqrt{5}}{5}$ and $\sqrt{20} = 2\sqrt{5}$

$\qquad\qquad\qquad = \left(\dfrac{1}{5} + 2\right)\sqrt{5}$ Factor out $\sqrt{5}$.

$\qquad\qquad\qquad = \left(\dfrac{11}{5}\right)\sqrt{5}$ $\dfrac{1}{5} + 2 = \dfrac{11}{5}$

$\qquad\qquad\qquad = \dfrac{11\sqrt{5}}{5}$

c) $\sqrt{2x^3} - \sqrt{4x^2} + \sqrt{18x^3} = \sqrt{x^2}\,\sqrt{2x} - 2x + \sqrt{9x^2} \cdot \sqrt{2x}$

$\qquad\qquad\qquad\qquad\quad = x\sqrt{2x} - 2x + 3x\sqrt{2x}$ Simplify each radical.

$\qquad\qquad\qquad\qquad\quad = -2x + 4x\sqrt{2x}$ Add like radicals only.

d) $\sqrt[3]{16} - \sqrt[3]{2} = \sqrt[3]{8}\,\sqrt[3]{2} - \sqrt[3]{2} = 2\sqrt[3]{2} - \sqrt[3]{2} = \sqrt[3]{2}$ ◀

Multiplying Radicals

We have already multiplied radicals in Section 5.2 when we rationalized denominators. The product rule for radicals allows multiplication of radicals with the same index, such as

$$\sqrt{5} \cdot \sqrt{3} = \sqrt{15}, \qquad \sqrt[3]{2} \cdot \sqrt[3]{5} = \sqrt[3]{10}, \qquad \text{and} \qquad \sqrt[5]{x^2} \cdot \sqrt[5]{x} = \sqrt[5]{x^3}.$$

Note that the product rule does not allow multiplication of radicals that have different indexes. We cannot use the product rule to multiply $\sqrt{2}$ and $\sqrt[3]{5}$.

EXAMPLE 3   Multiply and simplify the following expressions. Assume that the variables represent positive numbers.

a) $5\sqrt{6} \cdot 4\sqrt{3}$ **b)** $\sqrt{3a^2} \cdot \sqrt{6a}$ **c)** $\sqrt[3]{4} \cdot \sqrt[3]{4}$ **d)** $\sqrt[4]{x^3} \cdot \sqrt[4]{x^2}$

Solution

a) $5\sqrt{6} \cdot 4\sqrt{3} = 5 \cdot 4 \cdot \sqrt{6} \cdot \sqrt{3}$

$\qquad\qquad\quad = 20\sqrt{18}$ By the product rule

$\qquad\qquad\quad = 20 \cdot 3\sqrt{2}$ $\sqrt{18} = \sqrt{9} \cdot \sqrt{2} = 3\sqrt{2}$

$\qquad\qquad\quad = 60\sqrt{2}$

b) $\sqrt{3a^2} \cdot \sqrt{6a} = \sqrt{18a^3}$ By the product rule
$$= \sqrt{9a^2} \cdot \sqrt{2a}$$
$$= 3a\sqrt{2a} \qquad \text{Simplify.}$$

c) $\sqrt[3]{4} \cdot \sqrt[3]{4} = \sqrt[3]{16} = \sqrt[3]{8} \cdot \sqrt[3]{2} = 2\sqrt[3]{2}$

d) $\sqrt[4]{x^3} \cdot \sqrt[4]{x^2} = \sqrt[4]{x^5} = \sqrt[4]{x^4} \cdot \sqrt[4]{x} = x\sqrt[4]{x}$ ◀

To find a product such as $3\sqrt{2}(\sqrt{6} + \sqrt{2})$, in which addition is involved, we use the distributive property, $a(b + c) = ab + ac$.

$$3\sqrt{2}(\sqrt{6} + \sqrt{2}) = 3\sqrt{2} \cdot \sqrt{6} + 3\sqrt{2} \cdot \sqrt{2} \qquad \text{By the distributive property}$$
$$= 3\sqrt{12} + 3 \cdot 2 \qquad \begin{array}{l}\text{Because } \sqrt{2} \cdot \sqrt{6} = \sqrt{12} \\ \text{and } \sqrt{2} \cdot \sqrt{2} = 2\end{array}$$
$$= 3\sqrt{4} \cdot \sqrt{3} + 6 \qquad \sqrt{12} = \sqrt{4} \cdot \sqrt{3}$$
$$= 3 \cdot 2\sqrt{3} + 6 \qquad \text{Simplify } \sqrt{12}.$$
$$= 6\sqrt{3} + 6$$

A product such as $(\sqrt{2} + \sqrt{5})(2 - \sqrt{3})$ is a product of two binomials where the variables have been replaced by numbers. We can use FOIL to find the product:

$$(\sqrt{2} + \sqrt{5})(2 - \sqrt{3}) = 2\sqrt{2} - \sqrt{3} \cdot \sqrt{2} + 2\sqrt{5} - \sqrt{3} \cdot \sqrt{5}$$
$$= 2\sqrt{2} - \sqrt{6} + 2\sqrt{5} - \sqrt{15} \qquad \begin{array}{l}\text{There are no} \\ \text{like radicals.}\end{array}$$

EXAMPLE 4 Multiply and simplify.

a) $3\sqrt{2}(4\sqrt{2} - \sqrt{3})$ **b)** $(2\sqrt{3} + \sqrt{5})(3\sqrt{3} - 2\sqrt{5})$

c) $\sqrt[3]{a}(\sqrt[3]{a} - \sqrt[3]{a^2})$

Solution

a) $3\sqrt{2}(4\sqrt{2} - \sqrt{3}) = 3\sqrt{2} \cdot 4\sqrt{2} - 3\sqrt{2} \cdot \sqrt{3} \qquad \text{Distributive property}$
$$= 12 \cdot 2 - 3\sqrt{6} \qquad \text{Product rule}$$
$$= 24 - 3\sqrt{6}$$

b) $(2\sqrt{3} + \sqrt{5})(3\sqrt{3} - 2\sqrt{5})$
$$= 2\sqrt{3} \cdot 3\sqrt{3} - 2\sqrt{3} \cdot 2\sqrt{5} + \sqrt{5} \cdot 3\sqrt{3} - \sqrt{5} \cdot 2\sqrt{5}$$
$$= 18 - 4\sqrt{15} + 3\sqrt{15} - 10$$
$$= 8 - \sqrt{15} \qquad \text{Combine like radicals.}$$

c) $\sqrt[3]{a}(\sqrt[3]{a} - \sqrt[3]{a^2}) = \sqrt[3]{a^2} - \sqrt[3]{a^3} = \sqrt[3]{a^2} - a$ ◀

Conjugates

Consider the following product of a sum and a difference:

$$(4 - \sqrt{3})(4 + \sqrt{3}) = 16 - 3 = 13$$

The product of the irrational numbers $4 - \sqrt{3}$ and $4 + \sqrt{3}$ is the rational number 13. For this reason the expressions $4 - \sqrt{3}$ and $4 + \sqrt{3}$ are called **conjugates** of one another. We will use conjugates in Section 5.4 to rationalize some denominators.

EXAMPLE 5 Find the products of the conjugates.

a) $(2 + 3\sqrt{5})(2 - 3\sqrt{5})$

b) $(\sqrt{3} - \sqrt{2})(\sqrt{3} + \sqrt{2})$

Solution

a) $(2 + 3\sqrt{5})(2 - 3\sqrt{5}) = 4 - 45 = -41$
b) $(\sqrt{3} - \sqrt{2})(\sqrt{3} + \sqrt{2}) = 3 - 2 = 1$ ◄

Warm-ups

True or false?

1. $\sqrt{3} + \sqrt{3} = \sqrt{6}$
3. $2\sqrt{3} \cdot 3\sqrt{3} = 6\sqrt{3}$
5. $2\sqrt{5} \cdot 3\sqrt{2} = 6\sqrt{10}$
7. $\sqrt{2}(\sqrt{3} - \sqrt{2}) = \sqrt{6} - 2$
9. $(\sqrt{2} + \sqrt{3})^2 = 2 + 3$

2. $\sqrt{8} + \sqrt{2} = 3\sqrt{2}$
4. $\sqrt[3]{2} \cdot \sqrt[3]{2} = 2$
6. $2\sqrt{5} + 3\sqrt{5} = 5\sqrt{10}$
8. $\sqrt{12} = 2\sqrt{6}$
10. $(\sqrt{3} - \sqrt{2})(\sqrt{3} + \sqrt{2}) = 1$

5.3 EXERCISES

All variables in the following exercises represent positive numbers.

Simplify the sums and differences. Write the answers exactly. See Example 1.

1. $\sqrt{3} - 2\sqrt{3}$
4. $3\sqrt{6a} + 7\sqrt{6a}$
7. $\sqrt{3} - \sqrt{5} + 3\sqrt{3} - \sqrt{5}$
9. $\sqrt[3]{2} + \sqrt[3]{x} - \sqrt[3]{2} + 4\sqrt[3]{x}$
11. $\sqrt[3]{x} - \sqrt[3]{2x} + \sqrt[3]{x}$

2. $\sqrt{5} - 3\sqrt{5}$
5. $2\sqrt[3]{2} + 3\sqrt[3]{2}$

3. $5\sqrt{7x} + 4\sqrt{7x}$
6. $\sqrt[3]{4} + 4\sqrt[3]{4}$

8. $\sqrt{2} - 5\sqrt{3} - 7\sqrt{2} + 9\sqrt{3}$
10. $\sqrt[3]{5y} - 4\sqrt[3]{5y} + \sqrt[3]{x} + \sqrt[3]{x}$
12. $\sqrt[3]{ab} + \sqrt{a} + 5\sqrt{a} + \sqrt[3]{ab}$

Simplify each expression. Give exact answers. See Example 2.

13. $\sqrt{2} + \sqrt{8}$
16. $\dfrac{\sqrt{3}}{3} - \sqrt{3}$
19. $\sqrt{45} - \sqrt{18} + \sqrt{50} - \sqrt{20}$
22. $\sqrt{32} + \sqrt{\dfrac{1}{2}}$
25. $\sqrt[3]{24} + \sqrt[3]{375}$
28. $\sqrt[5]{64} + \sqrt[5]{2}$

14. $\sqrt{20} - \sqrt{125}$
17. $\sqrt{8} + \sqrt{28}$
20. $\sqrt{12} - \sqrt{18} - \sqrt{300} + \sqrt{98}$
23. $\sqrt[3]{54t} - \sqrt[3]{16t}$
26. $\sqrt[3]{2000w^2} - \sqrt[3]{16w^2}$

15. $\dfrac{\sqrt{2}}{2} + \sqrt{2}$
18. $\sqrt{12} + \sqrt{24}$
21. $\sqrt{80} + \sqrt{\dfrac{1}{5}}$
24. $\sqrt[3]{24} + \sqrt[3]{81}$
27. $\sqrt[4]{48} - \sqrt[4]{243}$

Simplify the products. Write the answers exactly. See Examples 3 and 4.

29. $\sqrt{3} \cdot \sqrt{5}$

30. $\sqrt{5} \cdot \sqrt{7}$

31. $(2\sqrt{5}) \cdot (3\sqrt{10})$

32. $(2\sqrt{7a}) \cdot (3\sqrt{2a})$

33. $(2\sqrt{5c})(5\sqrt{5})$

34. $(3\sqrt{2})(-4\sqrt{10})$

35. $(\sqrt[4]{9})(\sqrt[4]{27})$

36. $(\sqrt[3]{5})(\sqrt[3]{100})$

37. $(2\sqrt{3})^2$

38. $(-4\sqrt{2})^2$

39. $(\sqrt[3]{4x^2})(\sqrt[3]{2x^2})$

40. $(\sqrt[4]{4x^2})(\sqrt[4]{4x})$

41. $2\sqrt{3}(\sqrt{6} + 3\sqrt{3})$

42. $2\sqrt{5}(\sqrt{3} + 3\sqrt{5})$

43. $\sqrt{5}(\sqrt{10} - 2)$

44. $\sqrt{6}(\sqrt{15} - 1)$

45. $\sqrt{3t}(\sqrt{6t} + 3)$

46. $\sqrt{2}(\sqrt{x} - \sqrt{2x})$

47. $(\sqrt{3} + 2)(\sqrt{3} - 5)$

48. $(\sqrt{5} + 2)(\sqrt{5} - 6)$

49. $(\sqrt{11} - \sqrt{3})(\sqrt{11} + \sqrt{3})$

50. $(\sqrt{2} + 5)(\sqrt{2} + 5)$

51. $(2\sqrt{5} - 7)(2\sqrt{5} + 4)$

52. $(2\sqrt{6} - 3)(2\sqrt{6} + 4)$

53. $(2\sqrt{3} - \sqrt{6})(\sqrt{3} + 2\sqrt{6})$

54. $(3\sqrt{3} - \sqrt{2})(\sqrt{2} + \sqrt{3})$

55. $(\sqrt{2} + 3)^2$

56. $(\sqrt{2} - \sqrt{3})^2$

Find the product of each pair of conjugates. See Example 5.

57. $(\sqrt{3} - 2)(\sqrt{3} + 2)$

58. $(7 - \sqrt{3})(7 + \sqrt{3})$

59. $(\sqrt{5} + \sqrt{2})(\sqrt{5} - \sqrt{2})$

60. $(\sqrt{6} + \sqrt{5})(\sqrt{6} - \sqrt{5})$

61. $(2\sqrt{5} + 1)(2\sqrt{5} - 1)$

62. $(3\sqrt{2} - 4)(3\sqrt{2} + 4)$

63. $(3\sqrt{2} + \sqrt{5})(3\sqrt{2} - \sqrt{5})$

64. $(2\sqrt{3} - \sqrt{7})(2\sqrt{3} + \sqrt{7})$

65. $(5 - 3\sqrt{6})(5 + 3\sqrt{6})$

66. $(4\sqrt{2} + 3\sqrt{5})(4\sqrt{2} - 3\sqrt{5})$

Simplify each expression.

67. $\sqrt{300} + \sqrt{3}$

68. $\dfrac{1}{\sqrt{2}} + \sqrt{2}$

69. $2\sqrt{5} \cdot 5\sqrt{6}$

70. $3\sqrt{6} \cdot 5\sqrt{10}$

71. $2\sqrt{7}(\sqrt{7} - 2)$

72. $(2 + \sqrt{7})(\sqrt{7} - 2)$

73. $(2\sqrt{5} + \sqrt{2})(3\sqrt{5} - \sqrt{2})$

74. $(3\sqrt{2} - \sqrt{3})(2\sqrt{2} + 3\sqrt{3})$

75. $\dfrac{\sqrt{2}}{3} + \dfrac{\sqrt{2}}{5}$

76. $\dfrac{\sqrt{2}}{4} + \dfrac{\sqrt{3}}{5}$

77. $(5 + 2\sqrt{2})(5 - 2\sqrt{2})$

78. $(3 - 2\sqrt{7})(3 + 2\sqrt{7})$

79. $(3 + \sqrt{x})^2$

80. $(1 - \sqrt{x})^2$

81. $(5\sqrt{x} - 3)^2$

82. $(3\sqrt{a} + 2)^2$

83. $(1 + \sqrt{x + 2})^2$

84. $(\sqrt{x - 1} + 1)^2$

85. $(\sqrt{x - 3} + 2)^2$

86. $(1 - \sqrt{x + 3})^2$

87. $2\sqrt{a} + 3\sqrt{a}$

88. $5\sqrt{wy} - 7\sqrt{wy} + 6\sqrt{wy}$

89. $\sqrt{4w} - \sqrt{9w}$

90. $10\sqrt{m} - \sqrt{16m}$

91. $\sqrt{50a} + \sqrt{18a} - \sqrt{2a}$

92. $\sqrt{200z} + \sqrt{128z} - \sqrt{8z}$

93. $\sqrt{x^5} + 2x\sqrt{x^3}$

94. $\sqrt{8x^3} + \sqrt{50x^3} - x\sqrt{2x}$

95. $3a - \sqrt{a^3} + \sqrt{a^2}$

96. $(\sqrt{a} + 2)(\sqrt{a} - 2)$

97. $(\sqrt{x - 3} + 4)^2$

98. $(\sqrt{a} + 3)^2$

99. $(\sqrt{2x - 1} + 3)^2$

100. $(3 + \sqrt{x - 4})^2$

101. $(\sqrt{w} - 2)(\sqrt{w} + 2)$

102. $(2 - \sqrt{x + 1})^2$

103. $(\sqrt{x} - 5)^2$

104. $\sqrt{w^2} + \sqrt{w^6} - \sqrt{4w^2} + \sqrt{9w^6}$

105. $3\sqrt{m} \cdot 5\sqrt{m}$

106. $4\sqrt{w} \cdot 4\sqrt{w}$

107. $\sqrt{3x^3} \cdot \sqrt{6x^2}$

108. $\sqrt{2t} \cdot \sqrt{10t}$

5.4 More Operations with Radicals

IN THIS SECTION:

- Dividing Radicals
- Rationalizing the Denominator
- Powers of Radical Expressions

In this section we will continue studying operations with radicals. We will learn to rationalize some denominators that are different from those rationalized in Section 5.2.

Dividing Radicals

We know that

$$2 \div 6 = \frac{2}{6} = \frac{1}{3}.$$

We can divide radicals in the same way that we did the division of 2 and 6 above: we place one radical above the other and simplify. In some cases we simplify by rationalizing the denominator, and in others we actually divide the radicals. For example, the quotient rule for radicals in Section 5.2 says that the quotient of two square roots is the square root of the quotient. Thus,

$$\sqrt{10} \div \sqrt{5} = \frac{\sqrt{10}}{\sqrt{5}} = \sqrt{\frac{10}{5}} = \sqrt{2}. \qquad \text{Note that } \sqrt{10} \div 5 \neq \sqrt{2}.$$

There is no need to write all of the steps we have shown; we can simply write $\sqrt{10} \div \sqrt{5} = \sqrt{2}$. *Note that to apply the quotient rule, the radicals must have the same index.*

EXAMPLE 1 Divide and simplify. Assume that the variables represent positive numbers.

a) $\sqrt{5} \div \sqrt{3}$
c) $12\sqrt{6} \div 3\sqrt{2}$

b) $(3\sqrt{2}) \div (2\sqrt{3})$
d) $\sqrt[3]{10x^2} \div \sqrt[3]{5x}$

Solution

a) Because 5 is not evenly divisible by 3, we must rationalize the denominator.

$$\sqrt{5} \div \sqrt{3} = \frac{\sqrt{5}}{\sqrt{3}} = \frac{\sqrt{5}}{\sqrt{3}} \frac{\sqrt{3}}{\sqrt{3}} = \frac{\sqrt{15}}{3} \qquad \text{Note that } \frac{\sqrt{15}}{3} \neq \sqrt{5}.$$

b) $3\sqrt{2} \div 2\sqrt{3} = \dfrac{3\sqrt{2}}{2\sqrt{3}} = \dfrac{3\sqrt{2}\,\sqrt{3}}{2\sqrt{3}\,\sqrt{3}} = \dfrac{3\sqrt{6}}{2\cdot3} = \dfrac{\sqrt{6}}{2}$ Note that $\sqrt{6} \div 2 \neq \sqrt{3}$.

c) $12\sqrt{6} \div 3\sqrt{2} = \dfrac{12\sqrt{6}}{3\sqrt{2}} = 4\sqrt{3}$ $\sqrt{6} \div \sqrt{2} = \sqrt{3}$

d) $\sqrt[3]{10x^2} \div \sqrt[3]{5x} = \sqrt[3]{2x}$ By the quotient rule   ◄

In Chapter 6 it will be necessary to simplify expressions like

$$\frac{2 + \sqrt{8}}{2}.$$

In Chapter 4 we learned to reduce rational expressions by dividing out factors that are common to the numerator and denominator. To simplify $\dfrac{2 + \sqrt{8}}{2}$, we simplify $\sqrt{8}$, factor the numerator using the distributive property, and then divide out the common factors.

$$\frac{2 + \sqrt{8}}{2} = \frac{2 + 2\sqrt{2}}{2} = \frac{\cancel{2}(1 + \sqrt{2})}{\cancel{2}} = 1 + \sqrt{2}$$

EXAMPLE 2 Simplify the following radical expressions.

a) $\dfrac{4 - \sqrt{12}}{4}$ 　　　　　　　　**b)** $\dfrac{-6 + \sqrt{20}}{-2}$

Solution

a) $\dfrac{4 - \sqrt{12}}{4} = \dfrac{4 - 2\sqrt{3}}{4} = \dfrac{\cancel{2}(2 - \sqrt{3})}{\cancel{2}\cdot 2} = \dfrac{2 - \sqrt{3}}{2}$

Note that the remaining 2 in the numerator is not a factor of the numerator and cannot be divided out with the 2 in the denominator.

b) $\dfrac{-6 + \sqrt{20}}{-2} = \dfrac{-6 + 2\sqrt{5}}{-2} = \dfrac{-\cancel{2}(3 - \sqrt{5})}{-\cancel{2}} = 3 - \sqrt{5}$ ◄

Rationalizing the Denominator

We learned in Section 5.2 that a simplified expression involving radicals does not have radicals in the denominator. If an expression such as $4 - \sqrt{3}$ appears in a denominator, we can multiply both numerator and denominator by its conjugate, $4 + \sqrt{3}$, to get the rational number 13 in the denominator.

EXAMPLE 3 Simplify each expression by rationalizing the denominator.

a) $\dfrac{2 + \sqrt{3}}{4 - \sqrt{3}}$ 　　　　　　　　**b)** $\dfrac{\sqrt{5}}{\sqrt{6} + \sqrt{2}}$

Solution

a) $\dfrac{2 + \sqrt{3}}{4 - \sqrt{3}} = \dfrac{(2 + \sqrt{3})}{(4 - \sqrt{3})} \cdot \dfrac{(4 + \sqrt{3})}{(4 + \sqrt{3})}$ Multiply numerator and denominator by $4 + \sqrt{3}$.

$\qquad = \dfrac{8 + 6\sqrt{3} + 3}{13}$ $(4 - \sqrt{3})(4 + \sqrt{3}) = 13$

$\qquad = \dfrac{11 + 6\sqrt{3}}{13}$

b) $\dfrac{\sqrt{5}}{\sqrt{6} + \sqrt{2}} = \dfrac{\sqrt{5}}{(\sqrt{6} + \sqrt{2})} \cdot \dfrac{(\sqrt{6} - \sqrt{2})}{(\sqrt{6} - \sqrt{2})}$ Multiply numerator and denominator by $\sqrt{6} - \sqrt{2}$.

$\qquad = \dfrac{\sqrt{30} - \sqrt{10}}{4}$ $(\sqrt{6} + \sqrt{2})(\sqrt{6} - \sqrt{2}) = 6 - 2 = 4$ ◀

Powers of Radical Expressions

We can use the power of a product rule and the power of a power rule to simplify a radical expression raised to a power.

EXAMPLE 4 Simplify. Assume that the variables represent positive numbers.

a) $(5\sqrt{2})^3$ **b)** $(2\sqrt{x^3})^4$ **c)** $(3w\sqrt[3]{2w})^3$ **d)** $(2t\sqrt[4]{3t})^3$

Solution

a) $(5\sqrt{2})^3 = 5^3(\sqrt{2})^3$ By the power of a product rule

$\qquad = 125\sqrt{8}$ $(\sqrt{2})^3 = \sqrt{2^3} = \sqrt{8}$

$\qquad = 125 \cdot \sqrt{4} \cdot \sqrt{2}$

$\qquad = 250\sqrt{2}$

b) $(2\sqrt{x^3})^4 = 2^4(\sqrt{x^3})^4 = 16\sqrt{x^{12}} = 16x^6$

c) $(3w\sqrt[3]{2w})^3 = 3^3 \cdot w^3(\sqrt[3]{2w})^3 = 27w^3(2w) = 54w^4$

d) $(2t\sqrt[4]{3t})^3 = 2^3 \cdot t^3(\sqrt[4]{3t})^3 = 8t^3\sqrt[4]{27t^3}$ ◀

Warm-ups

True or false?

1. $\dfrac{\sqrt{6}}{\sqrt{2}} = \sqrt{3}$ **2.** $\dfrac{2}{\sqrt{2}} = \sqrt{2}$

3. $\dfrac{4 - \sqrt{10}}{2} = 2 - \sqrt{10}$ **4.** $\dfrac{1}{\sqrt{3}} = \dfrac{\sqrt{3}}{3}$

5. $\dfrac{8\sqrt{7}}{2\sqrt{7}} = 4\sqrt{7}$ **6.** $\dfrac{2(2 + \sqrt{3})}{(2 - \sqrt{3})(2 + \sqrt{3})} = 4 + 2\sqrt{3}$

7. $\dfrac{\sqrt{12}}{3} = \sqrt{4}$ **8.** $\dfrac{\sqrt{20}}{\sqrt{5}} = 2$

9. $(2\sqrt{4})^2 = 16$ **10.** $(3\sqrt{5})^3 = 27\sqrt{125}$

5.4 EXERCISES

All variables in the following exercises represent positive numbers. Divide and simplify. See Example 1.

1. $\sqrt{3} \div \sqrt{5}$

2. $\sqrt{5} \div \sqrt{7}$

3. $\sqrt[3]{20} \div \sqrt[3]{2}$

4. $\sqrt[4]{48} \div \sqrt[4]{3}$

5. $(3\sqrt{3}) \div (5\sqrt{6})$

6. $(2\sqrt{2}) \div (4\sqrt{10})$

7. $\dfrac{2\sqrt{3}}{3\sqrt{6}}$

8. $\dfrac{5\sqrt{12}}{4\sqrt{6}}$

9. $\sqrt[3]{x^7} \div \sqrt[3]{x}$

10. $\sqrt[4]{a^{10}} \div \sqrt[4]{a^2}$

Simplify. See Example 2.

11. $\dfrac{6 + \sqrt{45}}{3}$

12. $\dfrac{10 + \sqrt{50}}{5}$

13. $\dfrac{-2 + \sqrt{12}}{-2}$

14. $\dfrac{-6 + \sqrt{72}}{-6}$

15. $\dfrac{8 - \sqrt{32}}{20}$

16. $\dfrac{4 - \sqrt{28}}{6}$

17. $\dfrac{5 + \sqrt{75}}{10}$

18. $\dfrac{3 + \sqrt{18}}{6}$

Simplify each expression by rationalizing the denominator. See Example 3.

19. $\dfrac{\sqrt{2}}{\sqrt{3} - 1}$

20. $\dfrac{\sqrt{3}}{\sqrt{2} + \sqrt{6}}$

21. $\dfrac{\sqrt{2}}{\sqrt{6} + \sqrt{3}}$

22. $\dfrac{5}{\sqrt{7} - \sqrt{5}}$

23. $\dfrac{2\sqrt{3}}{3\sqrt{2} - \sqrt{5}}$

24. $\dfrac{3\sqrt{5}}{5\sqrt{2} + \sqrt{6}}$

25. $\dfrac{1 + 3\sqrt{2}}{2\sqrt{6} + 3\sqrt{10}}$

26. $\dfrac{3\sqrt{3} + 1}{4 - 5\sqrt{3}}$

Simplify. See Example 4.

27. $(\sqrt{2})^5$

28. $(\sqrt{3})^4$

29. $(2\sqrt{3})^3$

30. $(-2\sqrt{5})^3$

31. $(2\sqrt[3]{3})^3$

32. $(-2\sqrt[3]{4})^3$

33. $(2\sqrt[3]{5})^2$

34. $(3\sqrt[3]{4})^2$

35. $(\sqrt[3]{x})^5$

36. $(2\sqrt{y})^3$

37. $(\sqrt[3]{x^2})^6$

38. $(-2\sqrt[4]{y^3})^3$

Perform the indicated operations and simplify.

39. $\dfrac{\sqrt{3}}{\sqrt{2}} + \dfrac{2}{\sqrt{2}}$

40. $\dfrac{\sqrt{3}}{\sqrt{2}} + \dfrac{3\sqrt{6}}{2}$

41. $\dfrac{\sqrt{6}}{2} \cdot \dfrac{1}{\sqrt{3}}$

42. $\dfrac{\sqrt{6}}{\sqrt{7}} \cdot \dfrac{\sqrt{14}}{\sqrt{3}}$

43. $\dfrac{2}{\sqrt{7}} + \dfrac{5}{\sqrt{7}}$

44. $\dfrac{\sqrt{3}}{2\sqrt{2}} + \dfrac{\sqrt{5}}{3\sqrt{2}}$

45. $2\sqrt{w} \div 3\sqrt{w}$

46. $2 \div 3\sqrt{a}$

47. $\sqrt{a}(\sqrt{a} - 3)$

48. $3\sqrt{m}(2\sqrt{m} - 6)$

49. $4\sqrt{a}(a + \sqrt{a})$

50. $\sqrt{3ab}(\sqrt{3a} + \sqrt{3})$

51. $(2\sqrt{3m})^2$

52. $(-3\sqrt{4y})^2$

53. $(-2\sqrt{xy^2z})^2$

54. $(5a\sqrt{ab})^2$

55. $\sqrt[3]{m}(\sqrt[3]{m^2} - \sqrt[3]{m^5})$

56. $\sqrt[4]{w}(\sqrt[4]{w^3} - \sqrt[4]{w^7})$

57. $\sqrt[3]{8x^4} + \sqrt[3]{27x^4}$

58. $\sqrt[3]{16a^4} + a\sqrt[3]{2a}$

59. $(2m\sqrt[3]{2m^2})^3$

60. $(-2t\sqrt[6]{2t^2})^5$

61. $\dfrac{5}{\sqrt{2} - 1} + \dfrac{3}{\sqrt{2} + 1}$

62. $\dfrac{\sqrt{3}}{\sqrt{6} - 1} - \dfrac{\sqrt{3}}{\sqrt{6} + 1}$

*63. $\dfrac{1}{\sqrt{2}} + \dfrac{1}{\sqrt{3}}$

*64. $\dfrac{4}{2\sqrt{3}} + \dfrac{1}{\sqrt{5}}$

*65. $\dfrac{3}{\sqrt{2} - 1} + \dfrac{4}{\sqrt{2} + 1}$

*66. $\dfrac{3}{\sqrt{5} - \sqrt{3}} - \dfrac{2}{\sqrt{5} + \sqrt{3}}$

*67. $\dfrac{\sqrt{x}}{\sqrt{x} + 2} + \dfrac{3\sqrt{x}}{\sqrt{x} - 2}$

*68. $\dfrac{\sqrt{5}}{3 - \sqrt{y}} - \dfrac{\sqrt{5y}}{3 + \sqrt{y}}$

*69. $\dfrac{1}{\sqrt{x}} + \dfrac{1}{1 - \sqrt{x}}$

*70. $\dfrac{\sqrt{x}}{\sqrt{x} - 3} + \dfrac{5}{\sqrt{x}}$

Replace the question mark by an expression that makes the equation correct. Equations involving variables are to be identities.

71. $\dfrac{\sqrt{2}}{\sqrt{3}} = \dfrac{\sqrt{6}}{?}$

72. $\dfrac{2}{?} = \sqrt{2}$

73. $\dfrac{1}{\sqrt{2} - 1} = \dfrac{\sqrt{2} + 1}{?}$

74. $\dfrac{1}{1 + \sqrt{3}} = \dfrac{1 - \sqrt{3}}{?}$

75. $\dfrac{1}{3\sqrt{2}} = \dfrac{?}{6}$

76. $\dfrac{2}{5\sqrt{3}} = \dfrac{?}{15}$

77. $\dfrac{\sqrt{3}}{\sqrt{5} - 1} = \dfrac{?}{4}$

78. $\dfrac{\sqrt{6}}{\sqrt{6} + 2} = \dfrac{?}{2}$

*79. $\dfrac{1}{\sqrt{x} - 1} = \dfrac{?}{x - 1}$

*80. $\dfrac{5}{3 - \sqrt{x}} = \dfrac{?}{9 - x}$

*81. $\dfrac{3}{\sqrt{2} + x} = \dfrac{?}{2 - x^2}$

*82. $\dfrac{4}{2\sqrt{3} + a} = \dfrac{?}{12 - a^2}$

*83. $\dfrac{4}{\sqrt{x} - 1} = \dfrac{?}{x - 1}$

*84. $\dfrac{2}{\sqrt{3} - x} = \dfrac{?}{3 - x}$

Use the table of Appendix A or a calculator to find a decimal approximation for each radical expression. Round your answers to three decimal places.

85. $\sqrt{3} + \sqrt{5}$

86. $\sqrt{5} + \sqrt{7}$

87. $2\sqrt{3} + 5\sqrt{3}$

88. $7\sqrt{3}$

89. $\dfrac{-1 + \sqrt{6}}{2}$

90. $\dfrac{-1 - \sqrt{6}}{2}$

91. $\sqrt{5}(\sqrt{5} + \sqrt{3})$

92. $5 + \sqrt{15}$

93. $\dfrac{4 - \sqrt{10}}{-2}$

94. $\dfrac{4 + \sqrt{10}}{-2}$

95. $(2\sqrt{3})(3\sqrt{2})$

96. $6\sqrt{6}$

5.5 Solving Equations with Radicals and Exponents

IN THIS SECTION:

- The Odd-root Property
- The Even-root Property
- Raising Each Side to a Power
- Equations Involving Rational Exponents
- Summary of Methods
- Applications

In this section we will use our knowledge of radicals and exponents to solve equations involving radicals and exponents. When solving equations in Chapter 2, we wrote down a sequence of equivalent equations, each one simpler than the last, but each one having the same solution set. To solve some of the equations of this section, we must use equations that have different solution sets.

The Odd-root Property

Because $(-2)^3 = -8$ and $2^3 = 8$, the equation $x^3 = 8$ is equivalent to $x = 2$. The equation $x^3 = -8$ is equivalent to $x = -2$. Since there is only one real odd root of each real number, there is a simple rule for writing an equivalent equation in this situation.

Odd-root Property

> If n is an odd positive integer,
>
> $$x^n = k \text{ is equivalent to } x = \sqrt[n]{k},$$
>
> for any value of k.

EXAMPLE 1 Solve each equation.

a) $x^3 = 27$ **b)** $x^5 = -32$ **c)** $(x - 2)^3 = 24$

Solution

a) $x^3 = 27$

    $x = \sqrt[3]{27}$ Odd-root property

    $x = 3$

The solution set is $\{3\}$. Check.

b) $x^5 = -32$

    $x = \sqrt[5]{-32}$ Odd-root property

    $x = -2$

The solution set is $\{-2\}$. Check.

c) $(x - 2)^3 = 24$

    $x - 2 = \sqrt[3]{24}$ Odd-root property

       $x = 2 + 2\sqrt[3]{3}$ $\sqrt[3]{24} = \sqrt[3]{8}\sqrt[3]{3} = 2\sqrt[3]{3}$

The solution set is $\{2 + 2\sqrt[3]{3}\}$. Check.   ◀

The Even-root Property

In solving the equation $x^2 = 4$, we might be tempted to write the answer as $x = 2$, but $x = 2$ is *not* equivalent to the original equation. Because $2^2 = 4$ and $(-2)^2 = 4$, the solution set to $x^2 = 4$ is $\{2, -2\}$. So $x^2 = 4$ is equivalent to the compound sentence $x = 2$ or $x = -2$. We can shorten this by writing $x = \pm 2$ and read this as "x equals positive or negative 2."

    Equations involving other even powers are handled like the squares. Because $2^4 = 16$ and $(-2)^4 = 16$, the equation $x^4 = 16$ is equivalent to $x = \pm 2$. Note that the equation $x^4 = -16$ has no real solutions. The equation $x^6 = 5$ is equivalent to $x = \pm\sqrt[6]{5}$. We can now state a general rule.

Even-root Property

Suppose n is a positive even integer.

If $k > 0$, the equation $x^n = k$ is equivalent to the compound sentence $x = \pm \sqrt[n]{k}$ ($x = \sqrt[n]{k}$ or $x = -\sqrt[n]{k}$).

If $k = 0$, the equation $x^n = k$ is equivalent to $x = 0$.

If $k < 0$, the equation $x^n = k$ has no real solution.

EXAMPLE 2 Solve each equation.

a) $x^2 = 10$ 　　　　b) $(x - 3)^2 = 4$ 　　　　c) $2(x - 5)^2 - 7 = 0$

d) $x^4 = -4$ 　　　　e) $(x + 2)^6 = 0$ 　　　　f) $x^4 - 1 = 80$

Solution

a) $x^2 = 10$

$\quad x = \pm \sqrt{10}$ 　　Even-root property

The solution set is $\{-\sqrt{10},\ \sqrt{10}\}$.

b) $(x - 3)^2 = 4$

$\quad x - 3 = 2 \quad$ or $\quad x - 3 = -2 \quad$ Even-root property

$\qquad x = 5 \quad$ or $\qquad x = 1 \quad$ Add 3 to each side.

The solution set is $\{1,\ 5\}$.

c) $2(x - 5)^2 - 7 = 0$

$\quad 2(x - 5)^2 = 7 \qquad$ Add 7 to each side.

$\quad (x - 5)^2 = \dfrac{7}{2} \qquad$ Divide each side by 2.

$\quad x - 5 = \sqrt{\dfrac{7}{2}} \qquad$ or $\qquad x - 5 = -\sqrt{\dfrac{7}{2}} \qquad$ Even-root property

$\quad x = 5 + \dfrac{\sqrt{14}}{2} \qquad$ or $\qquad x = 5 - \dfrac{\sqrt{14}}{2} \qquad \sqrt{\dfrac{7}{2}} = \dfrac{\sqrt{7}}{\sqrt{2}} = \dfrac{\sqrt{14}}{2}$

$\quad x = \dfrac{10 + \sqrt{14}}{2} \qquad$ or $\qquad x = \dfrac{10 - \sqrt{14}}{2}$

The solution set is $\left\{ \dfrac{10 + \sqrt{14}}{2},\ \dfrac{10 - \sqrt{14}}{2} \right\}$.

d) This equation has no real solution because no real number has a fourth power that is negative.

e) $(x + 2)^6 = 0$ 　　　　　　　　　f) $x^4 - 1 = 80$

$\quad x + 2 = 0 \qquad$ Even-root property $\qquad\qquad x^4 = 81$

$\qquad x = -2 \qquad\qquad\qquad\qquad\qquad\qquad x = \pm \sqrt[4]{81} = \pm 3$

The solution set is $\{-2\}$. 　　　　　　　The solution set is $\{-3,\ 3\}$. ◄

The equations of parts (a), (b), and (c) of Example 1 are called **quadratic equations.** In a quadratic equation the highest power of the variable is 2. We will learn the precise definition of a quadratic equation in Chapter 6. The quadratic equations that we encounter here can be solved by using the even-root property. In Chapter 6 we will learn general methods for solving any quadratic equation.

Raising Each Side to a Power

If we start with the equation $x = 3$ and square both sides, we get $x^2 = 9$. The solution set to $x^2 = 9$ is $\{3, -3\}$, while the solution set to the original equation is $\{3\}$. Squaring both sides of an equation does not necessarily produce an equation that is equivalent to the original equation. The new equation may have additional solutions. We call these additional solutions **extraneous roots.** However, any solution or root of the original must be among the solutions to the new equation.

When we solve equations involving radicals, we raise each side to a power to eliminate the radical. We must check our answers because the new equations might not be equivalent to the old equations. Raising each side to an odd power will always give an equivalent equation; raising each side to an even power might not.

EXAMPLE 3 Solve each equation.

a) $\sqrt{2x - 3} = 5$

b) $\sqrt[3]{3x + 5} = \sqrt[3]{x - 1}$

c) $\sqrt{3x + 18} = x$

Solution

a)
$$\sqrt{2x - 3} = 5$$
$$(\sqrt{2x - 3})^2 = 5^2 \qquad \text{Square both sides.}$$
$$2x - 3 = 25$$
$$2x = 28$$
$$x = 14$$

Check: $\sqrt{2(14) - 3} = \sqrt{28 - 3} = \sqrt{25} = 5$
The solution set is $\{14\}$.

b)
$$\sqrt[3]{3x + 5} = \sqrt[3]{x - 1}$$
$$(\sqrt[3]{3x + 5})^3 = (\sqrt[3]{x - 1})^3 \qquad \text{Cube each side.}$$
$$3x + 5 = x - 1$$
$$2x = -6$$
$$x = -3$$

Check: $\sqrt[3]{3(-3) + 5} = \sqrt[3]{-3 - 1}$
$$\sqrt[3]{-4} = \sqrt[3]{-4} \qquad \text{Note that } \sqrt[3]{-4} \text{ is a real number.}$$

The solution set is $\{-3\}$. Note that in this example we checked for arithmetic mistakes. There was no possibility of extraneous roots here because we raised each side to an odd power.

c)

$$
\begin{array}{ll}
(\sqrt{3x + 18})^2 = x^2 & \text{Square both sides.} \\
3x + 18 = x^2 & \text{Simplify.} \\
-x^2 + 3x + 18 = 0 & \text{Subtract } x^2 \text{ from each side to get zero on one side.} \\
x^2 - 3x - 18 = 0 & \text{Multiply each side by } -1 \text{ for easier factoring.} \\
(x - 6)(x + 3) = 0 & \text{Factor.} \\
x - 6 = 0 \text{ or } x + 3 = 0 & \text{Zero-factor property} \\
x = 6 \quad\quad \text{ or } x = -3 &
\end{array}
$$

Because we squared both sides, we must check for extraneous roots. The equation

$$\sqrt{3(-3) + 18} = -3$$

is false, but

$$\sqrt{3(6) + 18} = 6$$

is true. The solution set is $\{6\}$.   ◀

In the next example the radicals are not eliminated after squaring both sides of the equation. In this case we must square both sides a second time. Note that to square the side with two terms, we must proceed as in squaring a binomial.

EXAMPLE 4   Solve $\sqrt{5x - 1} - \sqrt{x + 2} = 1$.

Solution   It is easier to square both sides if the two radicals are not on the same side, so we first rearrange the equation.

$$
\begin{array}{ll}
\sqrt{5x - 1} = 1 + \sqrt{x + 2} & \\
(\sqrt{5x - 1})^2 = (1 + \sqrt{x + 2})^2 & \text{Square both sides.} \\
5x - 1 = 1 + 2\sqrt{x + 2} + x + 2 & \text{Square the right side like a binomial.} \\
5x - 1 = 3 + x + 2\sqrt{x + 2} & \text{Combine like terms on the right side.} \\
4x - 4 = 2\sqrt{x + 2} & \text{Isolate the square root.} \\
2x - 2 = \sqrt{x + 2} & \text{Divide each side by 2.} \\
(2x - 2)^2 = (\sqrt{x + 2})^2 & \text{Square both sides.} \\
4x^2 - 8x + 4 = x + 2 & \text{Square the binomial on the left side.} \\
4x^2 - 9x + 2 = 0 & \\
(4x - 1)(x - 2) = 0 & \\
4x - 1 = 0 \text{ or } x - 2 = 0 & \\
x = \dfrac{1}{4} \quad\quad \text{ or } x = 2 &
\end{array}
$$

Because we squared both sides, we must check for extraneous roots.

$$\sqrt{5\left(\frac{1}{4}\right) - 1} - \sqrt{\left(\frac{1}{4}\right) + 2} = \sqrt{\frac{1}{4}} - \sqrt{\frac{9}{4}}$$

$$= \frac{1}{2} - \frac{3}{2}$$

$$= -1$$

$$\neq 1$$

$$\sqrt{5(2) - 1} - \sqrt{2 + 2} = \sqrt{9} - \sqrt{4}$$

$$= 3 - 2 = 1$$

Since 1/4 does not satisfy the original equation but 2 does, the solution set is {2}.   ◄

Equations Involving Rational Exponents

Equations involving rational exponents can be solved by combining the methods that we just learned for eliminating radicals and integral exponents. *For equations involving rational exponents we will always eliminate the root first and the power second.*

EXAMPLE 5   Solve each equation.

a) $x^{2/3} = 4$ **b)** $(w - 1)^{-2/5} = 4$ **c)** $(2t - 3)^{-2/3} = -1$

Solution

a) $x^{2/3} = 4$

$(x^{2/3})^3 = 4^3$ Cube each side.

$x^2 = 64$ $(2/3)(3) = 2$

$x = 8$ or $x = -8$ Even-root property

All of the equations are equivalent. The solution set is $\{-8, 8\}$. Check each of these solutions in the original equation.

b) $(w - 1)^{-2/5} = 4$

$[(w - 1)^{-2/5}]^{-5} = 4^{-5}$ Raise each side to the −5 to eliminate the negative exponent.

$$(w - 1)^2 = \frac{1}{1024}$$ $(-2/5)(-5) = 2$

$$w - 1 = \pm \sqrt{\frac{1}{1024}}$$ Even-root property

$$w - 1 = \frac{1}{32} \text{ or } w - 1 = -\frac{1}{32}$$

$$w = \frac{33}{32} \quad \text{or } w = \frac{31}{32}$$

The solution set is $\{31/32, 33/32\}$.

c) $(2t - 3)^{-2/3} = -1$

$[(2t - 3)^{-2/3}]^{-3} = (-1)^{-3}$ **Raise each side to the −3 power.**

$(2t - 3)^2 = -1$ $(-2/3)(-3) = 2$

By the even-root property, this equation has no solution. ◀

Summary of Methods

The three most important rules for solving equations with exponents and radicals are restated here.

Solving Equations with Exponents and Radicals

1. In raising each side of an equation to an even power, we can create an equation that gives extraneous roots. We must check all possible solutions in the original equation.

2. When applying the even-root property, remember that there is a positive root and a negative root.

3. For equations with rational exponents, raise each side to a positive or negative integral power first, then apply the even- or odd-root property. (Positive fraction: raise to a positive power; negative fraction: raise to a negative power.)

Applications

Applications of exponents and roots often involve the Pythagorean Theorem, which was introduced in Section 3.9. This theorem says that in any right triangle, the sum of the squares of the lengths of the legs is equal to the hypotenuse squared. We will use the Pythagorean Theorem to solve the following geometric problem.

EXAMPLE 6 A baseball diamond is actually a square, 90 feet on each side. What is the distance from third base to first base?

Solution First make a sketch as in Fig. 5.1. The distance from third base to first base is the length of the diagonal of the square shown in Fig. 5.1. Let x be the length of the diagonal of the square. The Pythagorean Theorem can be applied to the right triangle formed from the diagonal and two sides of the square. The sum of the squares of the sides is equal to the diagonal squared.

$$x^2 = 90^2 + 90^2$$
$$x^2 = 8100 + 8100$$
$$x^2 = 16200$$
$$x = \pm \sqrt{16200} = \pm 90\sqrt{2}$$

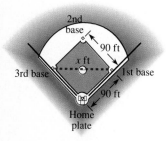

Figure 5.1

The length of the diagonal of a square must be positive, so we disregard the negative solution. Checking the answer in the original equation verifies that the exact length of a side is $90\sqrt{2}$ feet. ◀

Warm-ups

True or false?

1. The equations $x^2 = 4$ and $x = 2$ are equivalent.
2. The equation $x^2 = -25$ has no real solution.
3. There is no solution to the equation $x^2 = 0$.
4. The equation $x^3 = 8$ is equivalent to $x = \pm 2$.
5. The equation $-\sqrt{x} = 16$ has no solution.
6. To solve $\sqrt{x - 3} = \sqrt{2x + 5}$, the first step is to apply the even-root property.
7. Extraneous roots are roots that cannot be found.
8. If we square both sides of $\sqrt{x} = -7$, we will get an equation that has an extraneous root.
9. The equations $x^2 - 6 = 0$ and $x = \pm\sqrt{6}$ are equivalent.
10. We never get an equation that has extraneous roots from cubing each side of an equation.

5.5 EXERCISES

Solve each equation. See Example 1.

1. $x^3 = -1000$
2. $y^3 - 100 = 25$
3. $32m^5 - 1 = 0$
4. $243a^5 + 1 = 0$

5. $(y - 3)^3 = -8$
6. $(x - 1)^3 = -1$
7. $\frac{1}{2}x^3 + 4 = 0$
8. $3(x - 9)^7 = 0$

Solve each equation. See Example 2.

9. $x^2 = 25$
10. $x^2 - 36 = 0$
11. $x^2 - 20 = 0$
12. $a^2 = 40$
13. $x^2 = -9$
14. $w^2 + 49 = 0$
15. $(x - 3)^2 = 16$
16. $(a - 2)^2 = 25$
17. $(x + 1)^2 = 8$

18. $(w + 3)^2 = 12$
19. $\frac{1}{2}x^2 = 5$
20. $\frac{1}{3}x^2 = 6$

21. $4(a + 1)^2 = 1$
22. $9(m - 3)^2 = 1$
23. $(y - 3)^4 = 0$
24. $(2x - 3)^6 = 0$
25. $2x^6 = 128$
26. $3y^4 = 48$

27. $81a^4 - 16 = 0$
28. $64w^6 = 1$

Solve each equation and check for extraneous roots. See Example 3.

29. $\sqrt{x-3} = 7$

30. $\sqrt{a-1} = 6$

31. $2\sqrt{w+4} = 5$

32. $3\sqrt{w+1} = 6$

33. $\sqrt{2x+3} = \sqrt{x+12}$

34. $\sqrt{a+3} = \sqrt{2a-7}$

35. $\sqrt{2t+4} = \sqrt{t-1}$

36. $\sqrt{w-3} = \sqrt{4w+15}$

37. $\sqrt{4x^2+x-3} = 2x$

38. $\sqrt{x^2-5x+2} = x$

39. $\sqrt{x^2+2x-6} = 3$

40. $\sqrt{x^2-x-4} = 4$

41. $\sqrt{2x^2-1} = x$

42. $\sqrt{2x^2-3x-10} = x$

43. $\sqrt{2x^2+5x+6} = x$

44. $\sqrt{5x^2-9} = 2x$

45. $\sqrt{2x^2+6x+4} = x+1$

46. $\sqrt{2x^2+2x} = x+1$

Solve each equation and check for extraneous roots. See Example 4.

47. $\sqrt{x+3} - \sqrt{x-2} = 1$

48. $\sqrt{2x+1} - \sqrt{x} = 1$

49. $\sqrt{2x+2} - \sqrt{x-3} = 2$

50. $\sqrt{3x} + \sqrt{x-2} = 4$

51. $\sqrt{4-x} - \sqrt{x+6} = 2$

52. $\sqrt{6-x} - \sqrt{x-2} = 2$

Solve the equations. See Example 5.

53. $x^{2/3} = 4$

54. $a^{2/3} = 2$

55. $y^{-2/3} = 9$

56. $w^{-2/3} = 4$

57. $w^{1/3} = 8$

58. $a^{1/3} = 27$

59. $t^{-1/2} = 9$

60. $w^{-1/4} = \dfrac{1}{2}$

61. $a^{-1/3} = \dfrac{1}{2}$

62. $s^{1/2} = \dfrac{1}{2}$

63. $(3a-1)^{-2/5} = 1$

64. $(r-1)^{-2/3} = 1$

65. $(t-1)^{-2/3} = 2$

66. $(w+3)^{-1/3} = \dfrac{1}{3}$

Solve each word problem by writing an equation and solving it. Find the exact answer and simplify it using the rules for radicals. See Example 5.

67. Find the length of the side of a square whose diagonal is 8 feet.

68. Find the length of the side of a square whose diagonal measures 1 meter.

69. Find the length of the diagonal of a square painting whose sides measure 2 feet.

70. Find the length of the diagonal of a square patio with an area of 40 square meters.

71. Find the length of the side of a square sign whose area is 50 square feet.

72. Find the length of the side of a square whose area is 18 square kilometers.

73. Find the length of the side of a cube whose volume is 80 cubic feet.

74. Find the length of the side of a cube whose volume is 125 cubic meters.

75. In the bridge problem stated in the beginning of this chapter, the sides of a rectangle were 30 feet and 40 feet. Find the length of the diagonal of that rectangle.

76. What is the length of the diagonal of a rectangular billboard whose sides are 5 meters and 12 meters?

*77. Find the length of the diagonal of a side of a cubic packing crate whose volume is 2 cubic meters.

*78. Find the volume of a cube where the diagonal of a side measures 2 feet.

79. An architect designs a public park in the shape of a trape-zoid. Find the length of the diagonal road marked *a* in the figure.

80. Find the length of the border of the park marked *b* in the trapezoid shown in the figure.

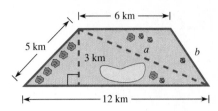

Figure for Exercises 79 and 80

Solve each equation.

81. $2x^2 + 3 = 7$

82. $3x^2 - 5 = 16$

83. $\sqrt[3]{2w + 3} = \sqrt[3]{w - 2}$

84. $\sqrt[3]{2 - w} = \sqrt[3]{2w - 28}$

85. $9x^2 - 1 = 0$

86. $4x^2 - 1 = 0$

87. $(w + 1)^{2/3} = -3$

88. $(x - 2)^{3/4} = 2$

89. $(a + 1)^{1/3} = -2$

90. $(a - 1)^{1/3} = -3$

91. $(4y - 5)^7 = 0$

92. $(5x)^9 = 0$

93. $\sqrt{x^2 + 5x} = 6$

94. $\sqrt{x^2 - 8x} = -3$

95. $\sqrt{4x^2} = x + 2$

96. $\sqrt{9x^2} = x + 6$

97. $(t + 2)^4 = 32$

98. $(w + 1)^4 = 48$

99. $\sqrt{x^2 - 3x} = x$

100. $\sqrt[4]{4x^4 - 48} = -x$

101. $x^{-3} = 8$

102. $x^{-2} = 4$

103. $a^{-2} = 3$

104. $w^{-2} = 18$

105. $\sqrt{x + 1} - \sqrt{2x + 9} = -2$

106. $\sqrt{x - 4} + \sqrt{2x - 1} = 4$

Use a calculator to find approximate solutions to the following equations. Round your answers to three decimal places.

107. $x^2 = 3.24$

108. $(x + 4)^3 = 7.51$

109. $\sqrt{x - 2} = 1.73$

110. $\sqrt[3]{x - 5} = 3.7$

111. $x^{2/3} = 8.86$

112. $(x - 1)^{-3/4} = 7.065$

5.6 Scientific Notation

IN THIS SECTION:

- Basic Ideas
- Converting from Scientific Notation
- Converting to Scientific Notation
- Computations with Scientific Notation

Many of the numbers that are encountered in science are either very large or very small. The distance from the earth to the sun is 93,000,000 miles. The hydrogen atom has a diameter of .00000001 centimeters and weighs .00000000000000000000000017 grams. Scientific notation is a convenient way of

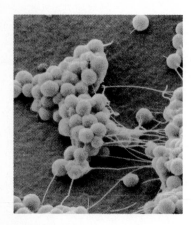

Staphylococcus Epidermides
bacteria magnified 3.145×10^3

writing very large and very small numbers. In this section we will learn to use positive and negative exponents to represent numbers in scientific notation, and we will use the rules of exponents to do computations with numbers in scientific notation.

Basic Ideas

To understand scientific notation, we need to understand the integral powers of 10.

$$10^1 = 10 \qquad 10^{-1} = \frac{1}{10} = .1$$

$$10^2 = 100 \qquad 10^{-2} = \frac{1}{100} = .01$$

$$10^3 = 1000 \qquad 10^{-3} = \frac{1}{1000} = .001$$

$$10^4 = 10,000 \qquad 10^{-4} = \frac{1}{10,000} = .0001$$

Now consider the result of multiplying a number by a positive power of 10. Multiplying a number by 10, 100, or 1000 just moves the decimal point either one, two, or three places to the right. For example,

$$
\begin{array}{ccc}
7.4 & 7.4 & 7.4 \\
\times\ 10 & \times\ 100 & \times\ 1000 \\
\hline
74 & 740 & 7400 \\
\text{1 place} & \text{2 places} & \text{3 places}
\end{array}
$$

Multiplying a number by a negative power of 10 moves the decimal point to the left. For example,

$$
\begin{array}{ccc}
7.4 & 7.4 & 7.4 \\
\times\ .1 & \times\ .01 & \times\ .001 \\
\hline
.74 & .074 & .0074 \\
\text{1 place} & \text{2 places} & \text{3 places}
\end{array}
$$

Notice that $1000 = 10^3$, and the decimal point moved three places to the right when we multiplied by 10^3. When we multiplied by 10^{-3} or .001, we moved the decimal point three places to the left.

The numbers 9.37×10^4 and 3.7×10^{-5} are examples of numbers written in scientific notation. In scientific notation the times symbol, $\times$, is used to indicate multiplication. *A number in scientific notation is written as a product of a number between 1 and 10, and a power of 10.* Numbers in scientific notation are written with only one digit to the left of the decimal point.

Converting from Scientific Notation

In scientific notation a number larger than 10 is written with a positive power of 10, and a number smaller than 1 is written with a negative power of 10. To convert a number in scientific notation to one in standard notation, we merely move the decimal point because we are multiplying by a power of 10. For example,

$$9.37 \times 10^4 = 93700.$$ Positive 4 indicates a large number, so move to the right.

4 places to the right

If the power of 10 is negative, the decimal point is moved to the left:

$$3.7 \times 10^{-5} = .000037$$ Negative 5 indicates a small number, so move to the left.

5 places to the left

In general, we use the following strategy to convert from scientific notation to standard notation.

STRATEGY
Converting from Scientific Notation

1. Determine the number of places to move the decimal point by examining the exponent on the 10.
2. Move to the right for a positive exponent and to the left for a negative exponent.

EXAMPLE 1   Write the following numbers using standard notation.

a) 7.62×10^5 **b)** 6.35×10^{-4} **c)** -3.897×10^4 **d)** -1.2×10^{-3}

Solution

a) $7.62 \times 10^5 = 762000. = 762,000$ Move the decimal point to the right since the exponent is positive.

5 places

b) $6.35 \times 10^{-4} = .000635$ Move the decimal point to the left since the exponent is negative.

4 places

c) Convert any negative number as if it were a positive number and then attach a negative sign to the result.

$$-3.897 \times 10^4 = -38,970$$

d) $-1.2 \times 10^{-3} = -.0012$ ◀

Converting to Scientific Notation

Scientific notation is used for writing large and small numbers. Numbers between 1 and 10 are not written in scientific notation. To convert a positive number to scientific notation, we just reverse the strategy for converting from scientific notation.

STRATEGY
Converting to Scientific
Notation

> **1.** Count the number of places (n) that the decimal point must be moved so that it will follow the first nonzero digit of the number.
> **2.** If the original number is larger than 10, use 10^n.
> **3.** If the original number is smaller than 1, use 10^{-n}.

To convert a negative number to scientific notation, ignore the negative sign, convert the number, and then attach the negative sign to the result.

EXAMPLE 2   Convert each number to scientific notation.

a) 834,000,000   **b)** .0000025   **c)** −3400   **d)** −.000000000164

Solution

a) $834,000,000 = 8.34 \times 10^8$ Use a positive 8 because the number is larger than 10.

    8 places, exponent 8

b) $.0000025 = 2.5 \times 10^{-6}$ Use −6 because this number is smaller than 1.

    6 places, exponent −6

c) Convert 3400 and attach the negative sign to the result.

$$-3400 = -3.4 \times 10^3$$

d) Convert .000000000164 and attach the negative sign to the result.

$$-.000000000164 = -1.64 \times 10^{-10}$$ ◄

Computations with Scientific Notation

We can perform computations with numbers in scientific notation by using the rules of exponents on the powers of 10.

EXAMPLE 3   Perform the indicated operations. Write the answers in scientific notation.

a) $(3 \times 10^5)(2 \times 10^4)$ **b)** $\dfrac{4 \times 10^5}{8 \times 10^{-3}}$

c) $\dfrac{(9 \times 10^{140})(8 \times 10^{-102})}{6 \times 10^{-105}}$ **d)** $\sqrt{9 \times 10^{36}}$

Solution

a) $(3 \times 10^5)(2 \times 10^4) = 3 \cdot 2 \cdot 10^5 \cdot 10^4 = 6 \times 10^9$

b) $\dfrac{4 \times 10^5}{8 \times 10^{-3}} = \dfrac{4}{8} \cdot \dfrac{10^5}{10^{-3}} = \dfrac{1}{2} \cdot 10^8$

$= (.5)10^8$

$= 5 \times 10^{-1} \cdot 10^8$ Convert .5 into scientific notation.

$= 5 \times 10^7$ Add the exponents.

c) $\dfrac{(9 \times 10^{140})(8 \times 10^{-102})}{6 \times 10^{-105}} = \dfrac{9 \cdot 8}{6} \cdot \dfrac{10^{140} 10^{-102}}{10^{-105}}$

$\qquad\qquad\qquad\qquad = 12 \cdot 10^{143}$

$\qquad\qquad\qquad\qquad = 1.2 \times 10^1 \times 10^{143}$ Convert 12 into 1.2×10^1.

$\qquad\qquad\qquad\qquad = 1.2 \times 10^{144}$

d) $\sqrt{9 \times 10^{36}} = \sqrt{9} \cdot \sqrt{10^{36}} = 3 \times 10^{18}$ ◀

EXAMPLE 4 Perform these computations by first converting each number to scientific notation.

a) $\dfrac{(40,000)(.000025)}{.005}$

b) $\dfrac{(60,000)^2(.00012)}{(2400)(.001)^3}$

Solution

a) $\dfrac{(40,000)(.000025)}{.005} = \dfrac{(4 \times 10^4)(2.5 \times 10^{-5})}{5 \times 10^{-3}}$

$\qquad\qquad\qquad\qquad = \dfrac{(4)(2.5)}{5} \cdot \dfrac{10^4 10^{-5}}{10^{-3}}$

$\qquad\qquad\qquad\qquad = 2 \times 10^2$

b) $\dfrac{(60,000)^2(.00012)}{(2400)(.001)^3} = \dfrac{(6 \times 10^4)^2(1.2 \times 10^{-4})}{(2.4 \times 10^3)(10^{-3})^3}$

$\qquad\qquad\qquad\qquad = \dfrac{(36)(10^8)(1.2)(10^{-4})}{(2.4)(10^3)(10^{-9})}$

$\qquad\qquad\qquad\qquad = 18 \times 10^{10}$

$\qquad\qquad\qquad\qquad = 1.8 \times 10^1 \times 10^{10}$

$\qquad\qquad\qquad\qquad = 1.8 \times 10^{11}$ ◀

In the next example we perform a computation with the aid of a scientific calculator. Refer to your calculator manual to see how to enter numbers in scientific notation.

EXAMPLE 5 Determine the number of hydrogen atoms in one kilogram of hydrogen.

Solution At the beginning of this section, we stated that the weight of a hydrogen atom is .0000000000000000000000017 grams. In scientific notation this is 1.7×10^{-24} grams. One kilogram is 1000, or 1×10^3, grams. So the number of hydrogen atoms in one kilogram of hydrogen is

$$\frac{1 \times 10^3}{1.7 \times 10^{-24}} = 5.882 \times 10^{26}.$$ ◀

If a number is too large to be displayed in standard notation, then a scientific calculator automatically displays the number in scientific notation. Most calculators are limited to powers of 10 from -99 to 99. Use a calculator with scientific notation to do Exercises 45–62.

Warm-ups

True or false?

1. $10^{-3} = .0001$

2. $234.7 = 2.347 \times 10^3$

3. $.000054 = 54 \times 10^{-6}$

4. $46.7 \times 10^5 = 4.67 \times 10^6$

5. $.512 \times 10^{-3} = 5.12 \times 10^{-4}$

6. $(4 \times 10^5)(5 \times 10^8) = 2.0 \times 10^{14}$

7. $\dfrac{8 \times 10^{30}}{2 \times 10^{-5}} = 4 \times 10^{25}$

8. $(4 \times 10^3) + (3 \times 10^2) = 43 \times 10^2$

9. $\dfrac{9 \times 10^{-5}}{3 \times 10^4} = 3 \times 10^{-9}$

10. $(3 \times 10^5) + (2 \times 10^{-3}) = 5 \times 10^2$

5.6 EXERCISES

Write each number in standard notation. See Example 1.

1. 4.86×10^8

2. 2.37×10^{-6}

3. -1.62×10^{-3}

4. -3.80×10^{-2}

5. 4.132×10^3

6. 5×10^{-6}

7. 1×10^6

8. 1×10^{-4}

9. 496×10^3

10. 48×10^{-3}

Write each number in scientific notation. See Example 2.

11. $320,000$

12. $43,298,000$

13. $.00000071$

14. $.00000894$

15. $-358,000,000$

16. $-.00237$

17. $.00007$

18. $8,295,100$

19. 235×10^5

20. $.43 \times 10^{-9}$

Perform the computations. Write each answer in scientific notation. See Example 3.

21. $(3 \times 10^5)(6 \times 10^9)$

22. $(-4 \times 10^3)(5 \times 10^{-6})$

23. $\dfrac{6 \times 10^{40}}{2 \times 10^{18}}$

24. $\dfrac{4.6 \times 10^{12}}{2.3 \times 10^5}$

25. $\dfrac{1 \times 10^{24}}{2 \times 10^8}$

26. $\dfrac{3 \times 10^{-5}}{4 \times 10^{-9}}$

27. $\dfrac{(4 \times 10^{-11})(6 \times 10^5)}{(3 \times 10^{10})(2 \times 10^{-7})}$

28. $\dfrac{(-4 \times 10^5)(6 \times 10^{-9})}{2 \times 10^{-16}}$

29. $\dfrac{(2.2 \times 10^{-5})(6 \times 10^{12})}{(3 \times 10^8)(1.1 \times 10^{-6})}$

30. $\dfrac{(4.8 \times 10^{-3})(5 \times 10^{-8})}{(1.2 \times 10^{-6})(2 \times 10^{12})}$

31. $(5 \times 10^3) + (4 \times 10^4)$

32. $(3 \times 10^{-6}) + (2 \times 10^{-7})$

33. $(4 \times 10^7)^3$

34. $(-9 \times 10^{-5})^2$

35. $\sqrt{1.6 \times 10^{13}}$

36. $\sqrt{2.5 \times 10^{-9}}$

37. $(8 \times 10^{15})^{1/3}$

38. $(-8 \times 10^{-12})^{-2/3}$

Perform the following computations by first converting each number to scientific notation. See Example 4.

39. $\dfrac{(5,000,000)(.0003)}{2000}$

40. $\dfrac{(6000)(.00004)}{(30,000)(.002)}$

41. $\dfrac{(-20,000)^3(.0005)^2}{(.0004)(.0001)^2}$

42. $\dfrac{(-400)^3(2000)^2}{(.0002)^2(800)}$

43. $\dfrac{\sqrt{.00000009}}{6,000,000}$

44. $\dfrac{20,000,000}{\sqrt{160,000,000,000}}$

Perform the following computations with the aid of a calculator. Write answers in scientific notation. Round the decimal part to three decimal places. See Example 5.

45. $(4.3 \times 10^9)(3.67 \times 10^{-5})$

46. $(2.34 \times 10^6)(8.7 \times 10^5)$

47. $(4.37 \times 10^{-6}) + (8.75 \times 10^{-5})$

48. $(6.72 \times 10^5) + (8.98 \times 10^6)$

49. $(9.27 \times 10^{80})(6.43 \times 10^{76})$

50. $(1.35 \times 10^{66})(2.7 \times 10^{74})$

51. $(5 \times 10^{99}) + (6 \times 10^{99})$

52. $(7 \times 10^{-99}) + (8 \times 10^{-99})$

53. $\dfrac{(-5.6 \times 10^{14})^2(3.2 \times 10^{-6})}{(6.4 \times 10^{-3})^3}$

54. $\dfrac{(3.51 \times 10^{-6})^3(4000)^5}{2\pi}$

55. $\dfrac{(-3.7 \times 10^{-8})(4.6 \times 10^7)}{(3.2 \times 10^5)(9.63 \times 10^4)}$

56. $\dfrac{(1.06 \times 10^3)(-3.1 \times 10^{-5})}{(4.6 \times 10^8)(-4.9 \times 10^{-13})}$

57. The distance from the earth to the sun is 93 million miles. Express this distance in feet (1 mile = 5280 feet).

58. The speed of light is 9.83569×10^8 feet per second. How long does it take light to get from the sun to the earth?

59. How long does it take a spacecraft traveling 1.2×10^5 kilometers per second to travel 4.6×10^{12} kilometers?

60. If the circumference of a very small circle is 2.35×10^{-8} meters, then what is the diameter of the circle?

61. If the area of a very small circle is 3.49×10^{-7} square centimeters, then what is the radius?

62. If the volume of a very small cubic crystal is 8.52×10^{-4} cubic inches, then what is the length of a side?

Wrap-up

CHAPTER 5

SUMMARY

	Definitions	Examples
Definition of *n*th roots	We say that a is an *n*th root of b if $a^n = b$.	$2^4 = 16$, $(-2)^4 = 16$ Both 2 and -2 are 4th roots of 16.
Definition of $1/n$ as an exponent	If n is a positive *even* integer and a is positive, then $a^{1/n}$ denotes the positive real *n*th root of a. If n is a positive *odd* integer and a is any real number, then $a^{1/n}$ is the real *n*th root of a.	$16^{1/4} = 2$ $(-8)^{1/3} = -2$ $8^{1/3} = 2$

Definition of nth root of zero	If n is a positive integer, then $0^{1/n} = 0$.	$0^{1/6} = 0$
Definition of rational exponents	If m and n are positive integers, then $a^{m/n} = (a^{1/n})^m$, provided that $a^{1/n}$ is defined. We can take the root first or the power first.	$8^{2/3} = (8^{1/3})^2$ $= 2^2 = 4$ $(-16)^{3/4}$ is not a real number.

Definition of negative rational exponents

If m and n are positive integers, then

$$a^{-m/n} = \frac{1}{a^{m/n}},$$

provided that $a^{1/n}$ is defined and nonzero.

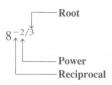

Rules for Rational Exponents

If a and b are nonzero real numbers and m and n are any rational numbers, then the following rules hold.

Examples

Product rule	$a^m \cdot a^n = a^{m+n}$	$3^{1/4} \cdot 3^{1/2} = 3^{3/4}$
Sign-change rule	$a^n = \dfrac{1}{a^{-n}}$	$x^{1/2} = \dfrac{1}{x^{-1/2}}, \quad a^{-2/3} = \dfrac{1}{a^{2/3}}$
Quotient rule	$\dfrac{a^m}{a^n} = a^{m-n}$	$\dfrac{x^{3/4}}{x^{1/4}} = x^{1/2}$
Power of a power rule	$(a^m)^n = a^{mn}$	$(2^{1/2})^{-1/2} = 2^{-1/4}$ $(x^{3/4})^4 = x^3$
Power of a product rule	$(ab)^n = a^n b^n$	$(a^2 b^6)^{1/2} = ab^3$
Power of a quotient rule	$\left(\dfrac{a}{b}\right)^n = \dfrac{a^n}{b^n}$	$\left(\dfrac{8}{x^6}\right)^{2/3} = \dfrac{4}{x^4}$

Radicals

Examples

Definition of radicals	If n is a positive integer and a is a number for which $a^{1/n}$ is defined, then $$\sqrt[n]{a} = a^{1/n}.$$ If $n = 2$, omit the 2 and write $\sqrt[2]{a}$ as $\sqrt{a}$.	$8^{1/3} = \sqrt[3]{8}$ $4^{1/2} = \sqrt{4}$
Product rule for radicals	Provided that all roots are defined, $$\sqrt[n]{ab} = \sqrt[n]{a} \cdot \sqrt[n]{b}.$$	$\sqrt{2} \cdot \sqrt{3} = \sqrt{6}$
Quotient rule for radicals	Provided that all roots are defined, $$\sqrt[n]{\dfrac{a}{b}} = \dfrac{\sqrt[n]{a}}{\sqrt[n]{b}}.$$	$\sqrt{4x} = 2\sqrt{x}$ $\sqrt{\dfrac{5}{9}} = \dfrac{\sqrt{5}}{3}$ $\sqrt{10} \div \sqrt{5} = \sqrt{2}$

Rules for simplifying radicals of index n	A simplified radical of index n has **1.** *no* perfect nth powers as factors of the radicand, **2.** *no* fractions inside the radical, and **3.** *no* radicals in the denominator.	$\sqrt{20} = \sqrt{4 \cdot 5} = 2\sqrt{5}$ $\sqrt{\dfrac{3}{2}} = \dfrac{\sqrt{3}}{\sqrt{2}}$ $\dfrac{\sqrt{3}}{\sqrt{2}} = \dfrac{\sqrt{3}}{\sqrt{2}}\dfrac{\sqrt{2}}{\sqrt{2}} = \dfrac{\sqrt{6}}{2}$

	Equations	**Examples**
Equations with radicals and exponents	**1.** In raising each side of an equation to an even power, we can create an equation that gives extraneous roots. We must check. **2.** When applying the even-root property, remember that there is a positive and a negative root. **3.** For equations with rational exponents, raise each side to a positive or negative integral power first, then apply the even- or odd-root property.	$\sqrt{x} = -3$ $x = 9$ $x^2 = 36$ $x = \pm 6$ $x^{-2/3} = 4$ $(x^{-2/3})^{-3} = 4^{-3}$ $x^2 = 1/64$ $x = \pm 1/8$

	Scientific Notation	**Examples**
Converting from scientific notation	**1.** Determine the number of places to move the decimal point by examining the exponent on the 10. **2.** Move the decimal point to the right for a positive exponent and to the left for a negative exponent.	$4 \times 10^3 = 4000$ $3 \times 10^{-4} = .0003$
Converting to scientific notation (positive numbers)	**1.** Count the number of places (n) that the decimal point must be moved so that it will follow the first nonzero digit of the number. **2.** If the original number was larger than 10, use 10^n. **3.** If the original number was smaller than 1, use 10^{-n}. To convert a negative number, ignore the negative sign, convert the number, and then attach the negative sign to the result.	$67,000 = 6.7 \times 10^4$ $.009 = 9 \times 10^{-3}$

REVIEW EXERCISES

5.1 *Simplify the expressions involving rational exponents. Assume that all variables represent positive real numbers. Write your answers with positive exponents.*

1. $(-27)^{-2/3}$ **2.** $-25^{3/2}$ **3.** $(2^6)^{1/3}$ **4.** $(5^2)^{1/2}$

5. $100^{-3/2}$ **6.** $1000^{-2/3}$ **7.** $\dfrac{3x^{-1/2}}{3^{-2}x^{-1}}$ **8.** $\dfrac{(x^2y^{-3}z)^{1/2}}{x^{1/2}yz^{-1/2}}$

9. $(a^{1/2}b)^3(ab^{1/4})^2$ **10.** $(t^{-1/2})^{-2}(t^{-2}v^2)$ **11.** $(x^{1/2}y^{1/4})(x^{1/4}y)$ **12.** $(a^{1/3}b^{1/6})^2(a^{1/3}b^{2/3})$

5.2 *Simplify the radical expressions. Assume that all variables represent positive real numbers.*

13. $\sqrt{72x^5}$

14. $\sqrt{90y^9z^4}$

15. $\sqrt[3]{72x^5}$

16. $\sqrt[3]{81a^8b^9}$

17. $\sqrt{2^6}$

18. $\sqrt{\dfrac{2}{5}}$

19. $\sqrt{\dfrac{1}{6}}$

20. $\sqrt[3]{\dfrac{2}{3}}$

21. $\sqrt[3]{\dfrac{1}{9}}$

22. $\dfrac{2}{\sqrt{3}}$

23. $\dfrac{3}{\sqrt{2}}$

24. $\sqrt{3^7}$

25. $\dfrac{\sqrt{10}}{\sqrt{6}}$

26. $\dfrac{\sqrt{5}}{\sqrt{8}}$

27. $\dfrac{3}{\sqrt{2a}}$

28. $\dfrac{a}{\sqrt{a}}$

29. $\dfrac{5}{\sqrt[3]{3x^2}}$

30. $\dfrac{b}{\sqrt[3]{b}}$

31. $\sqrt[4]{48x^5y^{12}}$

32. $\sqrt[5]{32x^{10}y^{12}}$

5.3 *Perform the operations and simplify.*

33. $\sqrt{27} + \sqrt{45} - \sqrt{75}$

34. $\sqrt{12} - \sqrt{50} + \sqrt{72}$

35. $\sqrt{\dfrac{1}{3}} + \sqrt{27}$

36. $\sqrt{\dfrac{1}{2}} - \sqrt{\dfrac{1}{8}}$

37. $3\sqrt{2}(5\sqrt{2} - 7\sqrt{3})$

38. $-2\sqrt{3}(\sqrt{3} - \sqrt{2})$

39. $(2 - \sqrt{3})(3 + \sqrt{2})$

40. $(2\sqrt{3} - \sqrt{5})(\sqrt{3} + \sqrt{5})$

41. $\sqrt[3]{40} - \sqrt[3]{5}$

42. $\sqrt[3]{54} + \sqrt[3]{16}$

43. $\sqrt[3]{13} \cdot \sqrt[3]{13}$

44. $\sqrt[3]{14} \cdot \sqrt[3]{14} \cdot \sqrt[3]{14}$

5.4 *Perform the operations and simplify.*

45. $5 \div \sqrt{2}$

46. $10\sqrt{3} \div 3\sqrt{2}$

47. $(\sqrt{3})^4$

48. $(\sqrt{2})^9$

49. $\dfrac{2 - \sqrt{8}}{2}$

50. $\dfrac{-3 - \sqrt{18}}{-6}$

51. $\dfrac{\sqrt{6}}{1 - \sqrt{3}}$

52. $\dfrac{\sqrt{15}}{2 + \sqrt{5}}$

53. $\dfrac{2\sqrt{3}}{3\sqrt{6} - \sqrt{12}}$

54. $\dfrac{-\sqrt{6}}{3\sqrt{2} + \sqrt{6}}$

5.5 *Solve each equation.*

55. $x^2 = 16$

56. $w^2 = 100$

57. $(a - 5)^2 = 4$

58. $(m - 7)^2 = 25$

59. $(a + 1)^2 = 5$

60. $(x + 5)^2 = 3$

61. $(m + 1)^2 = -8$

62. $(w + 4)^2 = 16$

63. $\sqrt{m - 1} = 3$

64. $3\sqrt{x + 5} = 12$

65. $\sqrt[3]{2x + 9} = 3$

66. $\sqrt[4]{2x - 1} = 2$

67. $w^{2/3} = 4$

68. $m^{-4/3} = 16$

69. $(m + 1)^{1/3} = 5$

70. $(w - 3)^{-2/3} = 4$

71. $\sqrt{x - 3} = \sqrt{x + 2} - 1$

72. $\sqrt{x^2 + 3x + 6} = 4$

73. $\sqrt{5x - x^2} = \sqrt{6}$

74. $\sqrt{x + 4} - 2\sqrt{x - 1} = -1$

75. $\sqrt{x + 7} - 2\sqrt{x} = -2$

76. $\sqrt{x} - \sqrt{x - 1} = 1$

77. $2\sqrt{x} - \sqrt{x - 3} = 3$

78. $1 + \sqrt{x + 7} = \sqrt{2x + 7}$

Solve each problem.

79. If we neglect air resistance, the number of feet *s* that an object falls from rest during *t* seconds is given by the equation $s = 16t^2$. How long would it take the landing gear of an airplane to reach the earth if it fell off the airplane at 12,000 feet?

80. Anne is pulling on a 60-foot rope attached to the top of a 48-foot tree, while Walter is cutting the tree at its base. How far from the base of the tree is Anne standing?

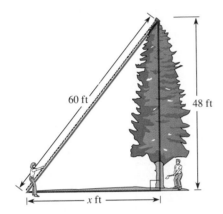

60 ft   48 ft

x ft

Figure for Exercise 80

81. If a guy wire of length 40 feet is attached to an antenna at a height of 30 feet, then how far from the base of the antenna is the wire attached to the ground?

82. Suppose that at the kickoff of a football game, the receiver catches the football at the left side of the goal line and runs for a touchdown diagonally across the field. How many yards would he run? (See Exercise 72 of Section 2.3.)

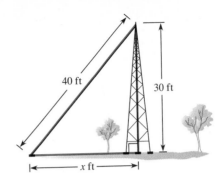

40 ft   30 ft

x ft

Figure for Exercise 81

83. The manufacturer of an antenna recommends that guy wires from the top of the antenna to the ground be attached to the ground at a distance from the base equal to the height of the antenna. How long would the guy wires be for a 200-foot antenna?

84. Betty observed that the lamp post in front of her house casts a shadow of length 8 feet when the angle of inclination of the sun is 60 degrees. How tall is the lamp post? (In a 30-60-90 right triangle, the side opposite 30 is one-half the length of the hypotenuse.)

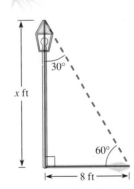

30°

x ft

60°

8 ft

Figure for Exercise 84

5.6 *Write each of the following numbers in standard notation.*

85. 8.36×10^6 **86.** -3.4×10^7 **87.** -5.7×10^{-4} **88.** 4×10^{-3}

Write each of the following numbers in scientific notation.

89. 8,070,000   **90.** $-90,000$ **91.** $-.000709$ **92.** $.0000005$

Perform each computation. Write the answer in scientific notation.

93. $(5 \times 10^9)(-3 \times 10^{-40})$ **94.** $\dfrac{1.2 \times 10^{32}}{4 \times 10^{-7}}$ **95.** $(5 \times 10^{12})^3$ **96.** $\dfrac{(2 \times 10^{-4})^3 \cdot 3 \times 10^{-9}}{4 \times 10^{-6}}$

Miscellaneous

Determine whether each of the following equations is true or false. An equation involving variables should be marked true only if it is an identity. Do not use a calculator.

97. $2^3 \cdot 3^2 = 6^5$

98. $16^{1/4} = 4^{1/2}$

99. $(\sqrt{2})^3 = 2\sqrt{2}$

100. $\sqrt[3]{9} = 3$

101. $8^{200} \cdot 8^{200} = 64^{200}$

102. $\sqrt{295} \cdot \sqrt{295} = 295$

103. $4^{1/2} = \sqrt{2}$

104. $\sqrt{a^2} = |a|$

105. $5^2 \cdot 5^2 = 25^4$

106. $\sqrt{6} \div \sqrt{2} = \sqrt{3}$

107. $\sqrt{w^{10}} = w^5$

108. $\sqrt{a^{16}} = a^4$

109. $\sqrt{x^6} = x^3$

110. $\sqrt[6]{16} = \sqrt[3]{4}$

111. $\sqrt{x^8} = x^4$

112. $\sqrt[9]{2^6} = 2^{2/3}$

113. $\sqrt{16} = 2$

114. $2^{1/2} \cdot 2^{1/4} = 2^{3/4}$

115. $2^{600} = 4^{300}$

116. $\sqrt{2} \cdot \sqrt[4]{2} = \sqrt[6]{2}$

117. $\dfrac{2 + \sqrt{6}}{2} = 1 + \sqrt{6}$

118. $\dfrac{4 + 2\sqrt{3}}{2} = 2 + \sqrt{3}$

119. $\sqrt{\dfrac{4}{6}} = \dfrac{2}{3}$

120. $8^{200} \cdot 8^{200} = 8^{400}$

121. $81^{2/4} = 81^{1/2}$

122. $(-64)^{2/6} = (-64)^{1/3}$

CHAPTER 5 TEST

Simplify each expression. Assume all variables represent positive numbers.

1. $8^{2/3}$

2. $4^{-3/2}$

3. $\sqrt{21} \div \sqrt{7}$

4. $2\sqrt{5} \cdot 3\sqrt{5}$

5. $\sqrt{20} + \sqrt{5}$

6. $\sqrt{5} + \dfrac{1}{\sqrt{5}}$

7. $2^{1/2} \cdot 2^{1/2}$

8. $\sqrt{72}$

9. $\sqrt{\dfrac{5}{12}}$

10. $\dfrac{6 + \sqrt{18}}{6}$

11. $(2\sqrt{3} + 1)(\sqrt{3} - 2)$

12. $\sqrt[3]{54}$

13. $\sqrt[4]{32}$

14. $\sqrt[5]{20} \div \sqrt[5]{10}$

15. $\dfrac{1}{\sqrt[3]{2}}$

16. $\sqrt{a^9}$

17. $\sqrt[3]{x^9}$

18. $\sqrt{20m^3}$

19. $x^{1/2} \cdot x^{1/4}$

20. $(x^{1/2})^{1/2}$

21. $\sqrt[3]{40x^7}$

22. $(4 + \sqrt{3})^2$

Rationalize the denominator and simplify.

23. $\dfrac{2}{5 - \sqrt{3}}$

24. $\dfrac{\sqrt{6}}{4\sqrt{3} + \sqrt{2}}$

Solve each equation.

25. $(x - 2)^2 = 49$

26. $2\sqrt{x + 4} = 3$

27. $w^{2/3} = 4$

28. $\sqrt{x - 1} + \sqrt{x + 4} = 5$

Show a complete solution to each problem.

29. Find the exact length of the side of a square whose diagonal is 3 feet.

30. Two positive numbers differ by 11 and their square roots differ by 1. Find the numbers.

31. If the perimeter of a rectangle is 20 feet and the diagonal is $2\sqrt{13}$ feet, then what are the length and width?

Convert to scientific notation.

32. 324,000

33. .0000867

Perform each computation by converting to scientific notation. Give the answer in scientific notation.

34. $\dfrac{(80,000)(.0006)}{2,000,000}$

35. $\dfrac{(.00006)^2(500)}{(30,000)^2(.01)}$

Tying It All Together

CHAPTERS 1–5

Find the solution set to each equation or inequality. For the inequalities, also sketch the graph of the solution set.

1. $3(x - 2) + 5 = 7 - 4(x + 3)$

2. $\sqrt{6x + 7} = 4$

3. $|2x + 5| > 1$

4. $8x^3 - 27 = 0$

5. $2x - 3 > 3x - 4$

6. $\sqrt{2x - 3} - \sqrt{3x + 4} = 0$

7. $\dfrac{w}{3} + \dfrac{w - 4}{2} = \dfrac{11}{2}$

8. $2(x + 7) - 4 = x - (10 - x)$

9. $(x + 7)^2 = 25$

10. $a^{-1/2} = 4$

11. $x - 3 > 2$ or $x < 2x + 6$

12. $a^{-2/3} = 16$

13. $3x^2 - 1 = 0$

14. $5 - 2(x - 2) = 3x - 5(x - 2) - 1$

15. $|3x - 4| < 5$

16. $3x - 1 = 0$

17. $\sqrt{y - 1} = 9$

18. $|5(x - 2) + 1| = 3$

19. $.06x - .04(x - 20) = 2.8$

20. $|3x - 1| > -2$

21. $\dfrac{3\sqrt{2}}{x} = \dfrac{\sqrt{3}}{4\sqrt{5}}$

22. $\dfrac{\sqrt{x} - 4}{x} = \dfrac{1}{\sqrt{x} + 5}$

23. $\dfrac{3\sqrt{2} + 4}{\sqrt{2}} = \dfrac{x\sqrt{18}}{3\sqrt{2} + 2}$

24. $\dfrac{x}{2\sqrt{5} - \sqrt{2}} = \dfrac{2\sqrt{5} + \sqrt{2}}{x}$

25. $\dfrac{\sqrt{2x} - 5}{x} = \dfrac{-3}{\sqrt{2x} + 5}$

26. $\dfrac{\sqrt{6} + 2}{x} = \dfrac{2}{\sqrt{6} + 4}$

27. $\dfrac{x - 1}{\sqrt{6}} = \dfrac{\sqrt{6}}{x}$

28. $\dfrac{x + 3}{\sqrt{10}} = \dfrac{\sqrt{10}}{x}$

The expression $\dfrac{-b + \sqrt{b^2 - 4ac}}{2a}$ *will be used in Chapter 6 to solve quadratic equations. Evaluate this expression for the following choices of a, b, and c.*

29. $a = 1$, $b = 2$, $c = -15$

30. $a = 1$, $b = 8$, $c = 12$

31. $a = 2$, $b = 5$, $c = -3$

32. $a = 6$, $b = 7$, $c = -3$

CHAPTER **6**

Quadratic Equations and Inequalities

On a recent bicycle trip across the desert, Erin traveled 60 miles before lunch and 46 miles after lunch. She cycled one hour more after lunch than before lunch. However, her speed was 4 miles per hour slower after lunch. What was Erin's speed before lunch and after lunch?

This problem is very similar to the uniform-motion problems that we solved in Chapters 2 and 5. However, to solve it we need to solve a quadratic equation that cannot be solved by any of the techniques that we have used previously. In this chapter we will learn to solve any quadratic equation. We will have the necessary skills to solve this problem when we see it again as Exercise 23 of Section 6.3.

6.1 Factoring and Completing the Square

IN THIS SECTION:

- Review of Factoring
- Review of the Even-root Property
- Completing the Square

In this section we first review the types of quadratic equations that we solved in Chapters 3, 4, and 5. Then we introduce the method of completing the square. This new method can be used to solve any quadratic equation.

Review of Factoring

Since Chapter 3, we have solved equations such as

$$x^2 = 16, \qquad (x - 1)^2 = 9, \qquad \text{and} \qquad 3x^2 - 4x - 15 = 0.$$

These are examples of quadratic equations. The definition of the quadratic equation was given in Chapter 3, but we will repeat it here.

DEFINITION
Quadratic Equation

A quadratic equation is an equation that can be written in the form

$$ax^2 + bx + c = 0,$$

where a, b, and c are real numbers with $a \neq 0$.

Recall that in Chapter 3 we used the following strategy to solve quadratic equations by factoring.

STRATEGY
Solving Equations
by Factoring

1. Write the equation with 0 on the right side.
2. Factor the left side.
3. Use the zero factor property to get simpler equations. (Set each factor equal to 0.)
4. Solve the simpler equations.
5. Check the answers in the original equation.

EXAMPLE 1   Solve by factoring $3x^2 - 4x - 15 = 0$.

Solution

$$(3x + 5)(x - 3) = 0$$

$$3x + 5 = 0 \quad \text{or} \quad x - 3 = 0 \quad \text{Zero-factor property}$$

$$3x = -5 \quad \text{or} \quad x = 3$$

$$x = -\frac{5}{3}$$

The solution set is $\{-5/3, 3\}$. Be sure to check the solutions in the original equation. ◄

Review of the Even-root Property

Recall that it is not necessary to write a quadratic equation in the form of the definition in order to solve it. In Chapter 3, we solved simple quadratics by using the even-root property.

EXAMPLE 2 Solve $(x - 1)^2 = 9$.

Solution

$$x - 1 = \pm\sqrt{9} \quad \text{The even-root property}$$

$$x - 1 = 3 \quad \text{or} \quad x - 1 = -3$$

$$x = 4 \quad \text{or} \quad x = -2$$

The solution set is $\{-2, 4\}$. Check these solutions in the original equation. ◄

Completing the Square

We cannot solve every quadratic by factoring, because not all quadratic polynomials can be factored. However, we can write any quadratic equation in the form of Example 2 and then apply the even-root property to solve it. This method is called **completing the square.**

The essential part of completing the square is to recognize a perfect-square trinomial when given its first two terms. For example, the perfect-square trinomial whose first two terms are $x^2 + 6x$ is $x^2 + 6x + 9$. The trinomial $x^2 + 6x + 9$ is a perfect-square trinomial because it is the square of the binomial $x + 3$:

$$x^2 + 6x + 9 = (x + 3)^2$$

The binomial $x^2 - 10x$ consists of the first two terms of the perfect-square trinomial $x^2 - 10x + 25$. Note that

$$x^2 - 10x + 25 = (x - 5)^2.$$

If the coefficient of x^2 is 1, there is a simple rule for identifying the last term in a perfect-square trinomial.

Rule for Finding the Last Term

The last term of a perfect-square trinomial is the square of one-half of the coefficient of the middle term. In symbols, the perfect-square trinomial whose first two terms are $x^2 + bx$ is $x^2 + bx + \left(\dfrac{b}{2}\right)^2$.

EXAMPLE 3 Find the perfect-square trinomial whose first two terms are given.

a) $x^2 + 8x$ **b)** $x^2 - 4x$ **c)** $x^2 + 5x$ **d)** $x^2 - \dfrac{3}{2}x$

Solution

a) One-half of 8 is 4, and 4 squared is 16. Therefore, the perfect-square trinomial is $x^2 + 8x + 16$.

b) One-half of -4 is -2, and -2 squared is 4. Therefore, the perfect-square trinomial is $x^2 - 4x + 4$.

c) One-half of 5 is 5/2, and 5/2 squared is 25/4. Therefore, the perfect-square trinomial is $x^2 + 5x + 25/4$.

d) One-half of 3/2 is 3/4, and $(3/4)^2 = 9/16$. Therefore, the perfect-square trinomial is $x^2 - \dfrac{3}{2}x + \dfrac{9}{16}$. ◀

The following examples show how to solve a quadratic equation by completing the square.

EXAMPLE 4 Solve by completing the square $x^2 + 6x + 5 = 0$.

Solution The perfect-square trinomial whose first two terms are $x^2 + 6x$ is $x^2 + 6x + 9$. We therefore move 5 to the right side of the equation, then add 9 to each side:

$$x^2 + 6x \quad = -5 \qquad \text{Subtract 5 from each side.}$$

$$x^2 + 6x + 9 = -5 + 9 \qquad \text{Add 9 to each side to complete the square.}$$

$$(x + 3)^2 = 4 \qquad \text{Factor the left side.}$$

$$x + 3 = \pm\sqrt{4} \qquad \text{The even-root property}$$

$$x + 3 = 2 \qquad \text{or} \qquad x + 3 = -2$$

$$x = -1 \qquad \text{or} \qquad x = -5$$

The solution set is $\{-1, -5\}$. Check these solutions in the original equation. ◀

Note that all of the perfect-square trinomials that we used so far had a leading coefficient of 1. If the leading coefficient is not 1, then we must divide each side of the equation by the leading coefficient to get an equation that does have a leading coefficient of 1. The strategy for solving a quadratic equation by completing the square is stated in the following box.

STRATEGY
Solving Quadratic Equations by Completing the Square

1. The coefficient of x^2 must be 1.
2. Place only the x^2 and the x terms on the left side.
3. Add to each side the square of one-half the coefficient of x.
4. Factor the left side as the square of a binomial.
5. Apply the even-root property.
6. Solve for x.
7. Simplify.

EXAMPLE 5 Solve by completing the square.

a) $2x^2 + 3x - 2 = 0$ **b)** $x^2 - 3x - 6 = 0$

Solution

a) For completing the square, the coefficient of x^2 must be 1. Thus, our first step is to divide each side of the equation by 2.

$$\frac{2x^2 + 3x - 2}{2} = \frac{0}{2} \qquad \text{Divide each side by 2.}$$

$$x^2 + \frac{3}{2}x - 1 = 0 \qquad \text{Simplify.}$$

$$x^2 + \frac{3}{2}x = 1 \qquad \text{Add 1 to each side.}$$

$$x^2 + \frac{3}{2}x + \frac{9}{16} = 1 + \frac{9}{16} \qquad \begin{array}{l}\text{One-half of 3/2 is 3/4, and}\\ \text{3/4 squared is 9/16.}\end{array}$$

$$\left(x + \frac{3}{4}\right)^2 = \frac{25}{16} \qquad \text{Factor the left side.}$$

$$x + \frac{3}{4} = \pm\sqrt{\frac{25}{16}} \qquad \text{The even-root property}$$

$$x + \frac{3}{4} = \frac{5}{4} \quad \text{or} \quad x + \frac{3}{4} = -\frac{5}{4}$$

$$x = \frac{2}{4} \quad \text{or} \quad x = -\frac{8}{4}$$

$$x = \frac{1}{2} \quad \text{or} \quad x = -2$$

The solution set is $\{-2, 1/2\}$. Check.

b)

$$x^2 - 3x - 6 = 0$$

$$x^2 - 3x = 6 \qquad \text{Add 6 to each side.}$$

$$x^2 - 3x + \frac{9}{4} = 6 + \frac{9}{4} \qquad \begin{array}{l}\text{One-half of } -3 \text{ is } -3/2\text{, and} \\ -3/2 \text{ squared is } 9/4.\end{array}$$

$$\left(x - \frac{3}{2}\right)^2 = \frac{33}{4} \qquad 6 + \frac{9}{4} = \frac{24}{4} + \frac{9}{4} = \frac{33}{4}$$

$$x - \frac{3}{2} = \pm\sqrt{\frac{33}{4}} \qquad \text{Even-root property}$$

$$x = \frac{3}{2} \pm \frac{\sqrt{33}}{2} \qquad \text{Add 3/2 to each side.}$$

$$x = \frac{3 \pm \sqrt{33}}{2}$$

The solution set is $\left\{ \dfrac{3 + \sqrt{33}}{2}, \dfrac{3 - \sqrt{33}}{2} \right\}$ ◀

The next two examples show equations that are not originally in the form of quadratic equations. However, after simplifying these equations we get quadratic equations to solve.

EXAMPLE 6   Solve $x + 3 = \sqrt{153 - x}$.

Solution

$$(x + 3)^2 = (\sqrt{153 - x})^2 \qquad \text{Square each side.}$$

$$x^2 + 6x + 9 = 153 - x \qquad \text{Simplify.}$$

$$x^2 + 7x - 144 = 0$$

$$(x - 9)(x + 16) = 0 \qquad \text{Factor.}$$

$$x - 9 = 0 \quad \text{or} \quad x + 16 = 0 \qquad \text{The zero factor property}$$

$$x = 9 \quad \text{or} \quad x = -16$$

Because we squared each side of the original equation, we must check for extraneous roots. Letting $x = 9$ in the original equation, we get

$$9 + 3 = \sqrt{153 - 9}$$

$$12 = \sqrt{144} \qquad \text{Correct.}$$

Letting $x = -16$ in the original equation, we get

$$-16 + 3 = \sqrt{153 - (-16)}$$

$$-13 = \sqrt{169} \qquad \text{Incorrect.}$$

Because -16 is an extraneous root, the solution set is $\{9\}$. ◀

EXAMPLE 7   Solve $\dfrac{1}{x} + \dfrac{3}{x - 2} = \dfrac{5}{8}$.

Solution

$$8x(x-2)\frac{1}{x} + 8x(x-2)\frac{3}{x-2} = 8x(x-2)\frac{5}{8}$$ Multiply each side by $8x(x-2)$.

$$8x - 16 + 24x = 5x^2 - 10x$$

$$32x - 16 = 5x^2 - 10x$$

$$-5x^2 + 42x - 16 = 0$$

$$5x^2 - 42x + 16 = 0$$ Multiply each side by -1 for easier factoring.

$$(5x - 2)(x - 8) = 0$$ Factor.

$$5x - 2 = 0 \quad \text{or} \quad x - 8 = 0$$

$$x = \frac{2}{5} \quad \text{or} \quad x = 8$$

The solution set is $\{2/5, 8\}$. Check.   ◄

Warm-ups

True or false?

1. Completing the square means drawing the fourth side.
2. $(x - 3)^2 = 12$ is equivalent to $x - 3 = 2\sqrt{3}$.
3. Every quadratic equation can be solved by factoring.
4. The trinomial $x^2 + \dfrac{4}{3}x + \dfrac{16}{9}$ is a perfect-square trinomial.
5. Every quadratic equation can be solved by completing the square.
6. To complete the square for the equation $2x^2 + 6x = 4$, we add 9 to each side.
7. $(2x - 3)(3x + 5) = 0$ is equivalent to $x = 3/2$ or $x = -5/3$.
8. In completing the square for $x^2 - 3x = 4$, we should add 9/4 to each side.
9. The equation $(x - 3)(x + 2) = 1$ is equivalent to the compound equation $x - 3 = 1$ or $x + 2 = 1$.
10. All quadratic equations have two distinct solutions.

6.1 EXERCISES

Solve by factoring. See Example 1.

1. $x^2 - x - 6 = 0$
2. $x^2 + 6x + 8 = 0$
3. $a^2 + 2a = 15$
4. $w^2 - 2w = 15$
5. $2x^2 - x - 3 = 0$
6. $6x^2 - x - 15 = 0$

7. $x^2 = x + 12$

8. $x^2 = x + 20$

9. $x^2 + 14x + 49 = 0$

10. $a^2 - 6a + 9 = 0$

11. $a^2 - 16 = 0$

12. $4w^2 - 25 = 0$

Solve by using the even-root property. See Example 2.

13. $x^2 = 81$

14. $x^2 = \dfrac{9}{4}$

15. $x^2 = \dfrac{16}{9}$

16. $a^2 = 32$

17. $(x - 3)^2 = 16$

18. $(x + 5)^2 = 4$

19. $4x^2 = 25$

20. $5w^2 = 3$

21. $(x + 1)^2 = 5$

22. $(a - 2)^2 = 8$

23. $\left(x - \dfrac{3}{2}\right)^2 = \dfrac{7}{4}$

24. $\left(w + \dfrac{2}{3}\right)^2 = \dfrac{5}{9}$

25. $\left(x + \dfrac{1}{2}\right)^2 = \dfrac{9}{4}$

26. $\left(x - \dfrac{2}{3}\right)^2 = \dfrac{4}{9}$

27. $\left(x - \dfrac{1}{2}\right)^2 = \dfrac{9}{2}$

28. $\left(x - \dfrac{1}{3}\right)^2 = \dfrac{1}{3}$

29. $\left(a + \dfrac{2}{3}\right)^2 = \dfrac{32}{9}$

30. $\left(w + \dfrac{1}{2}\right)^2 = 6$

Write a perfect-square trinomial that has its first two terms the same as those given. See Example 3.

31. $x^2 + 2x$

32. $m^2 + 14m$

33. $x^2 - 6x$

34. $w^2 - 5w$

35. $w^2 - 3w$

36. $x^2 - 10x$

37. $m^2 + 12m$

38. $x^2 + \dfrac{x}{4}$

39. $x^2 + \dfrac{x}{2}$

40. $m^2 + m$

41. $x^2 + \dfrac{2}{3}x$

42. $p^2 + \dfrac{3}{4}p$

Solve by completing the square. See Examples 4 and 5.

43. $x^2 - 2x - 15 = 0$

44. $x^2 + 2x - 24 = 0$

45. $x^2 + 6x - 7 = 0$

46. $x^2 - 6x - 7 = 0$

47. $x^2 + 8x = 20$

48. $x^2 + 10x = -9$

49. $2x^2 - 4x = 70$

50. $3x^2 - 6x = 24$

51. $x^2 - x - 20 = 0$

52. $x^2 - 3x - 10 = 0$

53. $x^2 + 5x = 14$

54. $x^2 + x = 2$

55. $2x^2 - x - 3 = 0$

56. $2x^2 - x - 15 = 0$

57. $3x^2 + 5x = 2$

58. $3x^2 - 7x = 6$

59. $x^2 + 4x = 6$

60. $x^2 + 6x - 8 = 0$

61. $x^2 + 8x - 4 = 0$

62. $x^2 + 10x - 3 = 0$

63. $2x^2 + 3x - 4 = 0$

64. $2x^2 + 5x - 1 = 0$

65. $5x^2 + 4x - 3 = 0$

66. $3x^2 + 4x - 1 = 0$

67. $-x^2 + x + 6 = 0$

68. $-x^2 + x + 12 = 0$

Solve each equation by an appropriate method. See Examples 6 and 7.

69. $x = \dfrac{\sqrt{x + 1}}{2}$

70. $x - 1 = \dfrac{\sqrt{x + 1}}{2}$

71. $\dfrac{4(2x - 3)^2}{9} = 1$

72. $\dfrac{(3 - 2x)^2}{25} = 1$

73. $\dfrac{x}{x - 2} = \dfrac{2x - 3}{x}$

74. $\dfrac{x}{x + 3} = \dfrac{3x}{5x - 1}$

75. $\dfrac{2}{x^2} + \dfrac{4}{x} + 1 = 0$

76. $\dfrac{1}{x^2} + \dfrac{3}{x} + 1 = 0$

77. $\dfrac{1}{x} + \dfrac{1}{x-1} = \dfrac{1}{4}$

78. $\dfrac{1}{x} - \dfrac{2}{1-x} = \dfrac{1}{2}$

79. $x^2 + 6x + k = 0$

80. $x^2 - 4x + c = 0$

6.2 The Quadratic Formula

IN THIS SECTION:
- Developing the Formula
- Using the Formula
- Number of Solutions
- Applications

In Section 6.1 we learned to solve quadratic equations by completing the square. The method of completing the square can be applied to any quadratic equation. Because it works on any quadratic, we can use completing the square to develop a formula that can be used to solve any quadratic equation.

Developing the Formula

Start with the general form of the quadratic equation,

$$ax^2 + bx + c = 0.$$

Assume a is positive for now and divide each side by a.

$$\frac{ax^2 + bx + c}{a} = \frac{0}{a}$$

$$x^2 + \frac{b}{a}x + \frac{c}{a} = 0$$

$$x^2 + \frac{b}{a}x = -\frac{c}{a} \qquad \text{Subtract } \tfrac{c}{a} \text{ from each side.}$$

One-half of b/a is $b/(2a)$. To complete the square on the left side, we add $\dfrac{b^2}{4a^2}$ to each side.

$$x^2 + \frac{b}{a}x + \frac{b^2}{4a^2} = \frac{b^2}{4a^2} - \frac{c}{a}$$

Factor the left side and get a common denominator for the right side.

$$\left(x + \frac{b}{2a}\right)^2 = \frac{b^2}{4a^2} - \frac{4ac}{4a^2} \qquad \frac{c}{a} \frac{(4a)}{(4a)} = \frac{4ac}{4a^2}$$

$$\left(x + \frac{b}{2a}\right)^2 = \frac{b^2 - 4ac}{4a^2}$$

$$x + \frac{b}{2a} = \pm\sqrt{\frac{b^2 - 4ac}{4a^2}} \qquad \text{Even-root property}$$

$$x = \frac{-b}{2a} \pm \frac{\sqrt{b^2 - 4ac}}{2a} \qquad \text{Because } a > 0, \sqrt{4a^2} = 2a$$

$$x = \frac{-b \pm \sqrt{b^2 - 4ac}}{2a}$$

We assumed a was positive so that $\sqrt{4a^2} = 2a$ would be correct. If a is negative, then $\sqrt{4a^2} = -2a$, and we get

$$x = \frac{-b}{2a} \pm \frac{\sqrt{b^2 - 4ac}}{-2a}.$$

The negative sign can be omitted in the term $-2a$ because of the $\pm$ symbol preceding it. For example, the results of $5 \pm (-3)$ and 5 ± 3 are the same. Thus, when a is negative we get the same formula as when a is positive. It is called the **quadratic formula.**

The Quadratic Formula

> The solution to the equation $ax^2 + bx + c = 0$, with $a \neq 0$, is given by the formula
>
> $$x = \frac{-b \pm \sqrt{b^2 - 4ac}}{2a}.$$

The quadratic formula is generally used instead of completing the square to solve a quadratic equation that cannot be factored. Even though we will no longer use completing the square to solve quadratic equations, it is an important skill to learn and will be used in the study of conic sections in Chapter 11.

Using the Formula

EXAMPLE 1 Solve each equation by using the quadratic formula.

a) $x^2 + 2x - 15 = 0$
c) $2x^2 + 6x + 3 = 0$

b) $4x^2 = 12x - 9$
d) $x^2 + x + 5 = 0$

Solution

a) To use the formula we first identify the values of a, b, and c.

$$1 \cdot x^2 + 2x - 15 = 0$$

$$\uparrow \qquad \uparrow \qquad \uparrow$$
$$a \qquad b \qquad c$$

The coefficient of x^2 is 1, so $a = 1$. The coefficient of $2x$ is 2, so $b = 2$. The constant term is -15, so $c = -15$. Substitute these values into the quadratic formula.

$$x = \frac{-2 \pm \sqrt{2^2 - 4(1)(-15)}}{2(1)}$$

$$= \frac{-2 \pm \sqrt{4 + 60}}{2}$$

$$= \frac{-2 \pm \sqrt{64}}{2} = \frac{-2 \pm 8}{2}$$

$$x = \frac{-2 + 8}{2} = 3 \qquad \text{or} \qquad x = \frac{-2 - 8}{2} = -5$$

The solution set is $\{-5, 3\}$. Check in the original equation.

b) Rewrite the equation in the form $ax^2 + bx + c = 0$ before identifying a, b, and c.

$$4x^2 - 12x + 9 = 0$$

$$a = 4, \qquad b = -12, \qquad c = 9$$

$$x = \frac{12 \pm \sqrt{(-12)^2 - 4(4)(9)}}{2(4)} \qquad \text{Because } b = -12,\ -b = 12.$$

$$= \frac{12 \pm \sqrt{144 - 144}}{8}$$

$$= \frac{12 \pm 0}{8} = \frac{12}{8} = \frac{3}{2}$$

The solution set is $\{3/2\}$. Check.

c) $2x^2 + 6x + 3 = 0 \qquad a = 2, \qquad b = 6, \qquad c = 3$

$$x = \frac{-6 \pm \sqrt{(6)^2 - 4(2)(3)}}{2(2)}$$

$$= \frac{-6 \pm \sqrt{36 - 24}}{4}$$

$$= \frac{-6 \pm \sqrt{12}}{4} = \frac{-6 \pm 2\sqrt{3}}{4} = \frac{2(-3 \pm \sqrt{3})}{2 \cdot 2} = \frac{-3 \pm \sqrt{3}}{2}$$

The solution set is $\left\{ \dfrac{-3 \pm \sqrt{3}}{2} \right\}$. Note that in this case the solutions are two irrational numbers.

d) $x^2 + x + 5 = 0$ $a = 1,$ $b = 1,$ $c = 5$

$$x = \frac{-1 \pm \sqrt{(1)^2 - 4(1)(5)}}{2(1)} = \frac{-1 \pm \sqrt{-19}}{2}$$

Because $\sqrt{-19}$ is not a real number, the solution set is the empty set, $\varnothing$. ◄

We have learned to solve quadratic equations by four different methods: the even-root property, factoring, completing the square, and the quadratic formula. Remember that any quadratic equation can be solved by completing the square or the quadratic formula, but the quadratic formula is easier to use than completing the square. Only certain types of quadratic equations can be solved by the even-root property or factoring.

Methods for Solving $ax^2 + bx + c = 0$

Method	Comments	Examples
Even-root property	Use when $b = 0$.	$x^2 = 3,$ $(x - 2)^2 = 8$
Factoring	Use when the polynomial can be factored.	$x^2 + 5x + 6 = 0$ $(x + 2)(x + 3) = 0$
Quadratic formula	Use when the first two methods do not apply.	$x^2 + 5x + 3 = 0$
Completing the square	Use the quadratic formula instead.	

Number of Solutions

The quadratic equations in Examples 1(a) and 1(c) each had two solutions. In each of those examples the value of $b^2 - 4ac$ was positive. In Example 1(b) the quadratic equation had only one solution, because the value of $b^2 - 4ac$ was zero. In Example 1(d) the quadratic equation had no real solutions, because $b^2 - 4ac$ had a negative value. Since the value of $b^2 - 4ac$ determines the number of solutions to a quadratic equation, we call $b^2 - 4ac$ the **discriminant.**

Number of Solutions to a Quadratic Equation

The equation $ax^2 + bx + c = 0$ with $a \neq 0$ has

two real solutions if $b^2 - 4ac > 0$,

one real solution if $b^2 - 4ac = 0$, and

no real solution if $b^2 - 4ac < 0$.

EXAMPLE 2 Use the discriminant to determine the number of solutions to each quadratic equation.

a) $x^2 - 3x - 5 = 0$ **b)** $x^2 = 3x - 9$ **c)** $4x^2 - 12x + 9 = 0$

Solution

a) $x^2 - 3x - 5 = 0$ $a = 1,$ $b = -3,$ $c = -5$

$$b^2 - 4ac = (-3)^2 - 4(1)(-5) = 9 + 20 = 29$$

Since the discriminant is positive, there are two real solutions to this quadratic equation.

b) $x^2 - 3x + 9 = 0$ $a = 1,$ $b = -3,$ $c = 9$

$$b^2 - 4ac = (-3)^2 - 4(1)(9) = 9 - 36 = -27$$

Since the discriminant is negative, there are no real solutions to this equation.

c) $4x^2 - 12x + 9 = 0$ $a = 4,$ $b = -12,$ $c = 9$

$$b^2 - 4ac = (-12)^2 - 4(4)(9) = 144 - 144 = 0$$

Since the discriminant is 0, there is only one real solution to this quadratic equation.   ◀

Applications

There are many types of problems whose solutions require the quadratic formula. When we need the quadratic formula, it is usually because the solutions are not rational numbers. Of course, this makes checking more difficult.

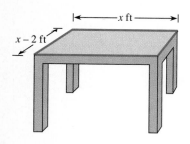

Figure 6.1

EXAMPLE 3 The area of a rectangular table top is to be 6 square feet. If the width is 2 feet shorter than the length, then what are the dimensions? (See Fig. 6.1.)

Solution Let x be the length and $x - 2$ be the width. Because the area is 6 square feet, we can write the equation

$$x(x - 2) = 6$$

or

$$x^2 - 2x - 6 = 0.$$

We must use the quadratic formula since this quadratic cannot be factored.

$$a = 1,   b = -2,   c = -6$$

$$x = \frac{2 \pm \sqrt{(-2)^2 - 4(1)(-6)}}{2(1)}$$

$$= \frac{2 \pm \sqrt{28}}{2} = \frac{2 \pm 2\sqrt{7}}{2} = 1 \pm \sqrt{7}$$

Because $1 - \sqrt{7}$ is a negative number, it cannot be the length of a table top. If $x = 1 + \sqrt{7}$, then $x - 2 = 1 + \sqrt{7} - 2 = \sqrt{7} - 1$. Checking the product of these expressions we get

$$(\sqrt{7} + 1)(\sqrt{7} - 1) = 7 - 1 = 6.$$

The length is $\sqrt{7} + 1$ feet, and the width is $\sqrt{7} - 1$ feet.   ◄

Warm-ups

True or false?

1. Completing the square is used to develop the quadratic formula.
2. In using the quadratic formula on the equation $3x^2 = 4x - 7$, we let $a = 3$, $b = 4$, and $c = -7$.
3. If $dx^2 + ex + f = 0$ and $d \neq 0$, then $x = \dfrac{-e \pm \sqrt{e^2 - 4df}}{2d}$.
4. The quadratic formula will not work on the equation $x^2 - 3 = 0$.
5. If $a = 2$, $b = -3$, and $c = -4$, then $b^2 - 4ac = 41$.
6. If the discriminant is equal to zero, then the quadratic equation has no real solutions because you cannot divide by zero.
7. If the discriminant is positive, then the quadratic equation has 2 distinct real numbers in its solution set.
8. If we use the quadratic formula on $2x - x^2 = 0$, we let $a = -1$, $b = 2$, and $c = 0$.
9. Two positive numbers that have a sum of 6 can be represented by x and $x + 6$.
10. Some quadratic equations have no real numbers in their solution sets.

6.2 EXERCISES

Solve each equation by using the quadratic formula. See Example 1.

1. $x^2 + 5x + 6 = 0$
2. $x^2 - 7x + 12 = 0$
3. $x^2 + x = 6$
4. $x^2 + 2x = 8$
5. $6x^2 - 7x - 3 = 0$
6. $8x^2 + 2x - 1 = 0$
7. $4x^2 - 4x + 1 = 0$
8. $4x^2 - 12x + 9 = 0$
9. $3x^2 + 2x + 5 = 0$
10. $-2x^2 + 3x = 6$
11. $x^2 + 5x - 1 = 0$
12. $x^2 + 3x - 5 = 0$
13. $x^2 + 8x + 6 = 0$
14. $x^2 + 6x + 4 = 0$
15. $2x^2 - 6x + 1 = 0$
16. $3x^2 - 8x + 2 = 0$
17. $\dfrac{1}{3}x^2 + \dfrac{1}{2}x = \dfrac{1}{3}$
18. $\dfrac{1}{2}x^2 + x = 1$

Calculate $b^2 - 4ac$ and determine the number of real solutions to each quadratic equation. See Example 2.

19. $x^2 - 6x + 2 = 0$
20. $x^2 + 6x + 9 = 0$
21. $2x^2 - 5x + 6 = 0$

22. $-x^2 + 3x - 4 = 0$

23. $4x^2 + 25 = 20x$

24. $x^2 = 3x + 5$

25. $x^2 - \frac{1}{2}x + \frac{1}{4} = 0$

26. $\frac{1}{2}x^2 - \frac{1}{3}x + \frac{1}{4} = 0$

27. $-3x^2 + 5x + 6 = 0$

28. $9x^2 + 16 = 24x$

29. $9 - 24x + 16x^2 = 0$

30. $12 - 7x + x^2 = 0$

Solve each problem. See Example 3.

31. Find two positive real numbers that differ by 1 and have a product of 16.

32. Find two positive real numbers that differ by 2 and have a product of 10.

33. Find two real numbers that have a sum of 6 and a product of 4.

34. Find two real numbers that have a sum of 8 and a product of 2.

35. The length of a bulletin board is 1 foot more than the width. The diagonal has length $\sqrt{3}$ feet. Find the length and width of the bulletin board.

36. The length of a rectangle is 2 meters longer than the width. The diagonal measures $\sqrt{6}$ meters. Find the length and width.

37. The length of a rectangular window is 4 feet longer than the width, and its area is 10 square feet. Find the length and width.

38. The diagonal of a square is 2 meters longer than a side. Find the length of a side.

If an object is given an initial velocity of v_0 feet per second from a height of s_0 feet, then its altitude S after t seconds is given by the formula $S = -16t^2 + v_0 t + s_0$.

39. If a pine cone is projected upward at a velocity of 16 feet per second from the top of a 96-foot pine tree, then how long does it take for it to reach the earth?

40. If a pine cone falls from the top of a 96-foot pine tree, then how long does it take to reach the earth?

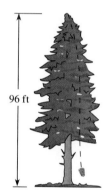

96 ft

Figure for Exercise 40

41. If a penny is thrown downward at 30 feet per second from the bridge at Royal Gorge, Colorado, how long does it take to reach the Arkansas River, 1000 feet below?

42. If a baseball is hit upward at 150 feet per second from a height of 5 feet, then how long will it take for the ball to reach the earth?

Find the solution set to each equation. Use the quadratic formula and a calculator. Round answers to three decimal places.

43. $x^2 + 3.2x - 5.7 = 0$

44. $x^2 + 7.15x + 3.24 = 0$

45. $5.2x^2 - 6.8x + 4.05 = 0$

46. $4.3x^2 - 9.86x - 3.75 = 0$

47. $x^2 - 7.4x + 13.69 = 0$

48. $1.44x^2 + 5.52x + 5.29 = 0$

49. $1.85x^2 + 6.72x + 3.6 = 0$

50. $3.67x^2 + 4.35x - 2.13 = 0$

51. $3x^2 + 14,379x + 243 = 0$

52. $x^2 + 12,347x + 6,741 = 0$

53. $x^2 + .00075x - .0062 = 0$

54. $x^2 - .00035x + 24 = 0$

6.3 Applications

IN THIS SECTION:

- **Equations Quadratic in Form**
- **Verbal Problems**

In this section we will use the methods for solving quadratic equations to solve equations that are not quadratic, and we will see more applications of the quadratic formula in solving problems.

Equations Quadratic in Form

Equations that are not quadratic equations, but that can be solved by the techniques used for quadratic equations, are said to be **quadratic in form.** We can change an equation that is quadratic in form to a quadratic equation by making a substitution. The following examples will illustrate the techniques for solving equations that are quadratic in form.

EXAMPLE 1 Solve $x^4 - 6x^2 + 8 = 0$.

Solution If we let $w = x^2$, then $w^2 = x^4$. Substitute these expressions into the original equation.

$$x^4 - 6x^2 + 8 = 0$$

$$\underset{w^2}{\uparrow} \qquad \underset{w}{\uparrow}$$

$$w^2 - 6w + 8 = 0$$

$$(w - 2)(w - 4) = 0 \qquad \text{Factor.}$$

$$w - 2 = 0 \qquad \text{or} \qquad w - 4 = 0$$

$$w = 2 \qquad \text{or} \qquad w = 4$$

$$x^2 = 2 \qquad \text{or} \qquad x^2 = 4 \qquad \text{Substitute } x^2 \text{ for } w.$$

$$x = \pm\sqrt{2} \qquad \text{or} \qquad x = \pm 2 \qquad \text{Even-root property}$$

There are four solutions to the original equation. The solution set is $\{-2, -\sqrt{2}, \sqrt{2}, 2\}$. Check. ◀

EXAMPLE 2 Solve $(x + 15)^2 - 3(x + 15) - 18 = 0$.

Solution   Let $a = x + 15$, and substitute a for $x + 15$ in the equation.

$$a^2 - 3a - 18 = 0$$
$$(a - 6)(a + 3) = 0 \qquad \text{Factor.}$$

$$\begin{array}{lll}
a - 6 = 0 & \text{or} & a + 3 = 0 \\
a = 6 & \text{or} & a = -3 \\
x + 15 = 6 & \text{or} & x + 15 = -3 \qquad \text{Replace } a \text{ by } x + 15. \\
x = -9 & \text{or} & x = -18
\end{array}$$

The solution set is $\{-18, -9\}$. Check in the original equation.   ◄

EXAMPLE 3   Solve $(x^2 + 2x)^2 - 11(x^2 + 2x) + 24 = 0$.

Solution   Let $a = x^2 + 2x$ and substitute.

$$a^2 - 11a + 24 = 0$$
$$(a - 8)(a - 3) = 0 \qquad \text{Factor.}$$

$$\begin{array}{lll}
a - 8 = 0 & \text{or} & a - 3 = 0 \\
a = 8 & \text{or} & a = 3 \\
x^2 + 2x = 8 & \text{or} & x^2 + 2x = 3 \qquad \begin{array}{l}\text{Replace } a \text{ by} \\ x^2 + 2x.\end{array} \\
x^2 + 2x - 8 = 0 & \text{or} & x^2 + 2x - 3 = 0 \\
(x - 2)(x + 4) = 0 & \text{or} & (x + 3)(x - 1) = 0 \\
x - 2 = 0 \text{ or } x + 4 = 0 & \text{or} & x + 3 = 0 \quad \text{or } x - 1 = 0 \\
x = 2 \text{ or } \quad x = -4 & \text{or} & x = -3 \text{ or} \quad x = 1
\end{array}$$

The solution set is $\{-4, -3, 1, 2\}$. Check.   ◄

EXAMPLE 4   Solve $x - 9x^{1/2} + 14 = 0$.

Solution   Let $w = x^{1/2}$, then $w^2 = (x^{1/2})^2 = x$. Now substitute w and w^2 into the original equation.

$$w^2 - 9w + 14 = 0$$
$$(w - 7)(w - 2) = 0$$

$$\begin{array}{lll}
w - 7 = 0 & \text{or} & w - 2 = 0 \\
w = 7 & \text{or} & w = 2 \\
x^{1/2} = 7 & \text{or} & x^{1/2} = 2 \qquad \text{Replace } w \text{ by } x^{1/2}. \\
x = 49 & \text{or} & x = 4 \qquad \text{Square each side.}
\end{array}$$

Because we squared each side, we must check. If $x = 49$, we get

$$49 - 9 \cdot 49^{1/2} + 14 = 49 - 9 \cdot 7 + 14 = 49 - 63 + 14 = 0.$$

If $x = 4$, we get

$$4 - 9 \cdot 4^{1/2} + 14 = 4 - 9 \cdot 2 + 14 = 4 - 18 + 14 = 0.$$

Since each solution checks, the solution set is $\{4, 49\}$. ◀

Verbal Problems

Applied problems often result in quadratic equations that cannot be factored. For equations that cannot be factored, it is necessary to use the quadratic formula to find the exact solutions. It is usually helpful also to find the decimal numbers that approximate the exact solutions.

EXAMPLE 5 Marvin's flower bed is presently rectangular in shape with a length of 10 feet and a width of 5 feet. He wants to increase the length and width by the

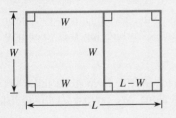

Math at Work

Is it possible to measure beauty?

Artists and philosophers have been trying to answer this question for over two thousand years. Looking at works of art—sculpture, painting, architecture—as well as natural objects, they have searched to find whether consistent numerical patterns can be found within beautiful objects.

Today, artists and architects still use concepts of beauty established by the ancient Greeks. One principle, called the Golden Rectangle, concerns the most pleasing proportions of a rectangle. The proportions of a Golden Rectangle are such that when a square is removed from one end, the sides of the remaining rectangle are proportional to the sides of the original rectangle. (See the figure to the left.) To discover the secret of the Golden Rectangle, we can set up a proportion between the ratio of the width to the length in the original rectangle and the width to the length in the remaining rectangle. The result is

$$\frac{L}{W} = \frac{W}{L - W}.$$

The front face of the Parthenon, built in Athens in the fifth century B. C., uses the principle of the Golden Rectangle. If an architect is designing a building that is to have a base width of 100 feet, then what height should the building be to form a Golden Rectangle? (*Hint:* Use $L = 100$ in the equation above.)

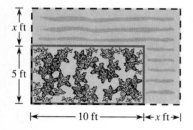

Figure 6.2

same amount to obtain a flower bed with an area of 75 square feet. What should the amount of increase be?

Solution Let x be the amount of increase. See Fig. 6.2.

The length and width of the new flower bed are $x + 10$ feet and $x + 5$ feet. Because the area is to be 75 square feet, we can write

$$(x + 10)(x + 5) = 75$$
$$x^2 + 15x + 50 = 75$$
$$x^2 + 15x - 25 = 0 \qquad \text{Get 0 on the right.}$$
$$x = \frac{-15 \pm \sqrt{225 - 4(1)(-25)}}{2(1)}$$
$$= \frac{-15 \pm \sqrt{325}}{2} = \frac{-15 \pm 5\sqrt{13}}{2}$$

Because the value of x must be positive, the exact increase is

$$x = \frac{-15 + 5\sqrt{13}}{2} \text{ feet.}$$

Using a calculator, we can find that x is approximately 1.51 feet. If $x = 1.51$ feet, then the new length is 11.51 feet, and the new width is 6.51 feet. The area of a rectangle with these dimensions is 74.93 square feet. Of course, the approximate dimensions do not give exactly 75 square feet. ◄

EXAMPLE 6 It takes Carla one hour longer to mow the lawn than it takes Sharon to mow the lawn. If they can mow the lawn in 5 hours working together, then how long would it take each girl by herself?

Solution Let x be the number of hours for Sharon to mow the lawn alone and $x + 1$ be the number of hours for Carla to mow the lawn alone. We can use a table to list all of the important quantities.

	Sharon	Carla	Together
Time in hours	x	$x + 1$	5
Portion per hour	$\dfrac{1}{x}$	$\dfrac{1}{x + 1}$	$\dfrac{1}{5}$

The equation expresses the fact that the sum of the portions done by each person in one hour is equal to the portion that they complete in one hour working together.

$$\frac{1}{x} + \frac{1}{x+1} = \frac{1}{5}$$

$$5x(x+1)\frac{1}{x} + 5x(x+1)\frac{1}{x+1} = 5x(x+1)\frac{1}{5} \qquad \text{Multiply by the LCD.}$$

$$5x + 5 + 5x = x^2 + x$$

$$10x + 5 = x^2 + x$$

$$-x^2 + 9x + 5 = 0$$

$$x^2 - 9x - 5 = 0 \qquad \text{Multiply each side by } -1.$$

$$x = \frac{9 \pm \sqrt{(-9)^2 - 4(1)(-5)}}{2(1)}$$

$$= \frac{9 \pm \sqrt{101}}{2}$$

If we use a calculator or the square-root table of Appendix A to get $\sqrt{101} \approx 10.05$, we can see that $(9 - \sqrt{101})/2$ is negative. Thus, Sharon's time alone must be $(9 + \sqrt{101})/2$ hours. To find Carla's time, we add one hour to Sharon's time:

$$\frac{9 + \sqrt{101}}{2} + 1 = \frac{9 + \sqrt{101}}{2} + \frac{2}{2} = \frac{11 + \sqrt{101}}{2}$$

Thus Carla's time alone is $(11 + \sqrt{101})/2$ hours. To find their approximate times, we use $\sqrt{101} \approx 10.05$. Thus, Sharon's time alone is approximately 9.525 hours, and Carla's time alone is approximately 10.525 hours. ◀

Warm-ups

True or false?

1. To solve the equation $x^4 - 5x^2 + 6 = 0$ by substitution, we should let $w = x^2$.

2. To solve $x^6 - 3x^3 - 10 = 0$ by substitution, we should let $w = x^2$.

3. We always use the quadratic formula on equations that are quadratic in form.

4. If $x = 4^{1/3}$, then $x^2 = 4^{2/3}$.

5. To solve $x - 7\sqrt{x} + 10 = 0$ by substitution, we let $\sqrt{w} = x$.

6. If $y = 2^{1/2}$, then $y^2 = 2^{1/4}$.

7. If John takes x hours to peel a 100-pound bag of potatoes, then in one hour he peels $100/x$ of the bag.

8. If Elvia drives 300 miles in $x + 2$ hours, then her average speed is $300/(x + 2)$ miles per hour.

9. If Ann's boat goes 10 miles per hour in still water, then against a 5-mile-per-hour current, it will go 2 miles per hour.

10. If squares with sides of length x inches are cut from the corners of an 11-inch by 14-inch rectangular piece of sheet metal, and the sides are folded up to form a box, then the dimensions of the bottom will be $11 - x$ by $14 - x$.

6.3 EXERCISES

Solve each equation. See Examples 1–4.

1. $x^4 - 14x^2 + 45 = 0$

2. $x^4 + 2x^2 = 15$

3. $x^6 + 7x^3 = 8$

4. $a^6 + 6a^3 = 16$

5. $(x^2 + 2x)^2 - 7(x^2 + 2x) + 12 = 0$

6. $(x^2 + 3x)^2 + (x^2 + 3x) - 20 = 0$

7. $(2a - 1)^2 + 2(2a - 1) - 8 = 0$

8. $(3a + 2)^2 - 3(3a + 2) = 10$

9. $(w - 1)^2 + 5(w - 1) + 5 = 0$

10. $(2x - 1)^2 - 4(2x - 1) + 2 = 0$

11. $x^{1/2} - 5x^{1/4} + 6 = 0$

12. $2x - 5\sqrt{x} + 2 = 0$

13. $2x - 5x^{1/2} - 3 = 0$

14. $x^{1/4} + 2 = x^{1/2}$

15. $x^{1/6} - x^{1/3} + 2 = 0$

16. $x^{2/3} - x^{1/3} - 20 = 0$

17. $\left(\dfrac{1}{y-1}\right)^2 + \left(\dfrac{1}{y-1}\right) = 6$

18. $\left(\dfrac{1}{w+1}\right)^2 - \left(\dfrac{2}{w+1}\right) - 24 = 0$

19. $2x^2 - 3 - 6\sqrt{2x^2 - 3} + 8 = 0$

20. $x^2 + x + \sqrt{x^2 + x} - 2 = 0$

Find the exact solution to each word problem. If the exact solution is an irrational number, then also find an approximate decimal solution. See Examples 5 and 6.

21. Harry and Gary are traveling to Nashville to make their fortunes. Harry leaves on the train at 8 A.M. and Gary travels by car, starting at 9 A.M. In order to complete the 300-mile trip and arrive at the same time as Harry, Gary travels 10 miles per hour faster than the train. At what time will they both arrive in Nashville?

22. Debbie traveled 5 miles upstream to fish in her favorite spot. It took her 20 minutes longer to get there than to return. If the current in the river is 4 miles per hour, then how fast will her boat go in still water?

Photo for Exercise 22

23. Erin was traveling across the desert on her bicycle. Before lunch she traveled 60 miles and after lunch 46 miles. She put in one hour more after lunch than before lunch, but her speed was 4 miles per hour slower than before. What was her speed before lunch and after lunch?

24. Kim starts to walk 3 miles to school at 7:30 A.M. Her brother Bryan starts at 7:45 A.M. on his bicycle, traveling 10 miles per hour faster than Kim. If they get to school at the same time, then how fast is each one traveling?

25. John takes 3 hours longer than Andrew to peel 500 pounds of apples. If together they can peel 500 pounds of apples in 8 hours, then how long would it take each one working alone?

26. It takes Brent one hour longer than Calvin to shuck a sack of oysters. If together they shuck a sack of oysters in 45 minutes, then how long would it take each one working alone?

27. Eric's garden is presently 20 feet by 30 feet. He wants to increase the length and width by the same amount in order to have a 1000-square-foot garden. What should the new dimensions of the garden be?

28. Thomas is going to make an open-top box by cutting equal squares from the four corners of an 11-inch by 14-inch sheet of cardboard and folding up the sides. If the area of the base is to be 80 square inches, then what size square should be cut from each corner?

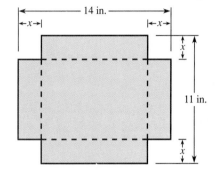

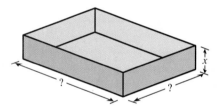

Figure for Exercise 28

6.4 Quadratic Inequalities

IN THIS SECTION:

- Solving Quadratic Inequalities with a Sign Graph
- Solving Rational Inequalities with a Sign Graph
- Solving Quadratic Inequalities That Cannot Be Factored
- Applications

We first solved inequalities in Chapter 2. In this section we will solve inequalities involving quadratic polynomials. We will use a new technique involving the rules for the sign of the product of a pair of real numbers.

Solving Quadratic Inequalities with a Sign Graph

An inequality involving a quadratic polynomial is called a **quadratic inequality.**

DEFINITION
Quadratic Inequality

A **quadratic inequality** is an inequality that can be written in the form

$$ax^2 + bx + c > 0,$$

where a, b, and c are real numbers with $a \neq 0$. The inequality symbols $<$, $\leq$, and $\geq$ may also be used.

Consider the quadratic inequality

$$x^2 + 3x - 10 > 0.$$

Since the left-hand side can be factored, we can write it as

$$(x + 5)(x - 2) > 0.$$

This inequality says that the product of $x + 5$ and $x - 2$ is positive. If both factors are negative or both positive, the product is positive. To analyze the signs of each factor carefully we make a **sign graph,** showing on the number line where the factor is positive, negative, or zero. Consider the possible values of the factor $x + 5$:

Value	Where	On the Number Line
$x + 5 = 0$	if $x = -5$	Put a 0 above -5.
$x + 5 > 0$	if $x > -5$	Put + signs to the right of -5.
$x + 5 < 0$	if $x < -5$	Put $-$ signs to the left of -5.

The sign graph shown in Fig. 6.3 for the factor $x + 5$ is made from the information in the preceding table.

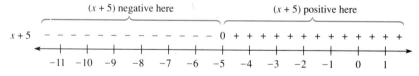

Figure 6.3

Now consider the possible values of the factor $x - 2$:

Value	Where	On the Number Line
$x - 2 = 0$	if $x = 2$	Put a 0 above 2.
$x - 2 > 0$	if $x > 2$	Put + signs to the right of 2.
$x - 2 < 0$	if $x < 2$	Put $-$ signs to the left of 2.

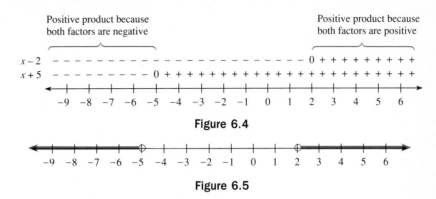

Figure 6.4

Figure 6.5

We put the information for the factor $x - 2$ on the sign graph for the factor $x + 5$ as shown in Fig. 6.4. We can see from Fig. 6.4 that the product is positive if $x < -5$ and the product is positive if $x > 2$. The solution set for the quadratic inequality is $\{x \mid x < -5 \text{ or } x > 2\}$ and its graph is shown in Fig. 6.5. Note that -5 and 2 are not included in the graph, because for those values of x the product is 0.

EXAMPLE 1   Solve $2x^2 + 5x \leq 3$ and graph the solution set.

Solution   Rewrite the inequality with 0 on one side.

$$2x^2 + 5x - 3 \leq 0$$
$$(2x - 1)(x + 3) \leq 0 \qquad \text{Factor.}$$

Examine the signs of each factor.

$$
\begin{array}{ll}
2x - 1 = 0 \quad \text{if } x = 1/2 & \qquad x + 3 = 0 \quad \text{if } x = -3 \\
2x - 1 > 0 \quad \text{if } x > 1/2 & \qquad x + 3 > 0 \quad \text{if } x > -3 \\
2x - 1 < 0 \quad \text{if } x < 1/2 & \qquad x + 3 < 0 \quad \text{if } x < -3
\end{array}
$$

Make a sign graph as shown in Fig. 6.6. The product is negative between -3 and $1/2$, when one factor is negative and the other positive. The product is 0 at -3 and at $1/2$. So the solution set is $\{x \mid -3 \leq x \leq 1/2\}$. The graph of the solution set is shown in Fig. 6.7.

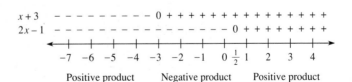

Figure 6.6

Figure 6.7

We summarize the strategy used for solving a quadratic inequality as follows.

STRATEGY
Solving a Quadratic
Inequality with a Sign Graph

1. Write the inequality with 0 on the right.
2. Factor the quadratic polynomial on the left.
3. Make a sign graph showing where each factor has a positive, negative, and zero value.
4. Use the rules for multiplying signed numbers to determine which regions satisfy the original inequality.

Solving Rational Inequalities with a Sign Graph

The inequalities $\dfrac{x+2}{x-3} \le 2$, $\qquad \dfrac{2x-3}{x+5} \le 0$, $\qquad$ and $\qquad \dfrac{2}{x+4} \ge \dfrac{1}{x+1}$

are called **rational inequalities.** When we solve equations that involve rational expressions, we usually multiply each side by the LCD. However, if we multiply each side of any inequality by a negative number we must reverse the inequality, and when we multiply by a positive number we do not. For this reason, we generally *do not multiply inequalities by variables.* The variables might be positive or negative. The next two examples show how to use a sign graph to solve rational inequalities that have variables in the denominator.

EXAMPLE 2 Solve $\dfrac{x+2}{x-3} \le 2$ and graph the solution set.

Solution We *do not* multiply each side by $x - 3$. Instead, subtract 2 from each side to get 0 on the right.

$$\frac{x+2}{x-3} - 2 \qquad \le 0$$

$$\frac{x+2}{x-3} - \frac{2(x-3)}{x-3} \le 0 \qquad \text{Get common denominator.}$$

$$\frac{x+2}{x-3} - \frac{2x-6}{x-3} \le 0 \qquad \text{Simplify.}$$

$$\frac{x+2-2x+6}{x-3} \le 0 \qquad \text{Subtract the rational expressions.}$$

$$\frac{-x+8}{x-3} \le 0 \qquad \text{The quotient is less than or equal to 0.}$$

Examine the signs of the numerator and denominator.

$$
\begin{array}{ll}
x - 3 = 0 \quad \text{if } x = 3 & \qquad -x + 8 = 0 \quad \text{if } x = 8 \\
x - 3 > 0 \quad \text{if } x > 3 & \qquad -x + 8 > 0 \quad \text{if } x < 8 \\
x - 3 < 0 \quad \text{if } x < 3 & \qquad -x + 8 < 0 \quad \text{if } x > 8
\end{array}
$$

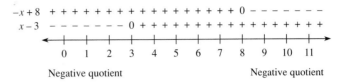

Figure 6.8

Make a sign graph as shown in Fig. 6.8. Using the rule for dividing signed numbers and the sign graph, we can identify where the quotient is negative or zero. The solution set is $\{x \mid x \geq 8 \text{ or } x < 3\}$. Note that 3 is not in the solution set because the quotient is undefined if $x = 3$. The graph of the solution set is shown in Fig. 6.9.

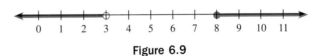

Figure 6.9   ◀

EXAMPLE 3 Solve $\dfrac{2}{x+4} \geq \dfrac{1}{x+1}$ and graph the solution set.

Solution We do not multiply by the LCD as we do in solving equations. Instead, subtract $1/(x+1)$ from each side.

$$\frac{2}{x+4} - \frac{1}{x+1} \geq 0$$

$$\frac{2(x+1)}{(x+4)(x+1)} - \frac{1(x+4)}{(x+1)(x+4)} \geq 0 \qquad \text{Get a common denominator.}$$

$$\frac{2x + 2 - x - 4}{(x+1)(x+4)} \geq 0 \qquad \text{Simplify.}$$

$$\frac{x - 2}{(x+1)(x+4)} \geq 0$$

Make a sign graph as shown in Fig. 6.10. The computation of

$$\frac{x - 2}{(x+1)(x+4)}$$

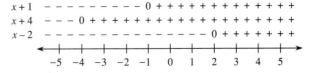

Figure 6.10

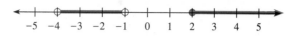

Figure 6.11

involves multiplication and division. The result of this computation is positive if all of the three binomials are positive, or if only one is positive and the other two are negative. The sign graph shows that this rational expression will have a positive value when x is between -4 and -1, and again when x is larger than 2. The solution set is

$$\{x \mid -4 < x < -1 \text{ or } x \geq 2\}.$$

Note that -1 and -4 are not in the solution set because they make the denominator zero. The graph of the solution set is shown in Fig. 6.11.   ◀

The strategy for solving rational inequalities with a sign graph is summarized in the following box.

STRATEGY
Solving Rational Inequalities with a Sign Graph

1. Rewrite the inequality with 0 on the right side.
2. Use only addition and subtraction to get an equivalent inequality.
3. Factor the numerator and denominator if possible.
4. Make a sign graph showing where each factor has a positive, negative, or zero value.
5. Use the rules for multiplying and dividing signed numbers to determine the regions that satisfy the original inequality.

Solving Quadratic Inequalities That Cannot Be Factored

The following example shows how to solve a quadratic inequality that involves a prime polynomial.

EXAMPLE 4   Solve $x^2 - 4x - 6 > 0$ and graph the solution set.

Solution   The quadratic polynomial is prime, but we can solve $x^2 - 4x - 6 = 0$ by the quadratic formula.

$$x = \frac{4 \pm \sqrt{16 - 4(1)(-6)}}{2(1)} = \frac{4 \pm \sqrt{40}}{2} = \frac{4 \pm 2\sqrt{10}}{2} = 2 \pm \sqrt{10}$$

We can see from the previous examples that the roots of the equation divide the number line into regions on which the quadratic polynomial is either positive valued or negative valued. To determine the sign of the polynomial, we test an arbitrary number in each region. These numbers are the **test points.** Because $2 + \sqrt{10} \approx 5.2$ and $2 - \sqrt{10} \approx -1.2$, we select -2, 0, and 7 for test points, as shown in Fig.

Figure 6.12

6.12. Now evaluate the polynomial $x^2 - 4x - 6$ at each of the test points.

Test Point	Value of $x^2 - 4x - 6$ at the Test Point	Sign of $x^2 - 4x - 6$ in Region of Test Point
-2	6	Positive
0	-6	Negative
7	15	Positive

We now make a sign graph showing the signs $x^2 - 4x - 6$, as shown in Fig. 6.13. The solution set to the inequality

$$x^2 - 4x - 6 > 0$$

is $\{x \mid x > 2 + \sqrt{10} \text{ or } x < 2 - \sqrt{10}\}$ and its graph is shown in Fig. 6.14.

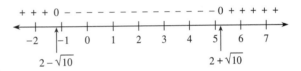

Figure 6.13

Figure 6.14

The test-point method used in Example 4 can be used also on inequalities that do factor. We summarize the strategy for solving inequalities using test points.

STRATEGY
Solving Quadratic Inequalities Using Test Points

1. Rewrite the inequality with 0 on the right.
2. Solve the quadratic equation that results from replacing the inequality symbol with the equals symbol.
3. Locate the solutions to the quadratic equation on a number line.
4. Select a test point in each region determined by the solutions to the quadratic equation.
5. Test each point in the original quadratic inequality to determine which regions satisfy the inequality.

Applications

The following example shows how a quadratic inequality can be used to solve a problem.

EXAMPLE 5   Charlene's daily profit P (in dollars) for selling x magazine subscriptions is determined by the formula $P = -x^2 + 80x - 1500$. For what values of x is her profit positive?

Solution   To find the values of x that make the profit positive, we must solve the inequality

$$-x^2 + 80x - 1500 > 0$$
$$x^2 - 80x + 1500 < 0 \qquad \text{Multiply each side by } -1.$$
$$(x - 30)(x - 50) < 0 \qquad \text{Factor.}$$

Make a sign graph as shown in Fig. 6.15. The product of the two factors is negative for x between 30 and 50. Because the last inequality is equivalent to the first, the profit is positive when the number of magazine subscriptions sold is greater than 30 and less than 50.

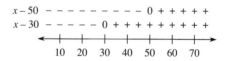

Figure 6.15   ◄

Warm-ups

True or false?

1. The solution set to $x^2 > 4$ is $\{x \mid x > 2\}$.

2. The inequality $\dfrac{x}{x - 3} > 2$ is equivalent to $x > 2x - 6$.

3. The inequality $(x - 1)(x + 2) < 0$ is equivalent to $x - 1 < 0$ or $x + 2 < 0$.

4. We cannot solve quadratic inequalities that do not factor.

5. The technique for solving quadratic inequalities is based on the rules for multiplying signed numbers.

6. Multiplying each side of an inequality by a variable should be avoided.

7. In solving quadratic or rational inequalities, we always get 0 on one side.

8. The inequality $\dfrac{x}{2} > 3$ is equivalent to $x > 6$.

9. The inequality $\dfrac{x-3}{x+2} < 1$ is equivalent to $\dfrac{x-3}{x+2} - 1 < 0$.

10. The solution set to $\dfrac{x+2}{x-4} \geq 0$ is $\{x \mid x \leq -2 \text{ or } x \geq 4\}$.

6.4 EXERCISES

Solve each inequality. State and graph the solution set. See Example 1.

1. $x^2 > 4$

2. $x^2 < 16$

3. $x^2 \leq 9$

4. $x^2 \geq 36$

5. $x^2 + x - 6 < 0$

6. $x^2 - 3x - 4 \geq 0$

7. $2x^2 + 5x \geq 12$

8. $2x^2 + 7x < 4$

9. $16 - x^2 > 0$

10. $9 - x^2 < 0$

11. $x^2 - 4x \geq 0$

12. $3x - x^2 > 0$

13. $5x - 10x^2 < 0$

14. $4x^2 - 8x \geq 0$

15. $x^2 + x > 0$

16. $4x^2 - 9 > 0$

17. $(2x + 4)(x - 1) < (x + 2)^2$

18. $x^2 + 6x + 9 \geq 0$

19. $3(2x^2 - 5) < x$

20. $6(x^2 - 2) + x < 0$

21. $x^2 \geq 4(x + 3)$

22. $x^2 + 25 < 10x$

23. $x^2 + 9 < 6x$

24. $(x + 2)^2 > 5x + 24$

25. $(x + 4)^2 > 10x + 31$

26. $\dfrac{1}{2}x^2 \geq 4 - x$

27. $\dfrac{1}{2}x^2 \leq x + 12$

28. $x^2 - 4x + 4 > 0$

29. $(x - 2)(x + 1)(x - 5) \geq 0$

30. $(x - 1)(x + 2)(2x - 5) < 0$

31. $x^3 + 3x^2 - x - 3 < 0$

32. $x^3 + 5x^2 - 4x - 20 \geq 0$

Solve each rational inequality. State and graph the solution set. See Examples 2 *and* 3.

33. $\dfrac{x}{x-3} > 0$

34. $\dfrac{a}{a+2} > 0$

35. $\dfrac{x+2}{x} \leq 0$

36. $\dfrac{w-6}{w} \leq 0$

37. $\dfrac{x-3}{x+6} > 0$

38. $\dfrac{x-2}{2x+5} < 0$

39. $\dfrac{x}{x+2} > -1$

40. $\dfrac{2}{x-5} > \dfrac{1}{x+4}$

41. $\dfrac{3}{x+2} > \dfrac{2}{x-1}$

42. $\dfrac{x+3}{x} \leq -2$

43. $\dfrac{x}{x-5} + \dfrac{3}{x-1} > 0$

44. $\dfrac{x}{x-16} + \dfrac{2}{x-6} \leq 0$

45. $\dfrac{x}{x-3} \leq \dfrac{-8}{x-6}$

46. $\dfrac{x}{x+20} > \dfrac{2}{x+8}$

Solve each inequality. State and graph the solution set. See Example 4.

47. $x^2 - 2x - 4 > 0$

48. $x^2 - 2x - 5 \leq 0$

49. $2x^2 - 6x + 3 \geq 0$

50. $2x^2 - 8x + 3 < 0$

51. $x^2 - 3x - 9 \leq 0$

52. $x^2 - 5x - 7 < 0$

53. $.23x^2 + 6.5x + 4.3 < 0$

54. $-.65x^2 + 3.2x + 5.1 > 0$

*55. $\dfrac{x}{x-2} > \dfrac{-1}{x+3}$

*56. $\dfrac{x}{3-x} > \dfrac{2}{x+5}$

Solve each problem by using a quadratic inequality. See Example 5.

57. The monthly profit P (in dollars) that Big Jim makes on the sale of x mobile homes is determined by the formula $P = x^2 + 5x - 50$. For what values of x is his profit positive?

58. Sharon's revenue R (in dollars) on the sale of x fruit cakes is determined by the formula $R = 50x - x^2$. Her cost C (in dollars) for producing x fruit cakes is given by the formula $C = 2x + 40$. For what values of x is Sharon's profit positive? (Profit = Revenue − Cost)

If an object is given an initial velocity of v_0 feet per second, from a height of s_0 feet, then its altitude S after t seconds is given by the formula $S = -16t^2 + v_0t + s_0$.

59. An arrow is shot straight upward with a velocity of 96 feet per second, from an altitude of 6 feet. For how many seconds is this arrow more than 86 feet high?

60. A cannon is fired straight upward from the surface of the earth. If the projectile leaves the barrel at 56 feet per second, then for what values of t is the projectile under 24 feet high?

61. An athlete in the shot-put projects the shot straight upward with a velocity of 30 feet per second from a height of 5 feet. For what values of t is the shot under 15 feet high?

62. A boy using a slingshot propels a stone straight into the air with a velocity of 44 feet per second from his tree house, 20 feet above the ground. For how long will this stone be more than 40 feet in the air?

6.5 Complex Numbers

IN THIS SECTION:

- **Definition**
- **Operations with Complex Numbers**
- **Square Roots of Negative Numbers**
- **Complex Solutions to Quadratic Equations**
- **Summary**

In Chapter 1 we discussed the real numbers and the various subsets of the real numbers. In this section we will define a set of numbers that has the real numbers as a subset.

Definition

The linear equation $2x = 1$ has no solution in the set of integers, but the set of integers is contained in the set of rational numbers. In the set of rational numbers, this equation has a solution. The situation is similar for the equation $x^2 = -4$. It has no solution in the set of real numbers, because the square of every real number is nonnegative. However, the set of real numbers is contained in another set of numbers called the **complex numbers,** where this equation has two solutions. The complex numbers were developed so that equations like $x^2 = -4$ would have solutions.

The complex numbers are based on the symbol $\sqrt{-1}$. In the real-number system this symbol has no meaning. In the set of complex numbers this symbol is given meaning. We call it i. We make the definition that

$$i = \sqrt{-1} \qquad \text{and} \qquad i^2 = -1.$$

DEFINITION
Complex Numbers

> The set of **complex numbers** is the set of all numbers of the form
>
> $$a + bi,$$
>
> where a and b are real numbers, $i = \sqrt{-1}$, and $i^2 = -1$.

In the complex number $a + bi$, a is called the **real part,** and b is called the **imaginary part.** If $b \neq 0$, the number $a + bi$ is called an **imaginary number.**

In dealing with complex numbers, we treat $a + bi$ as if it were a binomial, with i being a variable. Thus, we would write $2 + (-3)i$ as $2 - 3i$. We agree that $2 + i3$, $3i + 2$, and $i3 + 2$ are just different ways of writing $2 + 3i$. Some examples of complex numbers are

$$-3 - 5i, \qquad \frac{2}{3} - \frac{3}{4}i, \qquad 1 + i\sqrt{2}, \qquad 9 + 0i, \qquad \text{and} \qquad 0 + 7i.$$

For simplicity, we will write only $7i$ for $0 + 7i$. The complex number $9 + 0i$ is the real number 9, and $0 + 0i$ is the real number 0. Any complex number with $b = 0$ is a real number. For any real number a,

$$a + 0i = a.$$

Thus, the set of real numbers is a subset of the set of complex numbers. See Fig. 6.16.

Complex numbers	
Real numbers	Imaginary numbers
$3, \pi, \frac{5}{2}, 0, -9, \sqrt{2}$	$i, 2 + 3i, \sqrt{-5}, -3 - 8i$

Figure 6.16

Operations with Complex Numbers

We perform the operations of addition, subtraction, multiplication, and division of complex numbers as if the complex numbers were binomials with i being a variable. Whenever i^2 appears, we replace it by -1.

EXAMPLE 1   Find the following sums and differences.

a) $(2 + 3i) + (6 + i)$ 　　　　　　　　b) $(-2 + 3i) + (-2 - 5i)$
c) $(3 + 5i) - (1 + 2i)$ 　　　　　　　　d) $(-2 - 3i) - (1 - i)$

Solution

a) $(2 + 3i) + (6 + i) = 8 + 4i$
b) $(-2 + 3i) + (-2 - 5i) = -4 - 2i$
c) $(3 + 5i) - (1 + 2i) = 2 + 3i$
d) $(-2 - 3i) - (1 - i) = -3 - 2i$ ◄

The formal definition of addition and subtraction of complex numbers is given in the following box.

DEFINITION
Addition and Subtraction of Complex Numbers

The complex numbers $a + bi$ and $c + di$ are added and subtracted as follows:

$$(a + bi) + (c + di) = (a + c) + (b + d)i$$
$$(a + bi) - (c + di) = (a - c) + (b - d)i$$

Note that because $i^2 = -1$, we get

$$i^3 = i^2 \cdot i = -1 \cdot i = -i.$$

We can find the value of i^4 by using the known value of i^3:

$$i^4 = i^3 \cdot i = -i \cdot i = -i^2 = 1.$$

EXAMPLE 2   Find the following products.

a) $(2 + 3i)(4 + 5i)$ 　　　b) $2i(1 + i)$ 　　　c) $(2i)^2$
d) $(-2i)^2$ 　　　　　　　e) $(3 + i)(3 - i)$ 　　f) i^6

Solution

a) Use the FOIL method to find the product.

$$(2 + 3i)(4 + 5i) = 8 + 10i + 12i + 15i^2$$
$$= 8 + 22i + 15(-1) \quad \text{Replace } i^2 \text{ by } -1.$$
$$= 8 + 22i - 15$$
$$= -7 + 22i$$

b) $2i(1 + i) = 2i + 2i^2$ Distributive property

           $= 2i + 2(-1)$ Because $i^2 = -1$

           $= -2 + 2i$

c) $(2i)^2 = 2^2 i^2 = 4(-1) = -4$

d) $(-2i)^2 = (-2)^2 i^2 = 4i^2 = 4(-1) = -4$

e) This is the product of a sum and a difference.

$$(3 + i)(3 - i) = 9 - 3i + 3i - i^2$$
$$= 9 - (-1) \qquad i^2 = -1$$
$$= 10$$

f) $i^6 = i^2 \cdot i^4 = -1 \cdot 1 = -1$ ◄

The formal definition of multiplication of complex numbers is given in the following box.

DEFINITION
Multiplication of
Complex Numbers

> The complex numbers $a + bi$ and $c + di$ are multiplied as follows:
>
> $$(a + bi)(c + di) = (ac - bd) + (ad + bc)i$$

To divide a complex number by a real number, we divide each term by the real number. For example,

$$\frac{4 + 6i}{2} = 2 + 3i.$$

To understand division by a complex number, note that the product of the two imaginary numbers in Example 2(e) is a real number:

$$(3 + i)(3 - i) = 10.$$

We say that $3 + i$ and $3 - i$ are **complex conjugates** of each other.

DEFINITION
Complex Conjugates

> The complex numbers $a + bi$ and $a - bi$ are called **complex conjugates** of one another. Their product is the real number $a^2 + b^2$.

EXAMPLE 3 Find the product of the given complex number and its conjugate.

a) $2 + 3i$ **b)** $5 - 4i$

Solution

a) The conjugate of $2 + 3i$ is $2 - 3i$.

$$(2 + 3i)(2 - 3i) = 4 - 9i^2 = 4 + 9 = 13$$

b) The conjugate of $5 - 4i$ is $5 + 4i$.

$$(5 - 4i)(5 + 4i) = 25 + 16 = 41 \qquad \blacktriangleleft$$

We use the idea of complex conjugates to divide complex numbers. The process is similar to rationalizing the denominator. *Multiply the numerator and denominator of the quotient by the complex conjugate of the denominator.*

EXAMPLE 4 Find each quotient. Write the answer in the form $a + bi$.

a) $\dfrac{5}{3 - 4i}$ **b)** $\dfrac{i}{2 + i}$ **c)** $\dfrac{3 + 2i}{i}$

Solution

a) $\dfrac{5}{3 - 4i} = \dfrac{5(3 + 4i)}{(3 - 4i)(3 + 4i)}$ Multiply the numerator and denominator by $3 + 4i$, the conjugate of $3 - 4i$.

$$= \dfrac{15 + 20i}{9 - 16i^2}$$

$$= \dfrac{15 + 20i}{25}$$

$$= \dfrac{15}{25} + \dfrac{20}{25}i$$

$$= \dfrac{3}{5} + \dfrac{4}{5}i$$

b) Multiply the numerator and denominator by $2 - i$, the conjugate of $2 + i$.

$$\dfrac{i}{2 + i} = \dfrac{i(2 - i)}{(2 + i)(2 - i)} = \dfrac{2i - i^2}{4 - i^2} = \dfrac{2i + 1}{5} = \dfrac{1}{5} + \dfrac{2}{5}i$$

c) Multiply the numerator and denominator by $-i$, the conjugate of i.

$$\dfrac{3 + 2i}{i} = \dfrac{(3 + 2i)(-i)}{i(-i)} = \dfrac{-3i - 2i^2}{-i^2} = \dfrac{-3i + 2}{1} = 2 - 3i \qquad \blacktriangleleft$$

The formal definition of division of complex numbers is given in the following box.

DEFINITION
Division of
Complex Numbers

We divide the complex number $a + bi$ by the complex number $c + di$ as follows:

$$\dfrac{a + bi}{c + di} = \dfrac{(a + bi)(c - di)}{(c + di)(c - di)}$$

Square Roots of Negative Numbers

In Examples 2(c) and 2(d), we saw that both

$$(2i)^2 = -4 \quad \text{and} \quad (-2i)^2 = -4.$$

Because the square of each of these complex numbers is -4, both $2i$ and $-2i$ are square roots of -4. When we use the radical notation we write

$$\sqrt{-4} = 2i.$$

In the complex-number system the square root of a negative number is an imaginary number.

Square Root of a Negative Number

For any positive number b, $\sqrt{-b} = i\sqrt{b}$.

For example, $\sqrt{-9} = i\sqrt{9} = 3i$, and $\sqrt{-7} = i\sqrt{7}$. Note that the expression $\sqrt{7}i$ could easily be mistaken for the expression $\sqrt{7i}$, where i is under the radical. For this reason, when the coefficient of i is a radical, we write i preceding the radical.

EXAMPLE 5 Write each expression in the form $a + bi$, where a and b are real numbers.

a) $3 + \sqrt{-9}$ b) $\sqrt{-12} + \sqrt{-27}$ c) $\dfrac{-1 - \sqrt{-18}}{3}$

Solution

a) $3 + \sqrt{-9} = 3 + i\sqrt{9} = 3 + 3i$

b) $\sqrt{-12} + \sqrt{-27} = i\sqrt{12} + i\sqrt{27}$
$= 2i\sqrt{3} + 3i\sqrt{3}$ $\sqrt{12} = \sqrt{4}\sqrt{3} = 2\sqrt{3}$
$= 5i\sqrt{3}$ $\sqrt{27} = \sqrt{9}\sqrt{3} = 3\sqrt{3}$

c) $\dfrac{-1 - \sqrt{-18}}{3} = \dfrac{-1 - i\sqrt{18}}{3} = \dfrac{-1 - 3i\sqrt{2}}{3} = -\dfrac{1}{3} - i\sqrt{2}$ ◀

Complex Solutions to Quadratic Equations

Consider again the equation $x^2 = -4$. Although it has no solutions in the set of real numbers, it has imaginary solutions.

$$x^2 = -4$$
$$x = \pm\sqrt{-4}$$
$$x = \pm i\sqrt{4}$$
$$x = \pm 2i$$

Check: $(2i)^2 = 4i^2 = 4(-1) = -4$
$(-2i)^2 = 4i^2 = -4$
The solution set is $\{\pm 2i\}$.

If the domain of the variable in a quadratic equation is the set of complex numbers, then the equation has at least one solution. If the discriminant $b^2 - 4ac$ is negative, the solutions are imaginary numbers.

EXAMPLE 6 Find the complex solutions to the quadratic equations.

a) $x^2 + 2x + 5 = 0$ **b)** $2x^2 - 3x + 5 = 0$

Solution

a) $x = \dfrac{-2 \pm \sqrt{2^2 - 4(1)(5)}}{2(1)} = \dfrac{-2 \pm \sqrt{-16}}{2}$

$= \dfrac{-2 \pm i\sqrt{16}}{2}$

$= \dfrac{-2 \pm 4i}{2} = -1 \pm 2i$

The solution set is $\{-1 \pm 2i\}$. Check these solutions in the original equation.

b) $x = \dfrac{3 \pm \sqrt{(-3)^2 - 4(2)(5)}}{2(2)} = \dfrac{3 \pm \sqrt{9 - 40}}{4}$

$= \dfrac{3 \pm \sqrt{-31}}{4}$

$= \dfrac{3 \pm i\sqrt{31}}{4}$

The solution set is $\left\{ \dfrac{3 \pm i\sqrt{31}}{4} \right\}$ or $\left\{ \dfrac{3}{4} \pm i\dfrac{\sqrt{31}}{4} \right\}$ ◀

Summary

The following table summarizes the basic facts about complex numbers.

Complex Number Facts

1. Definition of i: $i = \sqrt{-1}$, and $i^2 = -1$.

2. A complex number has the form $a + bi$, where a and b are real numbers.

3. The complex number $a + 0i$ is the real number a.

4. If b is a positive real number, then $\sqrt{-b} = i\sqrt{b}$.

5. The numbers $a + bi$ and $a - bi$ are called complex conjugates of each other. Their product is the real number $a^2 + b^2$.

6. Definition of operations:

a) Addition $\qquad (a + bi) + (c + di) = (a + c) + (b + d)i$

b) Subtraction $\quad (a + bi) - (c + di) = (a - c) + (b - d)i$

c) Multiplication $\ (a + bi)(c + di) = (ac - bd) + (ad + bc)i$

d) Division $\qquad (a + bi) \div (c + di) = \dfrac{(a + bi)(c - di)}{(c + di)(c - di)}$

Warm-ups

True or false?

1. The set of real numbers is a subset of the set of complex numbers.
2. $2 - \sqrt{-6} = 2 - 6i$
3. $\sqrt{-9} = \pm 3i$
4. The solution set to the equation $x^2 = -9$ is $\{\pm 3i\}$.
5. $2 - 3i - (4 - 2i) = -2 - i$
6. $i^4 = 1$
7. $(2 - i)(2 + i) = 5$
8. $i^3 = i$
9. $i^{48} = 1$
10. When considering complex numbers, all quadratic equations have at least one solution.

6.5 EXERCISES

Perform the indicated operations with the complex numbers.

See Example 1.

1. $(2 + 3i) + (-4 + 5i)$
2. $(-1 + 6i) + (5 - 4i)$
3. $(2 - 3i) - (6 - 7i)$
4. $(2 - 3i) - (6 - 2i)$
5. $(-1 + i) + (-1 - i)$
6. $(-5 + i) + (-5 - i)$
7. $(-2 - 3i) - (6 - i)$
8. $(-6 + 4i) - (2 - i)$

See Example 2.

9. $3(2 + 5i)$
10. $4(1 - 3i)$
11. $2i(i - 5)$
12. $3i(2 - 6i)$
13. $-4i(3 - i)$
14. $-5i(2 + 3i)$
15. $(2 + 3i)(4 + 5i)$
16. $(2 + i)(3 + 4i)$
17. $(-1 + i)(2 - i)$
18. $(3 - 2i)(2 - 5i)$
19. $(-1 - 2i)(2 + i)$
20. $(1 - 3i)(1 + 3i)$
21. $(5 - 2i)(5 + 2i)$
22. $(4 + 3i)(4 + 3i)$
23. $(3 - 5i)^2$
24. $(5 - 2i)^2$
25. $(4 + 2i)(4 - 2i)$
26. $(4 - i)(4 + i)$
27. $(3i)^2$
28. $(5i)^2$
29. $(1 - i)(1 + i)$
30. $(2 + 6i)(2 - 6i)$

Find the product of the given complex number and its conjugate. See Example 3.

31. $1 - 2i$
32. $3 + 5i$
33. $-2 + i$
34. $-3 - 2i$
35. $4 - 6i$
36. $3 + i$
37. $2 - i\sqrt{3}$
38. $\sqrt{5} - 4i$

Find each quotient. See Example 4.

39. $\dfrac{2 + 6i}{2}$

40. $\dfrac{-4 + 8i}{4}$

41. $\dfrac{3}{4 + i}$

42. $\dfrac{6}{7 + 2i}$

43. $\dfrac{2 + i}{3 - 2i}$

44. $\dfrac{3 + 5i}{2 - i}$

45. $\dfrac{-5 + 2i}{i}$

46. $\dfrac{6 - 5i}{i}$

47. $\dfrac{4 + 3i}{-2i}$

48. $\dfrac{5 - 6i}{3i}$

Write each expression in the form a + bi, where a and b are real numbers. See Example 5.

49. $2 + \sqrt{-4}$

50. $3 - \sqrt{-9}$

51. $2\sqrt{-9} + 5$

52. $3\sqrt{-16} + 2$

53. $7 - \sqrt{-6}$

54. $\sqrt{-5} + 3$

55. $\sqrt{-8} + \sqrt{-18}$

56. $2\sqrt{-20} - \sqrt{-45}$

57. $\sqrt{16} + i$

58. $2i + \sqrt{25}$

59. $\dfrac{2 + \sqrt{-12}}{2}$

60. $\dfrac{-6 - \sqrt{-18}}{3}$

61. $\dfrac{-4 - \sqrt{-24}}{4}$

62. $\dfrac{8 + \sqrt{-20}}{-4}$

Find the complex solutions to each quadratic equation. See Example 6.

63. $x^2 = -9$

64. $x^2 + 4 = 0$

65. $x^2 = -12$

66. $x^2 = -25$

67. $2x^2 + 5 = 0$

68. $3x^2 + 4 = 0$

69. $3x^2 + 6 = 0$

70. $-3x^2 = 21$

71. $x^2 + 1 = 0$

72. $2x^2 + 1 = 0$

73. $3x^2 + 5x + 3 = 0$

74. $x^2 + 2x + 5 = 0$

75. $5x^2 - 4x + 1 = 0$

76. $2x^2 - 3x + 2 = 0$

77. $x^2 + 6x + 10 = 0$

78. $x^2 + 4x + 5 = 0$

79. $5x^2 + x + 5 = 0$

80. $-3x^2 - x + 1 = 0$

Write each expression in the form a + bi where a and b are real numbers.

81. $(2 - 3i)(3 + 4i)$

82. $(2 - 3i) + (3 + 4i)$

83. $\dfrac{2 - 3i}{3 + 4i}$

84. $i(2 - 3i)$

85. $(2 - 3i)(2 + 3i)$

86. $(-3i)^2$

87. $\sqrt{3} + \sqrt{-3}$

88. $\sqrt{25} - \sqrt{-25}$

Wrap-up

CHAPTER 6

SUMMARY

Quadratic Equations

		Examples
Quadratic equation	Any equation that can be written in the form $$ax^2 + bx + c = 0$$ where a, b, and c are real numbers, with $a \neq 0$.	$x^2 = 11$ $(x - 5)^2 = 99$ $x^2 + 3x - 20 = 0$
Methods for solving quadratic equations	Factoring	$x^2 + x - 6 = 0$ $(x + 3)(x - 2) = 0$
	The even-root property	$(x - 5)^2 = 10$ $x - 5 = \pm\sqrt{10}$
	Completing the square: take one-half of middle term, square it, then add it to each side.	$x^2 + 6x = -4$ $x^2 + 6x + 9 = -4 + 9$ $(x + 3)^2 = 5$
	Quadratic formula (works on any quadratic): $$x = \frac{-b \pm \sqrt{b^2 - 4ac}}{2a}$$	$2x^2 - 3x - 5 = 0$ $x = \dfrac{3 \pm \sqrt{9 - 4(2)(-5)}}{2(2)}$
Number of solutions	Determined by the discriminant $b^2 - 4ac$: $b^2 - 4ac > 0$ 2 real solutions $b^2 - 4ac = 0$ 1 real solution $b^2 - 4ac < 0$ no real solutions (2 complex solutions)	$x^2 + 6x - 12 = 0$ $6^2 - 4(1)(-12) > 0$ $x^2 + 10x + 25 = 0$ $10^2 - 4(1)(25) = 0$ $x^2 + 2x + 20 = 0$ $2^2 - 4(1)(20) < 0$
Equations quadratic in form	Use substitution to convert to a quadratic.	$x^4 + 3x^2 - 10 = 0$ Let $a = x^2$ $a^2 + 3a - 10 = 0$
Quadratic inequalities	Solved by making a sign graph, showing signs of the factors Use test points if the quadratic polynomial is prime.	$(x - 5)(x + 1) < 0$ $x + 1$ − − 0 + + + + + + + + $x - 5$ − − − − − − − − 0 + + $-3\ -2\ -1\ \ 0\ \ 1\ \ 2\ \ 3\ \ 4\ \ 5\ \ 6\ \ 7$

Complex Numbers		Examples

Complex numbers

Numbers of the form $a + bi$, where a and b are real numbers:
$$i = \sqrt{-1},\ i^2 = -1$$

$2 + 3i$
$-6i$
$\sqrt{2} + i$

Complex conjugates

$a + bi$ and $a - bi$ are called conjugates of one another. Their product is real.

$$(2 + 3i)(2 - 3i) = 4 - 9i^2$$
$$= 13$$

Complex number operations

Addition:
$$(a + bi) + (c + di) = (a + c) + (b + d)i$$
Subtraction:
$$(a + bi) - (c + di) = (a - c) + (b - d)i$$
Multiplication:
$$(a + bi)(c + di) = (ac - bd) + (ad + bc)i$$

Division—Multiply by conjugate:
$$(a + bi) \div (c + di) = \frac{(a + bi)(c - di)}{(c + di)(c - di)}$$

$$(2 + 5i) + (3 - 2i) = 5 + 3i$$

$$(2 + 5i) - (3 - 2i) = -1 + 7i$$

$$(2 + 5i)(3 - 2i) = 16 + 11i$$

$$(2 + 5i) \div (3 - 2i) = \frac{(2 + 5i)(3 + 2i)}{(3 - 2i)(3 + 2i)}$$
$$= \frac{-4 + 19i}{13}$$

REVIEW EXERCISES

6.1 *Solve by factoring.*

1. $x^2 - 2x - 15 = 0$

2. $x^2 - 2x - 24 = 0$

3. $2x^2 + x = 15$

4. $2x^2 + 7x = 4$

5. $w^2 - 25 = 0$

6. $a^2 - 121 = 0$

7. $4x^2 - 12x + 9 = 0$

8. $x^2 - 12x + 36 = 0$

Solve by using the even-root property.

9. $x^2 = 12$

10. $x^2 = 20$

11. $(x - 1)^2 = 9$

12. $(x + 4)^2 = 4$

13. $(x - 2)^2 = \dfrac{3}{4}$

14. $(x - 3)^2 = \dfrac{1}{4}$

15. $4x^2 = 9$

16. $2x^2 = 3$

Solve by completing the square.

17. $x^2 - 6x + 8 = 0$

18. $x^2 + 4x + 3 = 0$

19. $x^2 - 5x + 6 = 0$

20. $x^2 - x - 6 = 0$

21. $2x^2 - 7x + 3 = 0$

22. $2x^2 - x = 6$

23. $x^2 + 4x + 1 = 0$

24. $x^2 + 2x - 2 = 0$

6.2 *Solve by the quadratic formula.*

25. $x^2 - 3x - 10 = 0$

26. $x^2 - 5x - 6 = 0$

27. $6x^2 - 7x = 3$

28. $6x^2 = x + 2$

29. $x^2 + 4x + 2 = 0$

30. $x^2 + 6x = 2$

31. $3x^2 + 1 = 5x$

32. $2x^2 + 3x - 1 = 0$

6.3 *Find all of the real-number solutions to the following. Solve by any method.*

33. $x^6 + 7x^3 - 8 = 0$

34. $8x^6 + 63x^3 - 8 = 0$

35. $x^4 - 13x^2 + 36 = 0$

36. $x^4 + 7x^2 + 12 = 0$

37. $(x^2 + 3x)^2 - 28(x^2 + 3x) + 180 = 0$

38. $(x^2 + 1)^2 - 8(x^2 + 1) + 15 = 0$

39. $x^2 - 6x + 6\sqrt{x^2 - 6x} - 40 = 0$

40. $x^2 - 3x - 3\sqrt{x^2 - 3x} + 2 = 0$

41. $t^{-2} + 5t^{-1} - 36 = 0$

42. $a^{-2} + a^{-1} - 6 = 0$

43. $w - 13\sqrt{w} + 36 = 0$

44. $4a - 5\sqrt{a} + 1 = 0$

Find the exact answer and an approximate answer to the following problems.

45. Find two positive real numbers that differ by 4 and have a product of 4.

46. Find two positive real numbers that differ by 1 and have a product of 1.

47. On a 19-inch diagonal-measure television picture screen, the height is 4 inches less than the width. Find the height and width.

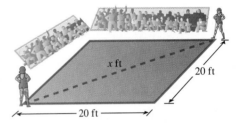

Figure for Exercise 48

48. A pit for mud wrestling is in the shape of a square, 20 feet on each side. How far apart are the wrestlers when they are in opposite corners of the pit?

49. A radical student group plans to print a message on an 8-inch by 10-inch paper. If the typed message requires 24 square inches of paper and they want an equal border on all sides, then how wide should the border be?

50. Winston can mow his dad's lawn in one hour less than it takes his brother Willie. If they take 2 hours to mow it when working together, then how long would it take Winston working alone?

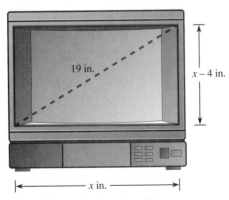

Figure for Exercise 47

6.4 *Solve each inequality. State and graph the solution set for each inequality.*

51. $a^2 + a > 6$

52. $x^2 - 5x + 6 > 0$

53. $x^2 - x - 20 \leq 0$

54. $a^2 + 2a \leq 15$

55. $w^2 - w < 0$

56. $x - x^2 \leq 0$

57. $\dfrac{x - 4}{x + 2} \geq 0$

58. $\dfrac{x - 3}{x + 5} < 0$

59. $\dfrac{x - 2}{x + 3} < 1$

60. $\dfrac{x - 3}{x + 4} > 2$

61. $\dfrac{3}{x + 2} > \dfrac{1}{x + 1}$

62. $\dfrac{1}{x + 1} < \dfrac{1}{x - 1}$

6.5 *Perform the indicated operations. Write answers in the form a + bi.*

63. $(2 - 3i)(-5 + 5i)$

64. $(2 + i) + (3 - 6i)$

65. $(1 - i) - (2 - 3i)$

66. $(3 - 2i) - (1 - i)$

67. $(2 + i)(5 - 2i)$

68. $(2 + i)(5 - 4i)$

69. $\dfrac{6 + 3i}{3}$ **70.** $\dfrac{8 + 12i}{4}$ **71.** $\dfrac{4 - \sqrt{-12}}{2}$

72. $\dfrac{6 + \sqrt{-18}}{3}$ **73.** $\dfrac{2 - 3i}{4 + i}$ **74.** $\dfrac{3 + i}{2 - 3i}$

Find the complex solutions to the quadratic equations.

75. $2x^2 - 4x + 3 = 0$ **76.** $2x^2 - 6x + 5 = 0$ **77.** $2x^2 + 3 = 3x$

78. $x^2 + x + 1 = 0$ **79.** $3x^2 + 2x + 2 = 0$ **80.** $x^2 + 2 = 2x$

81. $\dfrac{1}{2}x^2 + 3x + 8 = 0$ **82.** $\dfrac{1}{2}x^2 - 5x + 13 = 0$

CHAPTER 6 TEST

Calculate the value of $b^2 - 4ac$ and state how many real solutions each equation has.

1. $2x^2 - 3x + 2 = 0$ **2.** $-3x^2 + 5x - 1 = 0$ **3.** $4x^2 - 4x + 1 = 0$

Solve by using the quadratic formula.

4. $2x^2 + 5x - 3 = 0$ **5.** $x^2 + 6x + 6 = 0$

Solve by completing the square.

6. $x^2 + 10x + 25 = 0$ **7.** $2x^2 + x - 6 = 0$

Solve by any method.

8. $x(x + 1) = 12$ **9.** $a^4 - 5a^2 + 4 = 0$

10. $x - 2 - 8\sqrt{x - 2} + 15 = 0$

Solve each inequality. State and graph the solution set.

11. $w^2 + 3w < 18$ **12.** $\dfrac{2}{x - 2} < \dfrac{3}{x + 1}$

Find the exact solution to the problem.

13. The length of a rectangle is 2 feet longer than the width. If the area is 16 square feet, find the length and width.

Perform the computations. Write answers in the form $a + bi$.

14. $(3 - 5i) + (6 + 2i)$ **15.** $(2 - 8i) - (3 - 4i)$

16. $(2 - i)(-3 + 2i)$ **17.** $(2 - 3i)(2 + 3i)$

18. $(1 - 2i) \div (3 + i)$

Find the complex solutions to the quadratic equations.

19. $x^2 + 6x + 10 = 0$ **20.** $3x^2 - x + 1 = 0$

Tying It All Together

CHAPTERS 1–6

Solve each equation.

1. $2x - 15 = 0$
2. $2x^2 - 15 = 0$
3. $2x^2 + x - 15 = 0$
4. $2x^2 + 4x - 15 = 0$
5. $|4x + 11| = 3$
6. $|4x^2 + 11x| = 3$
7. $\sqrt{x} = x - 6$
8. $(2x - 5)^{2/3} = 4$

Solve each equation for y.

9. $2x - 3y = 9$
10. $\dfrac{y - 3}{x + 2} = -\dfrac{1}{2}$
11. $3y^2 + cy + d = 0$
12. $my^2 - ny = w$
13. $\dfrac{1}{3}x - \dfrac{2}{5}y = \dfrac{5}{6}$
14. $y - 3 = -\dfrac{2}{3}(x - 4)$

Let $m = \dfrac{y_1 - y_2}{x_1 - x_2}$. *Find the value of m for each of the following choices of* x_1, x_2, y_1, *and* y_2.

15. $x_1 = 2,\ x_2 = 5,\ y_1 = 3,\ y_2 = 7$
16. $x_1 = -3,\ x_2 = 4,\ y_1 = 5,\ y_2 = -6$
17. $x_1 = 13,\ x_2 = -12,\ y_1 = 6,\ y_2 = 6$
18. $x_1 = .3,\ x_2 = .5,\ y_1 = .8,\ y_2 = .4$
19. $x_1 = -1,\ x_2 = -5,\ y_1 = -3,\ y_2 = -8$
20. $x_1 = 1/2,\ x_2 = 1/3,\ y_1 = 3/5,\ y_2 = -4/3$

CHAPTER **7**

Linear Equations in Two Variables

The residents of Birdview have observed that the town has been growing by 5 people each year for many years. The algebraic description of this situation involves two variables: the population, P, and the time in years, t. The population depends on time in a very simple way. When the time increases by 1, the population increases by 5. We say that P is a linear function of t.

In 1980, Birdview had 204 residents. What will the population be in the year 2000? We could certainly answer this question by using just arithmetic, but a more satisfying answer would be to get a formula for P in terms of t. Once we have a formula relating the two variables, we can answer any questions about the value of one variable when given the value of the other. In this chapter we will learn how to write linear equations in two variables, and we will find a formula relating P and t when we do Problem 85 of Section 7.3.

7.1 The Rectangular Coordinate System

IN THIS SECTION:

- Ordered Pairs
- Plotting Points
- Graphing a Linear Equation
- Using X-intercept and Y-intercept for Graphing
- Distance between Two Points
- Applications

In Chapter 1 we learned to graph numbers on a number line. We used number lines to illustrate the solution sets to inequalities in Chapters 2 and 6. In this section we learn to graph pairs of numbers in a coordinate system made up of a pair of number lines. We will use this coordinate system to illustrate the solution sets to equations and inequalities in two variables.

Ordered Pairs

The equation $y = 2x + 3$ is an equation in two variables. This equation is satisfied if we choose a value for x and a value for y that make it true. For example, let $x = 4$ and $y = 11$, then the equation becomes

$$11 = 2 \cdot 4 + 3,$$

a true statement. A convenient way to write

$$x = 4 \quad \text{and} \quad y = 11$$

is as the **ordered pair** (4, 11). Another ordered pair that satisfies $y = 2x + 3$ is (5, 13), because $13 = 2 \cdot 5 + 3$. Note that (13, 5) does not satisfy $y = 2x + 3$, since $5 \neq 2 \cdot 13 + 3$. The format is always to write the value for x first and the value for y second.

To better understand ordered pairs of numbers, we need to study the rectangular coordinate system, or Cartesian coordinate system. The rectangular coordinate system consists of two number lines drawn at a right angle to one another, intersecting at zero on each number line. See Fig. 7.1.

The horizontal number line is called the **x-axis,** and the vertical number line is called the **y-axis.** The point at which they intersect is called the **origin.** The two number lines divide the plane into four regions called **quadrants.** The quadrants are numbered as shown in Fig. 7.1. They do not include any points on the axes.

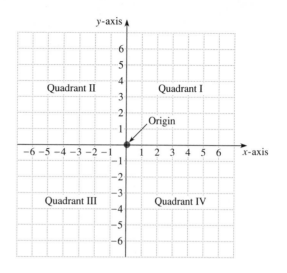

Figure 7.1

Plotting Points

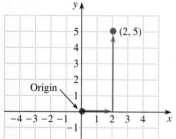

Figure 7.2

Just as every real number corresponds to a point on the number line, every pair of real numbers corresponds to a point in the rectangular coordinate system. For example, the pair $(2, 5)$ corresponds to the point that lies 2 units to the right of the origin and 5 units up. See Fig. 7.2. Locating a point in the rectangular coordinate system is referred to as **plotting,** or **graphing,** the point.

Because ordered pairs of numbers correspond to points in the plane, we frequently refer to an ordered pair as a point.

EXAMPLE 1 Plot the points $(2, 3)$, $(-1, 3)$, $(-2, -3)$, and $(4, -2)$ on a rectangular coordinate system.

Solution To plot $(2, 3)$, start at the origin, move 2 units to the right, then up 3. To graph $(-1, 3)$, start at the origin, move 1 to the left, then up 3. All 4 points are shown in Fig. 7.3.   ◀

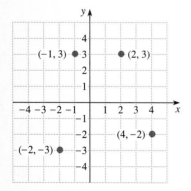

Figure 7.3

Graphing a Linear Equation

Consider again the equation

$$y = 2x + 3.$$

It is easy to find points that satisfy this equation. Choose any number for x, and then y must be $2x + 3$. If $x = 1$, then $y = 2(1) + 3 = 5$. Thus, $(1, 5)$ satisfies this equation. Figure 7.4 shows a table of values for x and y that satisfy the equation, and a rectangular coordinate system on which these points have been graphed.

Notice that the points graphed in Fig. 7.4 appear to lie along a straight line. If we choose x to be any real number and find the point (x, y) that satisfies the equation $y = 2x + 3$, we will get another point along this line. Plotting the points (x, y) for

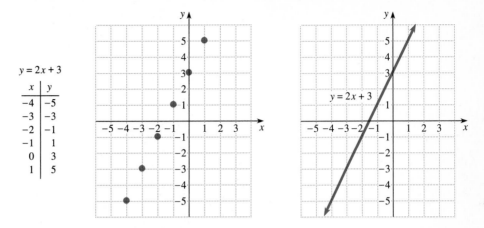

Figure 7.4   Figure 7.5

every real number x will give us the complete line shown in Fig. 7.5. The line is the graph of the equation $y = 2x + 3$. For this reason the equation is called a **linear equation.** More details about linear equations will be presented in Section 7.3. The arrows on the ends of the line indicate that it extends infinitely in each direction. The solution set to the equation is $\{(x, y) \mid y = 2x + 3\}$, and the line is a graph of this set.

EXAMPLE 2   Graph the linear equation $y + 2x = 1$. Plot at least 6 points.

Solution   If we solve the equation for y, we get $y = -2x + 1$. Now arbitrarily select 6 values for x, and then calculate the corresponding values for y. For example, if $x = 0$, then $y = -2 \cdot 0 + 1 = 1$. The table of values and the resulting line are shown in Fig. 7.6. ◄

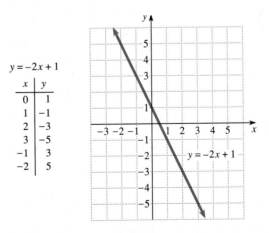

Figure 7.6

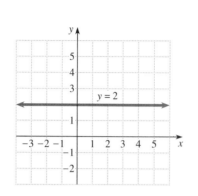

$y = 2$

x	y
-2	2
-1	2
0	2
1	2
2	2

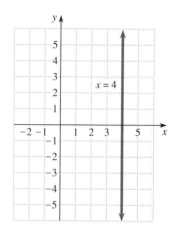

$x = 4$

x	y
4	-2
4	-1
4	0
4	1
4	2

Figure 7.7   **Figure 7.8**

EXAMPLE 3   Graph the linear equation $y = 2$. Plot at least 5 points.

Solution   We can think of the equation $y = 2$ as $y = 0 \cdot x + 2$. Because x is multiplied by 0, any ordered pair with a y-coordinate of 2 satisfies the equation. A table of values and the resulting horizontal line are shown in Fig. 7.7.   ◄

EXAMPLE 4   Graph the linear equation $x = 4$. Plot at least 5 points.

Solution   We think of the equation $x = 4$ as $0 \cdot y + x = 4$. Because y is multiplied by 0, the equation is satisfied by any ordered pair with an x-coordinate of 4. A table of values and the resulting vertical line are shown in Fig. 7.8.   ◄

Using *X*-intercept and *Y*-intercept for Graphing

Every nonvertical, nonhorizontal line has an x-intercept and a y-intercept. The **x-intercept** is the point where the line crosses the x-axis, and the **y-intercept** is the point where the line crosses the y-axis. The x-intercept has a y-coordinate of 0, and the y-intercept has an x-coordinate of 0. The intercepts can be used as two points that determine the location of the line. The next example shows how to graph a line by using the intercepts.

EXAMPLE 5   Use the intercepts to graph the line $3x - 4y = 6$.

Solution   Let $x = 0$ to find the y-intercept, and $y = 0$ to find the x-intercept.

If $x = 0$, then   If $y = 0$, then

$$3(0) - 4y = 6 \qquad\qquad\qquad 3x - 4(0) = 6$$

$$-4y = 6 \qquad\qquad\qquad\qquad 3x = 6$$

$$y = -\frac{3}{2} \qquad\qquad\qquad\qquad\quad x = 2$$

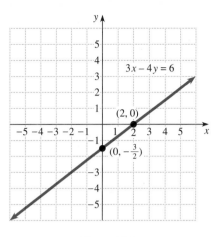

Figure 7.9

Figure 7.10

The y-intercept is $(0, -3/2)$, and the x-intercept is $(2, 0)$. The line containing the two intercepts is shown in Fig. 7.9. ◀

Distance between Two Points

Consider the two points $(1, 4)$ and $(-3, 1)$. The Pythagorean Theorem can be used to find the distance between the two points. As shown in Fig. 7.10, the distance between the points, d, is the length of the hypotenuse of a right triangle. We find the length of the two legs by subtracting. The length of side a is $4 - 1 = 3$, and the length of side b is $1 - (-3) = 4$. By the Pythagorean Theorem,

$$d^2 = 3^2 + 4^2$$

$$d^2 = 9 + 16$$

$$d^2 = 25 \qquad \text{There are two square roots of 25, but because distance}$$
$$\text{is always positive, we use only the positive root.}$$

$$d = 5.$$

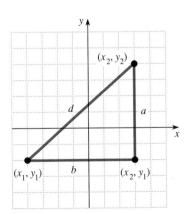

Figure 7.11

We can easily generalize this process to find a formula for the distance between any two points (x_1, y_1) and (x_2, y_2). As shown in Fig. 7.11, the distance between these points is again the length of the hypotenuse of a right triangle. The length of side a is $y_2 - y_1$, and the length of side b is $x_2 - x_1$. Using the Pythagorean Theorem we can write

$$d^2 = (x_2 - x_1)^2 + (y_2 - y_1)^2.$$

If we apply the even-root property and omit the negative square root, we can express this formula as follows.

Distance Formula

> The distance d between the points (x_1, y_1) and (x_2, y_2) is given by the formula
>
> $$d = \sqrt{(x_2 - x_1)^2 + (y_2 - y_1)^2}.$$

EXAMPLE 6 Find the distance between $(6, -4)$ and $(-8, -10)$.

Solution Let $(x_2, y_2) = (6, -4)$ and $(x_1, y_1) = (-8, -10)$. Now substitute the appropriate values into the distance formula.

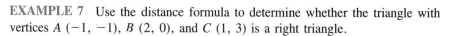

$$d = \sqrt{[6 - (-8)]^2 + [-4 - (-10)]^2}$$
$$= \sqrt{(14)^2 + (6)^2}$$
$$= \sqrt{196 + 36}$$
$$= \sqrt{232} = \sqrt{4 \cdot 58}$$
$$= 2\sqrt{58}$$

Note that we get the same answer if we let $(x_2, y_2) = (-8, -10)$ and $(x_1, y_1) = (6, -4)$.

The exact distance between the points is $2\sqrt{58}$. ◄

EXAMPLE 7 Use the distance formula to determine whether the triangle with vertices $A(-1, -1)$, $B(2, 0)$, and $C(1, 3)$ is a right triangle.

Solution A graph of this triangle is shown in Fig. 7.12. We use the distance formula to find the length of each side of the triangle.

$$\text{Length of } AB = \sqrt{(-1 - 2)^2 + (-1 - 0)^2} = \sqrt{9 + 1} = \sqrt{10}$$
$$\text{Length of } AC = \sqrt{(-1 - 1)^2 + (-1 - 3)^2} = \sqrt{4 + 16} = \sqrt{20}$$
$$\text{Length of } BC = \sqrt{(2 - 1)^2 + (0 - 3)^2} = \sqrt{1 + 9} = \sqrt{10}$$

The Pythagorean Theorem states that if the sum of the squares of the two shortest sides is equal to the square of the longest side, then the triangle is a right triangle. Because

$$(\sqrt{10})^2 + (\sqrt{10})^2 = 10 + 10 = 20 \quad \text{and} \quad (\sqrt{20})^2 = 20,$$

the triangle is a right triangle. ◄

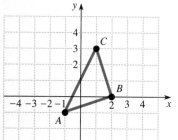

Figure 7.12

Applications

All of the equations that we graphed in this section are linear equations. Linear equations occur in many applications. Usually the equations are not as simple as our previous examples. We may have to rename the axes or adjust the scale to graph equations involving large numbers.

EXAMPLE 8 The cost per week, C, in dollars of producing n pairs of shoes for the Reebop Shoe Company is given by the equation $C = 2n + 8000$. Graph the equation for n between 0 and 800 inclusive ($0 \le n \le 800$).

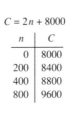

$C = 2n + 8000$

n	C
0	8000
200	8400
400	8800
800	9600

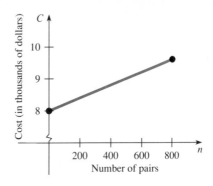

Figure 7.13

Solution   We let the horizontal axis be the n-axis and the vertical axis be the C-axis. The scales are adjusted to accommodate the large numbers. Let each unit on the n-axis represent 200 pairs, and each unit on the C-axis represent one thousand dollars. When graphing large numbers, the axes are often numbered so that the intersection of the two axes is not $(0, 0)$. Figure 7.13 shows a table of values for n and C, and the graph.   ◀

Warm-ups

True or false?

1. The point $(2, 5)$ satisfies the equation $3y - 2x = -4$.
2. The vertical axis is usually called the x-axis.
3. The point $(0, 0)$ is in quadrant I.
4. The point $(0, 1)$ is on the y-axis.
5. The graph of $x = 7$ is a vertical line.
6. The graph of $8 - y = 0$ is a horizontal line.
7. The distance between (x_1, y_1) and (x_2, y_2) is $\sqrt{(x_2 - x_1) + (y_2 - y_1)}$.
8. The distance between $(3, -2)$ and $(3, 5)$ is -7.
9. The distance between $(-2, 6)$ and $(3, 6)$ is 5.
10. The y-intercept for the line $y = 2x - 3$ is $(0, -3)$.

7.1 EXERCISES

Plot the following points on a rectangular coordinate system. For each point name the quadrant that it lies in or the axis that it lies on. See Example 1.

1. $(2, 5)$
2. $(-5, 1)$
3. $(-3, -1/2)$
4. $(-2, -6)$
5. $(0, 4)$
6. $(0, 2)$
7. $(\pi, -1)$
8. $(\pi/3, 0)$

9. $(-4, 3)$ **10.** $(0, -3)$ **11.** $(\pi/2, 0)$ **12.** $(\pi, 2)$

13. $(0, -1)$ **14.** $(4, -3)$ **15.** $(0, 0)$ **16.** $(-2, -1/2)$

Graph each linear equation. Plot at least 5 points for each. See Examples 2–4.

17. $y = x + 1$ **18.** $y = x - 1$ **19.** $y = -2x + 3$ **20.** $y = 2x - 3$

21. $y = x$ **22.** $y = -x$ **23.** $y = 3$ **24.** $y = 2 - x$

25. $y = 1 - x$ **26.** $y = -2$ **27.** $x = 2$ **28.** $x = 2y$

29. $y = \dfrac{1}{2}x - 1$ **30.** $x = -3$ **31.** $3x + y = 5$ **32.** $y = \dfrac{1}{2}x - 2$

33. $x + 2y = 4$ **34.** $2x + y = 1$ **35.** $x - y = 2$ **36.** $x = y - 2$

37. $x = y - 3$ **38.** $x - 2y = 1$ **39.** $2x + 2y = 1$ **40.** $2x - 2y = 1$

▦ **41.** $y = 1.35x - 4.27$ ▦ **42.** $y = -.26x + 3.86$

Use the x-intercept and y-intercept to graph each line. See Example 5.

43. $4x - 3y = 12$ **44.** $2x + 5y = 20$ **45.** $x - y + 5 = 0$ **46.** $x + y + 7 = 0$

47. $2x + 3y = 5$ **48.** $3x - 4y = 7$ **49.** $y = \dfrac{3}{5}x + \dfrac{2}{3}$ **50.** $y = -\dfrac{2}{3}x - \dfrac{5}{4}$

Use the distance formula to find the distance between each of the following pairs of points. See Example 6.

51. $(6, 5), (4, 2)$ **52.** $(7, 3), (5, 1)$ **53.** $(3, 5), (1, -3)$

54. $(6, 2), (3, -5)$ **55.** $(4, -2), (-3, -6)$ **56.** $(-2, 3), (1, -4)$

57. $(2, -7), (-5, -7)$ **58.** $(-3, 6), (-3, -1)$ **59.** $(0, -3), (4, 0)$

60. $(4, 0), (0, 3)$ **61.** $(0, 0), (-6, 8)$ **62.** $(0, 0), (-9, -12)$

63. $(-3, 2), (-5, 6)$ **64.** $(-3, -3), (2, 2)$ ▦ **65.** $(3.5, -6.2), (-1.4, 3.7)$

▦ **66.** $(-2.4, 1.5), (5.3, -1.6)$ ▦ **67.** $(\pi/2, 0), (\pi, 1)$ ▦ **68.** $(-\pi/2, 0), (\pi, -1)$

*69. $(1/2, 1/4), (1/3, 1/2)$ ***70.** $(3\sqrt{2}, \sqrt{3}), (5\sqrt{2}, -\sqrt{3})$

Use the distance formula to solve the geometric figure problems. See Example 7 and the front endpapers for a review of some geometric terms.

71. Determine whether the points $(-1, 1)$, $(0, 6)$, and $(5, 5)$ are the vertices of an isosceles triangle.

72. Determine whether the points $(-2, 10)$, $(4, 3)$, and $(2, -6)$ are the vertices of an isosceles triangle.

73. Determine whether the points $(5, -4)$, $(-5, -4)$, and $(-1, 4)$ are the vertices of an equilateral triangle.

74. Determine whether the points $(-4, -2)$, $(2, 6)$, and $(6, -2)$ are the vertices of an equilateral triangle.

75. Determine whether the triangle with vertices $(4, 4)$, $(0, 8)$, and $(-2, -2)$ is a right triangle.

76. Determine whether the triangle with vertices $(-1, 7)$, $(2, 1)$, and $(-4, -2)$ is a right triangle.

77. Determine whether the points $(-2, -13)$, $(1, -4)$, and $(3, 2)$ lie on the same straight line. (If the sum of the lengths of the two shortest line segments is equal to the length of the longest line segment, then the three points lie on a line.)

78. Determine whether the points $(-2, 13)$, $(1, 1)$, and $(3, -6)$ lie on the same straight line.

79. Determine whether the points $(1, 5)$, $(2, 3)$, $(3, 9)$, and $(4, 7)$ are the vertices of a parallelogram. (If the opposite sides of a quadrilateral are equal in length, then the quadrilateral is a parallelogram.)

80. Determine whether the points $(-2, -2)$, $(-1, 4)$, $(4, -1)$, and $(5, 5)$ are the vertices of a rhombus. (If all sides of a quadrilateral are the same length, then the quadrilateral is a rhombus.)

*81. Determine whether the points $(-2, 4)$, $(-1, 6)$, $(4, 1)$, and $(5, 3)$ are the vertices of a rectangle. (A parallelogram with at least one right angle is a rectangle.)

*82. Determine whether the points $(-7, -1)$, $(-4, 5)$, $(-1, -4)$, and $(2, 2)$ are the vertices of a square. (A rhombus with at least one right angle is a square.)

Solve the following. See Example 8.

83. In the eighties, Midwest Electric had a profit per share, P, that was determined by the equation $P = .25x + 3.70$, where x ranges from 0 to 9 corresponding to the years 1980 to 1989. What was the profit per share in 1983? Sketch the graph of this equation for x ranging from 0 to 9.

84. For the first 5 years, the value of a \$25,000 automobile is determined by the formula $V = -3,000x + 25,000$, where x is the age in years of the automobile. What is the value of this automobile when it is 4 years old? Sketch the graph of this equation for x between 0 and 5 inclusive.

85. For a one-day car rental, the X-press Car Company charges C dollars, where C is determined by the formula $C = .26m + 42$, and m is the number of miles driven. What is the charge for a car driven 400 miles? Sketch a graph of the equation for m ranging from 0 to 1000.

86. The Friendly Bob Loan Company gives each applicant a rating, t, from 0 to 10 according to the applicant's ability to repay, with higher rating indicating higher risk. The interest rate, r, is then determined by the formula $r = .02t + .15$. What interest rate would a person with a rating of 8 pay? Sketch the graph of the equation for t ranging from 0 to 10.

7.2 Slope of a Line

IN THIS SECTION:

- **Slope**
- **Formal Definition**
- **Perpendicular Lines**
- **Parallel Lines**
- **Geometric Applications of Slope**

In Section 7.1 we saw that many different equations had graphs that were straight lines. In this section we will look at graphs of straight lines in more detail and develop the concept of slope of a line.

Slope

Slope is a measurement of the steepness of a line. To calculate the **slope** of a line we use a ratio to compare the change in y-coordinate to the change in x-coordinate when moving from one point on the line to another point on the line.

DEFINITION
Slope

$$\text{Slope} = \frac{\text{change in } y\text{-coordinate}}{\text{change in } x\text{-coordinate}}$$

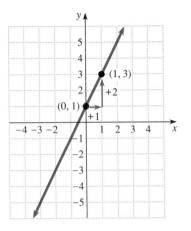

Figure 7.14

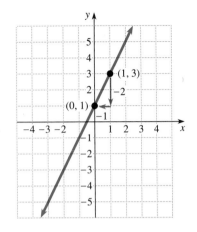

Figure 7.15

Consider the line in Fig. 7.14. In going from $(0, 1)$ to $(1, 3)$, there is a change of $+1$ in the x-coordinate and a change of $+2$ in the y-coordinate. Thus the slope of that line is

$$\frac{+2}{+1} = 2.$$

If we move from the point $(1, 3)$ to the point $(0, 1)$, as in Fig. 7.15, we say that there is a change of -2 in the y-coordinate and a change of -1 in the x-coordinate. We get

$$\text{slope} = \frac{-2}{-1} = 2.$$

We call the change in y-coordinate the **rise** and the change in x-coordinate the **run.** Moving up is a positive rise, and moving down is a negative rise. Moving to the right is a positive run, and moving to the left is a negative run. We usually use the letter m to stand for slope.

Another Definition of Slope

$$\text{Slope} = m = \frac{\text{change in } y\text{-coordinate}}{\text{change in } x\text{-coordinate}} = \frac{\text{rise}}{\text{run}}$$

EXAMPLE 1 Find the slopes of the lines in Fig. 7.16 by going from point A to point B.

Solution

a) A is located at $(0, 3)$ and B at $(2, 0)$. In going from A to B, the change in y is -3 and the change in x is 2. So $m = -3/2$.

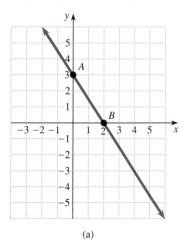

(a)

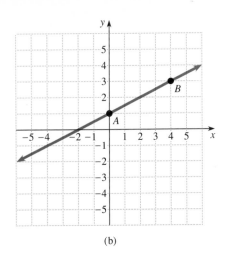

(b)

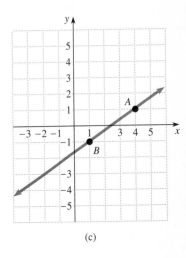

(c)

Figure 7.16

b) In going from A to B, we must rise 2 and run 4. Thus $m = 2/4 = 1/2$.

c) In going from A to B, the rise is -2 and the run is -3. Thus $m = -2/-3 = 2/3$. ◀

The ratio of rise to run is the ratio of the lengths of the two legs of a right triangle whose hypotenuse is on the line. See Fig. 7.17. As long as one leg is vertical and the other leg is horizontal, all such triangles for a certain line have the same shape—they are similar triangles. Because ratios of corresponding sides in similar triangles are equal, the ratio of rise to run has the same value no matter which two points of the line are used to find it.

EXAMPLE 2 Find the slope of the line shown in Fig. 7.18, using

a) points A and B **b)** points A and C **c)** points B and C.

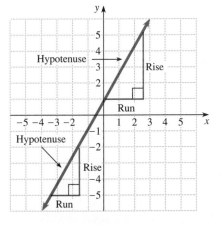

Figure 7.17

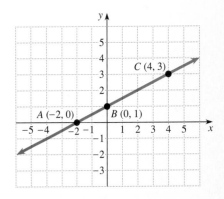

Figure 7.18

Solution

a) $m = \dfrac{\text{rise}}{\text{run}} = \dfrac{1}{2}$　　　　**b)** $m = \dfrac{\text{rise}}{\text{run}} = \dfrac{3}{6} = \dfrac{1}{2}$　　　　**c)** $m = \dfrac{2}{4} = \dfrac{1}{2}$　◀

Formal Definition

We can obtain the rise and run from a graph, or we can get them without a graph by subtracting the y-coordinates and the x-coordinates for two points on the line. See Fig. 7.19.

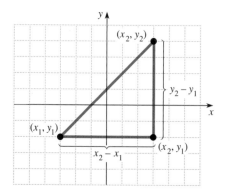

Figure 7.19

Formal Definition of Slope

The slope of the line containing the points (x_1, y_1) and (x_2, y_2) is denoted by the letter m, where

$$m = \frac{y_2 - y_1}{x_2 - x_1}, \text{ provided } x_2 - x_1 \neq 0.$$

EXAMPLE 3　Find the slope of each line.
a) The line through $(2, 5)$ and $(6, 3)$
b) The line through $(-2, 3)$ and $(-5, -1)$
c) The line through $(-6, 4)$ and the origin

Solution

a) Let $(x_1, y_1) = (2, 5)$ and $(x_2, y_2) = (6, 3)$. The assignment of (x_1, y_1) and (x_2, y_2) is arbitrary.

$$m = \frac{y_2 - y_1}{x_2 - x_1} = \frac{3 - 5}{6 - 2} = \frac{-2}{4} = -\frac{1}{2}$$

b) Let $(x_1, y_1) = (-5, -1)$ and $(x_2, y_2) = (-2, 3)$.

$$m = \frac{y_2 - y_1}{x_2 - x_1} = \frac{3 - (-1)}{-2 - (-5)} = \frac{4}{3}$$

c) $m = \dfrac{4 - 0}{-6 - 0} = \dfrac{4}{-6} = -\dfrac{2}{3}$ ◀

If we use the points $(-3, 2)$ and $(4, 2)$ to find the slope of the horizontal line in Fig. 7.20, we get

$$m = \frac{2 - 2}{-3 - 4} = \frac{0}{-7} = 0.$$

On any horizontal line, all points have the same y-coordinate, so the slope of any horizontal line is 0.

If we use the points $(1, -4)$ and $(1, 2)$ to calculate the slope of the vertical line in Fig. 7.21, we get

$$m = \frac{-4 - 2}{1 - 1} = \frac{-6}{0}.$$

Division by 0 is undefined; therefore, the slope of this line is undefined. For any vertical line, all points have the same x-coordinates, so $x_2 - x_1 = 0$. *Slope is undefined for any vertical line.*

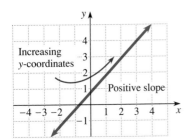

Increasing y-coordinates

Positive slope

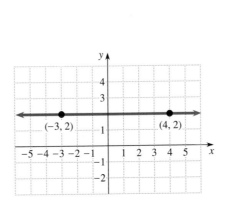

Figure 7.20

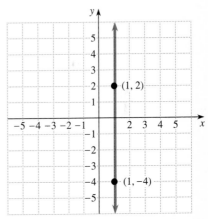

Figure 7.21

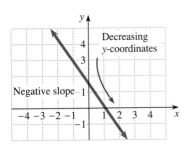

Decreasing y-coordinates

Negative slope

Figure 7.22

If we move from left to right along a line with positive slope, then we will be rising, or the y-coordinates will be increasing. If we move from left to right along a line with negative slope, then we will be falling, or the y-coordinates will be decreasing. See Fig. 7.22.

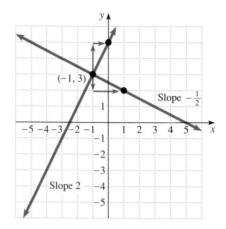

Figure 7.23

Perpendicular Lines

Consider two lines through the point $(-1, 3)$, one with slope $-1/2$ and the other with slope 2. Because slope is the ratio of rise to run, a slope of $-1/2$ means we can locate a second point on the line by starting at $(-1, 3)$ and going down 1 and 2 to the right. For the line with slope 2, we start at $(-1, 3)$ and go up 2 and 1 to the right. See Fig. 7.23.

The two lines that we sketched in Fig. 7.23 appear to be perpendicular to each other. This example illustrates a general situation. A line is perpendicular to another line if its slope is the negative of the reciprocal of the slope of the other. For example, lines with slopes $2/3$ and $-3/2$ are perpendicular.

Perpendicular Lines

> Two lines with slopes m_1 and m_2 are perpendicular if
> $$m_1 = -\left(\frac{1}{m_2}\right).$$

EXAMPLE 4 Find the slope of the line that contains the point $(1, 6)$ and is perpendicular to the line through the points $(-4, 1)$ and $(3, -2)$.

Solution The line through $(-4, 1)$ and $(3, -2)$ has slope

$$m = \frac{1 - (-2)}{-4 - 3} = \frac{3}{-7} = -\frac{3}{7}.$$

The line through $(1, 6)$ that is perpendicular to this line has slope $7/3$. ◀

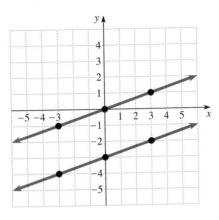

Figure 7.24

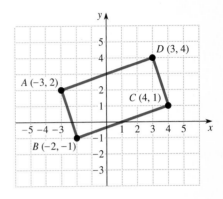

Figure 7.25

Parallel Lines

Consider the two lines shown in Fig. 7.24. Each of these lines has a slope of $1/3$, and these lines are parallel. In general we have the following fact.

Parallel Lines

Nonvertical parallel lines have equal slopes.

Geometric Applications of Slope

EXAMPLE 5 Use slope to determine whether the four points $(-3, 2)$, $(-2, -1)$, $(3, 4)$, and $(4, 1)$ are the vertices of a rectangle.

Solution First we sketch the figure determined by these points. See Fig. 7.25. It appears to be a rectangle, but we must prove that it is. Calculate the slope of each side.

$$m_{AB} = \frac{2 - (-1)}{-3 - (-2)} = \frac{3}{-1} = -3$$

$$m_{BC} = \frac{-1 - 1}{-2 - 4} = \frac{-2}{-6} = \frac{1}{3}$$

$$m_{CD} = \frac{4 - 1}{3 - 4} = \frac{3}{-1} = -3$$

$$m_{AD} = \frac{2 - 4}{-3 - 3} = \frac{-2}{-6} = \frac{1}{3}$$

Since the opposite sides have the same slope, they are parallel and the figure is a parallelogram. Because $1/3$ is the opposite of the reciprocal of -3, the intersecting lines are perpendicular. Therefore the figure is a rectangle. ◀

Warm-ups

True or false?

1. Slope is a measurement of the steepness of a line.
2. Slope is run divided by rise.
3. The line through (4, 5) and (−3, 5) has undefined slope.
4. The line through (−2, 6) and (−2, −5) has undefined slope.
5. Slope cannot be negative.
6. The slope of the line that crosses the *y*-axis at (0, −2) and the *x*-axis at (5, 0) is −2/5.
7. The line through (4, 4) and (5, 5) has slope 5/4.
8. Positive slope corresponds to an increasing graph.
9. Lines with slope 2/3 and −2/3 are perpendicular to each other.
10. Parallel lines have equal slopes.

7.2 EXERCISES

Determine the slope of each line. See Examples 1 and 2.

1.

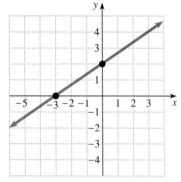

2.

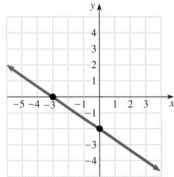

3.

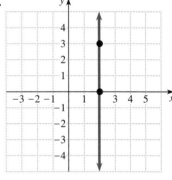

4.

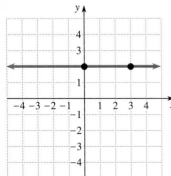

5.

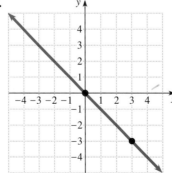

6.

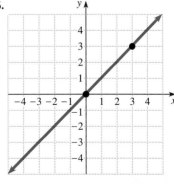

7.

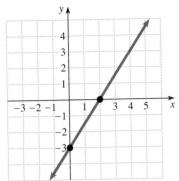

8.

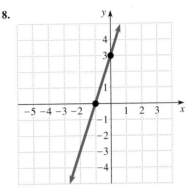

9.

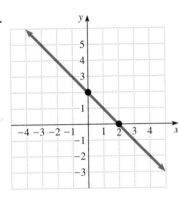

10.

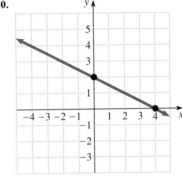

Find the slope of the line that contains each of the following pairs of points. See Example 3.

11. (2, 6), (5, 1)

12. (3, 4), (6, 10)

13. (−3, −1), (4, 3)

14. (−2, −3), (1, 3)

15. (−2, 2), (−1, 7)

16. (−3, 5), (1, −6)

17. (3, −5), (−1, −1)

18. (−3, 6), (−2, −1)

19. (0, 3), (5, 0)

20. (3, 0), (0, 10)

21. (3/4, −1), (−1/2, −1/2)

22. (1/2, 2), (1/4, 1/2)

23. (6, 212), (7, 209)

24. (1988, 306), (1990, 315)

25. (4, 7), (−12, 7)

26. (5, −3), (9, −3)

27. (2, 6), (2, −6)

28. (−3, 2), (−3, 0)

29. (24.3, 11.9), (3.57, 8.4)

30. (−2.7, 19.3), (5.46, −3.28)

31. ($\sqrt{2}$, 3), ($\sqrt{2}/2$, 5)

32. ($\sqrt{3}$, −1), (−$\sqrt{5}$, $\sqrt{2}$)

33. ($\pi/4$, 1), ($\pi/2$, 0)

34. ($\pi/3$, −1), ($\pi/6$, 0)

Find the slope of each of the lines described below. Make a sketch for each problem. See Example 4.

35. A line perpendicular to a line with slope 4/5.

36. A line perpendicular to a line with slope −5.

37. The line containing the point (3, 4) that is perpendicular to the line through (−5, 1) and (3, −2).

38. The line through (−3, −5) that is perpendicular to the line through (−2, 6) and (7, 3).

39. The line that goes through (2, 5) and runs parallel to the line through (−3, −2) and (4, 1).

40. The line through the origin that is parallel to the line through the points (−3, −5) and (4, −1).

Solve each geometric figure problem. See Example 5.

41. Use slope to determine whether the points $(-6, 1)$, $(-2, -1)$, $(0, 3)$, and $(4, 1)$ are the vertices of a parallelogram.

42. Use slope to determine whether the points $(-7, 0)$, $(-1, 6)$, $(-1, -2)$, and $(6, 5)$ are the vertices of a parallelogram.

43. Use slope to determine whether the points $(-3, 2)$, $(-1, -1)$, $(3, 6)$, and $(6, 4)$ are the vertices of a trapezoid.

44. Use slope to determine whether the points $(0, 0)$, $(0, 4)$, $(3, 6)$, and $(6, 4)$ are the vertices of a trapezoid.

45. Determine whether the points $(-4, 4)$, $(-1, -2)$, $(0, 6)$, and $(3, 0)$ are the vertices of a rectangle.

46. Determine whether the points $(-3, 2)$, $(1, -2)$, $(5, 2)$ and $(0, 6)$ are the vertices of a square.

47. Use slope to determine whether the points $(-3, 3)$, $(-1, 6)$, and $(0, 0)$ are the vertices of a right triangle.

48. Use slope to determine whether the points $(0, -1)$, $(2, 5)$, and $(5, 4)$ are the vertices of a right triangle.

49. What is the slope of a line that is perpendicular to a line with slope .247?

50. What is the slope of a line that is perpendicular to the line through $(3.27, -1.46)$ and $(-5.48, 3.61)$?

*51. Show that the diagonals of the rhombus with vertices $(-3, -1)$, $(0, 3)$, $(2, -1)$, and $(5, 3)$ are perpendicular to each other.

*52. Show that the diagonals of the square with vertices $(-5, 3)$, $(-3, -3)$, $(1, 5)$, and $(3, -1)$ are perpendicular.

7.3 Equations of Lines

IN THIS SECTION:

- **Point-slope Form**
- **Slope-intercept Form**
- **Standard Form**
- **Using Slope-intercept Form for Graphing**
- **Linear Functions**

In Section 7.1 we learned how to graph a straight line corresponding to a linear equation. The line contains all of the points that satisfy the equation. In this section we will start with a line or a description of a line, and we will write an equation corresponding to the line.

Point-slope Form

Figure 7.26 shows the line that has slope 2/3 and contains the point $(3, 5)$. The slope is the same no matter which two points of the line we use to calculate it. Thus, if we find the slope of this line using an arbitrary point of the line, say (x, y), and the specific point $(3, 5)$, we get

$$m = \frac{y - 5}{x - 3}.$$

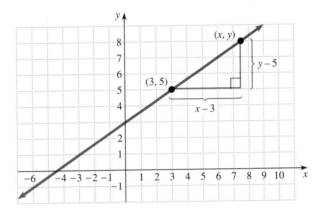

Figure 7.26

Because the slope of this line is 2/3, we can write

$$\frac{y - 5}{x - 3} = \frac{2}{3}.$$

Multiplying each side by $x - 3$, we get

$$y - 5 = \frac{2}{3}(x - 3).$$

Since (x, y) was an arbitrary point on the line, this equation is satisfied by every point on the line. It is called the **point-slope form** of the equation of the line. In general we have the following.

Point-slope Form

> The point-slope form of the equation of the line through the fixed point (x_1, y_1) with slope m is
>
> $$y - y_1 = m(x - x_1).$$

EXAMPLE 1 Find the equation of each line and solve it for y.

a) The line through $(-2, 5)$ with slope -3
b) The line through $(3, -2)$ and $(-1, 1)$
c) The line through $(2, 0)$ that is perpendicular to the line through $(5, -1)$ and $(-1, 3)$

Solution

a) Use $x_1 = -2$, $y_1 = 5$, and $m = -3$ in the point-slope form.

$$y - 5 = -3[x - (-2)]$$

Solving for y, we get

$$y - 5 = -3[x + 2]$$
$$y - 5 = -3x - 6$$
$$y = -3x - 1.$$

b) We are not given the slope, but we can obtain it from the points $(3, -2)$ and $(-1, 1)$.

$$m = \frac{1 - (-2)}{-1 - 3} = \frac{3}{-4} = -\frac{3}{4}$$

Now we use this slope and one of the points, say $(3, -2)$, to write the equation in point-slope form. We would get the same equation if we used the other point $(-1, 1)$.

$$y - (-2) = -\frac{3}{4}(x - 3)$$

Solving for y, we get

$$y + 2 = -\frac{3}{4}x + \frac{9}{4}$$

$$y = -\frac{3}{4}x + \frac{1}{4}. \qquad \frac{9}{4} - 2 = \frac{1}{4}$$

c) First find the slope of the line through $(5, -1)$ and $(-1, 3)$.

$$m = \frac{3 - (-1)}{-1 - 5} = \frac{4}{-6} = -\frac{2}{3}$$

A line perpendicular to this line has slope $3/2$. Now use the point $(2, 0)$ and a slope of $3/2$ in the point-slope formula.

$$y - 0 = \frac{3}{2}(x - 2)$$

Solving for y, we get

$$y = \frac{3}{2}x - 3. \qquad \blacktriangleleft$$

Slope-intercept Form

When we solve a linear equation for y, we are writing the equation in **slope-intercept form**. If we look at each part of Example 1, we see that after solving for y the slope is the coefficient of x. For example, the line $y = -3x - 1$ has slope -3.

The y-intercept is found by letting $x = 0$ in the equation. If $x = 0$ in the equation $y = -3x - 1$, then

$$y = -3(0) - 1 = -1.$$

So the y-intercept is $(0, -1)$. Notice that the y-coordinate of the y-intercept $(0, -1)$ appears in the equation:

$$y = -3x - 1$$
$$\uparrow \qquad \uparrow$$
$$\text{slope} \quad y\text{-intercept } (0, -1)$$

In general we have the following.

Slope-intercept Form

> The equation $y = mx + b$ is the equation of a line with y-intercept $(0, b)$ and slope m.

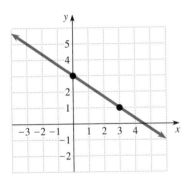

Figure 7.27

EXAMPLE 2 Write the slope-intercept form of the equation of the line shown in Fig. 7.27.

Solution From Fig. 7.27 we see that the y-intercept is $(0, 3)$. If we start at the y-intercept and move down 2 and 3 to the right, we get to another point on the line. So the slope is $-2/3$. The equation of this line in slope-intercept form is

$$y = -\frac{2}{3}x + 3.$$ ◄

Standard Form

If we arrange a linear equation so that the x and y terms are on the left side, it is in **standard form**.

Standard Form

> If A, B, and C are real numbers with A and B not both zero, then
> $$Ax + By = C$$
> is called the **standard form** of the equation of a line.

EXAMPLE 3 Write the equation

$$y = \frac{1}{2}x - \frac{3}{4}$$

in standard form with integral coefficients.

Solution

$$-\frac{1}{2}x + y = -\frac{3}{4} \qquad \text{Subtract } \frac{1}{2}x \text{ from each side.}$$

$$4\left(-\frac{1}{2}x + y\right) = 4\left(-\frac{3}{4}\right) \qquad \begin{array}{l}\text{Multiply by the LCD to get}\\\text{integral coefficients.}\end{array}$$

$$-2x + 4y = -3 \qquad \text{Simplify.}$$

$$2x - 4y = 3 \qquad \begin{array}{l}\text{Multiply by } -1 \text{ to make the}\\\text{coefficient of } x \text{ positive.}\end{array}$$

Either of these last two equations is acceptable for standard form with integral coefficients. We will generally write our answers with a positive coefficient for x.

◄

To find the slope and y-intercept of a line written in standard form, we convert the equation to slope-intercept form.

Math at Work

No matter what happens in the world of show business, "The show must go on." Singers and clowns, dancers and actors know this is true. Not only does a canceled or poorly performed show disappoint the audience but, as in any business, a concert hall or theater can operate only when a profit is being made.

To calculate whether a show is profitable, theater managers and concert organizers analyze the revenues versus the costs of the performance. The revenue from a performance can be expressed by the formula $R = px$, where p is the price per ticket and x is the number of tickets sold. There are two kinds of costs involved with every performance: fixed costs, which are the same no matter how many tickets are sold; and variable costs, which depend on the number of tickets sold. The total cost for a performance can be expressed by the formula $T = a + bx$, where a is the fixed cost, b is the variable cost per ticket, and x is the number of tickets sold.

For a performance to make a profit, the revenue in ticket sales must be greater than the total cost for the show. The number of tickets that must be sold can be found by graphing the two equations. The point where the two equations cross is called the "break-even" point. Any sales greater than that point will be a profit, but fewer sales than that point will result in a loss for the show.

A promoter is planning to book the Heebie Jeebies. The fixed costs for renting the auditorium, utilities, and labor are $50,000, and the variable costs are $5 per ticket sold. If the tickets are $15 each, then how many tickets must be sold for the show to break even?

EXAMPLE 4 Find the slope and y-intercept of the line $3x - 2y = 5$.

Solution Solve for y to get slope-intercept form.

$$3x - 2y = 5$$
$$-2y = -3x + 5$$
$$y = \frac{3}{2}x - \frac{5}{2} \qquad \text{Slope-intercept form}$$

The slope is $3/2$, and the y-intercept is $(0, -5/2)$. ◄

We learned in Section 7.1 that the graph of the equation $x = 4$ is a vertical line. Because slope is undefined for vertical lines, the equation of this line cannot be written in slope-intercept form or point-slope form. Only nonvertical lines can be written in those forms. However, the standard form includes the vertical line

$$x = 4$$

because it can be written as

$$1 \cdot x + 0 \cdot y = 4.$$

Every line has an equation in standard form.

EXAMPLE 5 Write an equation in standard form with integral coefficients for the line through $(2, 5)$ that is perpendicular to the line $2x + 3y = 1$.

Solution First solve the equation $2x + 3y = 1$ for y to find its slope.

$$2x + 3y = 1$$
$$3y = -2x + 1$$
$$y = -\frac{2}{3}x + \frac{1}{3} \qquad \text{The slope is } -2/3.$$

The slope of the line we are interested in is the opposite of the reciprocal of $-2/3$. We want the equation of a line with slope $3/2$ that contains $(2, 5)$. Use point-slope form.

$$y - 5 = \frac{3}{2}(x - 2) \qquad \text{Point-slope form}$$

$$y - 5 = \frac{3}{2}x - 3$$

$$y = \frac{3}{2}x + 2$$

$$-\frac{3}{2}x + y = 2$$

$$3x - 2y = -4 \qquad \text{Multiply each side by } -2.$$

The last equation is the standard form of the equation of the line through $(2, 5)$ that is perpendicular to $2x + 3y = 1$. ◄

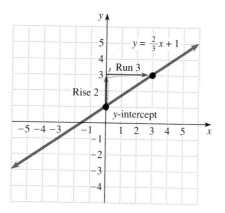

Figure 7.28

Using Slope-intercept Form for Graphing

In the slope-intercept form, a point of the line (the y-intercept) and the slope are readily available. As we did in Section 7.2, we start at the y-intercept and count off the rise and run to get a second point on the line.

EXAMPLE 6   Graph the line $2x - 3y = -3$.

Solution   First write the equation in slope-intercept form.

$$2x - 3y = -3$$
$$-3y = -2x - 3$$
$$y = \frac{2}{3}x + 1$$

The slope is 2/3, and the y-intercept is (0, 1). Start at 1 on the y-axis, then rise 2 and run 3 to locate a second point on the line. Since there is only one line containing any two given points, these two points determine the line. See Fig. 7.28.   ◀

The three methods that we have learned for graphing linear equations are summarized below.

Methods for Graphing a Linear Equation

1. Arbitrarily select some points that satisfy the equation and draw a line through them.
2. Find the x- and y-intercepts (provided they are not the origin) and draw a line through them.

3. Start at the y-intercept and use the slope to locate a second point, then draw a line through the two points.

If the y-coordinate of the y-intercept is an integer and the slope is a rational number, then the easiest method is usually to use the y-intercept and the slope.

Linear Functions

The linear equation $y = mx + b$ can be thought of as a formula that shows how to determine a value of y from a value of x. This formula is a specific example of a **function.** Functions in general will be discussed in Chapter 8. If $y = mx + b$, we say that y is a **linear function** of x. The next example shows how to use the point-slope formula to find the relationship between Fahrenheit and Celsius temperature.

EXAMPLE 7 Fahrenheit temperature F is a linear function of Celsius temperature C. Water freezes at 0° C or 32° F and boils at 100° C or 212° F. Find the linear equation that expresses F as a linear function of C.

Solution We want the equation of the line that contains the points (0, 32) and (100, 212). Use C as the independent (x) variable and F as the dependent (y) variable. The slope of the line is

$$m = \frac{212 - 32}{100 - 0} = \frac{180}{100} = \frac{9}{5}.$$

Using a slope of 9/5 and the point (100, 212) in the point-slope formula, we get

$$F - 212 = \frac{9}{5}(C - 100).$$

We can solve this equation for F to get the familiar formula relating Celsius and Fahrenheit temperature:

$$F = \frac{9}{5}C + 32 \qquad \blacktriangleleft$$

Warm-ups

True or false?

1. There is exactly one line that goes through a given point with a given slope.

2. The point-slope form of the equation of the line through (a, b) with slope m is $y - a = m(x - b)$.

3. The equation of the line through (a, b) with slope m is $y = mx + b$.

4. To find the y-intercept of a nonvertical line, we let $x = 0$ in the equation of the line.

5. To find the x-intercept of a nonhorizontal line, we let $y = 0$ in the equation of the line.

6. Every straight line in the coordinate plane has an equation in slope-intercept form.

7. The line $2y + 3x = 7$ has slope $-3/2$.

8. The line $y = 3x - 1$ is perpendicular to the line $y = (1/3)x - 1$.

9. The line $2y = 3x + 5$ has a y-intercept of $(0, 5)$.

10. Every straight line in the coordinate plane has an equation in standard form.

7.3 EXERCISES

Find the equation of each line and solve it for y. See Example 1.

1. The line containing the point $(2, -3)$ with slope -2.

2. The line containing the point $(-2, 5)$ with slope 6.

3. The line containing $(-2, 3)$ with slope $-1/2$.

4. The line containing $(3, 5)$ with slope $2/3$.

5. The line through $(2, 3)$ and $(-5, 6)$.

6. The line through $(-2, 1)$ and $(3, -4)$.

7. The line through $(3, 4)$ that is perpendicular to the line through $(-3, 1)$ and $(5, -1)$.

8. The line through $(0, 0)$ that is perpendicular to the line through $(0, 6)$ and $(-5, 0)$.

9. The line through $(0, 0)$ that is parallel to the line through $(9, -3)$ and $(-3, 6)$.

10. The line through $(2, -4)$ that is parallel to the line through $(6, 2)$ and $(-2, 6)$.

Write an equation in slope-intercept form (if possible) for each of the lines shown. See Example 2.

11.

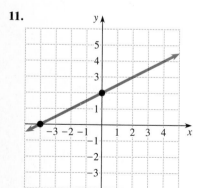

12.

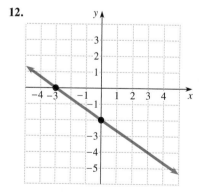

13.

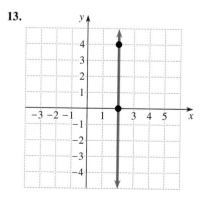

14.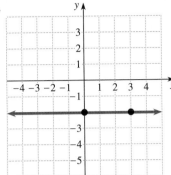

15.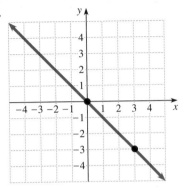

16.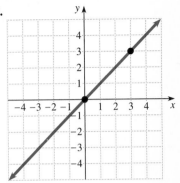

17.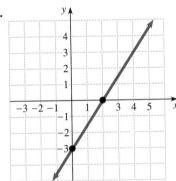

18.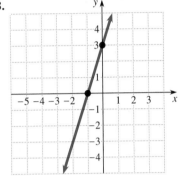

19.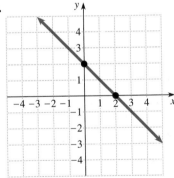

20.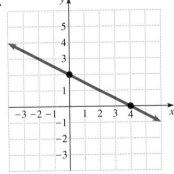

Write each equation in standard form with integral coefficients. See Example 3.

21. $y = \dfrac{1}{3}x - 2$

22. $y = \dfrac{1}{2}x + 7$

23. $y - 5 = \dfrac{1}{2}(x + 3)$

24. $y - 1 = \dfrac{1}{4}(x - 6)$

25. $y + \dfrac{1}{2} = \dfrac{1}{3}(x - 4)$

26. $y + \dfrac{1}{3} = \dfrac{1}{4}(x - 3)$

27. $.05x + .06y - 8.9 = 0$
29. $y + 55 = -.02(x - 600)$

28. $.03x - .07y = 2$
30. $y - 60 = -.08(x + 200)$

Write each of the equations in slope-intercept form, and identify the slope and y-intercept. See Example 4.

31. $2x + 5y = 1$
33. $3x - y - 2 = 0$
35. $2y - 3x + 4 = 0$
37. $y - 2 = 3(x - 1)$
39. $y + 5 = \frac{1}{2}(x - 2)$
41. $\frac{y - 5}{x + 4} = \frac{3}{2}$
43. $y - \frac{1}{2} = \frac{1}{3}\left(x + \frac{1}{4}\right)$
45. $y - 6000 = .01(x + 5700)$
47. $y + 3 = 5$
49. $.3x - .2y = 10$

32. $3x - 3y = 2$
34. $5 - x - 2y = 0$
36. $3y - x + 1 = 0$
38. $y + 4 = -2(x - 5)$
40. $y - 3 = -\frac{2}{3}(x + 7)$
42. $\frac{y - 6}{x - 2} = -\frac{3}{5}$
44. $y - \frac{1}{3} = -\frac{1}{2}\left(x - \frac{1}{4}\right)$
46. $y - 5000 = .05(x - 1990)$
48. $y - 9 = 0$
50. $.1y - .2x = 3.6$

Write an equation in standard form with integral coefficients for each of the lines described. See Example 5. In each case make a sketch.

51. The line with slope 1/2, going through (0, 5).
52. The line with slope 5, going through (0, 1/2).
53. The line with x-intercept (2, 0) and y-intercept (0, 4).
54. The line with y-intercept (0, 5) and x-intercept (4, 0).
55. The line through (−2, 1), parallel to $y = 2x + 6$.
56. The line through (1, −3), parallel to $y = -3x - 5$.
57. The line parallel to $2x + 4y = 1$ and containing (−3, 5).
58. The line parallel to $3x - 5y = -7$ and containing (2, 4).
59. The line through (1, 1), perpendicular to $y = (1/2)x - 3$.

60. The line through (−1, −2), perpendicular to $y = -3x + 7$.
61. The line through (−2, 3), perpendicular to $x + 3y = 4$.
62. The line perpendicular to $2y + 5 - 3x = 0$, containing (2, 7).
63. The line through (2, 5), running parallel to the x-axis.
64. The line through (−1, 6), running parallel to the y-axis.
65. The line parallel to $x = 3$, containing the point (−3, −2).
66. The line parallel to $y = 4$, containing the point (−7, 3).

Use the slope and y-intercept to graph each line. See Example 6.

67. $y = \frac{1}{2}x$
71. $y = -\frac{2}{3}x + 2$
75. $y - x + 3 = 0$
79. $y + 2 = 0$

68. $y = -\frac{2}{3}x$
72. $y = 3x - 4$
76. $x - y = 4$
80. $y - 3 = 0$

69. $y = 2x - 3$
73. $3y + x = 0$
77. $3x - 2y = 6$
81. $x + 3 = 0$

70. $y = -x + 1$
74. $4y - x = 0$
78. $3x + 5y = 10$
82. $x - 5 = 0$

Solve each problem. See Example 7.

83. Suppose the temperature, t, of a cup of water is a linear function of the number of seconds, s, it is in the microwave. If the temperature at $s = 0$ seconds is $t = 60°F$, and the temperature at $s = 120$ seconds is $200°F$, find the linear equation that expresses t as a function of s. What should the temperature be after 30 seconds? (Hint: Write the equation of the line containing the points (0, 60) and (120, 200) in the form $t = ms + b$.)

84. Suppose that the body temperature, t, of a person who has been lost overboard in the North Atlantic in the winter time is a linear function of the number of hours, h, that person has been in the water. If the body temperature for $h = .01$ hours is $98°F$ and the body temperature for $h = .03$ hours is $95°F$, find the linear equation that expresses t as a function of h. If hypothermia sets in at a body temperature of $88°F$, then how long would it take for this person to develop hypothermia?

85. The town of Birdview had a population of 204 in the 1980 census. In 1987, the whole town, all 239 people, attended the Firefighters' Benefit Picnic. If the population, P, is a linear function of the year, t, find the linear equation that expresses this relationship. What would be a prediction for the population in the year 2000?

86. The accountant at Apollo Manufacturing has determined that the cost, C, per week in dollars for making circuit boards is a linear function of the number, n, of circuit boards produced in a week. If $C = \$1500$ when $n = 1000$, and $C = \$2000$ when $n = 2000$, find the linear equation that expresses C in terms of n. What is the cost if they produce only one circuit board in a week?

7.4 Linear Inequalities

IN THIS SECTION:

- Definition
- Graph of a Linear Inequality
- Using a Test Point to Graph an Inequality
- Graphing Compound Inequalities

In the first three sections of this chapter we have been studying linear equations. We now turn our attention to linear inequalities.

Definition

A linear inequality is simply a linear equation with the equal sign replaced by an inequality symbol.

DEFINITION
Linear Inequality

> If A, B, and C are real numbers with A and B not both zero, then
>
> $$Ax + By \le C$$
>
> is called a **linear inequality**. In place of $\le$, we can also use $\ge$, $<$, or $>$.

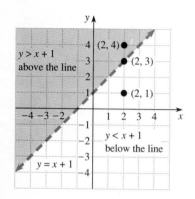

(2, 4) satisfies $y > x + 1$

(2, 3) satisfies $y = x + 1$

(2, 1) satisfies $y < x + 1$

Figure 7.29

Graph of a Linear Inequality

Consider the inequality

$$y > x + 1.$$

Which points in the rectangular coordinate system (xy-plane) satisfy this inequality? We want the points where the y-coordinate is larger than the x-coordinate plus 1. If we locate a point on the line $y = x + 1$, say (2, 3), the y-coordinate is equal to the x-coordinate plus 1. If we move vertically from that point to, say, (2, 4), the y-coordinate gets bigger but the x-coordinate does not change. See Fig. 7.29. For points above the line, the y-coordinate is greater than the x-coordinate plus 1. Thus the points above the line satisfy $y > x + 1$, and likewise points below the line satisfy $y < x + 1$.

The ideas just introduced can be used in graphing any linear inequality.

STRATEGY
Graphing a Linear Inequality

> **1.** Solve the inequality for y, then graph $y = mx + b$.
>
> $y > mx + b$ is the region above the line.
>
> $y = mx + b$ is the line itself.
>
> $y < mx + b$ is the region below the line.
>
> **2.** If the inequality involves x and not y, then graph the vertical line $x = k$.
>
> $x > k$ is the region to the right of the line.
>
> $x = k$ is the line itself.
>
> $x < k$ is the region to the left of the line.

EXAMPLE 1   Graph the inequalities.

a) $y < \dfrac{1}{2}x - 1$ **b)** $y \ge -2x + 1$ **c)** $3x - 2y < 6$

Solution

a) The set of points satisfying this inequality is the region below the line $y = (1/2)x - 1$. To show this region, we first graph the boundary line. The slope of

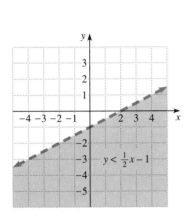

Figure 7.30

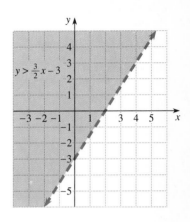

Figure 7.31

Figure 7.32

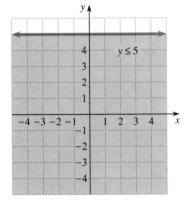

Figure 7.33

the line is $1/2$, and the y-intercept is $(0, -1)$. Start at -1 on the y-axis, then rise 1 and run 2 to get a second point of the line. We draw the line dashed because points on the line do not satisfy this inequality. The solution set to the inequality is the shaded region shown in Fig. 7.30.

b) Because the inequality symbol is $\geq$, every point on or above the line satisfies this inequality. To show that the line $y = -2x + 1$ is included, we make it a solid line. See Fig. 7.31.

c) First solve for y.

$$3x - 2y < 6$$
$$-2y < -3x + 6$$
$$y > \frac{3}{2}x - 3 \qquad \text{Divide by } -2 \text{ and reverse the inequality.}$$

To graph this inequality use a dashed line for the boundary $y = (3/2)x - 3$, and shade the region above the line. Remember, "$<$" means below the line and "$>$" means above the line *only when the inequality is solved for* y. See Fig. 7.32 for the graph. ◀

EXAMPLE 2   Graph the inequalities.

a) $y \leq 5$ **b)** $x > 4$

Solution

a) The line $y = 5$ is the horizontal line with y-intercept $(0, 5)$. Draw a solid horizontal line and shade below it as in Fig. 7.33.

b) The inequality $x > 4$ has no y term but it is easy to graph. The solution set, shown in Fig. 7.34, is the region to the right of the vertical line $x = 4$. ◀

Figure 7.34

Using a Test Point to Graph an Inequality

The graph of a linear equation such as $3x - 4y = 7$ separates the coordinate plane into two regions. One region satisfies the inequality $3x - 4y > 7$, and the other region satisfies the inequality $3x - 4y < 7$. We can tell which region satisfies which inequality by testing a point in one region. With this method, it is not necessary to solve the inequality for y.

EXAMPLE 3   Graph the inequality $3x - 4y > 7$.

Solution   First graph the equation $3x - 4y = 7$ using the x-intercept and the y-intercept. If $x = 0$, then $y = -7/4$. If $y = 0$, then $x = 7/3$. Use the x-intercept $(7/3, 0)$ and the y-intercept $(0, -7/4)$ to graph the line as shown in Fig. 7.35. Select a point on one side of the line, say $(0, 0)$, to test in the inequality. Because $3(0) - 4(0) > 7$ is false, the region on the other side of the line satisfies the inequality. The graph of $3x - 4y > 7$ is shown in Fig. 7.36.

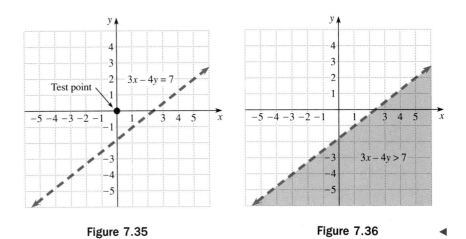

Figure 7.35   Figure 7.36   ◀

Graphing Compound Inequalities

We can write compound inequalities with two variables just as we do one variable. Consider the compound inequality

$$y > x - 3 \quad \text{and} \quad y < -\frac{1}{2}x + 2.$$

The solution set or graph of this compound inequality is the set of points that satisfy both inequalities. To determine the graph of this set, first graph the equations $y = x - 3$ and $y = (-1/2)x + 2$. The two lines divide the plane into four regions as shown in Fig. 7.37. Now test one point of each region to determine which region satisfies the compound inequality. Test the points $(3, 3)$, $(0, 0)$, $(4, -5)$, and $(5, 0)$.

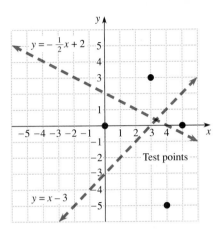

Figure 7.37

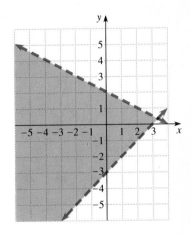

Figure 7.38

The only point that satisfies both inequalities is $(0, 0)$, because

$$0 > 0 - 3 \quad \text{and} \quad 0 < -\frac{1}{2} \cdot 0 + 2$$

are both true. Each of the other three points fails to satisfy at least one of the inequalities. Check them. To illustrate the solution set to the compound inequality, we shade the region containing $(0, 0)$ as shown in Fig. 7.38. ◄

EXAMPLE 4 Graph the compound inequality

$$2x - 3y \le -6 \quad \text{or} \quad x + 2y \ge 4.$$

Solution First graph the lines $2x - 3y = -6$ and $x + 2y = 4$ by using x- and y-intercepts. See Fig. 7.39. The graph of the compound inequality is the set of points that satisfy either one inequality or the other (or both). Test the points $(0, 0)$, $(4, 2)$, $(0, 5)$, and $(-3, 2)$; only $(0, 0)$ fails to satisfy at least one of the inequalities. Thus only the region containing the origin is left unshaded. See Fig. 7.40.

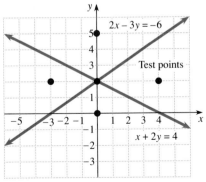

Figure 7.39

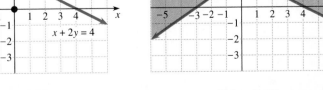

Figure 7.40 ◄

Note that the graph of a compound inequality involving the word *and* is the intersection of the solution sets of the two inequalities. If the compound inequality involves the word *or*, then the graph is the union of the graphs of the two inequalities. The graph of an absolute value inequality is found by graphing an equivalent compound inequality.

EXAMPLE 5   Graph each absolute value inequality.

a) $|y - 2x| \le 3$ **b)** $|x - y| > 1$

Solution

a) The inequality $|y - 2x| \le 3$ is equivalent to the compound inequality

$$y - 2x \le 3 \quad \text{and} \quad y - 2x \ge -3.$$

First graph the lines $y - 2x = 3$ and $y - 2x = -3$ as shown in Fig. 7.41. These lines divide the plane into three regions. Test a point from each region in the original inequality, say $(-5, 0)$, $(0, 0)$, and $(5, 0)$:

$$|0 - 2(-5)| \le 3 \qquad |0 - 2 \cdot 0| \le 3 \qquad |0 - 2 \cdot 5| \le 3$$

Only $(0, 0)$ satisfies the original inequality. So the region satisfying the absolute value inequality is the shaded region shown in Fig. 7.42, the region between the two lines. The boundary lines are solid because of the $\le$ symbol.

b) The inequality $|x - y| > 1$ is equivalent to

$$x - y > 1 \quad \text{or} \quad x - y < -1.$$

First graph the lines $x - y = 1$ and $x - y = -1$ as shown in Fig. 7.43. Test a

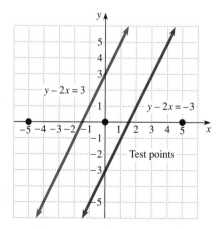

Figure 7.41

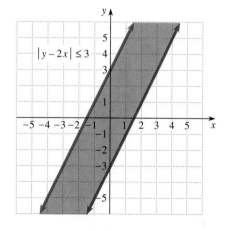

Figure 7.42

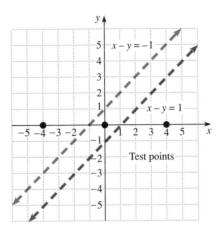

Figure 7.43   Figure 7.44

point from each region in the original inequality, say $(-4, 0)$, $(0, 0)$, and $(4, 0)$:

$$|-4 - 0| > 1 \qquad |0 - 0| > 1 \qquad |4 - 0| > 1$$

Because $(-4, 0)$ and $(4, 0)$ satisfy the inequality, we shade those regions as shown in Fig. 7.44. The boundary lines are dashed because of the $>$ symbol.

◄

Warm-ups

True or false?

1. The point $(2, -3)$ satisfies the inequality $y > -3x + 2$.
2. The graph of $3x - y > 2$ is the region above the line $3x - y = 2$.
3. The graph of $3x + y < 5$ is the region below the line $y = -3x + 5$.
4. The graph of $x < -3$ is the region to the left of the vertical line $x = 3$.
5. The graph of $y > x + 3$ and $y < 2x - 6$ is the intersection of the graphs of the two inequalities.
6. The graph of $y \leq 2x - 3$ or $y \geq 3x + 5$ is the union of the graphs of the two inequalities.
7. The point $(2, -5)$ is in the solution set to the compound inequality $y > -3x + 5$ and $y < 2x - 3$.
8. The point $(-3, 2)$ is in the solution set to the compound inequality $y \leq 3x - 6$ or $y \leq x + 5$.

9. The inequality $|2x - y| \leq 4$ is equivalent to the compound inequality $2x - y \leq 4$ and $2x + y \leq 4$.

10. The graph of $|x - y| > 3$ is the same as the graph of the compound inequality $x - y > 3$ or $x - y < -3$.

7.4 EXERCISES

Graph each linear inequality. See Examples 1 and 2.

1. $y < x + 2$	**2.** $y < x - 1$	**3.** $y \leq -2x + 1$	**4.** $y \geq -3x + 4$
5. $x + y > 3$	**6.** $x + y \leq -1$	**7.** $2x + 3y < 9$	**8.** $-3x + 2y > 6$
9. $3x - 4y \leq 8$	**10.** $4x - 5y > 10$	**11.** $x - y > 0$	**12.** $2x - y < 0$
13. $x \geq 1$	**14.** $x < 0$	**15.** $y < 3$	**16.** $y > -1$

Graph each linear inequality by using a test point. See Example 3.

17. $2x - 3y < 5$	**18.** $5x - 4y > 3$	**19.** $x + y + 3 \geq 0$
20. $x - y - 6 \leq 0$	**21.** $\frac{1}{2}x + \frac{1}{3}y < 1$	**22.** $2 - \frac{2}{5}y > \frac{1}{2}x$

Graph each compound inequality. See Example 4.

23. $y > x$ and $y > -2x + 3$	**24.** $y < x$ and $y < -3x + 2$
25. $y < x + 3$ or $y > -x + 2$	**26.** $y \geq x - 5$ or $y \leq -2x + 1$
27. $x + y \leq 5$ and $x - y \leq 3$	**28.** $2x - y < 3$ and $3x - y > 0$
29. $x - 2y \leq 4$ or $2x - 3y \leq 6$	**30.** $4x - 3y \leq 3$ or $2x + y \geq 2$
31. $y > 2$ and $x < 3$	**32.** $x \leq 5$ and $y \geq -1$
33. $y \geq x$ and $x \leq 2$	**34.** $y < x$ and $y > 0$
35. $2x < y + 3$ or $y > 2 - x$	**36.** $3 - x < y + 2$ or $x > y + 5$
37. $x - 1 < y < x + 3$	**38.** $x - 1 < y < 2x + 5$
**39. $0 \leq y \leq x$ and $x \leq 1$	**40. $x \leq y \leq 1$ and $x \geq 0$
**41. $1 \leq x \leq 3$ and $2 \leq y \leq 5$	**42. $-1 < x < 1$ and $-1 < y < 1$

Graph the absolute value inequalities. See Example 5.

43. $|x + y| < 2$	**44.** $|2x + y| < 1$	**45.** $|2x + y| \geq 1$	**46.** $|x + 2y| \geq 6$
47. $|x - y - 3| < 5$	**48.** $|x - 2y + 4| > 2$	**49.** $|x - 2y| \geq 4$	**50.** $|x - 3y| \leq 6$
51. $|x| > 2$	**52.** $|x| \leq 3$	**53.** $|y| < 1$	**54.** $|y| \geq 2$
**55. $y < |x|$		**56. $y < |x + 3|$	
**57. $|x| < 2$ and $|y| < 3$		**58. $|x| \geq 3$ or $|y| \geq 1$	
**59. $|x - 3| < 1$ and $|y - 2| < 1$		**60. $|x - 2| \geq 3$ or $|y - 5| \geq 2$	

Wrap-up

CHAPTER 7

SUMMARY

Concepts	Examples

Distance formula

The distance between (x_1, y_1) and (x_2, y_2) is

$$\sqrt{(x_2 - x_1)^2 + (y_2 - y_1)^2}$$

$(1, -2), (3, -4)$

$$\sqrt{(1-3)^2 + [-2-(-4)]^2} = \sqrt{4+4}$$
$$= \sqrt{8} = 2\sqrt{2}$$

X-intercept

The point where a nonhorizontal line intersects the x-axis

$2x + y = 6$
$(3, 0)$ x-intercept

Y-intercept

The point where a nonvertical line intersects the y-axis

$(0, 6)$ y-intercept

Slope of a line

(Idea) slope $= \dfrac{\text{change in } y\text{-coordinate}}{\text{change in } x\text{-coordinate}}$

$= \dfrac{\text{rise}}{\text{run}}$

(Definition) slope of line through (x_1, y_1), (x_2, y_2) is

$$m = \frac{y_2 - y_1}{x_2 - x_1} \qquad (x_2 - x_1 \neq 0)$$

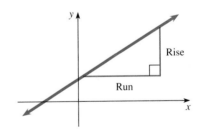

$(4, -2), (3, -6)$

$$m = \frac{-6 - (-2)}{3 - 4} = \frac{-4}{-1} = 4$$

Types of slope

 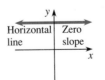

Perpendicular lines

The slope of one line is the negative of the reciprocal of the slope of the other.

A line with slope 3 is perpendicular to a line with slope $-1/3$.

Parallel lines	Nonvertical parallel lines have equal slopes.	$y = 2x - 3$ $y = 2x + 7$
Equations of lines	**Point-slope form:** m = slope, (x_1, y_1) is a point of the line $y - y_1 = m(x - x_1)$	Line through $(5, -3)$ with slope -2: $y + 3 = -2(x - 5)$
	Slope-intercept form: m = slope, $(0, b)$ is the y-intercept $y = mx + b$	Line with slope $1/2$ crossing y-axis at $(0, -3)$: $y = (1/2)x - 3$
	Standard form: A and B are not both zero $Ax + By = C$	$3x - 7y = 12$
	Every line can be expressed in standard form.	$x = 3$ (vertical) $y = -5$ (horizontal)
Methods for graphing a linear equation	1. Arbitrarily select some points that satisfy the equation, and draw a line through them. 2. Find the x- and y-intercepts (provided they are not the origin), and draw a line through them. 3. Start at the y-intercept and use the slope to locate a second point, then draw a line through the two points.	
Linear inequalities	A and B not both zero $Ax + By \leq C$	$2x - 3y \leq 7$
Graphing linear inequalities	1. Solve for y, then graph the line $y = mx + b$. $y > mx + b$ is above the line. $y = mx + b$ is on the line. $y < mx + b$ is below the line. 2. If the inequality involves x and not y, graph $x = k$. $x > k$ is to the right of the vertical line. $x = k$ is on the vertical line. $x < k$ is to the left of the vertical line.	$x - 2y > 6$ $-2y > -x + 6$ $y < (1/2)x - 3$ is below $y = (1/2)x - 3$ $3x + 4 > 7$ $3x > 3$ $x > 1$ is right of $x = 1$.
Test points	A linear inequality may also be graphed by graphing the equation and then testing a point to determine which region satisfies the inequality.	

Compound
inequalities

And—intersection:

$x \geq 1$ and $y \geq 2$

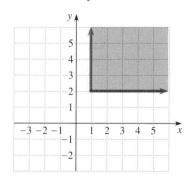

Or—union:

$x > 1$ or $y > 2$

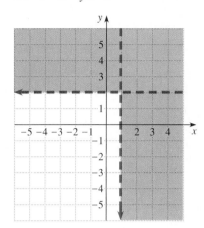

REVIEW EXERCISES

7.1 *For each point, name the quadrant that it lies in or the axis that it lies on.*

1. $(-3, -2)$ **2.** $(0, \pi)$ **3.** $(\pi, 0)$ **4.** $(-5, 4)$

5. $(0, -1)$ **6.** $(\pi/2, 1)$ **7.** $(\sqrt{2}, -3)$ **8.** $(6, -3)$

Find the distance between each pair of points.

9. $(2, 5), (1, 3)$ **10.** $(3, -1), (-5, 2)$ **11.** $(-6, 0), (0, -8)$ **12.** $(-6, 0), (0, -8)$

Complete the given ordered pairs so that each ordered pair satisfies the given equation.

13. $(0, \ ), (\ , 0), (4, \ ), (\ , -3), y = -3x + 2$ **14.** $(0, \ ), (\ , 0), (-6, \ ), (\ , 5), 2x + 3y = 5$

7.2 *Find the slope of the line through each pair of points.*

15. $(-5, 6), (-2, 9)$ **16.** $(-2, 7), (3, -4)$ **17.** $(4, 1), (-3, -2)$ **18.** $(6, 0), (0, -3)$

7.3 *Find the slope and y-intercept for each line.*

19. $y = -3x + 4$ **20.** $2y - 3x + 1 = 0$ **21.** $y - 3 = (2/3)(x - 1)$ **22.** $y - 3 = 5$

Write each equation in standard form with integral coefficients.

23. $y = \dfrac{2}{3}x - 4$ **24.** $y = -.05x + .26$ **25.** $y - 1 = \dfrac{1}{2}(x + 3)$ **26.** $\dfrac{1}{2}x - \dfrac{1}{3}y = \dfrac{1}{4}$

Write the equation of the line containing the given point and having the given slope. Put each equation into standard form with integral coefficients.

27. $(1, -3), m = 1/2$ **28.** $(0, 2), m = 3$ **29.** $(-2, 6), m = -3/4$ **30.** $(2, 1/2), m = 1/4$

31. $(3, 5), m = 0$ **32.** $(0, 0), m = -1$

Draw the graph of each line.

33. $y = 2x - 3$

34. $y = (2/3)x + 1$

35. $3x - 2y = -6$

36. $y - 3 = 10$

37. $2x = 8$

38. $4x + 5y = 10$

39. $5x - 3y = 7$

40. $3x + 4y = -1$

7.4 *Graph each linear inequality.*

41. $y > 3x - 2$

42. $y \leq 2x + 3$

43. $x - y \leq 5$

44. $2x + y > 1$

45. $3x > 2$

46. $4y \leq 0$

47. $4x - 2y \geq 6$

48. $-5x - 3y > 6$

49. $5x - 2y < 9$

50. $3x + 4y \leq -1$

Graph each compound or absolute value inequality.

51. $y > 3$ and $y - x < 5$

52. $x + y \leq 1$ or $y \leq 4$

53. $3x + 2y \geq 8$ or $3x - 2y \leq 6$

54. $x + 8y > 8$ and $x - 2y < 10$

55. $|x + 2y| < 10$

56. $|x - 3y| \geq 9$

57. $|x| \leq 5$

58. $|y| > 6$

59. $|y - x| > 2$

60. $|x - y| \leq 1$

Miscellaneous

Write an equation in standard form with integral coefficients for each line described.

61. The line that crosses the x-axis at 2 and the y-axis at -6

62. The line with an x-intercept of $(4, 0)$ and slope $-1/2$

63. The line with a y-intercept of $(0, 6)$ and slope 3

64. The line through the origin with slope -1

65. The horizontal line through $(2, 5)$

66. The vertical line through $(-3, -2)$

67. The line through $(-1, 4)$ with slope $-1/2$

68. The line through $(2, -3)$ with slope 0

69. The line through $(2, -6)$ and $(2, 5)$

70. The line through $(-3, 6)$ and $(4, 2)$

71. The line through $(0, 0)$ and perpendicular to $x = 5$

72. The line through $(2, -3)$ perpendicular to $y = -3x + 5$

73. The line through $(-1, 4)$ and parallel to $y = 2x + 1$

74. The line through $(2, 1)$ and perpendicular to $y = 10$

Use slope or distance to solve each geometric problem.

75. Show that the points $(-5, -5)$, $(-3, -1)$, $(6, 2)$, and $(4, -2)$ are the vertices of a parallelogram.

76. Show that the points $(-5, -5)$, $(4, -2)$, and $(3, 1)$ are the vertices of a right triangle.

77. Show that the points $(-4, -3)$, $(4, -1)$, and $(2, 7)$ are the vertices of an isosceles triangle.

78. Show that the points $(-2, 2)$, $(-1, -1)$, $(1, 3)$, and $(2, 0)$ are the vertices of a square.

79. Show that the points $(-1, 0)$, $(3, 0)$, and $(1, 2\sqrt{3})$ are the vertices of an equilateral triangle.

80. Show that the points $(-2, 2)$, $(0, 0)$, $(2, 6)$, and $(4, 4)$ are the vertices of a rectangle.

81. Find the perimeter of the triangle with vertices $(-4, -3)$, $(4, -1)$, and $(2, 7)$.

82. Determine whether the points $(2, 1)$, $(4, 7)$, and $(-3, -14)$ lie on a straight line.

Solve the following problems.

83. The charge, C, in dollars for renting an air hammer from the Tools R' Us Rental Company is determined from the formula $C = 26 + 17d$, where d is the number of days in the rental period. Graph this function for d from 1 to 30. If the air hammer is worth \$1080, then in how many days would the rental charge equal the value of the air hammer?

84. A subject is given 3 mg of an experimental drug, and a resting heart rate of 82 is recorded. Another subject is given 5 mg of the same drug, and a resting heart rate of 89 is recorded. If we assume the heart rate, h, is a linear function of the dosage, d, find the linear equation expressing h in terms of d. If a subject is given 10 mg of the drug, what would the expected heart rate be?

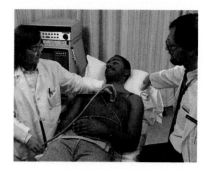

Photo for Exercise 84

CHAPTER 7 TEST

Complete each ordered pair so that it satisfies the given equation.

1. $(0, \ )$, $(\ , 0)$, $(4, \ )$, $(\ , -8)$, $2x + y = 5$

Use the distance formula to find the distance between each pair of points.

2. $(3, -2)$, $(5, -6)$

3. $(0, 3)$, $(-4, 0)$

Find the slope of the line that contains each pair of points.

4. $(-3, 7)$, $(2, 1)$

5. $(0, -6)$, $(4, 2)$

Determine the slope and y-intercept for each line.

6. $y = -\dfrac{1}{2}x - 2$

7. $8x - 5y = -10$

For each line described below, write its equation in standard form with integral coefficients.

8. The line with y-intercept $(0, 3)$ and slope $-1/2$.

9. The line through $(-3, 5)$ with slope -4.

10. The line through $(4, 1)$ and $(-2, -5)$.

11. The line through $(2, 3)$ that is perpendicular to the line $3x - 5y = 7$.

12. The line through $(-4, 6)$ parallel to $y = -1$.

13. The line shown in the graph:

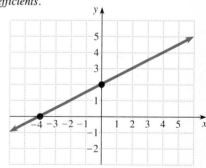

Figure for Exercise 13

Use the slope or distance formula to solve the following problems.

14. Show that $(-1, -2)$, $(2, -1)$, and $(1, 2)$ are the vertices of a right triangle.

15. Show that $(-1, -2)$, $(0, 0)$, $(6, 2)$, and $(5, 0)$ are the vertices of a parallelogram.

Sketch the graph of each line.

16. $y = \dfrac{1}{3}x - 2$

17. $5x - 2y = 7$

Sketch the graph of each inequality.

18. $y > -\dfrac{1}{2}x + 3$

19. $3x - 2y - 6 \le 0$

Sketch the graph of each compound or absolute value inequality.

20. $x > 2$ and $x + y > 0$

21. $|2x + y| \ge 3$

22. Suppose the value, V, in dollars of an automobile is a linear function of its age, a, in years. If an automobile was valued at \$22,000 brand new and \$16,000 when it is 3 years old, find the linear equation that expresses V in terms of a.

Tying It All Together

CHAPTERS 1–7

Simplify the following expressions.

1. 5^3
2. $125^{1/3}$
3. $2(7) + 4$
4. $\dfrac{18 - 4}{2}$
5. $\sqrt[3]{4 \cdot 6 + 3}$
6. $\dfrac{3^3 - 3}{6}$

7. $\dfrac{(x + h)^2 - x^2}{h}$

8. $\dfrac{3(x + h)^2 - 2(x + h) - (3x^2 - 2x)}{h}$

Sketch the graph of each of the following lines.

9. $y = \dfrac{1}{3}x$
10. $y = 3x$
11. $y = 2x + 5$
12. $y = \dfrac{x - 5}{2}$

Find all real numbers that solve each equation.

13. $2x - 5 = 9$
14. $3(y - 2) + 4 = -5$
15. $|2x - 3| = 0$
16. $\sqrt{x - 4} = 0$
17. $x^2 + 9 = 0$
18. $x^2 - 8x + 12 = 0$

Graph the solution set to each inequality or compound inequality on the number line.

19. $2x - 1 > 7$
20. $5 - 3x \le -1$
21. $x - 5 \le 4$ and $3x - 1 < 8$
22. $2x \le -6$ or $5 - 2x < -7$
23. $|x - 3| < 2$
24. $|1 - 2x| \ge 7$

Find the domain of each expression.

25. $\dfrac{x - 2}{x + 3}$
26. $\sqrt{x}$
27. $\dfrac{5x - 3}{x^2 + 1}$

28. $\dfrac{1}{|x - 1|}$
29. $\dfrac{x - 3}{x^2 - 10x + 9}$
30. $.3x$

Functions

Sheriff John Stone of Macon County has been fighting for a new jail for a long time. To support his arguments he has been studying the costs of housing prisoners in the Macon County Jail for the last 10 years. He has repeatedly argued that jail overcrowding is more expensive than building a new jail. When the jail is overcrowded the guard to prisoner ratio has to be higher than normal, and the meals must be catered because the jail kitchen is small.

From the cost data the sheriff has determined a formula (or function) that gives the cost in dollars per day per prisoner when n prisoners are in the jail, $C(n) = .015n^2 - 4.5n + 400$. The formula shows that the cost per prisoner is large when there are too few prisoners as well as when there are too many. What number of prisoners will make the cost per prisoner a minimum? We will be able to solve problems of this type after studying functions in this chapter. The number of prisoners that minimizes the cost per prisoner in the Macon County Jail will be found when we do Exercise 59 of Section 8.2.

8.1 Definitions and Notation

IN THIS SECTION:

- **Relations and Functions**
- **Domain and Range**
- **The *f*-Notation**

In Section 7.3 we studied linear functions. In this section we will introduce the general concepts of relations and functions.

Relations and Functions

A farmer in Illinois suspected that the amount of June rainfall had some relationship to the price he got for a bushel of corn in the fall. He checked his records for the previous five years and found the following.

Rainfall in inches	2.7	1.6	.4	8.1	2.7
Price in dollars	3.89	3.82	5.85	5.36	4.70

Using the notation of ordered pairs, we could describe this data as a set of ordered pairs where the first coordinate is rainfall and the second is price.

$$\{(2.7, 3.89), \quad (1.6, 3.82), \quad (.4, 5.85), \quad (8.1, 5.36), \quad (2.7, 4.70)\}$$

Any set of ordered pairs is called a **relation.** There seems to be some relationship between rainfall and price, but it is not clear-cut because the same rainfall has resulted in different prices. The ordered pairs (2.7, 3.89) and (2.7, 4.70) have the same first coordinate and different second coordinates. Therefore, the price cannot be determined from the rainfall.

If the farmer had discovered that the June rainfall determined the price of corn, we would say that the price of corn is a function of the amount of June rainfall. In general, when the value of one variable determines the value of a second variable, we say that the second variable is a function of the first. Exactly how the value of the second variable is determined may be unknown, so the definition of a function is simply stated in terms of ordered pairs.

DEFINITION
Function

> A **function** is a set of ordered pairs where no two ordered pairs have the same first coordinate and different second coordinates.

The first variable is called the **independent variable,** and the second variable is called the **dependent variable.** In this text we will allow the variables to represent real numbers only.

EXAMPLE 1   Determine whether each relation is a function.

a) {(1, 2), (1, 5), (3, 7)}   **b)** {(4, 5), (3, 5), (2, 6), (1,7)}

Solution

a) This relation is not a function because (1, 2) and (1, 5) have the same first coordinates but different second coordinates.
b) This relation is a function. Note that the same second coordinate with different first coordinates is permitted in a function.   ◀

We have seen and used many functions in the past, but we did not always call them functions. We usually have a formula or equation that shows us how the value of one variable determines the value of another. The formula

$$A = \pi r^2$$

shows us that the area of a circle is a function of the radius. The radius determines the area. The formula

$$V = L \cdot W \cdot H$$

expresses the fact that the volume of a box is a function of three variables: length, width, and height. The equation

$$y = 2x - 3$$

expresses y as a linear function of x. To say that an equation such as $y = 2x - 3$ defines y as a function of x, means that the set of ordered pairs

$$\{(x, y) \mid y = 2x - 3\}$$

is a function. Some equations define functions and some do not. To determine if an equation defines a function, we use the definition of function.

EXAMPLE 2   Determine whether each relation is a function.

a) $\{(x, y) \mid y = 5x^2 - 7x + 2\}$ **b)** $\{(x, y) \mid y^2 = x\}$
c) $\{(x, y) \mid 2x + 3y = 6\}$

Solution

a) This relation is a function, because with each value we select for x there is only one value determined for y.
b) This relation is not a function, because if we select $x = 4$, then $y = \pm 2$. Both (4, 2) and (4, −2) belong to this set.
c) If we solve $2x + 3y = 6$ for y, we get $y = (-2/3)x + 2$. Because each value of x determines only one value for y, this relation is a function.   ◀

We often omit the set notation and refer to an equation as a relation or a function. Any equation with two variables defines a relation, but not necessarily a function.

EXAMPLE 3   Determine which of these relations defines y as a function of x.

a) $x = |y|$ **b)** $y = x^2$ **c)** $x^2 + y^2 = 16$

Solution

a) If $x = 5$, then $y = \pm 5$. Both $(5, 5)$ and $(5, -5)$ satisy this equation, so y is not a function of x.

b) Since each value of x determines only one value of y, y is a function of x.

c) If we let $x = 0$, then $y = \pm 4$. Both $(0, 4)$ and $(0, -4)$ satisfy this equation, so y is not a function of x. ◄

Domain and Range

The set of all possible numbers that can be used for the independent variable (x) is called the **domain** of the relation. The set of all values of the dependent variable (y) is called the **range** of the relation. Often we will clearly state the domain, and at times the domain will be understood to be only those values that can be used in the expression defining the relation.

EXAMPLE 4   State the domain and range of each relation.

a) $\{(2, 5), (2, 7), (4, 3)\}$ **b)** $y = \sqrt{x}$
c) $w = v + 1$ for $v > 2$

Solution

a) The domain is the set of numbers used as first coordinates, $\{2, 4\}$. The range is the set of second coordinates, $\{3, 5, 7\}$.

b) Because $\sqrt{x}$ is a real number only for $x \geq 0$, the domain is $\{x \mid x \geq 0\}$, the nonnegative real numbers. The range is the set of numbers that results from taking the principal square root of every nonnegative real number. Thus, the range is also the set of nonnegative real numbers, $\{y \mid y \geq 0\}$.

c) The condition $v > 2$ specifies the domain of the function. The domain is $\{v \mid v > 2\}$, the real numbers greater than 2. The range is the set of values of w, but if $v > 2$ then certainly $w > 3$. The range is the set of real numbers greater than 3, $\{w \mid w > 3\}$. ◄

In Section 8.2 we will use the graph of a relation to determine the range.

The f-Notation

If the variable y is a function of x, we write $y = f(x)$ to emphasize this relationship. We read $f(x)$ as "f of x." If there is a formula for determining y from x, we can be

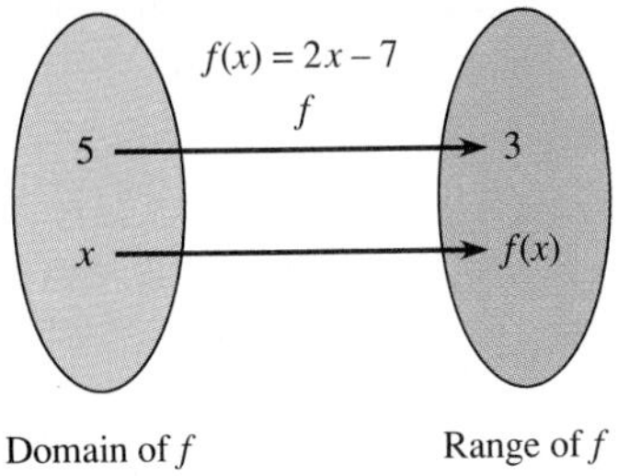

$f(x) = 2x - 7$
f

5 → 3

x → $f(x)$

Domain of f   Range of f

Figure 8.1

more specific. For example, if $y = 2x - 7$, we write $f(x) = 2x - 7$. We use y and $f(x)$ interchangeably. Letters other than f may also be used.

If $f(x) = 2x - 7$ and $x = 5$, then

$$f(5) = 2(5) - 7 = 10 - 7 = 3, \quad \text{or} \quad f(5) = 3.$$

We read the last equation as "f of 5 equals 3." When x is 5 the value of this function (or the value of the dependent variable) is 3.

A function is a rule for pairing numbers in one set (the domain) with numbers in another set (the range). The diagram shown in Fig. 8.1 is helpful in understanding the idea of a function.

The x in notation such as $f(x) = x^2 + 2$ is sometimes called a dummy variable because the letter used is unimportant. The notation $f(t) = t^2 + 2$ or even $f(\ ) = (\ )^2 + 2$ could be used to define the same function. This function is a rule for determining the second coordinate when given the first coordinate of an ordered pair. The rule in this case is to square the first coordinate and then add 2. If the first coordinate is, say, $x + 3$, the second coordinate is $(x + 3)^2 + 2$, or $x^2 + 6x + 11$. Note that $x + 3$ is squared and not just x.

EXAMPLE 5 Suppose $f(x) = x^2 + 2$. Find the following.

a) $f(-2)$ **b)** $f(x + h)$ **c)** x, if $f(x) = 4$

Solution

a) Replace x by -2 in the equation defining the function f.

$$f(-2) = (-2)^2 + 2 = 4 + 2 = 6$$

So $f(-2) = 6$.

b) Replace x by $x + h$ in the equation defining the function f.

$$f(x + h) = (x + h)^2 + 2$$
$$= x^2 + 2xh + h^2 + 2$$

c) If $f(x) = 4$, we have

$$x^2 + 2 = 4$$
$$x^2 = 2$$
$$x = \pm\sqrt{2}.$$

Warm-ups

True or false?

1. Any set of ordered pairs is a function.

2. The circumference of a circle is a function of the diameter.

3. The set $\{(1, 2), (3, 2), (5, 2)\}$ is a function.

4. Every relation is a function.
5. The set $\{(1, 5), (3, 6), (1, 7)\}$ is a function.
6. The domain of $f(x) = \sqrt{x}$ is the set of positive real numbers.
7. The range of $g(x) = |x|$ is the set of nonnegative real numbers.
8. The set $\{(x, y) \mid x = 4y\}$ is a function.
9. The set $\{(x, y) \mid x = y^4\}$ is a function.
10. If $h(x) = x^2 - 3$, then $h(-2) = 1$.

8.1 EXERCISES

Determine whether each relation is a function. See Example 1.

1. $\{(2, 3), (3, 4), (4, 5)\}$
2. $\{(2, -5), (2, 5), (3, 10)\}$
3. $\{(-2, 4), (2, 4), (3, 6)\}$
4. $\{(3, 6), (6, 3)\}$
5. $\{(1, -1), (1, 1)\}$
6. $\{(-1, -1)\}$
7. $\{(2, 2)\}$
8. $\{(3, 7), (-3, 7), (6, 7)\}$

Determine whether each relation is a function. See Example 2.

9. $\{(x, y) \mid y = (x - 1)^2\}$
10. $\{(x, y) \mid y = x^2 - 12x + 1\}$
11. $\{(x, y) \mid x = |y|\}$
12. $\{(x, y) \mid x = y^2 + 2\}$
13. $\{(x, y) \mid y = x\}$
14. $\{(x, y) \mid y = 1/x\}$
15. $\{(x, y) \mid x = 5y + 2\}$
16. $\{(x, y) \mid x = 3y\}$

Determine whether each relation is a function. See Example 3.

17. $x = 2y$
18. $x = \sqrt{y}$
19. $y = \sqrt{x}$
20. $y = x^{2/3} + 3x^{1/3} + 5$
21. $y = x^{1/3}$
22. $x = 2|y|$
23. $x^2 + y^2 = 1$
24. $y = |x - 1|$

Determine the domain and range of each relation. See Example 4.

25. $\{(2, 3), (2, 5), (2, 7)\}$
26. $\{(3, 1), (5, 1), (4, 1)\}$
27. $y = |x|$
28. $y = 2x + 1$
29. $x = \sqrt{y}$
30. $y = \sqrt{x} + 1$
31. $y = x^{1/3}$ for $x \geq 8$
32. $x = y^2$ for $0 \leq x \leq 9$

Let $f(x) = 3x - 2$, $g(x) = x^2 - 3x + 2$, and $h(x) = |x + 2|$. Find the following. See Example 5.

33. $f(4)$
34. $f(100)$
35. $g(-2)$
36. $g(6)$
37. $h(-3)$
38. $h(-19)$

39. x, if $f(x) = 5$

40. x, if $f(x) = 49$

41. x, if $g(x) = 0$

42. x, if $g(x) = 2$

43. x, if $h(x) = 3$

44. x, if $h(x) = 7$

45. $f(a)$

46. $f(a + 1)$

47. $g(b + 2)$

48. $g(x + 2)$

49. $g(x + h)$

50. $g(2 + h)$

Solve each problem.

51. Express the area of a square, A, as a function of the length of a side, s.

52. Express the perimeter of a square, P, as a function of the length of a side, s.

53. If a certain fabric is priced at $3.98 per yard, express the cost of a purchase, C, as a function of the number of yards purchased, y.

54. If Mildred earns $14.50 per hour, express her total pay, P, as a function of the number of hours worked, h.

55. A pizza parlor charges $14.95 for a pizza plus $.50 for each topping. Express the cost of a pizza, C, as a function of the number of toppings, n.

56. A gravel dealer charges $120 for a minimum load of 9 cubic yards, and $10 more for each additional cubic yard. Express the total charge, C, as a function of the number of yards sold, n, where $n \geq 9$.

57. Suppose that the heart rate of a certain individual is a linear function of the number of minutes she spends on a treadmill. A heart rate of 78 was measured after 2 minutes, and a heart rate of 86 was measured after 4 minutes. Write the

Photo for Exercise 57

heart rate, h, as a linear function of the number of minutes, t, on the treadmill, for $0 \leq t \leq 8$.

58. To determine how much he charges for printing a book, a printer uses a linear function of the number of pages. If he charges $8.60 for a 400-page book and $12.20 for a 580-page book, then what is the linear function that he uses?

Let $f(x) = \sqrt{x + 2}$ and $g(x) = 3x^2 - 8x + 2$. Find the following. Round answers to three decimal places.

59. $f(3.46)$

60. $g(-1.37)$

61. $g(-3.5)$

62. $f(-1.2)$

63. x, if $f(x) = 5.6$

64. x, if $f(x) = 8.251$

65. x, if $g(x) = 0$

66. x, if $g(x) = 1$

Let $f(x) = x^2 - x + 2$ and $g(x) = 3x - 5$. Calculate and simplify the following.

*67. $g(x + 2) - g(x)$

*68. $g(a + 1) - g(a)$

*69. $\dfrac{g(x + h) - g(x)}{h}$

*70. $\dfrac{g(3 + h) - g(3)}{h}$

*71. $\dfrac{f(a + 2) - f(a)}{2}$

*72. $f(x - 1) - f(x)$

*73. $\dfrac{f(x + h) - f(x)}{h}$

*74. $\dfrac{f(5 + h) - f(5)}{h}$

8.2 Graphs of Functions

IN THIS SECTION:

- ● Linear Functions
- ● Absolute-value Functions
- ● Quadratic Functions
- ● Square-root Functions
- ● Graphs of Relations
- ● Vertical-line Test
- ● Applications

In Chapter 7 we learned that graphs are an important tool in understanding linear functions. In this section we will again use the graphs of functions as an aid to understanding the functions.

Linear Functions

In Chapter 7 we learned that if y is determined from x by a formula of the form $y = mx + b$, then y is a linear function of x. We can restate this definition using f-notation.

DEFINITION
Linear Function

> A function of the form
>
> $$f(x) = mx + b,$$
>
> where m and b are any real numbers, is called a **linear function.**

The graph of the linear function $f(x) = mx + b$ is exactly the same as the graph of the linear equation $y = mx + b$. If $m = 0$, the linear function $f(x) = b$ is called a **constant function.**

EXAMPLE 1 Graph the constant function $f(x) = 3$, and state the domain and range.

Solution When we use the f-notation, we can label the vertical axis as $f(x)$ rather than y. The graph of the function $f(x) = 3$ is the same as the graph of the horizontal line $y = 3$. See Fig. 8.2. The domain is the set of all real numbers. We can see from the graph that the only y-coordinate used is 3. So the range is {3}.   ◄

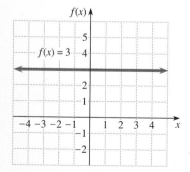

Figure 8.2

EXAMPLE 2 Graph the function $f(x) = 3x - 4$, and state the domain and range.

Solution   The *y*-intercept is $(0, -4)$, and the slope of the line is 3. The slope and intercept determine the position of the line, just as they did in Chapter 7. See Fig. 8.3. The domain is R. We can see from the graph that every real number is used as a *y*-coordinate. So the range is also R.   ◄

Absolute-value Functions

The concept of absolute value can be used to define a function.

DEFINITION
Absolute-value Function

> The function
> $$f(x) = |x|$$
> is called the **absolute-value function.**

EXAMPLE 3   Graph the function $f(x) = |x|$, and state the domain and range.

Solution   To graph this function, we find points that satisfy the equation $f(x) = |x|$. Fig. 8.4 shows a table of values for *x* and $f(x)$, along with a graph of these points. Since the domain of $f(x)$ is the set of real numbers, we use both positive and negative values for *x*. If we plotted all of the points satisfying $f(x) = |x|$, we would get the V-shaped graph shown in Fig. 8.5. The domain is R, but because the graph does not go below the *x*-axis, the range is the set of nonnegative real numbers.   ◄

Many functions involving absolute value have graphs that are V-shaped, as in Fig. 8.5. To graph other functions involving absolute value, we must choose points that determine the correct location of the V-shaped graph.

EXAMPLE 4   Graph each function, and state the domain and range.

a) $f(x) = |x| - 2$ **b)** $g(x) = |2x - 6|$

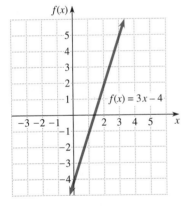

Figure 8.3

x	f(x)
−3	3
−2	2
−1	1
0	0
1	1
2	2
3	3

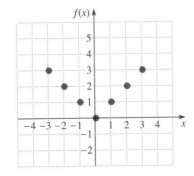

Figure 8.4

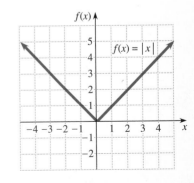

Figure 8.5

x	f(x)
-3	1
-2	0
-1	-1
0	-2
1	-1
2	0
3	1

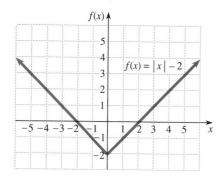

Figure 8.6

x	g(x)
-1	8
0	6
1	4
2	2
3	0
4	2
5	4

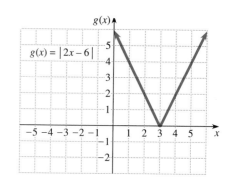

Figure 8.7

Solution

a) Choose values for x and find $f(x)$. A table of values and the graph are shown in Fig. 8.6. The domain is R, and the range is $\{y \mid y \geq -2\}$.

b) Make a table of values for x and $g(x)$, then draw the graph as shown in Fig. 8.7. The domain is R, and the range is $\{y \mid y \geq 0\}$. ◄

Quadratic Functions

The general quadratic polynomial, $ax^2 + bx + c$, can be used to define a function.

DEFINITION
Quadratic Function

A function of the form

$$f(x) = ax^2 + bx + c,$$

where a, b, and c are real numbers with $a \neq 0$, is called a **quadratic function.**

Without the term ax^2, this function would be a linear function. That is why we specify $a \neq 0$. The next example illustrates the graph of the simplest quadratic function.

EXAMPLE 5   Graph the function $h(x) = x^2$, and state the domain and range.

Solution   We must plot enough points to get the correct shape of the graph. See Fig. 8.8 for the table of values and the graph. The domain is R, and the range is $\{y \mid y \geq 0\}$. ◄

EXAMPLE 6   Graph the function $g(x) = 4 - x^2$, and state the domain and range.

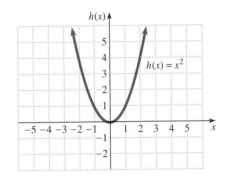

x	$h(x)$
-2	4
-1	1
0	0
1	1
2	4
$\pm\frac{1}{2}$	$\frac{1}{4}$

$h(x) = x^2$

$4 = (-2)^2$

$0 = 0^2$

Figure 8.8

Solution   We plot enough points to get the correct shape of the graph. See Fig. 8.9 for the table of values and the graph. The domain is R, and the range is $\{y \mid y \le 4\}$.

◀

   The graph of any quadratic function is called a **parabola.** All parabolas are similar to the ones in Figs. 8.8 and 8.9. The key point on a parabola is the point where the curve reaches its lowest point (Fig. 8.8) or its highest point (Fig. 8.9). This point is called the **vertex.** The y-coordinate of the vertex is called the **maximum** or **minimum value** of the function, depending on whether it is the highest or lowest point on the parabola. For $h(x) = x^2$ the vertex is (0, 0), and 0 is the minimum value of the function. For $g(x) = 4 - x^2$ the vertex is (0, 4), and 4 is the maximum value of the function. It will be shown in Chapter 11 that for the general quadratic function, the x-coordinate of the vertex is $-b/(2a)$.

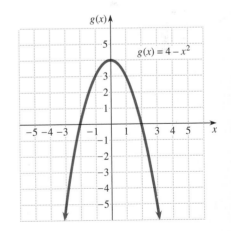

x	$g(x)$
-2	0
-1	3
0	4
1	3
2	0

$g(x) = 4 - x^2$

$0 = 4 - (-2)^2$

$4 = 4 - (0)^2$

$0 = 4 - (2)^2$

Figure 8.9

Vertex of a Parabola

For the parabola determined by the function
$$f(x) = ax^2 + bx + c \qquad (a \neq 0),$$
the x-coordinate of the vertex is $-b/(2a)$.

When we graph a parabola we should always locate the vertex. With the vertex and several nearby points, we can see the correct shape of the parabola.

EXAMPLE 7   Graph the function $f(x) = -x^2 - x + 2$, and state the domain and range.

Solution   First find the x-coordinate of the vertex.

$$x = \frac{-b}{2a} = \frac{-(-1)}{2(-1)} = \frac{1}{-2} = -\frac{1}{2}$$

Next calculate $f(x)$ for $x = -1/2$, and then find a few points on either side of the vertex. See Fig. 8.10. The domain is R. Because the graph goes no higher than 9/4, the range is $\{y \mid y \leq 9/4\}$.   ◀

Square-root Functions

EXAMPLE 8   Graph the equation $y = \sqrt{x}$, and state the domain and range.

Solution   The graph of the equation $y = \sqrt{x}$ and the graph of the function $f(x) = \sqrt{x}$ are the same. Because $\sqrt{x}$ is only defined for $x \geq 0$, the domain of this function is the set of nonnegative real numbers. Figure 8.11 shows a table of values and

x	$f(x)$
-2	0
-1	2
$-\frac{1}{2}$	$\frac{9}{4}$
0	2
1	0

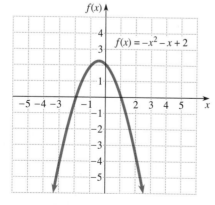

Figure 8.10

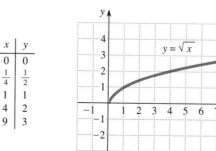

x	y
0	0
$\frac{1}{4}$	$\frac{1}{2}$
1	1
4	2
9	3

Figure 8.11

the graph. Note that x is chosen from the nonnegative numbers. The domain is $\{x \mid x \geq 0\}$, and the range is $\{y \mid y \geq 0\}$. ◀

There are infinitely many points on each of the graphs in the examples just presented. It is impossible to plot every point. We try to plot enough points to see the general shape of the graph. Even graphs drawn by computers are done by plotting only a finite number of points. We can get the correct graph of a function while plotting only a few points if we recall the shapes of the graphs of similar functions.

Graphs of Relations

EXAMPLE 9 Graph the relation $x = y^2$, and state the domain and range.

Solution We are to graph the set $\{(x, y) \mid x = y^2\}$. This relation is not a function. Figure 8.12 shows a table of values and the graph. The domain is $\{x \mid x \geq 0\}$, and the range is R. ◀

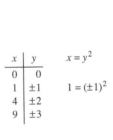

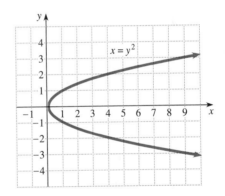

x	y
0	0
1	±1
4	±2
9	±3

$x = y^2$

$1 = (\pm 1)^2$

Figure 8.12

Vertical-line Test

The equation $x = y^2$ from Example 9 does not define a function, because for each positive x there are two values of y. If a vertical line crosses a graph twice, then we must have two points with the same x-coordinate and different y-coordinates. This means that the graph is not the graph of a function.

Vertical-line Test

If the graph of a relation can be crossed twice by a vertical line, then the graph is not the graph of a function.

EXAMPLE 10 Which of the graphs in Fig. 8.13 are graphs of functions?

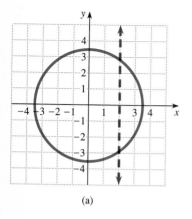

(a)

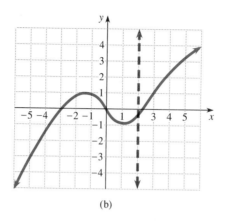

(b)

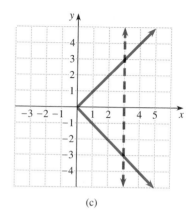

(c)

Figure 8.13

Solution Neither (*a*) nor (*c*) is the graph of a function, since we can draw vertical lines that cross these graphs twice. Graph (*b*) is the graph of a function, because no vertical line crosses this graph twice.   ◄

Applications

The maximum or minimum value of a function is important in applications. Many applications involve functions that determine quantities such as profit or cost. We may be interested in the maximum profit or the minimum cost. Maximums and minimums of functions are studied extensively in calculus. In the next example we will find the maximum altitude reached by a projectile.

EXAMPLE 11 A bullet is fired upward from the surface of the earth with an initial velocity of 96 feet per second. The altitude of the bullet in feet for any time, *t*, in seconds is given by the function $h(t) = -16t^2 + 96t$. Graph this function for $0 \leq t \leq 6$, and determine the maximum height reached by the bullet.

Solution The function $h(t)$ is a quadratic function. The *t*-coordinate of the vertex of the parabola is

$$t = \frac{-b}{2a} = \frac{-96}{2(-16)} = 3.$$

Because the maximum or minimum value of any quadratic function is reached at the vertex, the maximum value of this function is the second coordinate of the vertex.

$$h(3) = -16 \cdot 3^2 + 96 \cdot 3 = 144$$

The maximum height reached by the bullet is 144 feet. A table of values and the graph of this function are shown in Fig. 8.14.   ◄

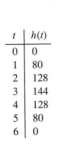

t	$h(t)$
0	0
1	80
2	128
3	144
4	128
5	80
6	0

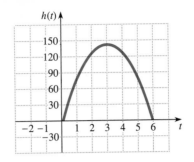

Figure 8.14

Warm-ups

True or false?

1. Any function can be graphed by calculating ordered pairs of the function and plotting them.

2. The graph of every linear function is a straight line.

3. We should look for a V-shaped graph when graphing an absolute-value function.

4. The domain of $f(x) = 1/x$ is R.

5. The vertex of a parabola is determined by $x = b/(2a)$.

6. The range of any quadratic function is R.

7. The y-axis and the $f(x)$-axis are the same.

8. The vertical-line test allows us to determine at a glance whether a graph is the graph of a function.

9. The domain of $f(x) = \sqrt{x-1}$ is the set of real numbers greater than one.

10. The domain of any quadratic function is R.

8.2 EXERCISES

Graph each linear function, and state its domain and range. See Examples 1 and 2.

1. $f(x) = 2x - 1$

2. $g(x) = x + 2$

3. $g(x) = -\dfrac{1}{2}x + 2$

4. $h(x) = -\dfrac{2}{3}x + 5$

5. $h(x) = 3$

6. $f(x) = -4$

7. $y = \dfrac{2}{3}x - 1$

8. $y = -\dfrac{3}{4}x + 4$

9. $y = -.3x + 6.5$

10. $y = .25x - .5$

Graph each absolute-value function, and state its domain and range. See Examples 3 and 4.

11. $h(x) = |x + 1|$

12. $f(x) = |x - 2|$

13. $g(x) = |3x|$

14. $g(x) = |x| - 3$

15. $f(x) = |2x - 1|$

16. $y = |2x - 3|$

17. $y = |x| + 1$

18. $f(x) = |x - 2| + 1$

19. $y = |x - 1| + 2$

20. $h(x) = |-2x|$

Graph each quadratic function, and state its domain and range. See Examples 5–7.

21. $f(x) = x^2 + 2$

22. $f(x) = 2x^2$

23. $g(x) = x^2 - x - 2$

24. $g(x) = -2x^2 + 3$

25. $h(x) = x^2 - 2x - 8$

26. $y = x^2 + 2x - 3$

27. $y = -x^2 + 2x$

28. $h(x) = -2x^2 - 3x + 2$

29. $y = -x^2 + 4x - 3$

30. $y = 9 - x^2$

Graph each square-root function, and state its domain and range. See Example 8.

31. $f(x) = \sqrt{x - 1}$

32. $f(x) = \sqrt{x + 1}$

33. $g(x) = 2\sqrt{x}$

34. $g(x) = \sqrt{x} - 1$

35. $h(x) = -\sqrt{x}$

36. $h(x) = -\sqrt{x - 1}$

37. $y = -\sqrt{x} + 2$

38. $y = 2\sqrt{x} + 1$

Graph each relation, and state its domain and range. See Example 9.

39. $x = |y|$

40. $x = -|y|$

41. $x = -y^2$

42. $x = 1 - y^2$

43. $x + 9 = y^2$

44. $x = |y + 1|$

45. $x = \sqrt{y}$

46. $x = (y - 1)^2$

47. $x = (y + 2)^2$

48. $x = -\sqrt{y}$

49. $x = 5$

50. $x = -3$

Use the vertical-line test to determine which of the graphs are graphs of functions. See Example 10.

51.

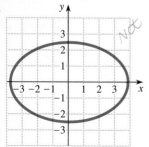

52.

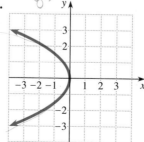

53.

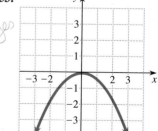

NO

54.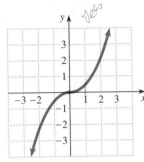
yes

55.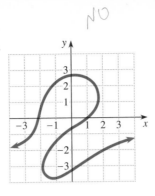

56.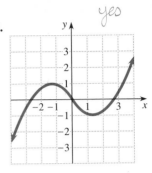
yes

Solve each problem. See Example 11.

57. If an object is projected upward with an initial velocity of 64 feet per second, then its height is a function of time, given by $h(t) = -16t^2 + 64t$. Graph this function for $0 \leq t \leq 4$. What is the maximum height reached by this object?

58. If a soccer ball is kicked straight up with an initial velocity of 32 feet per second, then its height above the earth is a function of time given by $h(t) = -16t^2 + 32t$. Graph this function for $0 \leq t \leq 2$. What is the maximum height reached by this ball?

59. The sheriff has determined that the cost in dollars per day per prisoner for housing prisoners in the Macon County Jail is a function of the number of prisoners, n. If this function is given by $C(n) = .015n^2 - 4.5n + 400$, then what is the minimum cost per prisoner? Graph this function for $1 \leq n \leq 200$.

60. A horticulturist has determined that the number of inches a young redwood tree grows in one year is a function of the annual rainfall, r, given by $g(r) = -.02r^2 + r + 1$. Graph this function for $0 \leq r \leq 50$. What is the maximum number of inches a young redwood can grow in a year?

Graph each function, and state the domain and range.

61. $f(x) = 1 - |x|$

62. $h(x) = \sqrt{x - 3}$

63. $y = (x - 3)^2$

64. $y = -x^3$

65. $g(x) = \sqrt{x^2}$

66. $f(x) = -2x + 4$

67. $y = x^3$

68. $y = 2|x|$

69. $h(x) = 1/x$

70. $y = -1/x$

71. $y = 3x - 5$

72. $g(x) = (x + 2)^2$

8.3 Combining Functions

IN THIS SECTION:

- Basic Operations with Functions
- Composition

In Sections 8.1 and 8.2 we learned the fundamental ideas about functions and their graphs. In this section we will learn how to combine functions to obtain new functions.

Basic Operations with Functions

A manufacturer of fashionable dog dishes can sell all of the dishes he produces. A financial analyst has determined that if x dishes are produced, the total cost in dollars is determined by the function

$$C(x) = 30x + 500.$$

If x dishes are sold, the total revenue in dollars is determined by the function

$$R(x) = 50x.$$

Because profit equals revenue minus cost, we can find a function for determining the profit by subtracting the functions for cost and revenue:

$$P(x) = R(x) - C(x)$$
$$= 50x - (30x + 500)$$
$$= 20x - 500$$

The function $P(x) = 20x - 500$ expresses the profit as a function of x.

In general, we define the sum, difference, product, and quotient of two functions as follows.

DEFINITION
Sum, Difference, Product, and Quotient Functions

If f and g are functions, the functions $f + g, f - g, f \cdot g$, and f/g are defined as follows:

$$(f + g)(x) = f(x) + g(x)$$
$$(f - g)(x) = f(x) - g(x)$$
$$(f \cdot g)(x) = f(x) \cdot g(x)$$
$$(f/g)(x) = f(x)/g(x) \qquad \text{provided } g(x) \neq 0$$

The domain of the function $f + g, f - g, f \cdot g$, or f/g is the set of numbers for which both f and g are defined. Thus, the domain is the intersection of the domain of f with the domain of g.

EXAMPLE 1   Let $f(x) = 4x - 12$ and $g(x) = x - 3$. Find the following.

a) $(f + g)(x)$ **b)** $(f - g)(x)$

c) $(f \cdot g)(x)$ **d)** $(f/g)(x)$

Solution

a) $(f + g)(x) = f(x) + g(x)$
$$= 4x - 12 + x - 3$$
$$= 5x - 15$$

b) $(f - g)(x) = f(x) - g(x) = 4x - 12 - (x - 3) = 3x - 9$

c) $(f \cdot g)(x) = f(x) \cdot g(x) = (4x - 12)(x - 3) = 4x^2 - 24x + 36$

d) $(f/g)(x) = f(x)/g(x) = \dfrac{4x - 12}{x - 3} = \dfrac{4(x - 3)}{x - 3} = 4$ ◄

EXAMPLE 2   Let $f(x) = 4x - 12$ and $g(x) = x - 3$. Find $(f + g)(2)$.

Solution   In Example 1(a) we found a general formula for the function $f + g$, namely, $(f + g)(x) = 5x - 15$. If we replace x by 2, we get

$$(f + g)(2) = 5(2) - 15 = -5.$$

We can also find $(f + g)(2)$ by evaluating each function separately and then adding the results. Because $f(2) = -4$ and $g(2) = -1$, we get

$$(f + g)(2) = f(2) + g(2) = -4 + (-1) = -5.$$ ◄

Composition

A salesperson's monthly salary is a function of the number of cars she sells: $1000 plus $50 for each car sold. If we let S be her salary and n be the number of cars sold, then S is a function of n:

$$S = \$1000 + \$50n$$

Each month the dealer contributes $100 plus 5% of her salary to a profit-sharing plan. If P represents the amount put into profit sharing, then P is a function of S:

$$P = \$100 + .05S$$

Now P is a function of S, and S is a function of n. Is P a function of n? The value of n certainly determines the value of P. In fact, we can easily write a formula for P in terms of n by substituting one formula into the other:

$$P = 100 + .05S$$
$$= 100 + .05(1000 + 50n) \qquad \text{Substitute } S = 1000 + 50n.$$
$$= 100 + 50 + 2.5n$$
$$= 150 + 2.5n$$

We have written P as a function of n, bypassing S. We call this idea **composition of functions.**

EXAMPLE 3   If $y = x^2 - 2x + 3$ and $z = 2y - 5$, write z as a function of x.

Solution

$$z = 2y - 5$$
$$= 2(x^2 - 2x + 3) - 5 \qquad \text{Replace } y \text{ by } x^2 - 2x + 3.$$
$$= 2x^2 - 4x + 1$$

The equation $z = 2x^2 - 4x + 1$ expresses z as a function of x.   ◀

We can identify the functions used in the salary and profit-sharing example above as follows:

$$S = g(n) = 1000 + 50n \qquad \text{Salary is function of the number of cars sold.}$$
$$P = f(S) = 100 + .05S \qquad \text{Profit sharing is a function of salary.}$$

Using this function notation, the composition of f and g is written as $f{\circ}g$. Read "$f{\circ}g$" as "f composite g." The function $f{\circ}g$ is defined by the equation

$$(f{\circ}g)(n) = f[g(n)].$$

To find a formula for $(f{\circ}g)(n)$, we proceed as follows:

$$(f{\circ}g)(n) = f[g(n)]$$
$$= f(1000 + 50n) \qquad \text{Because } g(n) = 1000 + 50n$$
$$= 100 + .05(1000 + 50n) \qquad \text{Replace } S \text{ by } 1000 + 50n.$$
$$= 150 + 2.5n$$

Thus, $(f{\circ}g)(n) = 150 + 2.5n$.

DEFINITION
Composition of Functions

If $f(x)$ and $g(x)$ are two functions, the **composition** of f and g, written $f{\circ}g$, is defined by the equation

$$(f{\circ}g)(x) = f[g(x)],$$

provided $g(x)$ is in the domain of f. The composition of g and f is written as $g{\circ}f$.

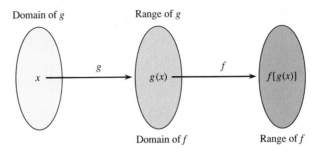

Figure 8.15

The diagram shown in Fig. 8.15 is helpful in understanding composition of functions.

EXAMPLE 4 Let $f(x) = 3x - 2$ and $g(x) = x^2 + 2x$. Find the following.

a) $(g \circ f)(2)$ b) $(f \circ g)(2)$ c) $(g \circ f)(x)$ d) $(f \circ g)(x)$

Solution

a) $(g \circ f)(2) = g[f(2)] = g[3(2) - 2] = g[4] = 4^2 + 2 \cdot 4 = 24$
So $(g \circ f)(2) = 24$.

b) $(f \circ g)(2) = f[g(2)] = f[8] = 22$
Thus, $(f \circ g)(2) = 22$.

c) $(g \circ f)(x) = g[f(x)] = g[3x - 2] = (3x - 2)^2 + 2(3x - 2)$
$$= 9x^2 - 6x$$
Thus, $(g \circ f)(x) = 9x^2 - 6x$.

d) $(f \circ g)(x) = f[g(x)] = f[x^2 + 2x] = 3(x^2 + 2x) - 2$
$$= 3x^2 + 6x - 2$$
Thus, $(f \circ g)(x) = 3x^2 + 6x - 2$.   ◄

It is often helpful to consider a complicated function as a composition of simpler functions. For example, the function $Q(x) = (x - 3)^2$ consists of two operations, subtracting 3 and squaring. So Q can be described as a composition of the functions $f(x) = x - 3$ and $g(x) = x^2$. To check this, we find $(g \circ f)(x)$:

$$(g \circ f)(x) = g[f(x)] = g(x - 3) = (x - 3)^2$$

Thus, $Q = g \circ f$.

EXAMPLE 5 Let $f(x) = x - 2$, $g(x) = 3x$, and $h(x) = \sqrt{x}$. Write each of the following functions as a composition, using f, g, and h.

a) $F(x) = \sqrt{x - 2}$ b) $H(x) = x - 4$ c) $K(x) = 3x - 6$

Solution

a) Because $F(x) = \sqrt{x - 2} = \sqrt{f(x)} = h[f(x)] = (h \circ f)(x)$, we have $F = h \circ f$.

b) Because $H(x) = x - 2 - 2 = f(x) - 2 = f[f(x)] = (f \circ f)(x)$, we have $H = f \circ f$.

c) Because $K(x) = 3(x - 2) = 3 \cdot f(x) = g[f(x)] = (g \circ f)(x)$, we have $K = g \circ f$. ◄

In Section 8.4 we will study pairs of functions f and g where

$$(f \circ g)(x) = x \quad \text{and} \quad (g \circ f)(x) = x.$$

Notice that we start with x, and after applying both functions, we end up with x. One function reverses what the other function does.

EXAMPLE 6   Show that $(f \circ g)(x) = x$ for the following pairs of functions.

a) $f(x) = 2x - 1$ and $g(x) = \dfrac{x + 1}{2}$

b) $f(x) = x^3 + 1$ and $g(x) = (x - 1)^{1/3}$

Solution

a) $(f \circ g)(x) = f[g(x)] = f\left(\dfrac{x + 1}{2}\right)$

$$= 2\left(\dfrac{x + 1}{2}\right) - 1$$

$$= x + 1 - 1$$

$$= x$$

b) $(f \circ g)(x) = f[g(x)] = f((x - 1)^{1/3})$

$$= ((x - 1)^{1/3})^3 + 1$$

$$= x - 1 + 1$$

$$= x$$ ◀

Warm-ups

True or false?

1. If $f(x) = x - 2$ and $g(x) = x + 3$, then $(f - g)(x) = -5$.

2. If $f(x) = x + 4$ and $g(x) = 3x$, then $(f/g)(2) = 1$.

3. The functions $f \circ g$ and $g \circ f$ are always the same.

4. If $f(x) = x^2$ and $g(x) = x + 2$, then $(f \circ g)(x) = x^2 + 2$.

5. The functions $f \circ g$ and $f \cdot g$ are always the same.

6. In finding the function $g \circ f$, it does not matter whether we use $g[f(x)]$ or $f[g(x)]$.

7. If $f(x) = 3x$ and $g(x) = x/3$, then $(f \circ g)(x) = x$.

8. If $a = 3b^2 - 7b$ and $c = a^2 + 3a$, then c is a function of b.

9. Complicated functions can sometimes be expressed as compositions of simpler functions.

10. If $F(x) = (x - 1)^2$, $h(x) = x - 1$, and $g(x) = x^2$, then $F = g \circ h$.

8.3 EXERCISES

Let $f(x) = 4x - 3$ and $g(x) = x^2 - 2x$. Find the following. See Examples 1 and 2.

1. $(f + g)(x)$ **2.** $(f - g)(x)$ **3.** $(f \cdot g)(x)$

4. $(f/g)(x)$ **5.** $(f + g)(3)$ **6.** $(f + g)(2)$

7. $(f - g)(-3)$ **8.** $(f - g)(-2)$ **9.** $(f \cdot g)(-1)$

10. $(f \cdot g)(-2)$ **11.** $(f/g)(4)$ **12.** $(f/g)(-2)$

Use the two functions given to write y as a function of x. See Example 3.

13. $y = 3a - 2, \quad a = 2x - 6$ **14.** $y = 2c + 3, \quad c = -3x + 4$

15. $y = 2d + 1, \quad d = (x + 1)/2$ **16.** $y = -3d + 2, \quad d = (2 - x)/3$

17. $y = m^2 - 1, \quad m = x + 1$ **18.** $y = \sqrt{n - 1}, \quad n = x + 2$

19. $y = \dfrac{a - 3}{a + 2}, \quad a = \dfrac{2x + 3}{1 - x}$ **20.** $y = \dfrac{w + 2}{w - 5}, \quad w = \dfrac{5x + 2}{x - 1}$

Let $f(x) = 2x - 3$, $g(x) = x^2 + 3x$, and $h(x) = (x + 3)/2$. Find the following. See Example 4.

21. $(g \circ f)(1)$ **22.** $(f \circ g)(-2)$ **23.** $(f \circ g)(1)$ **24.** $(g \circ f)(-2)$

25. $(f \circ f)(4)$ **26.** $(h \circ h)(3)$ **27.** $(h \circ f)(5)$ **28.** $(f \circ h)(0)$

29. $(f \circ h)(5)$ **30.** $(h \circ f)(0)$ **31.** $(g \circ h)(-1)$ **32.** $(h \circ g)(-1)$

33. $(f \circ g)(2.36)$ **34.** $(h \circ f)(23.761)$ **35.** $(g \circ f)(x)$ **36.** $(g \circ h)(x)$

37. $(f \circ g)(x)$ **38.** $(h \circ g)(x)$ **39.** $(f \circ f)(x)$ **40.** $(h \circ h)(x)$

41. $(h \circ f)(x)$ **42.** $(f \circ h)(x)$

Let $f(x) = \sqrt{x}$, $g(x) = x^2$, and $h(x) = x - 3$. Write each of the following functions as a composition using f, g, or h. See Example 5.

43. $F(x) = \sqrt{x - 3}$ **44.** $N(x) = \sqrt{x} - 3$ **45.** $G(x) = x^2 - 6x + 9$ **46.** $P(x) = x$

47. $H(x) = x^2 - 3$ **48.** $M(x) = x^{1/4}$ **49.** $J(x) = x - 6$ **50.** $R(x) = \sqrt{x^2 - 3}$

51. $K(x) = x^4$ **52.** $Q(x) = \sqrt{x^2 - 6x + 9}$

Show that $(f \circ g)(x) = x$ and $(g \circ f)(x) = x$ for each given pair of functions. See Example 6.

53. $f(x) = 3x + 5, \quad g(x) = (x - 5)/3$ **54.** $f(x) = 3x - 7, \quad g(x) = (x + 7)/3$

55. $f(x) = x^3 - 9, \quad g(x) = \sqrt[3]{x + 9}$ **56.** $f(x) = x^3 + 1, \quad g(x) = \sqrt[3]{x - 1}$

57. $f(x) = \dfrac{x - 1}{x + 1}, \quad g(x) = \dfrac{x + 1}{1 - x}$ **58.** $f(x) = \dfrac{x + 1}{x - 3}, \quad g(x) = \dfrac{3x + 1}{x - 1}$

59. $f(x) = \dfrac{1}{x}, \quad g(x) = \dfrac{1}{x}$ **60.** $f(x) = 2x^3, \quad g(x) = \left(\dfrac{x}{2}\right)^{1/3}$

Solve each problem.

61. A plastic-bag manufacturer has determined that he can sell as many bags as he can produce each month. If he produces x thousand bags in a month, his revenue is $R(x) = $ $x^2 - 10x + 30$ dollars, and his cost is $C(x) = 2x^2 - 30x + 200$ dollars. Use the fact that profit is revenue minus cost to write the profit as a function of x.

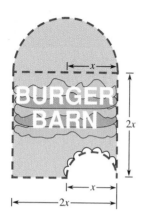

Figure for Exercise 62

62. A sign is in the shape of a square with a semicircle of radius x adjoining one side, and a semicircle of diameter x removed from the opposite side. If the sides of the square are length $2x$, then write the area of the sign as a function of x.

63. Suppose that the average family spends 25% of their income on food, $F = .25I$, and 10% of their food dollar on junk food, $J = .10F$. Write J as a function of I.

64. Write a formula for the area of a circle inscribed in a square of area M.

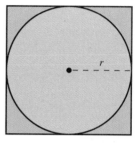

Area of square = M

Figure for Exercise 64

8.4   Inverse Functions

IN THIS SECTION:

- Inverse of a Function
- Inverse Functions Using *f*-Notation
- Finding the Inverse Function
- Procedure for Finding Inverses
- Graphs of *f* and f^{-1}

In Section 8.3 we introduced the idea of a pair of functions such that $(f \circ g)(x) = x$ and $(g \circ f)(x) = x$. Each function reverses what the other function does. In this section we will explore that idea further.

Inverse of a Function

A function is a set of ordered pairs where no two ordered pairs have the same first coordinate and different second coordinates. The set

$$f = \{(2, 6), (5, 15)\}$$

is a function. We use the letter f as a name for this set or function just as we use the letter f as a name for a function in the f-notation. The ordered pairs indicate how numbers in the domain $\{2, 5\}$ are paired with numbers in the range $\{6, 15\}$. The **inverse** of the function f, denoted f^{-1}, is a function whose ordered pairs are obtained from f by interchanging the x- and y-coordinates:

$$f^{-1} = \{(6, 2), (15, 5)\}$$

We read f^{-1} as "f inverse." Note that the -1 in f^{-1} is not an exponent. It does not mean $1/f$. The domain of f^{-1} is $\{6, 15\}$, and the range of f^{-1} is $\{2, 5\}$. The inverse function f^{-1} pairs numbers in the range of f with numbers in the domain of f. It reverses what the function f does.

If we interchange the x- and y-coordinates of the function $g = \{(3, 1), (4, 5), (6, 1)\}$, we get $\{(1, 3), (5, 4), (1, 6)\}$. This set of ordered pairs is not a function, since it contains ordered pairs with the same first coordinates and different second coordinates. So the function g does not have an inverse function. We say that the function f is **invertible** and the function g is not invertible. The reason g is not invertible is that the definition of function allows more than one number of the domain to be paired with the same number in the range. Of course when this pairing is reversed, the definition of function is violated.

DEFINITION
One-to-one Function

> If a function is such that no two ordered pairs have different x-coordinates and the same y-coordinate, then the function is called a **one-to-one** function.

In a one-to-one function each member of the domain corresponds to just one member of the range, and each member of the range corresponds to just one member of the domain. Functions that are one-to-one are invertible functions.

DEFINITION
Inverse Function

> The inverse of a one-to-one function f is the function f^{-1}, which is obtained from f by interchanging the coordinates in each ordered pair of f.

EXAMPLE 1 Determine whether each function is invertible. If it is invertible, then find the inverse function.
a) $f = \{(2, 4), (-2, 4), (3, 9)\}$
b) $g = \{(2, 1/2), (5, 1/5), (7, 1/7)\}$
c) $h = \{(3, 5), (7, 9)\}$

Solution

a) Because $(2, 4)$ and $(-2, 4)$ have the same y-coordinate, this function is not one-to-one, and it is not invertible.

b) This function is one-to-one, and therefore it is invertible.

$$g^{-1} = \{(1/2, 2), (1/5, 5), (1/7, 7)\}$$

c) This function is invertible, and $h^{-1} = \{(5, 3), (9, 7)\}$.   ◄

Inverse Functions Using f-Notation

Consider the function $f(x) = 3x$. Each y-coordinate is 3 times the x-coordinate. This function is one-to-one. If we reverse the coordinates to find the inverse function, then an ordered pair such as $(2, 6)$ becomes $(6, 2)$. In the inverse function the y-coordinate is one-third of the x-coordinate. So the inverse of $f(x) = 3x$ is $f^{-1}(x) = x/3$.

Consider the composition of $f(x) = 3x$ and $f^{-1}(x) = x/3$:

$$(f^{-1} \circ f)(x) = f^{-1}[f(x)] = f^{-1}(3x) = \frac{3x}{3} = x$$

$$(f \circ f^{-1})(x) = f[f^{-1}(x)] = f(x/3) = 3 \cdot \frac{x}{3} = x$$

Note that we started with x, and after applying both functions in either order, we ended up with x. The two functions are called inverses of each other because each one reverses the operation that the other performs. If two functions are written using f-notation, then the composition of the functions can be used to determine whether they are inverses of each other.

Identifying Inverse Functions

Two functions f and g are inverses of each other, if *both* of the following conditions are met.

1. $(g \circ f)(x) = x$ for every number x in the domain of f.
2. $(f \circ g)(x) = x$ for every number x in the domain of g.

EXAMPLE 2 Determine whether the functions f and g are inverses of each other.

a) $f(x) = 2x - 1$, $g(x) = \frac{1}{2}x + 1$ **b)** $f(x) = x - 5$, $g(x) = x + 5$

c) $f(x) = x^2$, $g(x) = \sqrt{x}$

Solution

a) Find the composition of g and f:

$$(g \circ f)(x) = g[f(x)] = g(2x - 1) = \frac{1}{2}(2x - 1) + 1 = x + \frac{1}{2}$$

So f and g are not inverses of each other.

b) Find $g \circ f$ and $f \circ g$:

$$(g \circ f)(x) = g[\,f(x)] = g(x - 5) = x - 5 + 5 = x$$
$$(f \circ g)(x) = f[\,g(x)] = f(x + 5) = x + 5 - 5 = x$$

Because each of these equations is true for any real number x, f and g are inverses of each other. We may write $g = f^{-1}$ or $f^{-1}(x) = x + 5$.

c) If x is any real number, we can write

$$(g \circ f)(x) = g[\,f(x)] = g(x^2) = \sqrt{x^2} = |x|.$$

The domain of f is R, and $|x| \neq x$ if x is negative. So g and f are not inverses of each other. Note that $f(x) = x^2$ is not a one-to-one function, because both $(3, 9)$ and $(-3, 9)$ are ordered pairs of this function. Thus $f(x) = x^2$ does not have an inverse. ◄

Finding the Inverse Function

The function $g(x) = \sqrt{x}$ is a one-to-one function, and therefore it has an inverse function. To reverse the operation of square root, we use squaring. The inverse function is

$$g^{-1}(x) = x^2 \text{ for } x \geq 0.$$

Note that by restricting the domain of g^{-1} to the nonnegative numbers, g^{-1} is one-to-one. With this restriction, it is true that $(g \circ g^{-1})(x) = x$ and $(g^{-1} \circ g)(x) = x$ for every nonnegative number x.

For functions involving one operation it is easy to find the inverse function. For example, the inverse of $f(x) = x^3$ (cubing) is $f^{-1}(x) = x^{1/3}$ (the cube root). The inverse of $h(x) = 6x$ is $h^{-1}(x) = x/6$. The inverse of $k(x) = x - 9$ is $k^{-1}(x) = x + 9$.

If a function involves two operations, the inverse function undoes those operations in the opposite order that the function does them. Consider the function $g(x) = 3x - 5$. Because $g(4) = 7$, this function pairs 4 with 7. The function g^{-1} should pair 7 with 4. See Fig. 8.16. To calculate $g(4)$, we multiply 4 by 3 and then subtract 5. To reverse these operations, we must add 5 and then divide by 3. We undo these two operations by reversing the last operation first.

$$g^{-1}(7) = \frac{7 + 5}{3} = 4$$

Thus, the inverse of g should be

$$g^{-1}(x) = \frac{x + 5}{3}.$$

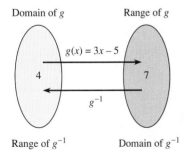

Domain of g Range of g

$g(x) = 3x - 5$

4   7

g^{-1}

Range of g^{-1} Domain of g^{-1}

Figure 8.16

To see that g^{-1} really does reverse g for any value of x, we must calculate $(g^{-1} \circ g)(x)$ and $(g \circ g^{-1})(x)$.

$$(g^{-1} \circ g)(x) = g^{-1}[g(x)] = g^{-1}[3x - 5] = \frac{3x - 5 + 5}{3} = \frac{3x}{3} = x$$

A similar computation shows that $(g \circ g^{-1})(x) = x$. Try it.

EXAMPLE 3 Determine whether each function has an inverse. If it does, then find the inverse.

a) $f(x) = x^5$ **b)** $g(x) = |x|$ **c)** $h(x) = 2x + 1$

Solution

a) The fifth root reverses the fifth power, so $f^{-1}(x) = \sqrt[5]{x}$. Check by finding $(f^{-1} \circ f)(x)$ and $(f \circ f^{-1})(x)$.

$$(f^{-1} \circ f)(x) = f^{-1}(x^5) = \sqrt[5]{x^5} = x$$
$$(f \circ f^{-1})(x) = f(\sqrt[5]{x}) = (\sqrt[5]{x})^5 = x$$

Since each equation is true for any real number, these functions are inverses of each other.

b) Because the ordered pairs $(2, 2)$ and $(-2, 2)$ belong to this function, the function is not one-to-one. The function does not have an inverse.

c) To reverse multiplication by 2 and addition by 1, we must subtract 1 and divide by 2. Thus

$$h^{-1}(x) = \frac{x - 1}{2}.$$

Note that dividing by 2 and then subtracting 1 does not invert the function h. Check. ◀

Procedure for Finding Inverses

There is a systematic procedure that we can use to find the formula for the inverse of a function. Because the inverse function interchanges the roles of x and y, that is what we do in the formula. For example, to find the inverse of $h(x) = 2x + 1$ or $y = 2x + 1$, we first interchange x and y.

$$y = 2x + 1$$
$$x = 2y + 1 \qquad \text{Interchange } x \text{ and } y.$$
$$x - 1 = 2y$$
$$\frac{x - 1}{2} = y \qquad \text{Solve for } y.$$
$$h^{-1}(x) = \frac{x - 1}{2} \qquad \text{Replace } y \text{ by } h^{-1}(x).$$

Note that this result is the same result as Example 1(c).

The next example shows a function whose inverse cannot be found without the systematic procedure.

EXAMPLE 4 If $f(x) = \dfrac{x + 1}{x - 3}$, find $f^{-1}(x)$.

Solution

$$y = \frac{x + 1}{x - 3} \qquad \text{Use } y \text{ in place of } f(x).$$

$$x = \frac{y + 1}{y - 3} \qquad \text{Switch } x \text{ and } y.$$

$$x(y - 3) = y + 1 \qquad \text{To solve for } y, \text{ first multiply each side by } y - 3.$$

$$xy - 3x = y + 1$$

$$xy - y = 3x + 1$$

$$y(x - 1) = 3x + 1 \qquad \text{Factor out } y.$$

$$y = \frac{3x + 1}{x - 1}$$

$$f^{-1}(x) = \frac{3x + 1}{x - 1} \qquad \text{Replace } y \text{ by } f^{-1}(x). \qquad \blacktriangleleft$$

The strategy for finding the inverse of a function $f(x)$ is summarized as follows.

STRATEGY
Finding f^{-1}

1. Replace $f(x)$ by y.	2. Interchange x and y.
3. Solve the equation for y.	4. Replace y by $f^{-1}(x)$.

Graphs of f and f^{-1}

Consider $f(x) = x^2$ for $x \geq 0$ and $f^{-1}(x) = \sqrt{x}$. Their graphs are shown in Fig. 8.17. Notice the symmetry. If we folded the paper along the line $y = x$, the two graphs would coincide.

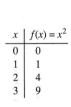

x	$f(x) = x^2$
0	0
1	1
2	4
3	9

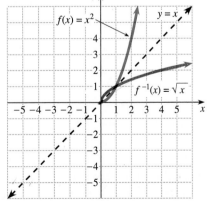

x	$f^{-1}(x) = \sqrt{x}$
0	0
1	1
4	2
9	3

Figure 8.17

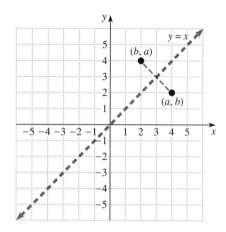

Figure 8.18

If a point (a, b) is on the graph of the function f, then (b, a) must be on the graph of f^{-1}. See Fig. 8.18. Points (b, a) and (a, b) lie on opposite sides of the diagonal line $y = x$ and are the same distance from it. For this reason, the graphs of f and f^{-1} are symmetric with respect to the line $y = x$.

EXAMPLE 5 Find the inverse of the function $f(x) = \sqrt{x - 1}$, and graph f and f^{-1} on the same pair of axes.

Solution To find f^{-1}, switch x and y in the formula $y = \sqrt{x - 1}$.

$$x = \sqrt{y - 1}$$
$$x^2 = y - 1 \qquad \text{Square both sides.}$$
$$x^2 + 1 = y$$

Because the range of f is the set of nonnegative real numbers, we must restrict the domain of f^{-1} to be the set of nonnegative real numbers. Thus $f^{-1}(x) = x^2 + 1$ for $x \geq 0$. The two graphs are shown in Fig. 8.19.

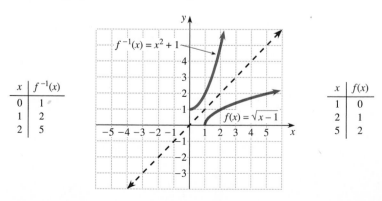

x	$f^{-1}(x)$
0	1
1	2
2	5

x	$f(x)$
1	0
2	1
5	2

Figure 8.19

Warm-ups

True or false?

1. The inverse of $\{(1, 3), (2, 5)\}$ is $\{(3, 1), (2, 5)\}$
2. The function $f(x) = 3$ is a one-to-one function.
3. If $g(x) = 2x$, then $g^{-1}(x) = 1/(2x)$.
4. Only one-to-one functions are invertible.
5. The domain of g is the same as the range of g^{-1}.
6. The function $f(x) = x^4$ is invertible.
7. If $f(x) = -x$, then $f^{-1}(x) = -x$.
8. If h is invertible and $h(7) = -95$, then $h^{-1}(-95) = 7$.
9. If $k(x) = 3x - 6$, then $k^{-1}(x) = (1/3)x + 2$.
10. If $f(x) = 3x - 4$, then $f^{-1}(x) = \dfrac{x}{3} + 4$.

8.4 EXERCISES

Determine whether each function is invertible. If it is invertible, then find the inverse. See Example 1.

1. $\{(-3, 3), (-2, 2), (0, 0), (2, 2)\}$
2. $\{(1, 1), (2, 8), (3, 27)\}$
3. $\{(16, 4), (9, 3), (0, 0)\}$
4. $\{(-1, 1), (-3, 81), (3, 81)\}$
5. $\{(0, 5), (5, 0), (6, 0)\}$
6. $\{(3, -3), (-2, 2), (1, -1)\}$
7. $\{(0,0)\ (2, 2), (9, 9)\}$
8. $\{(9, 1), (2, 1), (7, 1), (0, 1)\}$

Determine whether each pair of functions f and g are inverses of each other. See Example 2.

9. $f(x) = 2x,\quad g(x) = .5x$
10. $f(x) = 3x,\quad g(x) = .33x$
11. $f(x) = x^4,\quad g(x) = x^{1/4}$
12. $f(x) = |2x|,\quad g(x) = |x/2|$
13. $f(x) = 2x - 10,\quad g(x) = \dfrac{1}{2}x + 5$
14. $f(x) = 3x + 7,\quad g(x) = \dfrac{x - 7}{3}$
15. $f(x) = -x,\quad g(x) = -x$
16. $f(x) = \dfrac{1}{x},\quad g(x) = \dfrac{1}{x}$

Find the inverse mentally for each function that has an inverse. Check the compositions. See Example 3.

17. $f(x) = 5x$
18. $h(x) = -3x$
19. $g(x) = x - 9$
20. $j(x) = x + 7$
21. $k(x) = 5x - 9$
22. $r(x) = 2x - 8$
23. $m(x) = \dfrac{2}{x}$
24. $s(x) = \dfrac{-1}{x}$
25. $n(x) = x$
26. $t(x) = -5$
27. $p(x) = \sqrt[4]{x}$
28. $v(x) = \sqrt[6]{x}$
29. $u(x) = -x + 1$
30. $w(x) = -x + 6$

Determine f^{-1} for each function by using the procedure shown in Example 4.

31. $f(x) = 3x - 9$ **32.** $f(x) = -3x + 10$ **33.** $f(x) = \sqrt[3]{x - 4}$

34. $f(x) = \sqrt[3]{x + 2}$ **35.** $f(x) = \sqrt[3]{3x + 7}$ **36.** $f(x) = \sqrt[3]{7 - 5x}$

37. $f(x) = \dfrac{x + 1}{x - 2}$ **38.** $f(x) = \dfrac{1 - x}{x + 3}$ **39.** $f(x) = \dfrac{6}{x}$

40. $f(x) = \dfrac{-3}{x}$ **41.** $f(x) = \dfrac{3}{x - 4}$ **42.** $f(x) = \dfrac{2}{x + 1}$

43. $f(x) = x^2 + 3$ for $x \geq 0$ **44.** $f(x) = x^2 - 5$ for $x \geq 0$

45. $f(x) = \sqrt{x + 2}$ **46.** $f(x) = \sqrt{x - 4}$

Find the inverse of each function, and graph f and f^{-1} on the same pair of axes. See Example 5.

47. $f(x) = 2x + 3$ **48.** $f(x) = -3x + 2$

49. $f(x) = x^2 - 1$ for $x \geq 0$ **50.** $f(x) = x^2 + 3$ for $x \geq 0$

51. $f(x) = 5x$ **52.** $f(x) = \dfrac{x}{4}$

53. $f(x) = x^3$ **54.** $f(x) = 2x^3$

55. $f(x) = \sqrt{x - 2}$ **56.** $f(x) = \sqrt{x + 3}$

For each pair of functions, find $(f^{-1} \circ f)(x)$.

57. $f(x) = x^3 - 1$, $f^{-1}(x) = \sqrt[3]{x + 1}$ **58.** $f(x) = 2x^3 + 1$, $f^{-1}(x) = \sqrt[3]{\dfrac{x - 1}{2}}$

59. $f(x) = \dfrac{1}{2}x - 3$, $f^{-1}(x) = 2x + 6$ **60.** $f(x) = 3x - 9$, $f^{-1}(x) = \dfrac{1}{3}x + 3$

61. $f(x) = \dfrac{1}{x} + 2$, $f^{-1}(x) = \dfrac{1}{x - 2}$ **62.** $f(x) = 4 - \dfrac{1}{x}$, $f^{-1}(x) = \dfrac{1}{4 - x}$

*63. $f(x) = \dfrac{x + 1}{x - 2}$, $f^{-1}(x) = \dfrac{2x + 1}{x - 1}$ ***64.** $f(x) = \dfrac{3x - 2}{x + 2}$, $f^{-1}(x) = \dfrac{2x + 2}{3 - x}$

8.5 Variation

IN THIS SECTION:

- Direct Variation
- Finding the Constant
- Inverse Variation
- Joint Variation
- More Variation

If $y = 3x$, then as x varies so does y. Certain functions are customarily expressed in terms of variation. In this section we will learn to write formulas for those functions from verbal descriptions of the functions.

Direct Variation

In a community with an 8% sales-tax rate, the amount of tax, t (in dollars), is a function of the amount of the purchase, a (in dollars). This function is expressed by the formula

$$t = .08a.$$

If the amount increases, then the tax increases. If a decreases, then t decreases. In this situation we say that t *varies directly with* a, or t *is directly proportional to* a. The constant tax rate, .08, is called the **variation constant,** or **proportionality constant.** Notice that t is just a simple linear function of a. We are merely introducing some new terms to an old idea.

DEFINITION
Direct Variation

The statement y **varies directly as** x, or y **is directly proportional to** x, means that

$$y = kx$$

for some constant, k. The constant, k, is a fixed nonzero real number.

Finding the Constant

EXAMPLE 1 Joyce is traveling by car, and the distance she travels, d, varies directly with the amount of time, t, that she drives. In 3 hours she drove 120 miles. Find the constant of variation, and write d as a function of t.

Solution Because d varies directly as t, we must have a constant, k, such that

$$d = kt.$$

Because $d = 120$ when $t = 3$, we can write

$$120 = k \cdot 3,$$
or
$$40 = k.$$

Thus, $d = 40t$.

EXAMPLE 2 In a downtown office building, the monthly rent for an office is directly proportional to the size of the office. If a 420-square-foot office rents for $1260 per month, then what is the rent for a 900-square-foot office?

Solution Because the rent, R, varies directly with the area of the office, A, we have

$$R = kA.$$

Because a 420-square-foot office rents for $1260, we can substitute to find k.

$$1260 = k \cdot 420$$
$$3 = k$$

Thus $R = 3A$. To get the rent for a 900-square-foot office, insert 900 into the formula.

$$R = 3 \cdot 900$$
$$R = \$2700 \qquad \blacktriangleleft$$

Inverse Variation

In making a 500-mile trip by car, the time it takes is a function of the speed of the car. The greater the speed, the less time it will take. If you decrease the speed, the time increases. We say that the time is *inversely proportional* to the speed. Using the formula $D = RT$, or $T = D/R$, we can write

$$T = \frac{500}{R}.$$

Math at Work

It is easy to tell the difference between having your face slapped and having it caressed, between loud and soft music, between bright and dim lights, between hot and cold air. But how do we tell when a pleasurable sensation changes into a painful one?

Our senses of sight, smell, sound, taste, and touch are constantly bombarded by stimuli. Psychologists have been interested in how our brain makes judgments about these perceptions. One question that has been widely studied is the ability of people to judge changes or differences in the stimulus to their senses.

A nineteenth-century psychologist named Ernst Weber discovered that the smallest detectable difference in stimulus energy, known as the just-noticeable-difference (JND), is directly proportional to the intensity of the stimulus. The constant, K (the Weber Constant), differs for each sense—smell, sight, touch, etc.—and for different types of sensation within those senses. Expressed algebraically,

$$\text{JND} = Kx,$$

where x is the intensity of the stimulus.

The value of K for weight is .02. For example, the amount of weight that could be added to a 25-pound package before the mail carrier detects a difference is $.02(25) = .5$ pounds. For a 100-pound package the JND is $.02(100) = 2$ pounds. If the Weber Constant for brightness is .017, then what is the JND for a 60-watt light bulb? If the Weber Constant for loudness is .1, then what is the JND for a radio playing at 50 decibels?

In general, we make the following definition.

DEFINITION
Inverse Variation

The statement **y varies inversely as x,** or **y is inversely proportional to x,** means that

$$y = \frac{k}{x}$$

for some nonzero constant, k.

Be sure to understand the difference between direct and inverse variation. If y varies *directly* as x, then *as x increases y increases.* If y varies *inversely* as x, then *as x increases y decreases.*

EXAMPLE 3 Suppose a is inversely proportional to b, and when $b = 5$, $a = 1/2$. Find a when $b = 12$.

Solution Because a is inversely proportional to b, we have

$$a = \frac{k}{b}$$

for some constant, k. Since $a = 1/2$ when $b = 5$, we can find k by substituting these values into the formula.

$$\frac{1}{2} = \frac{k}{5}$$

$$\frac{5}{2} = k \qquad \text{Solve for } k.$$

To find a when $b = 12$, we use the formula with k replaced by 5/2.

$$a = \frac{5/2}{b}$$

$$a = \frac{5/2}{12} = \frac{5}{24}$$ ◀

Joint Variation

On a deposit of $5000 in a savings account, the interest earned, I, depends on the rate, r, and the time, t. Assuming that the interest is simple interest, we can use the formula $I = Prt$ to write

$$I = 5000rt.$$

The variable I is a function of two independent variables, r and t. In this case, we say that I varies jointly as r and t.

DEFINITION
Joint Variation

The statement *y* **varies jointly as** *x* **and** *z,* or *y* **is jointly proportional to** *x* **and** *z* means that

$$y = kxz$$

for some nonzero constant, *k*.

EXAMPLE 4   Suppose that *y* varies jointly with *x* and *z*, and *y* = 12 when *x* = 5 and *z* = 2. Find *y* when *x* = 10 and *z* = −3.

Solution   Because *y* varies jointly with *x* and *z*, we can write

$$y = kxz$$

for some constant, *k*. Now use *y* = 12, *x* = 5, and *z* = 2.

$$12 = k \cdot 5 \cdot 2$$
$$12 = 10k$$
$$\frac{6}{5} = k$$

Now that we know the value of *k*, we can write

$$y = \frac{6}{5}xz.$$

To find *y* when *x* = 10 and *z* = −3, substitute into the equation.

$$y = \frac{6}{5}(10)(-3)$$
$$y = -36 \qquad \blacktriangleleft$$

More Variation

We frequently combine the ideas of direct, inverse, and joint variation with powers and roots. Study the examples contained in the following box.

More Variation Examples

Statement	*Formula*
y varies directly as the square root of *x*.	$y = k\sqrt{x}$
y is directly proportional to the cube of *x*.	$y = kx^3$
y is inversely proportional to x^2.	$y = \dfrac{k}{x^2}$

y varies inversely as the square root of x.	$y = \dfrac{k}{\sqrt{x}}$
y varies jointly as x and the square of z.	$y = kxz^2$
y varies directly with x and inversely with the square root of z.	$y = \dfrac{kx}{\sqrt{z}}$

Notice that these variation terms *never* signify addition or subtraction. We always use multiplication, unless we see the word "inversely." Then we divide.

EXAMPLE 5   Suppose that m varies directly with a and inversely with b. If $m = 9$ when $a = 2$ and $b = 6$, then what is m when $a = 5$ and $b = 9$?

Solution   If m varies directly with a and inversely with b, then

$$m = \frac{ka}{b}.$$

Use $m = 9$, $a = 2$, and $b = 6$ to find k:

$$9 = \frac{k \cdot 2}{6}$$

$$9 = \frac{k}{3}$$

$$27 = k$$

Now use $k = 27$, $a = 5$, and $b = 9$ to find m:

$$m = \frac{27 \cdot 5}{9}$$

$$m = 15$$ ◄

Warm-ups

True or false?
1. If a varies directly as b, then $a = kb$.
2. If a is inversely proportional to b, then $a = bk$.
3. If a is jointly proportional to b and c, then $a = bc$.
4. If a is directly proportional to the square root of c, then $a = k\sqrt{c}$.
5. If b is directly proportional to a, then $b = ka^2$.
6. If a varies directly as b and inversely as c, then $a = \dfrac{kb}{c}$.
7. If a is jointly proportional to c and the square of b, then $a = \dfrac{kc}{b^2}$.

8. If a varies directly as c and inversely as the square root of b, then $a = \dfrac{kc}{\sqrt{b}}$.

9. If b is directly proportional to a and inversely proportional to the square of c, then $b = ka\sqrt{c}$.

10. If b varies inversely with the square of c, then $b = \dfrac{k}{c^2}$.

8.5 EXERCISES

Write a formula that expresses the relationship described by each statement. Use k as a constant of variation. See Examples 1–5.

1. a varies directly as m.

2. w varies directly with P.

3. d varies inversely with e.

4. y varies inversely with x.

5. I varies jointly as r and t.

6. q varies jointly as w and v.

7. m is directly proportional to the square of p.

8. g is directly proportional to the cube of r.

9. B is directly proportional to the cube root of w.

10. F is directly proportional to the square of m.

11. t is inversely proportional to the square of x.

12. y is inversely proportional to the square root of z.

13. v varies directly as m and inversely as n.

14. b varies directly as the square of n and inversely as the square root of v.

Find the variation constant, and write a formula that expresses the indicated variation. See Example 1.

15. y varies directly as x, and $y = 6$ when $x = 4$.

16. m varies directly as w, and $m = \dfrac{1}{3}$ when $w = \dfrac{1}{4}$.

17. A varies inversely as B, and $A = 10$ when $B = 3$.

18. c varies inversely as d, and $c = .31$ when $d = 2$.

19. m varies inversely as the square root of p, and $m = 12$ when $p = 9$.

20. s varies inversely as the square root of v, and $s = 6$ when $v = 3/2$.

21. A varies jointly as t and u, and $A = -6$ when $t = -5$ and $u = 3$.

22. N varies jointly as the square of p and the cube of q, and $N = 72$ when $p = 3$ and $q = 2$.

23. y varies directly as x and inversely as z, and $y = 2.37$ when $x = \pi$ and $z = \sqrt{2}$.

24. a varies directly as the square root of m and inversely as the square of n, and $a = 5.47$ when $m = 3$ and $n = 1.625$.

Solve each variation problem. See Examples 2–5.

25. If y varies directly as x, and $y = 7$ when $x = 5$, find y when $x = -3$.

26. If n varies directly as p, and $n = .6$ when $p = .2$, find n when $p = \sqrt{2}$.

27. If w varies inversely as z, and $w = 6$ when $z = 2$, find w when $z = -8$.

28. If p varies inversely as q and $p = 5$ when $q = \sqrt{3}$, find p when $q = 5$.

29. If A varies jointly as F and T, and $A = 6$ when $F = 3\sqrt{2}$ and $T = 4$, find A when $F = 2\sqrt{2}$ and $T = 1/2$.

30. If j varies jointly as the square of r and the cube of v, and $j = -3$ when $r = 2\sqrt{3}$ and $v = 1/2$, find j when $r = 3\sqrt{5}$ and $v = 2$.

31. If D varies directly with t and inversely with the square of s, and $D = 12.35$ when $t = 2.8$ and $s = 2.48$, find D when $t = 5.63$ and $s = 6.81$.

32. If M varies jointly with x and the square of v, and $M = 39.5$ when $x = \sqrt{10}$ and $v = 3.87$, find M when $x = \sqrt{30}$ and $v = 7.21$.

Solve each word problem.

33. At Larry's Lawn Service the cost of lawn maintenance varies directly with the size of the lawn. If the monthly maintenance on a 4000-square-foot lawn is $280, then what is the maintenance fee for a 6000-square-foot lawn?

34. The weight of a female iguana is directly proportional to her length. If a 4-foot female weighs 30 pounds, then how much should a 5-foot female weigh?

35. The volume of a gas in a cylinder at a fixed temperature is inversely proportional to the weight on the piston. If the gas has a volume of 6 cubic centimeters for a weight of 30 kilograms, then what would the volume be for a weight of 20 kilograms?

36. A software vendor sells a software package at a price inversely proportional to the number of packages sold per month. When they are selling 900 packages per month, the price is $80 each. If they sell 1000 packages per month, then what should the new price be?

37. The price of an aluminum culvert is jointly proportional to its radius and length. If a 12-foot culvert with a 6-inch radius costs $324, then what is the price of a 10-foot culvert with an 8-inch radius?

38. The cost of a piece of PVC water pipe varies jointly as its diameter and length. If a 20-foot pipe with a diameter of one inch costs $6.80, then what will be the cost of a 10-foot pipe with a 3/4-inch diameter?

39. The price of a steel rod varies jointly as the length and the square of the diameter. If an 18-foot rod with a 2-inch diameter costs $12.60, then what is the cost of a 12-foot rod with a 3-inch diameter?

40. The weight of a cylindrical can of pea soup varies jointly with the height and the square of the radius. If a 4-inch-high can with a 1.5-inch radius weighs 16 ounces, then what is the weight of a 5-inch-high can with a radius of 3 inches?

41. The distance an object falls in a vacuum varies directly as the square of the time it is falling. If an object falls .16 feet in .1 second, then how far does it fall in .5 second?

42. The cost of material used in making a Frisbee varies directly with the square of the diameter. If it costs the manufacturer $.45 for the material in a Frisbee with a 9-inch diameter, then what is the cost for the material in a 12-inch-diameter Frisbee?

43. The basic law of leverage is that the force required to lift an object is inversely proportional to the length of the lever. If a force of 2000 pounds applied 2 feet from the pivot point would lift a car, then what force would be required at 10 feet to lift the car?

44. The resistance of a wire varies directly with the length and inversely as the square of the diameter. If a wire of length 20 feet and diameter .1 inch, has a resistance of 2 ohms, then what is the resistance of a 30-foot wire with a diameter of .2 inches?

8.6 Synthetic Division and the Factor Theorem

IN THIS SECTION:
- **Synthetic Division**
- **The Factor Theorem**

In Section 3.7 we learned how to divide a polynomial by a binomial. In this section we will learn a faster method, called **synthetic division,** and we will see how it can be used to find the roots of a polynomial function.

Synthetic Division

If we compare the division of $x^3 - 5x^2 + 4x - 3$ by $x - 2$ as done by the method used in Section 3.7 and as done by synthetic division, synthetic division certainly looks easier.

Ordinary Division of Polynomials

$$
\begin{array}{r}
x^2 - 3x - 2 \leftarrow \text{Quotient} \\
x - 2 \overline{)x^3 - 5x^2 + 4x - 3} \\
\underline{x^3 - 2x^2} \\
-3x^2 + 4x \\
\underline{-3x^2 + 6x} \\
-2x - 3 \\
\underline{-2x + 4} \\
-7 \leftarrow \text{Remainder}
\end{array}
$$

Synthetic Division

$$
\begin{array}{r|rrr|r}
2 & 1 & -5 & 4 & -3 \\
  &   & 2 & -6 & -4 \\
\hline
  & 1 & -3 & -2 & -7
\end{array}
$$

Quotient   Remainder

In synthetic division we write down only the necessary parts of ordinary division; we eliminate repetitions. For example, count how many times x and x^2 are written in the ordinary division. In synthetic division we write down only the coefficients of the terms of the dividend, because it is the coefficients of the variables that determine the coefficients of the quotient. It is easy to identify the polynomials, because we always keep the terms of the polynomials in order of descending exponents. The numbers 1, -5, 4, and -3 are the coefficients of

$$1 \cdot x^3 - 5x^2 + 4x - 3.$$

For $x - 2$ we write only the 2. The bottom line gives us the quotient and the remainder. The remainder is always last, and the numbers to the left of the remainder are the coefficients of the quotient, $x^2 - 3x - 2$.

We should look more closely at synthetic division to see exactly how it is done. We start with the following arrangement of coefficients.

$$
\begin{array}{r|rrrr}
2 & 1 & -5 & 4 & -3 \\
\hline
\end{array}
$$

Next we bring the first coefficient, 1, straight down.

$$
\begin{array}{r|rrrr}
2 & 1 & -5 & 4 & -3 \\
  & \downarrow \text{ Bring down} \\
\hline
  & 1
\end{array}
$$

We then multiply the 1 by the 2 from the divisor, place the answer under the -5, and then add that column.

$$
\begin{array}{r|rrrr}
2 & 1 & -5 & 4 & -3 \\
  &   & 2 \text{ add} \\
\hline
  & 1 & -3
\end{array}
$$

Multiply

We then repeat this multiply-and-add step for each of the remaining columns.

$$
\begin{array}{r|rrrr}
2 & 1 & -5 & 4 & -3 \\
  &   & 2 & -6 & -4 \\
\hline
  & 1 & -3 & -2 & -7
\end{array}
$$

Multiply

The quotient is $1x^2 - 3x - 2$, and the remainder is -7.

Note that in ordinary division we multiply and subtract, while in synthetic division we multiply and add. We can add because we used 2 for the divisor when dividing by $x - 2$. Using 2 automatically changes the sign of the bottom number of each column, allowing us to add rather than subtract. To divide by $x + 3$ using synthetic division, we use -3 for the divisor in order to change the signs.

Synthetic division is used only for dividing a polynomial by the binomial $x - c$, where c is a constant. If the binomial is $x - 7$, we have $c = 7$. For the binomial $x + 7$, we must recognize that $x + 7 = x - (-7)$, and so $c = -7$. The strategy for getting the quotient and remainder by synthetic divison is simple if we follow these steps.

STRATEGY
Synthetic Division

1. List the coefficients of the polynomial (dividend), and include zeros for any missing terms.
2. For dividing by $x - c$, place c to the left.
3. Bring the first coefficient down.
4. Multiply by c and add for each column.

EXAMPLE 1 Find the quotient and remainder when the polynomial is divided by the binomial following it.

a) $x^3 + 3x^2 - 5x + 4$, $x - 3$ **b)** $2x^4 - 5x^2 + 6x - 9$, $x + 2$

Solution

a) When dividing by $x - 3$, use 3 for the divisor.

Bring
down   Add

$$
\begin{array}{r|rrrr}
3 & 1 & 3 & -5 & 4 \\
  &   & 3 & 18 & 39 \\
\hline
  & 1 & 6 & 13 & 43
\end{array}
$$

Coefficients of $x^3 + 3x^2 - 5x + 4$

Multiply

The degree of the quotient is one less than the degree of the dividend. The last number represents the remainder, 43, and the other three numbers tell us that the quotient is $x^2 + 6x + 13$.

b) If we divide by $x + 2$, we must think of $x + 2$ as $x - (-2)$ and use -2 for the divisor. We must also be careful to use a zero for the coefficient of x^3, because x^3 is missing.

$$
\begin{array}{r|rrrrr}
-2 & 2 & 0 & -5 & 6 & -9 \\
 &   & -4 & 8 & -6 & 0 \\
\hline
 & 2 & -4 & 3 & 0 & -9
\end{array}
$$

← Coefficients of $2x^4 + 0 \cdot x^3 - 5x^2 + 6x - 9$

← Quotient and remainder

Because the degree of the dividend is 4, the degree of the quotient is 3. The quotient is $2x^3 - 4x^2 + 3x$, and the remainder is -9. Note that we can also express the results of this division in the form quotient + remainder/divisor as follows.

$$\frac{2x^4 - 5x^2 + 6x - 9}{x + 2} = 2x^3 - 4x^2 + 3x + \frac{-9}{x + 2} \qquad \blacktriangleleft$$

The Factor Theorem

Consider the polynomial $x^2 + 2x - 15$. If we use the polynomial as a rule for defining a function, we get the polynomial function

$$P(x) = x^2 + 2x - 15.$$

The values of x for which $P(x) = 0$ are called the **roots** of the function. We can find the roots of the function by solving the equation

$$x^2 + 2x - 15 = 0$$

as follows.

$$(x + 5)(x - 3) = 0$$
$$x + 5 = 0 \quad \text{or} \quad x - 3 = 0$$
$$x = -5 \quad \text{or} \quad x = 3$$

Because $x + 5$ is a factor of $x^2 + 2x - 15$, -5 is a solution to the equation $x^2 + 2x - 15 = 0$ and a root of the function. We can check that -5 is a root of the polynomial function as follows:

$$P(-5) = (-5)^2 + 2(-5) - 15 = 25 - 10 - 15 = 0$$

Because $x - 3$ is a factor of the polynomial, 3 is also a solution to the equation $x^2 + 2x - 15 = 0$ and a root of the polynomial function. We can check by computing:

$$P(3) = 3^2 + 2 \cdot 3 - 15 = 9 + 6 - 15 = 0$$

There is a one-to-one correspondence between linear factors of the polynomial and roots of the polynomial function.

Now suppose that $P(x)$ represents an arbitrary polynomial. If $x - c$ is a factor of the polynomial $P(x)$, then c is a solution to the equation $P(x) = 0$, and so $P(c) = 0$.

If we divide $P(x)$ by $x - c$ and the remainder is 0, we must have

$$P(x) = (x - c)(\text{quotient}).$$ Dividend equals the divisor
times the quotient.

If the remainder is 0, then $x - c$ is a factor of $P(x)$.
The **factor theorem** summarizes these ideas.

The Factor Theorem

> Suppose $P(x)$ is a polynomial with variable x. The following statements
> are equivalent.
> **1.** The remainder is zero when $P(x)$ is divided by $x - c$.
> **2.** $x - c$ is a factor of $P(x)$.
> **3.** c is a solution to $P(x) = 0$.
> **4.** $P(c) = 0$ (c is a root of the function P).

To say that statements are equivalent means that the truth of any one of them implies
that the others are true.

EXAMPLE 2 Use synthetic division to determine whether 2 is a root of the
polynomial function $P(x) = x^3 - 3x^2 + 5x - 2$.

Solution Perform the synthetic division and check the remainder.

$$
\begin{array}{r|rrrr}
2 & 1 & -3 & 5 & -2 \\
  &   & 2 & -2 & 6 \\
\hline
  & 1 & -1 & 3 & \;4
\end{array}
$$

By the factor theorem, 2 is a root of the function if and only if the remainder is 0.
Because the remainder is 4, 2 is not a root of the function.   ◄

EXAMPLE 3 Use synthetic division to determine whether $x + 4$ is a factor of
$x^3 + 3x^2 + 16$.

Solution Perform the synthetic division and check the remainder.

$$
\begin{array}{r|rrrr}
-4 & 1 & 3 & 0 & 16 \\
   &   & -4 & 4 & -16 \\
\hline
   & 1 & -1 & 4 & \;0
\end{array}
$$

Because the remainder is 0, $x + 4$ is a factor, and we can factor the polynomial as
follows:

$$x^3 + 3x^2 + 16 = (x + 4)(x^2 - x + 4)$$

Note that $x^2 - x + 4$ is a prime polynomial, so the factoring is complete.   ◄

EXAMPLE 4   Suppose the equation $x^3 - 4x^2 - 17x + 60 = 0$ is known to have a solution that is an integer between -3 and 3 inclusive. Find the solution set.

Solution   We can use synthetic division to help us discover a solution. We try 2 first.

$$
\begin{array}{r|rrrr}
2 & 1 & -4 & -17 & 60 \\
  &   & 2 & -4 & -42 \\
\hline
  & 1 & -2 & -21 & 18
\end{array}
$$

Because the remainder is 18, 2 is not a solution to the equation. Next we try 3.

$$
\begin{array}{r|rrrr}
3 & 1 & -4 & -17 & 60 \\
  &   & 3 & -3 & -60 \\
\hline
  & 1 & -1 & -20 & 0
\end{array}
$$

If 3 did not work, we would then try another integer between -3 and 3.

The remainder is 0, so 3 is a solution to the equation, and $x - 3$ is a factor of the polynomial. The other factor is the quotient, $x^2 - x - 20$.

$$x^3 - 4x^2 - 17x + 60 = 0$$
$$(x - 3)(x^2 - x - 20) = 0 \qquad \text{Factor.}$$
$$(x - 3)(x - 5)(x + 4) = 0 \qquad \text{Factor completely.}$$
$$x - 3 = 0 \quad \text{or} \quad x - 5 = 0 \quad \text{or} \quad x + 4 = 0$$
$$x = 3 \quad \text{or} \qquad x = 5 \quad \text{or} \qquad x = -4$$

The solution set is $\{3, 5, -4\}$. Check each of these solutions in the original equation. ◀

Warm-ups

True or false?

1. When dividing a polynomial by $x - 3$, we use -3 for the synthetic division process.
2. We can use synthetic division to divide $x^3 - 5x^2 + 3x - 7$ by $x^2 - 5$.
3. If we divide $3x^5 - 4x^2 - 3$ by $x + 2$, the quotient will be a fourth-degree polynomial.
4. If we divide $x^3 - 8$ by $x - 2$, then the remainder will be zero.
5. If the remainder is zero when $x^4 - 3x^2 + 6x - 4$ is divided by $x - 1$, then $x - 1$ is a factor of $x^4 - 3x^2 + 6x - 4$.
6. If -2 is in the solution set to $x^4 - 8x^2 - 7x + 2 = 0$, then $x + 2$ is a factor of $x^4 - 8x^2 - 7x + 2$.

7. The binomial $x - 1$ is a factor of $x^{35} - 3x^{24} + 2x^{18}$.

8. The binomial $x + 1$ is a factor of $x^3 - 3x^2 + x + 5$.

9. The remainder of $x^3 - 5x + 4$ divided by $x - 1$ is 0.

10. If $P(x) = x^3 - 5x - 2$ and the remainder is zero when $P(x)$ is divided by $x + 2$, then $P(-2) = 0$.

8.6 EXERCISES

Use synthetic division to find the quotient and remainder when the first polynomial is divided by the second. See Example 1.

1. $x^3 - 5x^2 + 6x - 3$, $x - 2$

2. $x^3 + 6x^2 - 3x - 5$, $x - 3$

3. $2x^2 - 4x + 5$, $x + 1$

4. $3x^2 - 7x + 4$, $x + 2$

5. $3x^4 - 15x^2 + 7x - 9$, $x - 3$

6. $-2x^4 + 3x^2 - 5$, $x - 2$

7. $x^5 - 1$, $x - 1$

8. $x^6 - 1$, $x + 1$

9. $x^3 - 5x + 6$, $x + 2$

10. $x^3 - 3x - 7$, $x - 4$

▦ 11. $2.3x^2 - .14x + .6$, $x - .32$

▦ 12. $1.6x^2 - 3.5x + 4.7$, $x + 1.8$

Use synthetic division to write each of the following in the form quotient + remainder/divisor. See Example 1.

13. $\dfrac{x^2 - 4x + 9}{x + 2}$

14. $\dfrac{x^2 - 5x - 10}{x - 3}$

15. $\dfrac{2x - 3}{x + 6}$

16. $\dfrac{4x - 5}{x - 2}$

17. $\dfrac{3x^3 - 4x^2 + 7}{x - 1}$

18. $\dfrac{-2x^3 + x^2 - 3}{x + 2}$

Use synthetic division to determine whether the number following each function is a root of the function. See Example 2.

19. $P(x) = x^3 - x^2 + x - 1$,   1

20. $P(x) = -2x^3 - 5x^2 + 3x + 10$, -2

21. $P(x) = -x^4 - 3x^3 - 2x^2 + 18$, -3

22. $P(x) = x^4 - x^2 - 8x - 16$,   4

23. $P(x) = 2x^3 - 4x^2 - 5x + 9$,   2

24. $P(x) = x^3 + 5x^2 + 2x + 1$, -3

Use synthetic division to determine whether -3 is a solution to each of the following equations. See Example 2.

25. $x^3 + 5x^2 + 2x + 12 = 0$

26. $x^2 - 3x + 6 = 0$

27. $x^4 + 3x^3 - 5x^2 - 10x + 5 = 0$

28. $-x^3 - 4x^2 + x + 12 = 0$

29. $-2x^3 - 4x^2 + 5x - 3 = 0$

30. $x^4 + x^3 - 5x^2 + 7 = 0$

▦ 31. $.8x^2 - .3x - 8.1 = 0$

▦ 32. $6.2x^2 + 4.7x - 41.7 = 0$

Use synthetic division to determine whether the first polynomial is a factor of the second. If it is, then factor the polynomial completely. See Example 3.

33. $x - 3$, $x^3 - 6x - 9$

34. $x + 2$, $x^3 - 6x - 4$

35. $x + 5$, $x^3 + 9x^2 + 23x + 15$

36. $x - 3$, $x^4 - 9x^2 + x - 7$

37. $x - 2,$ $x^3 - 8x^2 + 4x - 6$

38. $x + 5,$ $x^3 + 125$

39. $x + 1,$ $x^4 + x^3 - 8x - 8$

40. $x - 2,$ $x^3 - 6x^2 + 12x - 8$

41. $x - .5,$ $2x^3 - 3x^2 - 11x + 6$

42. $x - \dfrac{1}{3},$ $3x^3 - 10x^2 - 27x + 10$

Solve each of the following equations, given that at least one of the solutions to each equation is an integer between −5 and 5. See Example 4.

43. $x^3 - 13x + 12 = 0$

44. $x^3 + 2x^2 - 5x - 6 = 0$

45. $2x^3 - 9x^2 + 7x + 6 = 0$

46. $6x^3 + 13x^2 - 4 = 0$

47. $2x^3 - 3x^2 - 50x - 24 = 0$

48. $x^3 - 7x^2 + 2x + 40 = 0$

49. $x^3 + 5x^2 + 3x - 9 = 0$

50. $x^3 + 6x^2 + 12x + 8 = 0$

*51. $x^4 - 4x^3 + 3x^2 + 4x - 4 = 0$

*52. $x^4 + x^3 - 7x^2 - x + 6 = 0$

Wrap-up

CHAPTER 8

SUMMARY

	Concepts	Examples
Relation	Any set of ordered pairs	$\{(1, 2), (1, 3)\}$
Function	A relation where no two ordered pairs have the same first coordinate and different second coordinates	$\{(1, 2), (3, 5)\}$ $\{(x, y) \mid y = x^2\}$
	If y is a function of x, then y is uniquely determined by x.	
Domain	The set of values of the independent variable	$y = x^2$ Domain: R
Range	The set of values of the dependent variable	Range: nonnegative real numbers
f-notation	For y is a function of x, we write $y = f(x)$.	$y = 2x + 3$ $f(x) = 2x + 3$

Linear function	$y = mx + b$, or $f(x) = mx + b$	

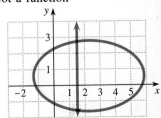

Absolute-value function $y = |x|$, or $f(x) = |x|$

Quadratic function

$$f(x) = ax^2 + bx + c$$

$\dfrac{-b}{2a}$ is the x-coordinate of the vertex.

Square-root function $f(x) = \sqrt{x}$

Vertical-line test   If a graph can be crossed twice by a vertical line, then it is not the graph of a function.

Not a function

Operations with functions	$(f + g)(x) = f(x) + g(x)$ $(f - g)(x) = f(x) - g(x)$ $(f \cdot g)(x) = f(x) \cdot g(x)$ $(f/g)(x) = f(x)/g(x)$	Let $f(x) = x^2 - 1$, $g(x) = x + 1$ $(f + g)(x) = x^2 + x$ $(f - g)(x) = x^2 - x - 2$ $(f \cdot g)(x) = (x^2 - 1)(x + 1)$ $\quad = x^3 + x^2 - x - 1$ $(f/g)(x) = (x^2 - 1)/(x + 1)$ $\quad = x - 1$
Composition of functions	$(g{\circ}f)(x) = g[f(x)]$	$(f{\circ}g)(x) = f(x + 1)$ $\quad = (x + 1)^2 - 1$ $\quad = x^2 + 2x$
One-to-one function	A function where no two ordered pairs have different x-coordinates and the same y-coordinate	$\{(2, 20), (3, 30)\}$
	One-to-one functions are invertible.	$\{(20, 2), (30, 3)\}$
f-notation for inverse	Two functions f and g are inverses of each other if *both* of the following conditions are met.	
	1. $(g{\circ}f)(x) = x$ for every number x in the domain of f. 2. $(f{\circ}g)(x) = x$ for every number x in the domain of g.	$g(x) = x + 1$ $g^{-1}(x) = x - 1$
Procedure for finding f^{-1}	Replace $f(x)$ by y. Switch x and y. Solve for y. Replace y by $f^{-1}(x)$.	$h(x) = 3x + 2$ $y = 3x + 2$ $x = 3y + 2$ $x - 2 = 3y$ $\dfrac{x - 2}{3} = y$ $h^{-1}(x) = \dfrac{x - 2}{3}$
Graphs of f and f^{-1}	Symmetric with respect to the line $y = x$	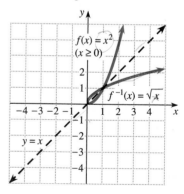
Variation	y varies directly as x. y varies inversely as x. y varies jointly as x and z.	$y = kx$ $y = k/x$ $y = kxz$

Synthetic
division

Use only for dividing by $x - c$.

$x + 2 = x - (-2)$

$$(x^2 + 4x - 7) \div (x + 2)$$

$$-2 \,|\, \begin{array}{ccc} 1 & 4 & -7 \\ & -2 & -4 \\ \hline 1 & 2 & -11 \end{array}$$

Factor
theorem

Suppose $P(x)$ is a polynomial with variable x. The
following statements are equivalent.

1. The remainder is zero when $P(x)$ is divided
by $x - c$.
2. $x - c$ is a factor of $P(x)$.
3. c is a solution to $P(x) = 0$.
4. $P(c) = 0$ (c is a root of the function P).

REVIEW EXERCISES

8.1 *Determine whether each relation is a function.*

1. $\{(5, 7),(5, 10)\}$ 2. $\{(10, 7),(6, 7)\}$ 3. $\{(x, y) \,|\, y = x^2\}$

4. $x^2 = 1 + y^2$ 5. $\{(x, y) \,|\, x = y^4\}$ 6. $y = \sqrt{x - 1}$

Determine the domain and range of each relation.

7. $\{(1, 3),(2, 3)\}$ 8. $\{(x, y) \,|\, y = -x^2\}$ 9. $f(x) = 2x - 7$

10. $x = y^{1/2}$ 11. $y = |x|$ 12. $y = \sqrt{x} - 1$

Let $f(x) = 2x - 5$ and $g(x) = x^2 + x - 6$. Find the following.

13. $f(0)$ 14. $f(-3)$ 15. $g(0)$ 16. $g(-2)$

17. $g(1/2)$ 18. $g(-1/2)$ 19. $f(a)$ 20. $f(x + 3)$

21. $g(a - 1)$ 22. $g(x + h)$ 23. a,   if $f(a) = 1$ 24. x,   if $g(x) = 0$

8.2 *Graph each function, and state the domain and range.*

25. $f(x) = 3x - 4$ 26. $y = .3x$ 27. $h(x) = |x| - 2$

28. $y = |x - 2|$ 29. $y = x^2 - 2x + 1$ 30. $g(x) = x^2 - 2x - 15$

31. $k(x) = \sqrt{x} + 2$ 32. $y = \sqrt{x - 2}$

33. $y = \dfrac{2}{x}$ 34. $y = 4 - x^2$

Graph each relation, and state its domain and range.

35. $x = 2$ 36. $x = y^2 - 1$ 37. $x = |y| + 1$ 38. $x = \sqrt{y - 1}$

8.3 *Let $f(x) = 3x + 5$, $g(x) = x^2 - 2x$, and $h(x) = \dfrac{x - 5}{3}$. Find the following.*

39. $f(-3)$ 40. $h(-4)$ 41. $(h \circ f)(\sqrt{2})$ 42. $(f \circ h)(\pi)$

43. $(g \circ f)(2)$ 44. $(g \circ f)(x)$ 45. $(f + g)(3)$ 46. $(f - g)(x)$

47. $(f \cdot g)(x)$ 48. $(f/g)(1)$ 49. $(f \circ f)(0)$ 50. $(f \circ f)(x)$

Let $f(x) = |x|$, $g(x) = x + 2$, and $h(x) = x^2$. Write each of the following functions as a composition of functions, using f, g, or h.

51. $F(x) = |x + 2|$

52. $G(x) = |x| + 2$

53. $H(x) = x^2 + 2$

54. $K(x) = x^2 + 4x + 4$

55. $I(x) = x + 4$

56. $J(x) = x^4 + 2$

8.4 *Determine whether each function is invertible. If it is invertible, find the inverse.*

57. $\{(-2, 4),(2, 4)\}$

58. $\{(1, 1),(3, 3)\}$

59. $f(x) = 8x$

60. $g(x) = 13x - 6$

61. $h(x) = \sqrt[3]{x - 6}$

62. $i(x) = -\dfrac{x}{3}$

63. $j(x) = \dfrac{x + 1}{x - 1}$

64. $k(x) = |x| + 7$

65. $m(x) = (x - 1)^2$

66. $n(x) = \dfrac{3}{x}$

Find the inverse of each function, and graph f and f^{-1} on the same pair of axes.

67. $f(x) = 3x - 1$

68. $f(x) = 2 - x^2$ for $x \geq 0$

69. $f(x) = \dfrac{x^3}{2}$

70. $f(x) = -\dfrac{1}{4}x$

8.5 *Solve each variation problem.*

71. If y varies directly as m and $y = -3$ when $m = 1/4$, find y when $m = -2$.

72. If a varies inversely as b and $a = 6$ when $b = -3$, find a when $b = 4$.

73. If c varies directly as m and inversely as n, and $c = 20$ when $m = 10$ and $n = 4$, find c when $m = 6$ and $n = -3$.

74. If V varies jointly as h and the square of r, and $V = 32$ when $h = 6$ and $r = 3$, find V when $h = 3$ and $r = 4$.

8.6 *Use synthetic division to find the quotient and remainder when the first polynomial is divided by the second.*

75. $x^3 - 5x^2 + 4x - 12$, $\quad x - 3$

76. $x^4 - 3x^2 + 5x - 7$, $\quad x + 2$

77. $x^4 - 4x + 2$, $\quad x + 1$

78. $x^3 - 3x^2 + 9x - 17$, $\quad x - 4$

Given that either 2 or -2 is a root to each of the following polynomial functions, find all of the roots.

79. $P(x) = x^3 - 4x^2 - 11x + 30$

80. $P(x) = x^3 - 2x^2 - 6x + 12$

81. $P(x) = x^3 - 5x^2 - 2x + 24$

82. $P(x) = x^3 + 7x^2 + 4x - 12$

Miscellaneous

83. The cost, C, of crop dusting is a linear function of the number of pounds of chemical, n, distributed over the fields. Betty paid \$1800 for 400 pounds and then paid \$2300 for 600 pounds. Write C as a linear function of n.

84. Evelyn's grade on a math test varies directly with the number of hours spent studying and inversely with the number of hours spent partying during the 24 hours preceding the test. If she scored a 90 on a test where she studied 10 hours and partied 2 hours, then what should she score after studying 4 hours and partying 6 hours?

Photo for Exercise 83

85. If $a = 3k + 2$ and $k = 5w - 6$, then what is the relationship between a and w?

86. Let B be the area of a square inscribed in a circle of radius r and area A. Write B as a function of A.

87. What is the vertex of the parabola $y = x^2 - 3x + 2$?

88. The profit that Luther makes on the sale of magazine subscriptions is determined by the formula $P(x) = 50x - x^2 - 400$, where x is the number of subscriptions he sells and $P(x)$ is in dollars. What value of x will maximize his profit? What is his maximum profit?

89. If $f(x) = 3x^2 - 4x + 2$, find and simplify the quantity
$$\frac{f(x + h) - f(x)}{h}.$$

90. A window is in the shape of a square of side s, with a semicircle of diameter s above it. Write a function that expresses the total area of the window as a function of s.

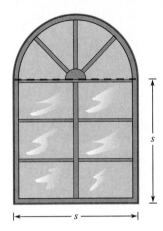

Figure for Exercise 90

CHAPTER 8 TEST

Determine the domain and range of each relation.

1. $f(x) = |x|$

2. $x = \sqrt{y}$

Let $f(x) = -2x + 5$ and $g(x) = x^2 + 4$. Find the following.

3. $f(-3)$

4. $g(a + 1)$

5. $(g \circ f)(x)$

6. $f^{-1}(x)$

7. $(g + f)(2)$

8. $(f \cdot g)(1)$

9. $(f^{-1} \circ f)(-1782)$

10. x,   if $g(x) = 9$

Sketch a graph of each function.

11. $g(x) = -\dfrac{2}{3}x + 1$

12. $y = |x| - 4$

13. $f(x) = x^2 + 2x - 8$

14. $h(x) = 2\sqrt{x} - 1$

Solve each problem.

15. Find the inverse of the function $f(x) = \dfrac{2x + 1}{x - 1}$.

16. The volume of a sphere varies directly as the cube of the radius. If a sphere with radius 3 feet has a volume of 36π cubic feet, then what is the volume of a sphere with a radius of 2 feet?

17. A mail-order firm charges its customers a shipping and handling fee of $3.00, plus $.50 per pound for each order shipped. Express the shipping and handling fee as a function of the weight of the order, n.

18. Suppose y varies directly as x and inversely as the square root of z. If $y = 12$ when $x = 7$ and $z = 9$, then what is the constant of variation?

19. Graph the relation $x = y^2$.

20. If a ball is tossed into the air from a height of 6 feet with a velocity of 40 feet per second, then its altitude at time t can be described by the quadratic function $A(t) = -16t^2 + 40t + 6$. For what value of t is the altitude at its maximum?

21. Use synthetic division to find the quotient and remainder when $2x^4 - 5x^3 + 3x^2 + 6x - 12$ is divided by $x + 1$.

22. Given that all of the solutions to the equation $x^3 - 12x^2 + 47x - 60 = 0$ are whole numbers smaller than 10, find the solution set.

Tying It All Together

CHAPTERS 1–8

Simplify each expression.

1. $125^{-2/3}$

2. $\left(\dfrac{8}{27}\right)^{-1/3}$

3. $\sqrt{18} - \sqrt{8}$

4. $x^5 \cdot x^3$

5. $16^{1/4}$

6. $\dfrac{x^{12}}{x^3}$

Find the solution set to each equation.

7. $x^2 = 9$

8. $3^x = 9$

9. $27^x = 3$

10. $x^{1/4} = 3$

11. $x^{1/6} = -2$

12. $x^3 = 8$

13. $8^x = 2$

14. $|x| = 8$

15. $8x = 2$

16. $x^2 = 8$

17. $x^2 - 4x - 6 = 0$

18. $2x^2 - x = 21$

19. $|5x - 4| = 21$

20. $\sqrt{2x - 3} = 9$

Sketch the graph of each set.

21. $\{(x, y) \mid y = 5\}$

22. $\{(x, y) \mid y = 2x - 5\}$

23. $\{(x, y) \mid x = 5\}$

24. $\{(x, y) \mid 3y = x\}$

25. $\{(x, y) \mid y = 5x^2\}$

26. $\{(x, y) \mid y = -2x^2\}$

Exponential and Logarithmic Functions

Why is disposing of nuclear waste a problem? Many types of waste that we generate are relatively harmless compared to nuclear waste. Nuclear waste does not decay over a short period of time like some other types of waste. In fact, nuclear waste never decays completely. The amount that decays is always only a fraction of the material present. This type of decay can be described by an exponential function.

The amount, A, of a certain radioactive substance remaining after t years is given by the formula $A = A_0 e^{-.0003t}$, where A_0 is the initial amount. If we have 218 grams of this substance today, then how much of it will be left 1000 years from now? From the formula relating the amount of radioactive material and time, we can easily answer any questions concerning how much material will be present at any time in the future. We will answer this question when we solve Exercise 87 of the Chapter 9 Review.

9.1 Exponential Functions

IN THIS SECTION:

- Evaluation
- Domain
- Graphing Exponential Functions
- One-to-one Property
- Compound Interest

In Chapter 8 we studied functions such as

$$f(x) = x^2,\ g(x) = x^3,\ \text{and}\ h(x) = x^{1/2}.$$

For these functions, the variable is the base. In this section we will discuss functions that have a variable as an exponent.

Evaluation

Functions that have the variable only in the position of an exponent are called **exponential functions**. For example, the functions

$$f(x) = 2^x,\ f(x) = (1/2)^x,\ \text{and}\ f(x) = 3^{2x+1}$$

are exponential functions. We evaluate exponential functions using our knowledge of exponents.

EXAMPLE 1   Let $f(x) = 2^x$, $g(x) = (1/3)^{1-x}$, and $h(x) = -3^x$. Evaluate the following.

a) $f(3/2)$ b) $g(3)$ c) $h(2)$

Solution

a) $f(3/2) = 2^{3/2} = \sqrt{2^3} = \sqrt{8} = 2\sqrt{2}$
b) $g(3) = (1/3)^{1-3} = (1/3)^{-2} = 3^2 = 9$
c) $h(2) = -3^2 = -9$ Note that $-3^2 \neq (-3)^2$.   ◄

Domain

Consider the function $f(x) = 2^x$. For what values of x is this function defined? We know that 2^x is defined for every rational number. For example,

$$2^{3.14} = 2^{314/100} = \sqrt[100]{2^{314}}.$$

Can we use an irrational number such as π for x? What does 2^π mean? As a decimal number, an irrational number is an infinite nonrepeating decimal. For example,

$$\pi = 3.141592654 \ldots$$

In doing computations we use only rational approximations for π. We use as many digits from the exact value of π as we like, depending on the accuracy we want. Using the x^y key on a calculator, we get the following results.

$$2^{3.1} = 8.5741877 \ldots$$
$$2^{3.14} = 8.8152409 \ldots$$
$$2^{3.141} = 8.8213533 \ldots$$
$$2^{3.1415} = 8.8244110 \ldots$$
$$2^{3.14159} = 8.8249615 \ldots$$

Notice that by choosing the exponent closer and closer to π, we get results that appear to be getting closer and closer to some number. We define 2^{π} to be that number. If we use a calculator to find 2^{π}, we get

$$2^{\pi} = 8.824977827 \ldots$$

So 2^{π} is the number that is approached by using better and better rational approximations for π as the exponent on a base of 2. We can define 2^x for any irrational exponent in the same manner. Thus, we can use the set of all real numbers as the domain of the function $f(x) = 2^x$.

Graphing Exponential Functions

EXAMPLE 2   Sketch the graph of $f(x) = 2^x$.

Solution   We make a table of values and plot the points as shown in Fig. 9.1. Notice that as x increases, 2^x increases. As x decreases, 2^x decreases but does not reach 0. Since the domain of the function is R, we can draw the graph as a smooth curve through the points that we found in the table. Note that the range is $\{y \mid y > 0\}$.

x	$f(x)$
0	1
$\frac{1}{2}$	1.414
1	2
2	4
3	8
-1	$\frac{1}{2}$
-2	$\frac{1}{4}$
-3	$\frac{1}{8}$

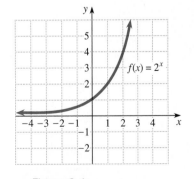

Figure 9.1

EXAMPLE 3   Graph the function $f(x) = (1/2)^x$.

Solution   Make a table of values and plot the points as shown in Fig. 9.2. Notice that as x gets larger and larger, $(1/2)^x$ gets closer and closer to 0. Also note that $(1/2)^x$ could be written as $1/2^x$ or as 2^{-x}.

x	$f(x)$
0	1
1	$\frac{1}{2}$
2	$\frac{1}{4}$
3	$\frac{1}{8}$
-1	2
-2	4

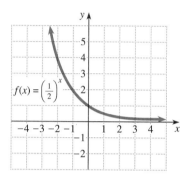

Figure 9.2   ◄

Math at Work

A shard of pottery is found in an archaeological dig in Arizona. The fossil of a sea creature is discovered in the mountains of Vermont. And a painting hanging in a New York museum is suspected of being a forgery.

Scientists use a method called carbon-14 dating to attach probable dates of origin to these objects. All living organisms contain radioactive carbon, carbon 14. The amount of carbon 14 remains constant in the organism while it is alive, but after the organism's death, the carbon 14 decays exponentially at a rate of one-half every 5700 years. By comparing the amount of carbon 14 remaining in the organism with the amount that has decayed, the age of the organism can be estimated.

A long-standing controversy of the art world recently was settled when carbon-14 dating was used to fix dates to the Shroud of Turin. Once thought to be the burial cloth of Christ, it is now known to have been made between 1260 and 1390 A.D.

The formula used to measure the decay of carbon 14 is

$$A = A_0 \cdot 2^{-t/5700}$$

where A is the amount of carbon 14 present after t years and A_0 is the original amount of carbon 14. If a fossilized bone originally contained 100 grams of carbon 14, how many grams will remain 5000 years later?

Notice the similarity between the graphs in Figs. 9.1 and 9.2. The graph of $f(x) = 2^x$ is rising as we go from left to right, and the graph of $f(x) = (1/2)^x$ is falling as we go from left to right. As x gets bigger 2^x grows, but $(1/2)^x$ decreases. We now give the definition of exponential functions.

DEFINITION
Exponential Function

A function of the form

$$f(x) = a^x,$$

where a is any positive real number other than 1, is called an **exponential function.**

We rule out the base 1 in the definition, because $f(x) = 1^x$ is the same as the linear function $f(x) = 1$. Zero is not used as a base because $0^x = 0$ for any positive x and nonpositive powers of 0 are undefined. Negative numbers are not used as bases because an expression such as $(-4)^x$ is not a real number if $x = 1/2$.

There are two bases for exponential functions that are used more frequently than the others. They are 10 and e. A scientific calculator usually has function keys for 10^x and e^x. The number e is an irrational number that is approximately equal to 2.718. The number e seems like an unusual number to use as a base. However, e, like π, occurs naturally in many applications. We will see one in Example 9 of this section.

All exponential functions have similar graphs. However, we can alter their shape and location by making changes in the exponent.

EXAMPLE 4   Sketch the graph of $f(x) = 3^{2x-1}$.

Solution   A table of values and the graph are shown in Fig. 9.3.

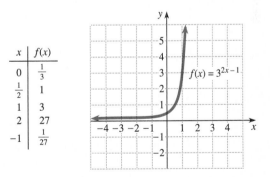

x	$f(x)$
0	$\frac{1}{3}$
$\frac{1}{2}$	1
1	3
2	27
-1	$\frac{1}{27}$

Figure 9.3

EXAMPLE 5   Sketch the graph of $y = -2^{-x}$.

Solution Because $-2^{-x} = -(2^{-x})$, all y-coordinates are negative. A table of values and the graph are shown in Fig. 9.4.

x	y
0	-1
1	$-\frac{1}{2}$
2	$-\frac{1}{4}$
-1	-2
-2	-4

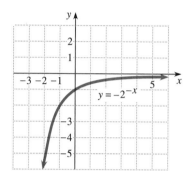

Figure 9.4   ◄

One-to-one Property

The exponential functions are **one-to-one** functions. This means that no two different values of x result in the same value for y. We state this property formally as follows.

One-to-one Property of Exponential Functions

For $a > 0$ and $a \neq 1$, $a^{x_1} = a^{x_2}$ if and only if $x_1 = x_2$.

The next example shows how this property is used in solving equations involving exponential functions.

EXAMPLE 6 Let $f(x) = 2^x$ and $g(x) = (1/2)^{1-x}$. Find x if:

a) $f(x) = 32$ **b)** $g(x) = 8$

Solution

a) Because $f(x) = 2^x$, we have $2^x = 32$. We also know that $2^5 = 32$. Thus,

$$2^x = 2^5.$$

Because the exponential functions are one-to-one, it must be that $x = 5$.

b) We must find x such that $(1/2)^{1-x} = 8$. To solve this equation we write each side as a power of the same base. Because $1/2 = 2^{-1}$ and $8 = 2^3$, we can write

$$(2^{-1})^{1-x} = 2^3$$

$$2^{x-1} = 2^3 \qquad -1 \cdot (1 - x) = x - 1$$

$$x - 1 = 3 \qquad \text{One-to-one property}$$

$$x = 4.$$
                                                                            ◄

EXAMPLE 7 Solve each equation.

a) $9^{|x|} = 3$

b) $\dfrac{1}{8} = 4^x$

Solution

a) Because $9 = 3^2$, we can write each side as a power of the same base.

$$(3^2)^{|x|} = 3$$
$$3^{2|x|} = 3^1$$
$$2|x| = 1 \qquad \text{One-to-one property}$$
$$|x| = \frac{1}{2}$$
$$x = \pm\frac{1}{2}$$

The solution set is $\{\pm 1/2\}$. Check.

b) Because $1/8 = 2^{-3}$ and $4 = 2^2$, we can write each side as a power of the same base.

$$2^{-3} = (2^2)^x$$
$$2^{-3} = 2^{2x}$$
$$2x = -3 \qquad \text{One-to-one property}$$
$$x = -\frac{3}{2}$$

The solution set is $\{-3/2\}$. Check. ◀

Compound Interest

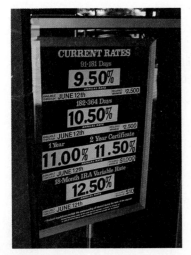

Exponential functions are used to describe phenomena such as population growth, radioactive decay, and compound interest. If a bank pays 6% compounded quarterly on an account, then the interest is computed every 3 months at 1.5% (one-quarter of 6%). Suppose an account has $5000 in it at the beginning of a quarter. We can apply the simple-interest formula $A = P + Prt$, with $r = 6\%$ and $t = 1/4$, to find how much is in the account at the end of the quarter.

$$A = P + Prt$$
$$= P(1 + rt)$$
$$= 5000\left(1 + .06 \cdot \frac{1}{4}\right)$$
$$= 5000(1.015)$$
$$= \$5075$$

To repeat this computation for another quarter, we multiply $5075 by 1.015. If S

represents the amount in the account at the end of n quarters, we can write S as an exponential function of n.

$$S = \$5000(1.015)^n$$

In general, the amount S is given by an exponential function.

Compound-interest Formula

If P represents the principal, i the interest rate per period, n the number of periods, and S the amount at the end of n periods, then

$$S = P(1 + i)^n.$$

EXAMPLE 8   If \$350 is deposited in an account paying 12% compounded monthly, then how much is in the account at the end of 6 years?

Solution   Interest is paid 12 times per year, so the account earns 1/12 of 12%, or 1% each month, for 72 months. Because the value of i is .01, we have

$$S = \$350(1.01)^{72} \qquad \text{Calculate } (1.01)^{72} \text{ on a calculator with an } x^y \text{ key.}$$
$$= \$716.48. \qquad\qquad\qquad\qquad\qquad\qquad\qquad\qquad \blacktriangleleft$$

If we shorten the length of the time period (yearly, quarterly, monthly, daily, hourly, etc.), the number of periods increases, while the interest rate for the period decreases. The limit to this process is called **continuous compounding.** When a savings institution pays interest compounded continuously, it is using the following formula.

Continuous-compounding Formula

If P is the principal, or beginning balance, r is the annual percentage rate compounded continuously, t is the time in years, and S is the amount, or ending balance, then

$$S = P \cdot e^{rt}.$$

EXAMPLE 9   If \$350 is deposited in an account paying 12% compounded continuously, then how much is in the account after 6 years?

Solution   Use $r = 12\%$, $t = 6$, and $P = \$350$ in the formula for compounding interest continuously.

$$S = P \cdot e^{rt}$$
$$= 350 \cdot e^{(.12)(6)}$$
$$= 350 \cdot e^{.72} \qquad \text{Calculate } e^{.72} \text{ on a calculator with an } e^x \text{ key.}$$
$$= \$719.05$$

Note that compounding continuously amounted to only a few dollars more than compounding monthly in Example 8. $\blacktriangleleft$

Warm-ups

True or false?

1. If $f(x) = 4^x$, then $f(-1/2) = -2$.
2. If $f(x) = (1/3)^x$, then $f(-1) = 3$.
3. The function $f(x) = x^4$ is an exponential function.
4. The functions $f(x) = (1/2)^x$ and $g(x) = 2^{-x}$ have the same graph.
5. If $f(x) = 2^x$, then $f^{-1}(x) = \sqrt{x}$.
6. $2^{1.4} = \sqrt[5]{2^7}$
7. $2^{1.41} = \sqrt[100]{2^{141}}$
8. The expression $2^{\sqrt{2}}$ is undefined.
9. The functions $f(x) = 2^{-x}$ and $g(x) = 1/(2^x)$ have the same graph.
10. If $f(x) = 3^x$, then $f^{-1}(x) = 3^{-x}$.

9.1 EXERCISES

Let $f(x) = 4^x$, $g(x) = (1/3)^{x+1}$, *and* $h(x) = 10^x$. *Find the following. See Example* 1.

1. $f(2)$
2. $f(-1)$
3. $f(1/2)$
4. $f(-1/2)$
5. $f(-2)$
6. $f(0)$
7. $f(3/2)$
8. $g(-1)$
9. $g(-2)$
10. $g(1)$
11. $g(0)$
12. $g(-3)$
13. $h(0)$
14. $h(-1)$
15. $h(2)$
16. $h(3)$

Sketch the graph of each function. See Examples 2 *and* 3.

17. $f(x) = 3^x$
18. $g(x) = 4^x$
19. $h(x) = (1/3)^x$
20. $i(x) = (1/4)^x$
21. $y = 10^x$
22. $y = (.1)^x$

Sketch the graph of each function. See Examples 4 *and* 5.

23. $f(x) = -2^x$
24. $g(x) = 2^{-x}$
25. $k(x) = -2^{x-2}$
26. $A(x) = 10^{1-x}$
27. $y = 10^{x+2}$
28. $y = 3^{2x+1}$
29. $H(x) = 10^{|x|}$
30. $s(x) = 2^{(x^2)}$
31. $P = 5000(1.05)^t$
32. $d = 800 \cdot 10^{-4t}$

Let $f(x) = 2^x$, $g(x) = (1/3)^x$, *and* $h(x) = 4^{2x-1}$. *Find x for each of the following. See Example* 6.

33. $f(x) = 4$
34. $f(x) = 1/4$
35. $f(x) = 4^{2/3}$
36. $f(x) = 1$
37. $g(x) = 9$
38. $g(x) = 1/9$
39. $g(x) = 1$
40. $g(x) = \sqrt{3}$
41. $h(x) = 16$
42. $h(x) = 1/2$
43. $h(x) = 1$
44. $h(x) = \sqrt{2}$

Solve each equation. See Example 7.

45. $2^x = 64$
46. $-3^x = -9$
47. $5^{-x} = 25$
48. $10^x = .001$
49. $10^{2x} = .1$
50. $2^x = 1/4$
51. $3^x = 1/9$
52. $(2/3)^{x-1} = 9/4$

53. $3^{2x-5} = 81$ **54.** $(1/4)^{3x} = 16$ **55.** $10^{|x|} = 1000$ **56.** $-2^{1-x} = 8$

57. $-3^{2-x} = -81$ **58.** $10^{x-1} = .01$

Solve each problem. See Example 8.

59. If $6000 is deposited in an account paying 5% compounded quarterly, then what amount will be in the account after 10 years?

60. If $400 is deposited in an account paying 10% compounded quarterly, then what amount will be in the account after 7 years?

61. The number of grams of a certain radioactive substance present at time t is given by the formula $A = 300 \cdot 3^{-6t}$, where t is the number of years. Find the amount present at time $t = 0$. Find the amount present after 20 years.

62. The population of a certain country appears to be growing according to the formula $P = 20 \cdot 3^{.1t}$, where P is the population in millions and t is the number of years since 1980. What was the population in 1980? What will be the population in the year 2000?

63. The value of a certain textbook seems to decrease according to the formula $V = 45 \cdot 2^{-.9t}$, where V is the value in dollars and t is the age of the book in years. What is the book worth when it is new? What is it worth when it is 2 years old?

64. In a Minnesota swamp in the springtime, the number of mosquitoes per acre appears to grow according to the formula $N = 10^{.1t+2}$, where t is the number of days since the last frost. What is the size of the mosquito population at times $t = 10$, $t = 20$, and $t = 30$?

Solve each problem. See Example 9.

65. If $500 is deposited in an account paying 7% compounded continuously, how much will be in the account after 3 years?

66. If $7000 is deposited in an account paying 8% compounded continuously, then what will it amount to after 4 years?

67. How much interest will be earned the first year on $80,000 on deposit in an account paying 7.5% compounded continuously?

68. If $7500 is deposited in an account paying 6.75% compounded continuously, then how much will be in the account after 5 years and 215 days?

Use a scientific calculator to find the value of each expression.

69. $10^{3.46}$ **70.** $10^{1.23}$ **71.** $2^{4.67}$ **72.** $2^{-1.48}$

73. e^2 **74.** $e^{-3.2}$ **75.** e^0 **76.** $e^{-.015}$

9.2 Logarithms

IN THIS SECTION:

- Evaluation
- Graphing Logarithmic Functions
- Inverses
- Solving Logarithmic Equations
- Applications

Logarithmic functions are the inverses of the exponential functions that we studied in Section 9.1. In this section we will learn to evaluate logarithms and graph logarithmic functions.

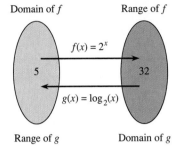

Domain of *f*   Range of *f*

$f(x) = 2^x$

5   32

$g(x) = \log_2(x)$

Range of *g*   Domain of *g*

Figure 9.5

DEFINITION
log_a x

Evaluation

The exponential function $f(x) = 2^x$ pairs a number with the result we get by using that number as an exponent on a base of 2. For example, because $2^5 = 32$, the function $f(x) = 2^x$ pairs the 5 with 32. The base 2 logarithm function is the inverse of the function $f(x) = 2^x$. The base 2 logarithm function pairs a number with the exponent that we use on 2 to obtain the number. Thus, the base 2 logarithm of 32 is 5, because $2^5 = 32$. This fact is written as $\log_2(32) = 5$. See Fig. 9.5. We can state the formal definition of logarithm as follows.

> If a is a positive number other than one, the statement $y = \log_a x$ is equivalent to the statement $a^y = x$.

EXAMPLE 1 Write each statement involving logarithms as a statement using exponents, and each statement involving exponents as a statement with logarithms.

a) $3 = \log_5 125$ **b)** $3^x = 7$ **c)** $5 = \log_a x$

Solution

a) The base 5 logarithm of 125 equals 3, which means that 3 is the exponent on 5 that gives 125. Thus, $5^3 = 125$.
b) $x = \log_3 7$
c) $a^5 = x$ ◄

In the next example we use the definition of logarithm to evaluate logarithms.

EXAMPLE 2 Evaluate each logarithm.

a) $\log_2(1/8)$ **b)** $\log_{1/2}(4)$ **c)** $\log_{10}(.001)$

Solution

a) The number $\log_2(1/8)$ is the power of 2 that gives us 1/8. Because $1/8 = 2^{-3}$, we have $\log_2(1/8) = -3$.
b) Because $4 = (1/2)^{-2}$, $\log_{1/2}(4) = -2$.
c) Because $.001 = 10^{-3}$, $\log_{10}(.001) = -3$.   ◄

There are two bases for logarithms that are used more frequently than the others; they are 10 and e. Most scientific calculators have function keys for $\log_{10} x$ and $\log_e x$. The base 10 logarithm is called the **common logarithm** and is usually written as $\log(x)$. The base e logarithm is called the **natural logarithm,** and we usually write $\ln(x)$ instead of $\log_e x$. The simplest way to obtain a common or natural logarithm is to use a scientific calculator. However, a table of common logarithms can be found in Appendix B. In the next example we find natural and common logarithms of certain numbers without a calculator or a table.

EXAMPLE 3   Evaluate each of the following logarithms.

a) $\log(1000)$ **b)** $\ln(e)$ **c)** $\log(1/10)$

Solution

a) Because $10^3 = 1000$, $\log(1000) = 3$.   **b)** Because $e^1 = e$, $\ln(e) = 1$.
c) Because $10^{-1} = 1/10$, $\log(1/10) = -1$.   ◀

Graphing Logarithmic Functions

EXAMPLE 4   Sketch the graph of the function $g(x) = \log_2 x$.

Solution   Figure 9.6 shows a table of values and the graph.

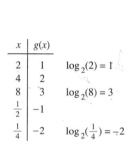

x	$g(x)$	
2	1	$\log_2(2) = 1$
4	2	
8	3	$\log_2(8) = 3$
$\frac{1}{2}$	-1	
$\frac{1}{4}$	-2	$\log_2(\frac{1}{4}) = -2$

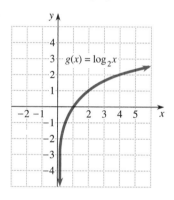

Figure 9.6   ◀

EXAMPLE 5   Graph the function $f(x) = \log_{1/2} x$.

Solution   For positive values of x, find $f(x)$. For example, if $x = 1$, then $f(1) = \log_{1/2} 1 = 0$, because $(1/2)^0 = 1$. A table of values and the graph are shown in Fig. 9.7.

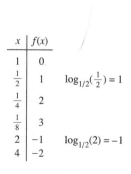

x	$f(x)$	
1	0	
$\frac{1}{2}$	1	$\log_{1/2}(\frac{1}{2}) = 1$
$\frac{1}{4}$	2	
$\frac{1}{8}$	3	
2	-1	$\log_{1/2}(2) = -1$
4	-2	

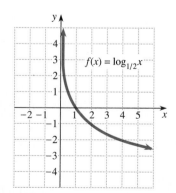

Figure 9.7   ◀

Inverses

The function $g(x) = \log_2 x$ is defined to be the inverse of the exponential function $f(x) = 2^x$. For example, the point $(3, 8)$ is on the graph of $f(x) = 2^x$, because $2^3 = 8$. The point $(8, 3)$ is on the graph of $g(x) = \log_2 x$, because $\log_2 8 = 3$. We have seen that the range of the exponential function $f(x) = 2^x$ is the set of positive real numbers. Since $g(x) = \log_2 x$ is the inverse of $f(x)$, its domain is the set of positive real numbers. In Chapter 8 we learned that the graph of the inverse to any function can be found by reflecting the graph of the function across the line $y = x$. In Fig. 9.8, we have the graphs of $f(x) = 2^x$ and $f^{-1}(x) = \log_2 x$ shown on the same pair of axes.

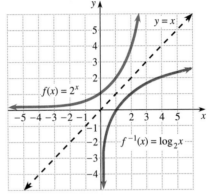

Figure 9.8

EXAMPLE 6   Find the inverse of each of the following functions.

a) $f(x) = 10^x$ **b)** $g(x) = \log_3 x$

Solution

a) The inverse of $f(x) = 10^x$ is $f^{-1}(x) = \log_{10} x$.
b) The inverse of $g(x) = \log_3 x$ is $g^{-1}(x) = 3^x$. ◀

Solving Logarithmic Equations

In Section 9.1 we learned that the exponential functions are one-to-one functions. The logarithmic functions are also one-to-one functions. This fact can be stated as follows.

One-to-one Property of Logarithms

If $\log_a m = \log_a n$, then $m = n$.

The one-to-one property of logarithms and the definition of logarithms are the two basic tools that we use to solve equations involving logarithms. We will use these ideas to solve equations in the following example.

EXAMPLE 7   Solve each equation.

a) $\log_3 x = -2$ **b)** $\log_x 8 = -3$ **c)** $\log(x^2) = \log(4)$

Solution

a) Use the definition of logarithms to write an equivalent equation:

$$3^{-2} = x$$

$$\frac{1}{9} = x$$

The solution set is $\{1/9\}$.

b) For this equation, the equivalent exponential statement leads us to an equation of the type that we solved in Chapter 5.

$$x^{-3} = 8$$
$$(x^{-3})^{-1} = 8^{-1} \qquad \text{Raise each side to the } -1 \text{ power.}$$
$$x^3 = \frac{1}{8}$$
$$x = \sqrt[3]{\frac{1}{8}} = \frac{1}{2} \qquad \text{Odd-root property}$$

The solution set is $\{1/2\}$.

c) Because each side of the equation is a logarithm, the one-to-one property of logarithms is used to get an equivalent equation.

$$\log(x^2) = \log(4)$$
$$x^2 = 4 \qquad \text{One-to-one property of logarithms}$$
$$x = \pm 2 \qquad \text{Even-root property}$$

The solution set is $\{\pm 2\}$.   ◄

Applications

When money is compounded continuously, the formula

$$t = \frac{1}{r}\ln(A/P)$$

expresses the relationship between the time in years, t, the annual interest rate, r, the principal, P, and the amount, A. This formula is used to determine how long it takes for a deposit to grow to a specific amount.

EXAMPLE 8   How long does it take $80 to grow to $240 at 12% compounded continuously?

Solution Use $r = .12$, $P = \$80$, and $A = \$240$ in the formula, and use a calculator to evaluate the logarithm.

$$t = \frac{1}{.12} \ln(240/80) = \frac{\ln(3)}{.12} = 9.155$$

It takes approximately 9.155 years, or 9 years and 57 days. ◀

Warm-ups

True or false?

1. $\log_{25} 5 = 2$
2. The statement $a^3 = 2$ is equivalent to the statement $\log_a 2 = 3$.
3. $\log(-10) = -1$
4. If $f(x) = a^x$, then $f^{-1}(x) = \log_a x$.
5. $\log(0) = 0$
6. $5^{\log_5 7} = 7$
7. $\log_{1/2}(32) = -5$
8. If $f(x) = \ln(x)$, then $f^{-1}(x) = e^x$.
9. The graphs of exponential functions and logarithmic functions are totally unrelated.
10. The domain of $f(x) = \log_6 x$ is R.

9.2 EXERCISES

Write each exponential equation as a logarithmic equation and each logarithmic equation as an exponential equation. See Example 1.

1. $\log_2 8 = 3$
2. $\log_{10} 10 = 1$
3. $10^2 = 100$
4. $5^3 = 125$
5. $y = \log_5 x$
6. $m = \log_b N$
7. $2^a = b$
8. $a^3 = c$
9. $\log_3 x = 10$
10. $\log_c t = 4$
11. $e^3 = x$
12. $m = e^x$

Evaluate the following. See Examples 2 and 3.

13. $\log_2 4$
14. $\log_2 1$
15. $\log_2 16$
16. $\log_4 16$
17. $\log_2 64$
18. $\log_8 64$
19. $\log_4 64$
20. $\log_{64} 64$
21. $\log_2(1/4)$

22. $\log_2(1/8)$

23. $\log(100)$

24. $\log(1)$

25. $\log(.01)$

26. $\log(10,000)$

27. $\log_{1/3}(1/3)$

28. $\log_{1/3}(1/9)$

29. $\log_{1/3}(27)$

30. $\log_{1/3}(1)$

31. $\ln(e^3)$

32. $\ln(1)$

33. $\ln(e^2)$

34. $\ln(1/e)$

35. $\log(5)$

36. $\log(.03)$

37. $\ln(6.238)$

38. $\ln(.23)$

Sketch the graph of each function. See Examples 4 and 5.

39. $f(x) = \log_3 x$

40. $g(x) = \log_{10} x$

41. $y = \log_4 x$

42. $y = \log_5 x$

43. $h(x) = \log_{1/4}(x)$

44. $y = \log_{1/3}(x)$

For each function f find f^{-1}. See Example 6.

45. $f(x) = 6^x$

46. $f(x) = 4^x$

47. $f(x) = \ln(x)$

48. $f(x) = \log(x)$

49. $f(x) = \log_{1/2}(x)$

50. $f(x) = \log_{1/4}(x)$

Solve each equation for x. See Example 7.

51. $x = (1/2)^{-2}$

52. $x = 3^2$

53. $5 = 25^x$

54. $.1 = 10^x$

55. $\log(x) = -3$

56. $\log(x) = 5$

57. $\log_x 36 = 2$

58. $\log_x 100 = 2$

59. $\log_x 5 = -1$

60. $\log_x 16 = -2$

61. $\log(x^2) = \log(9)$

62. $\ln(2x - 3) = \ln(x + 1)$

63. $3 = 10^x$

64. $10^x = .03$

65. $10^x = 1/2$

66. $75 = 10^x$

Solve each problem. See Example 8.

67. How long does it take $5000 to grow to $10,000 at 12% compounded continuously?

68. How long does it take $5000 to grow to $10,000 at 6% compounded continuously?

69. How long does it take to earn $1000 in interest on a deposit of $6000 at 8% compounded continuously?

70. How long does it take to earn $1000 interest on a deposit of one million dollars at 9% compounded continuously?

72. Find the pH of a solution with a hydrogen ion concentration of one mole per liter.

73. Find the pH of a solution with a hydrogen ion concentration of 3.16×10^{-12} moles per liter. Round to nearest hundredth.

74. Find the pH of a solution with a hydrogen ion concentration of 5.68×10^{-3} moles per liter.

In chemistry, the pH of a solution is defined to be

$$pH = -\log_{10}[H^+],$$

where H^+ is the hydrogen ion concentration of the solution in moles per liter.

71. Find the pH of a solution with a hydrogen ion concentration of 10^{-7} moles per liter.

9.3 Properties

IN THIS SECTION:

- Logarithm of a Product
- Logarithm of a Quotient
- Logarithm of a Power
- Inverse Properties
- Summary
- Using the Properties

The properties of logarithms are very similar to the properties of exponents, because logarithms are exponents. In this section we will use the properties of exponents to write some properties of logarithms. The properties will be used in solving logarithmic equations in Section 9.4.

Logarithm of a Product

If $M = a^x$ and $N = a^y$, then we can use the product rule for exponents to write

$$MN = a^x \cdot a^y = a^{x+y}.$$

The equation

$$MN = a^{x+y} \quad \text{is equivalent to} \quad \log_a(MN) = x + y.$$

Because $M = a^x$ and $N = a^y$ are equivalent to $\log_a M = x$ and $\log_a N = y$, we can replace x and y in the equation

$$\log_a(MN) = x + y$$

to get the product rule for logarithms.

Product Rule for Logarithms

$$\log_a(MN) = \log_a M + \log_a N$$

The product rule says that *the logarithm of a product is the sum of the logarithms*, provided all of the logarithms are defined.

EXAMPLE 1 Write each expression as a single logarithm.

a) $\log_2 7 + \log_2 5$

b) $\ln(\sqrt{2}) + \ln(\sqrt{3})$

Solution

a) $\log_2 7 + \log_2 5 = \log_2 35$ By the product rule for logarithms
b) $\ln(\sqrt{2}) + \ln(\sqrt{3}) = \ln(\sqrt{6})$ By the product rule for logarithms   ◄

Logarithm of a Quotient

If $M = a^x$ and $N = a^y$, then we can use the quotient rule for exponents to write

$$\frac{M}{N} = \frac{a^x}{a^y} = a^{x-y}.$$

The equation

$$\frac{M}{N} = a^{x-y}$$

is equivalent to

$$\log_a(M/N) = x - y.$$

Because $\log_a M = x$ and $\log_a N = y$, we can replace x and y in the equation

$$\log_a(M/N) = x - y$$

to get the quotient rule for logarithms.

**Quotient Rule
for Logarithms**

$$\log_a(M/N) = \log_a M - \log_a N$$

The quotient rule says that *the logarithm of a quotient is the difference of the logarithms,* provided all of the logarithms are defined.

EXAMPLE 2   Write each expression as a single logarithm.

a) $\log_2 3 - \log_2 7$ **b)** $\ln(w^8) - \ln(w^2)$

Solution

a) $\log_2 3 - \log_2 7 = \log_2(3/7)$ By the quotient rule for logarithms

b) $\ln(w^8) - \ln(w^2) = \ln\left(\dfrac{w^8}{w^2}\right)$ By the quotient rule for logarithms

$\qquad\qquad\qquad\quad = \ln(w^6)$ By the quotient rule for exponents   ◄

Logarithm of a Power

If $M = a^x$, we can use the power rule for exponents to write

$$M^N = (a^x)^N = a^{Nx}.$$

The equation

$$M^N = a^{Nx} \qquad \text{is equivalent to} \qquad \log_a(M^N) = Nx. \cdot$$

Because $\log_a M = x$, we get the power rule for logarithms.

Power Rule for Logarithms

$$\log_a(M^N) = N \cdot \log_a M$$

The power rule says that *the logarithm of a power of a number is the power times the logarithm of the number*, provided all of the logarithms are defined.

EXAMPLE 3 Rewrite each logarithm in terms of log2.

a) $\log(2^{10})$ **b)** $\log\sqrt{2}$ **c)** $\log(1/2)$

Solution

a) $\log(2^{10}) = 10 \cdot \log2$ By the power rule for logarithms

b) $\log\sqrt{2} = \log(2^{1/2}) = \dfrac{1}{2} \cdot \log2$ By the power rule for logarithms

c) $\log(1/2) = \log(2^{-1}) = -\log2$ By the power rule for logarithms   ◄

Inverse Properties

The exponential function with base a and the logarithm function with base a are inverses of each other. We can express this fact with the following two inverse properties.

Inverse Properties

$$\log_a a^M = M \quad \text{and} \quad a^{\log_a M} = M$$

For example,
$$\ln(e^5) = 5 \quad \text{and} \quad 2^{\log_2 8} = 8.$$

Summary

A summary of the properties of logarithms follows.

Properties of Logarithms

If M, N, and a are positive numbers, $a \neq 1$, then

1. $\log_a a = 1$ **2.** $\log_a 1 = 0$

3. $\log_a a^M = M$ **4.** $a^{\log_a M} = M$

5. $\log_a MN = \log_a M + \log_a N$ Product rule

6. $\log_a(M/N) = \log_a M - \log_a N$ Quotient rule

7. $\log_a(M^N) = N \cdot \log_a M$ Power rule

8. $\log_a(1/N) = -\log_a N$

Note that property 8 is property 6 with $M = 1$.

Using the Properties

The rules of logarithms are often used to write the logarithm of a complicated expression in terms of logarithms of simpler expressions.

EXAMPLE 4   Use the properties of logarithms to write each of the following expressions in terms of log2 and log3.

a) log6   **b)** log16   **c)** log(2/3)   **d)** log(9/2)   **e)** log(1/3)

Solution

a) $\log 6 = \log(2 \cdot 3) = \log 2 + \log 3$ By the product rule
b) $\log 16 = \log(2^4) = 4 \cdot \log 2$ By the power rule
c) $\log(2/3) = \log 2 - \log 3$ By the quotient rule
d) $\log(9/2) = \log 9 - \log 2 = \log(3^2) - \log 2$ By the quotient rule
$\qquad\qquad\quad = 2 \cdot \log 3 - \log 2$ By the power rule
e) $\log(1/3) = -\log 3$ By property 8   ◄

EXAMPLE 5   Rewrite each expression using a sum or difference of multiples of logarithms.

a) $\log\left[\dfrac{xz}{y}\right]$ **b)** $\log_3\left[\dfrac{(x-3)^{2/3}}{\sqrt{x}}\right]$

Solution

a) $\log\left[\dfrac{xz}{y}\right] = \log(xz) - \log(y) = \log(x) + \log(z) - \log(y)$

b) $\log_3\left[\dfrac{(x-3)^{2/3}}{\sqrt{x}}\right] = \log_3(x-3)^{2/3} - \log_3\sqrt{x}$ $\sqrt{x} = x^{1/2}$

$\qquad\qquad\qquad = \dfrac{2}{3}\log_3(x-3) - \dfrac{1}{2}\log_3 x$ ◄

In the next example we use the properties of logarithms to convert an expression involving several logarithms into a single logarithm. We are learning to rewrite expressions involving logarithms in order to solve equations involving logarithms in Section 9.4.

EXAMPLE 6   Rewrite each expression as a single logarithm.

a) $\dfrac{1}{2}\log(x) - 2 \cdot \log(x+1)$ **b)** $3 \cdot \log(y) + \dfrac{1}{2}\log(z) - \log(x)$

Solution

a) $\dfrac{1}{2}\log(x) - 2 \cdot \log(x+1) = \log(x)^{1/2} - \log(x+1)^2$ By the power rule

$\qquad\qquad\qquad\qquad = \log\left[\dfrac{\sqrt{x}}{(x+1)^2}\right]$ By the quotient rule

b) $3 \cdot \log(y) + \dfrac{1}{2}\log(z) - \log(x) = \log(y^3) + \log\sqrt{z} - \log(x)$ **By the power rule**

$$= \log(y^3 \cdot \sqrt{z}) - \log(x) \quad \text{By the product rule}$$

$$= \log\left[\frac{y^3 \cdot \sqrt{z}}{x}\right] \quad \text{By the quotient rule} \quad \blacktriangleleft$$

Warm-ups

True or false?

1. $\log_2(x^2/8) = \log_2(x^2) - 3$ **2.** $\dfrac{\log(100)}{\log(10)} = \log(100) - \log(10)$

3. $\ln\sqrt{2} = \dfrac{\ln(2)}{2}$ **4.** $3^{\log_3 17} = 17$

5. $\log_2(1/8) = \dfrac{1}{\log_2 8}$ **6.** $\ln(8) = 3\ln(2)$

7. $\ln(1) = e$ **8.** $\dfrac{\log(100)}{10} = \log(10)$

9. $\dfrac{\log_2 8}{\log_2 2} = \log_2 4$ **10.** $\ln(2) + \ln(3) - \ln(7) = \ln(6/7)$

9.3 EXERCISES

Assume that all variables involved in logarithms represent numbers for which the logarithms are defined.

Write each expression as a single logarithm. See Example 1.

1. $\log 3 + \log 7$ **2.** $\ln 5 + \ln 4$ **3.** $\log_3 5 + \log_3 x$

4. $\ln(x) + \ln(y)$ **5.** $\log(x^2) + \log(x^3)$ **6.** $\ln(a^3) + \ln(a^5)$

7. $\ln 2 + \ln 3 + \ln 5$ **8.** $\log_2 x + \log_2 y + \log_2 z$

Write each expression as a single logarithm. See Example 2.

9. $\log 8 - \log 2$ **10.** $\ln 3 - \ln 6$ **11.** $\log_2(x^6) - \log_2(x^2)$

12. $\ln(w^9) - \ln(w^3)$ **13.** $\log 7 - \log 9$ **14.** $\log_3 13 - \log_3 4$

Write each expression in terms of log3. See Example 3.

15. $\log(27)$ **16.** $\log(1/9)$ **17.** $\log(\sqrt{3})$

18. $\log(\sqrt[4]{3})$ **19.** $\log(3^x)$ **20.** $\log(3^{-99})$

Use the properties of logarithms to write each of the following in terms of log(3) and log(5). See Example 4.

21. $\log(15)$

22. $\log(9)$

23. $\log(5/3)$

24. $\log(3/5)$

25. $\log(25)$

26. $\log(1/27)$

27. $\log(75)$

28. $\log(.6)$

29. $\log(1/3)$

30. $\log(45)$

31. $\log(.2)$

32. $\log(9/25)$

Rewrite each expression as a sum or difference of multiples of logarithms. Assume that all variables are such that all expressions are defined. See Example 5.

33. $\log(xyz)$

34. $\log(3y)$

35. $\log_2 8x$

36. $\log_2 16y$

37. $\ln(x/y)$

38. $\ln(z/3)$

39. $\log(10x^2)$

40. $\log(100\sqrt{x})$

41. $\log_5\left[\dfrac{(x-3)^2}{\sqrt{w}}\right]$

42. $\log_3\left[\dfrac{(y+6)^3}{y-5}\right]$

43. $\ln\left[\dfrac{yz\sqrt{x}}{w}\right]$

44. $\ln\left[\dfrac{(x-1)\sqrt{w}}{x^3}\right]$

Rewrite each expression as a single logarithm. See Example 6.

45. $\ln(x) - \ln(w) + \ln(z)$

46. $\ln(x) - \ln(3) - \ln(7)$

47. $3 \cdot \ln(y) + 2 \cdot \ln(x) - \ln(w)$

48. $5 \cdot \ln(r) + 3 \cdot \ln(t) - 4 \cdot \ln(s)$

49. $\dfrac{1}{2}\log(x-3) - \dfrac{2}{3}\log(x+1)$

50. $\dfrac{1}{2}\log(y-4) + \dfrac{1}{2}\log(y+4)$

51. $\dfrac{2}{3}\log_2(x-1) - \dfrac{1}{4}\log_2(x+2)$

52. $\dfrac{1}{2}\log_3(y+3) + 6 \cdot \log_3(y)$

Determine whether each of the following equations is true or false.

53. $\log(56) = \log(7) \cdot \log(8)$

54. $\log(5/9) = \dfrac{\log(5)}{\log(9)}$

55. $\log_2(4^2) = (\log_2 4)^2$

56. $\ln(4^2) = (\ln 4)^2$

57. $\ln(25) = 2 \cdot \ln(5)$

58. $\ln(3e) = 1 + \ln(3)$

59. $\dfrac{\log_2 64}{\log_2 8} = \log_2 8$

60. $\dfrac{\log_2 16}{\log_2 4} = \log_2 4$

61. $\log(1/3) = -\log(3)$

62. $\log_2(8 \cdot 2^{59}) = 62$

63. $\log_2(16^5) = 20$

64. $(\log_2 5)(\log_5 2) = 1$

65. $\log(10^3) = 3$

66. $\log_3 3^7 = 7$

67. $\log(100 + 3) = 2 + \log 3$

68. $\dfrac{\log_7 32}{\log_7 8} = \dfrac{5}{3}$

9.4 Solving Equations

IN THIS SECTION:
- **Logarithmic Equations**
- **Exponential Equations**
- **Changing the Base**
- **Strategies for Equation Solving**
- **Applications**

We solved some equations involving exponents and logarithms in Sections 1 and 2 of this chapter. In this section we will use the properties of exponents and logarithms to solve more complex equations.

Logarithmic Equations

Our first example involves only one logarithm.

EXAMPLE 1   Solve $\log(x + 3) = 2$.

Solution   Write the equivalent exponential equation.

$$10^2 = x + 3$$
$$100 = x + 3$$
$$97 = x$$

The solution set is $\{97\}$. Check.   ◄

In the next example, we use the product rule for logarithms to write a sum of two logarithms as a single logarithm.

EXAMPLE 2   Solve $\log_2(x + 3) + \log_2(x - 3) = 4$.

Solution   Because the sum of two logarithms is the logarithm of the product, we can write

$$\log_2[(x + 3)(x - 3)] = 4$$
$$x^2 - 9 = 2^4 \qquad \text{Write as an exponential equation.}$$
$$x^2 - 9 = 16$$
$$x^2 = 25$$
$$x = \pm 5.$$

If we check $x = -5$, we get

$$\log_2(-5 + 3) + \log_2(-5 - 3) = 4$$
$$\log_2(-2) + \log_2(-8) = 4. \qquad \text{Incorrect.}$$

Because the domain of any logarithm function is the set of positive real numbers, neither of these logarithms is defined. If we check $x = 5$, we get

$$\log_2(5 + 3) + \log_2(5 - 3) = 4$$
$$\log_2(8) + \log_2(2) = 4$$
$$3 + 1 = 4. \qquad \text{Correct.}$$

The solution set is $\{5\}$.   ◄

EXAMPLE 3   Solve $\log(x) + \log(x - 1) = \log(8x - 12) - \log(2)$.

Solution   Apply the product rule to the left-hand side and the quotient rule to the right-hand side to get a single logarithm on each side.

$$\log[x(x - 1)] = \log\left[\frac{8x - 12}{2}\right]$$
$$x(x - 1) = \frac{8x - 12}{2} \qquad \text{By the one-to-one property of logarithms}$$
$$x^2 - x = 4x - 6 \qquad \text{Simplify.}$$
$$x^2 - 5x + 6 = 0$$
$$(x - 2)(x - 3) = 0$$
$$x - 2 = 0 \quad \text{or} \quad x - 3 = 0$$
$$x = 2 \quad \text{or} \qquad x = 3$$

Neither 2 nor 3 causes undefined terms in the original equation, so the solution set is $\{2, 3\}$.   ◄

Exponential Equations

If an equation has a single exponential expression, we can write the equivalent logarithmic equation.

EXAMPLE 4   Find the exact solution to the equation $2^x = 10$.

Solution   The equivalent logarithmic equation is

$$x = \log_2 10.$$

The solution set is $\{\log_2 10\}$. The number $\log_2 10$ is the exact solution to the equation. Later in this section we will learn how to use the change-of-base formula to find an approximate value for an expression of this type.   ◄

When solving exponential equations, we may be able to write each side as a power of the same base.

EXAMPLE 5 Solve $2^{(x^2)} = 4^{3x-4}$.

Solution Because $4 = 2^2$, we can write

$$2^{(x^2)} = (2^2)^{3x-4}.$$

Then
$$2^{(x^2)} = 2^{6x-8} \qquad 2(3x-4) = 6x - 8$$

$$x^2 = 6x - 8 \qquad \begin{array}{l} \text{By the one-to-one property} \\ \text{of exponential functions} \end{array}$$

$$x^2 - 6x + 8 = 0$$

$$(x - 4)(x - 2) = 0$$

$$x - 4 = 0 \quad \text{or} \quad x - 2 = 0$$

$$x = 4 \quad \text{or} \qquad x = 2.$$

The solution set is $\{2, 4\}$.   ◀

The following example shows how to solve an equation where the sides are powers of apparently unrelated bases.

EXAMPLE 6 Find the exact and approximate solution to $2^{2x-1} = 3^x$.

Solution We first take the base 10 logarithm of each side.

$$\log(2^{2x-1}) = \log(3^x)$$

$$(2x - 1)\log 2 = x \cdot \log 3 \qquad \text{By the power rule}$$

$$2x \cdot \log 2 - \log 2 = x \cdot \log 3 \qquad \begin{array}{l} \text{By using the distributive property} \\ \text{on the left-hand side} \end{array}$$

$$2x \cdot \log 2 - x \cdot \log 3 = \log 2 \qquad \text{Solve for } x.$$

$$x(2 \cdot \log 2 - \log 3) = \log 2 \qquad \text{Factor out } x.$$

$$x = \frac{\log 2}{2 \cdot \log 2 - \log 3}$$

This expression is the exact solution to the equation. To find an approximate answer, we use a calculator to find a decimal value for this expression. We get $x \approx 2.409$. Check this answer with a calculator. For the exact answer we could have taken the logarithm of each side using any base, but to find the approximate answer we use base 10, because it is available on the calculator.   ◀

Changing the Base

Scientific calculators have an x^y key for computing any power of any base, in addition to the function keys for computing 10^x and e^x. For logarithms we have the keys ln and log, but there are no function keys for logarithms using other bases. The following example shows how to find a decimal value for a base 3 logarithm.

EXAMPLE 7 Find a decimal approximation to $\log_3 7$.

Solution Let $y = \log_3 7$, and write the equivalent exponential equation.

$$3^y = 7$$

$$\log(3^y) = \log(7) \qquad \text{Take the base 10 logarithm of each side.}$$

$$y \cdot \log(3) = \log(7) \qquad \text{By the power rule}$$

$$y = \frac{\log(7)}{\log(3)} \qquad \text{Divide.}$$

$$\approx \frac{.8451}{.4771} \approx 1.7712 \qquad \begin{array}{l}\text{Use a calculator to find}\\ \log(7) \text{ and } \log(3).\end{array}$$

Thus $\log_3 7 \approx 1.7712$. ◀

We can easily find a general formula for changing bases on a logarithm. We proceed exactly as in Example 7. If

$$y = \log_a M,$$

then

$$a^y = M$$

$$\log_b(a^y) = \log_b M \qquad \text{Take the base } b \text{ logarithm of each side.}$$

$$y \cdot \log_b a = \log_b M \qquad \text{By the power rule}$$

$$y = \frac{\log_b M}{\log_b a}. \qquad \text{Divide.}$$

The last equation gives us a formula for writing a logarithm with one base in terms of logarithms with a different base.

Base-change Formula

If a and b are positive numbers, not equal to one, and M is positive, then

$$\log_a M = \frac{\log_b M}{\log_b a}.$$

In words, we take the logarithm with the new base and divide by the logarithm of the old base.

Strategies for Equation Solving

To solve equations involving logarithms and exponents, we need to keep certain strategies in mind. There is no formula that will solve every equation in this section. The following list summarizes the ideas that we used in solving these equations.

STRATEGY
Solving Exponential and
Logarithmic Equations

1. If the equation has a single logarithm, write the equivalent exponential equation (Example 1).
2. Use the properties of logarithms to combine logarithms as much as possible (Example 2).
3. If the equation has a single exponential expression, write the equivalent logarithmic equation (Example 4).
4. Use the one-to-one properties:
   a) If $\log_a M = \log_a N$, then $M = N$ (Example 3).
   b) If $a^M = a^N$, then $M = N$ (Example 5).
5. To get a decimal answer take the log or ln of each side of an exponential equation (Example 6).

Applications

EXAMPLE 8 If $500 is deposited into an account paying 8% compounded quarterly, then in how many quarters will the account have $1000 in it?

Solution We use the compound-interest formula, $S = P(1 + i)^n$, with a principal of $500, an amount of $1000, and an interest rate of 2% each quarter.

$$1000 = 500(1.02)^n$$

$$2 = (1.02)^n \qquad \text{Divide each side by 500.}$$

$$\ln(2) = \ln(1.02)^n \qquad \text{Take the ln of each side.}$$

$$\ln(2) = n \cdot \ln(1.02) \qquad \text{By the power rule}$$

$$\frac{\ln(2)}{\ln(1.02)} = n \qquad \text{Divide each side by ln(1.02).}$$

$$n = 35.0028 \qquad \text{Use a calculator to find an approximate answer.}$$

It takes approximately 35 quarters, or 8 years and 9 months, for the initial investment to be worth $1000. Note that we get the same answer using the base 10 logarithm, log. Try it. ◀

Warm-ups

True or false?

1. If $\log(x - 2) + \log(x + 2) = 7$, then $\log(x^2 - 4) = 7$.
2. If $\log(3x + 7) = \log(5x - 8)$, then $3x + 7 = 5x - 8$.
3. If $e^{x-6} = e^{x^2-5x}$, then $x - 6 = x^2 - 5x$.

4. If $2^{3x-1} = 3^{5x-4}$, then $3x - 1 = 5x - 4$.

5. If $\log_2(x^2 - 3x + 5) = 3$, then $x^2 - 3x + 5 = 8$.

6. $\log_3 5 = \dfrac{\ln(3)}{\ln(5)}$

7. $\dfrac{\ln(2)}{\ln(6)} = \dfrac{\log(2)}{\log(6)}$

8. $\log(5) = \ln(5)$

9. If $2^{2x-1} = 3$, then $2x - 1 = \log_2 3$.

10. If $5^x = 23$, then $x \cdot \ln(5) = \ln(23)$.

9.4 EXERCISES

Solve each equation. See Examples 1 and 2.

1. $\log_2(x + 1) = 3$

2. $\log_3(x^2) = 4$

3. $\log(x) + \log(5) = 1$

4. $\ln(x) + \ln(3) = 0$

5. $\log_2(x - 1) + \log_2(x + 1) = 3$

6. $\log_3(x - 4) + \log_3(x + 4) = 2$

7. $\log_2(x - 1) - \log_2(x + 2) = 2$

8. $\log_4(8x) - \log_4(x - 1) = 2$

9. $\log_2(x - 4) + \log_2(x + 2) = 4$

10. $\log_6(x + 6) + \log_6(x - 3) = 2$

Solve each equation. See Example 3.

11. $\ln(x) + \ln(x + 5) = \ln(x + 1) + \ln(x + 3)$

12. $\log(x) + \log(x + 5) = 2 \cdot \log(x + 2)$

13. $\log(x + 3) + \log(x + 4) = \log(x^3 + 13x^2) - \log(x)$

14. $\log(x^2 - 1) - \log(x - 1) = \log(6)$

15. $2 \cdot \log(x) = \log(20 - x)$

16. $2 \cdot \log(x) + \log(3) = \log(2 - 5x)$

Solve each equation. See Examples 4 and 5.

17. $3^x = 7$

18. $2^{x-1} = 5$

19. $2^{3x+4} = 4^{x-1}$

20. $9^{2x-1} = 27^{1/2}$

21. $(1/3)^x = 3^{1+x}$

22. $4^{3x} = (1/2)^{1-x}$

Find the exact solution and approximate solution to each equation. See Example 6.

23. $x \cdot \ln2 = \ln7$

24. $x \cdot \log3 = \log5$

25. $3x - x \cdot \ln2 = 1$

26. $2x + x \cdot \log5 = \log7$

27. $3^x = 5$

28. $2^x = 1/3$

29. $2^{x-1} = 9$

30. $10^{x-2} = 6$

31. $2^x = 3^{x+5}$

32. $e^x = 10^x$

33. $5^{x+2} = 10^{x-4}$

34. $3^{2x} = 6^{x+1}$

35. $8^x = 9^{x-1}$

36. $5^{x+1} = 8^{x-1}$

Use a scientific calculator and the base-change formula to find an approximation for each of the following. Round answers to four decimal places. See Example 7.

37. $\log_2 3$

38. $\log_3 5$

39. $\log_3(1/2)$

40. $\log_5(2.56)$

41. $\log_{1/2}(4.6)$

42. $\log_{1/3}(3.5)$

43. $\log_{.1}(.03)$

44. $\log_{.2}(1.06)$

Solve each of the following problems. See Example 8.

45. How many months does it take for $1000 to grow to $1500 in an account paying 12% compounded monthly?

46. How many years does it take for $25 to grow to $100 in an account paying 8% compounded annually?

47. In a certain country the number of people above the poverty level is presently 28 million and growing 5% annually. Assuming that the population is growing continuously, the population, P, (in millions), t years from now, is determined by the formula $P = 28e^{.05t}$. In how many years will there be 40 million people above the poverty level?

48. In the same country of Problem 47, the number of people below the poverty level is currently 20 million and growing 7% annually. This population (in millions), t years from now, is determined by the formula $P = 20e^{.07t}$. In how many years will there be 40 million people below the poverty level?

49. For this problem use the information given in Problems 47 and 48. In how many years will the number of people above the poverty level equal the number of people below the poverty level?

50. In a declining country there are currently 100 million workers and 40 million retired people. The population of workers is decreasing according to the formula $W = 100e^{-.01t}$, where t is in years and W is in millions. The population of retired people is increasing according to the formula $R = 40e^{.09t}$, where t is in years and R is in millions. In how many years will the number of workers equal the number of retired people?

51. What is the hydrogen ion concentration of a solution with a pH of 4.2? (See Exercises 71–74 of Section 9.2)

52. What is the hydrogen ion concentration of a solution with a pH of 8.6?

Find the exact solution to each equation.

53. $3^x = 20$

54. $2^x = 128$

55. $\log_3(x) + \log_3(5) = 1$

56. $\log(x) - \log(3) = \log(6)$

57. $8^x = 2^{x+1}$

58. $2^x = 5^{x+1}$

Wrap-up

CHAPTER 9

SUMMARY

	Concepts	Examples
Exponential function	Variable in the position of the exponent $f(x) = a^x \quad a > 0,\ a \neq 1$	$f(x) = 2^x$ Domain: R Range: $y > 0$
Logarithm function	Inverse of exponential function $f(x) = \log_a x$ or $y = \log_a x$ means $a^y = x$ Base 10: common logarithm (log) Base e: natural logarithm (ln)	$f(x) = \log_2 x$ Domain: $x > 0$ Range: R

Graphs

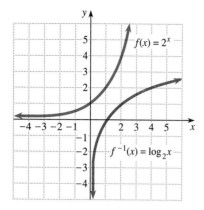

Properties of logarithms

If M, N, and a are positive numbers, $a \neq 1$, then

1. $\log_a a = 1$
2. $\log_a 1 = 0$
3. $\log_a(a^M) = M$
4. $a^{\log_a M} = M$
5. $\log_a MN = \log_a M + \log_a N$

6. $\log_a(M/N) = \log_a M - \log_a N$

7. $\log_a(M^N) = N \cdot \log_a M$

8. $\log_a(1/N) = -\log_a N$
9. $\log_a M = \dfrac{\log_b M}{\log_b a}$

$\log_5 5 = 1$, because $5^1 = 5$
$\log_3 1 = 0$, because $3^0 = 1$
$\log_2(2^{13}) = 13$
$10^{\log 19} = 19$
Product rule
$\quad \ln(3x) = \ln 3 + \ln(x)$
Quotient rule
$\quad \ln(1/3) = \ln(1) - \ln(3)$
Power rule
$\quad \log(x^3) = 3 \cdot \log(x)$
$\ln(1/3) = -\ln(3)$
Base change
$$\log_3 5 = \frac{\ln 5}{\ln 3}$$

Solving equations

1. Use the properties of logarithms to combine logarithms as much as possible.
2. If there is a single logarithm, write the equivalent exponential equation.
3. If there is a single exponential expression, write the equivalent logarithmic equation.
4. If $\log_a M = \log_a N$, then $M = N$.

5. If $a^M = a^N$, then $M = N$.
6. To get a decimal answer, take the log or ln of each side of an exponential equation.

$\log x + \log(x - 3) = 1$
$\log[x(x - 3)] = 1$
$\quad\quad x(x - 3) = 10^1$

$2^x = 3, \quad\quad x = \log_2 3$

$\ln(x) = \ln(2x - 5)$
$\quad\quad x = 2x - 5$
$2^{3x} = 2^{5x-7}, \quad\quad 3x = 5x - 7$
$\quad\quad 2^x = 3$
$\ln(2^x) = \ln 3$
$$x = \frac{\ln 3}{\ln 2}$$

REVIEW EXERCISES

9.1 *Let* $f(x) = 5^x$, $g(x) = 10^{x-1}$, *and* $h(x) = (1/4)^x$. *Find the following.*

1. $f(-2)$ **2.** $f(0)$ **3.** $f(3)$ **4.** $f(4)$

5. $g(1)$ **6.** $g(-1)$ **7.** $g(0)$ **8.** $g(3)$

9. $h(-1)$ **10.** $h(2)$ **11.** $h(1/2)$ **12.** $h(-1/2)$

13. x, if $f(x) = 25$ **14.** x, if $f(x) = -1/125$ **15.** x, if $g(x) = 1000$ **16.** x, if $g(x) = .001$

17. x, if $h(x) = 32$ **18.** x, if $h(x) = 8$ **19.** $f(1.34)$ **20.** $f(-3.6)$

21. $g(3.25)$ **22.** $g(4.87)$ **23.** $h(2.82)$ **24.** $h(\pi)$

25. $h(\sqrt{2})$ **26.** $h(1/3)$

Sketch the graph of each function.

27. $f(x) = 5^x$ **28.** $g(x) = e^x$ **29.** $y = (1/5)^x$ **30.** $y = 3^{-x}$

31. $f(x) = -3^{x-1}$ **32.** $y = 1 + 2^x$ **33.** $y = 1 - 2^x$ **34.** $y = e^{-x}$

9.2 *Let* $f(x) = \log_2 x$, $g(x) = \log(x)$, *and* $h(x) = \log_{1/2} x$. *Find the following.*

35. $f(1/8)$ **36.** $f(64)$ **37.** $g(.1)$ **38.** $g(1)$

39. $g(100)$ **40.** $h(1/8)$ **41.** $h(1)$ **42.** $h(4)$

43. x, if $f(x) = 8$ **44.** x, if $g(x) = 3$ **45.** $f(77)$ **46.** $g(88.4)$

47. $h(33.9)$ **48.** $h(.05)$ **49.** x, if $f(x) = 2.475$ **50.** x, if $g(x) = 1.426$

For each function f, find f^{-1} *and sketch the graphs of f and* f^{-1} *on the same set of axes.*

51. $f(x) = 10^x$ **52.** $f(x) = \log_8 x$ **53.** $f(x) = e^x$ **54.** $f(x) = \log_3 x$

9.3 *Rewrite each expression as a sum or difference of multiples of logarithms.*

55. $\log(x^2 y)$ **56.** $\log_3(x^2 + 2x)$ **57.** $\ln(16)$ **58.** $\log(y/\sqrt{x})$

59. $\log_5(1/x)$ **60.** $\ln(xy/z)$

Rewrite each expression as a single logarithm.

61. $\dfrac{1}{2}\log(x + 2) - 2 \cdot \log(x - 1)$ **62.** $3 \cdot \ln(x) + 2 \cdot \ln(y) - \dfrac{1}{3} \cdot \ln(z)$

9.4 *Find the exact solution to each equation.*

63. $\log_2 x = 8$ **64.** $\log_3 x = .5$ **65.** $\log_2 8 = x$

66. $3^x = 8$ **67.** $x^3 = 8$ **68.** $3^2 = x$

69. $\log_x 27 = 3$ **70.** $\log_x 9 = -1/3$ **71.** $x \cdot \ln 3 - x = \ln 7$

72. $x \cdot \log 8 = x \cdot \log 4 + \log 9$ **73.** $3^x = 5^{x-1}$ **74.** $5^{2x^2} = 5^{3-5x}$

75. $4^{2x} = 2^{x+1}$ **76.** $\log(12) = \log(x) + \log(7 - x)$ **77.** $\ln(x + 2) - \ln(x - 10) = \ln(2)$

78. $2 \cdot \ln(x + 3) = 3 \cdot \ln 4$ **79.** $\log(x) - \log(x - 2) = 2$ **80.** $\log_2 x = \log_2(x + 16) - 1$

Use a scientific calculator to find an approximate solution to each of the following. Round your answers to 4 decimal places.

81. $6^x = 12$ **82.** $5^x = 8^{3x+2}$ **83.** $3^{x+1} = 5$ **84.** $\log_3 x = 2.634$

Miscellaneous

85. What does $10,000 invested at 11.5% compounded annually amount to after 15 years?

86. How many years does it take for an investment to double at 6.5% compounded annually?

87. The amount, A, of a certain radioactive substance remaining after t years is given by the formula $A = A_0 e^{-.0003t}$, where A_0 is the initial amount. If we have 218 grams of this substance today, then how much of it will be left 1000 years from now?

88. The population of white-tail deer in a certain national forest is believed to be growing according to the formula $P = 517 + 10 \cdot \ln(8t + 1)$, where t is the time in years from the present. What is the current population? In how many years will there be 600 deer?

89. Melissa deposited $1000 into an account paying 5% annually, and on the same day, Frank deposited $900 into an account paying 7% compounded continuously. In how many years will they both have the same amount in their accounts?

90. The value of imported goods for a small South American country is believed to be growing according to the formula $I = 15 \cdot \log(16t + 33)$, and the value of exported goods appears to be growing according to the formula

$$E = 30 \cdot \log(t + 3),$$

where I and E are in millions of dollars, and t is the time in years from the present. What is the current value of imports and exports? In how many years will the value of imports equal the value of exports?

Photo for exercise 88

CHAPTER 9 TEST

Let $f(x) = 5^x$ and $g(x) = \log_5 x$. Find the following.

1. $f(2)$ **2.** $f(-1)$ **3.** $f(0)$ **4.** $g(125)$ **5.** $g(1)$ **6.** $g(1/5)$

Sketch the graph of each function.

7. $y = 2^x$ **8.** $f(x) = \log_2 x$ **9.** $y = (1/3)^x$ **10.** $g(x) = \log_{1/3} x$

Suppose $\log_a M = 6$ and $\log_a N = 4$. Find the following.

11. $\log_a (MN)$ **12.** $\log_a (M^2/N)$ **13.** $\dfrac{\log_a M}{\log_a N}$

14. $\log_a (a^3 M^2)$ **15.** $\log_a (1/N)$

Find the exact solution to each equation.

16. $3^x = 12$ **17.** $\log_3 x = 1/2$ **18.** $5^x = 8^{x-1}$

19. $\log(x) + \log(x + 15) = 2$ **20.** $2 \cdot \ln(x) = \ln(3) + \ln(6 - x)$

Use a scientific calculator to find an approximate solution to each of the following. Round your answers to four decimal places.

21. Solve $20^x = 5$.

22. Solve $\log_3 x = 2.75$.

23. The number of bacteria present in a culture at time t is given by the formula $N = 10e^{.4t}$, where t is in hours. How many bacteria are present initially? How many are present after 24 hours?

24. How many hours does it take for the bacteria population of Problem 23 to double?

Tying It All Together

CHAPTERS 1–9

Find the exact solution to each equation.

1. $(x - 3)^2 = 8$

2. $\log_2(x - 3) = 8$

3. $2^{x-3} = 8$

4. $2x - 3 = 8$

5. $|x - 3| = 8$

6. $\sqrt{x - 3} = 8$

7. $\log_2(x - 3) + \log_2 x = \log_2(18)$

8. $2\log_2(x - 3) = \log_2(5 - x)$

9. $\dfrac{1}{2}x - \dfrac{2}{3} = \dfrac{3}{4}x + \dfrac{1}{5}$

10. $3x^2 - 6x + 2 = 0$

Find the inverse of each function.

11. $f(x) = \dfrac{1}{3}x$

12. $g(x) = \log_3 x$

13. $f(x) = 2x - 4$

14. $h(x) = \sqrt{x}$

15. $j(x) = 1/x$

16. $k(x) = 5^x$

17. $m(x) = e^{x-1}$

18. $n(x) = \ln(x)$

Sketch the graph of each equation.

19. $y = 2x$

20. $y = 2^x$

21. $y = x^2$

22. $y = \log_2 x$

23. $y = \dfrac{1}{2}x - 4$

24. $3x + y = 5$

25. $y = 2 - x^2$

26. $y = e^2$

Systems of Equations

Alex is using a defective scale that makes a constant error. It always gives a reading that is some fixed amount too high. Alex knows that a can of soup and a can of tuna together weigh 24 ounces on this scale, and that 2 cans of tuna weigh 18 ounces on this scale. The only other information he has is that 4 identical cans of soup and 3 identical cans of tuna have a total weight of 80 ounces when placed on an accurate scale. Is it possible for Alex to figure out the correct weight of each type of can and the error in the scale from this information?

The problem presented here involves three unknowns. In this chapter we will solve problems like this by writing a system of three equations involving the three variables. We will learn several methods for solving different kinds of systems. This particular problem is Exercise 31 in Section 10.3.

10.1 Systems of Linear Equations in Two Variables

IN THIS SECTION:

- Solving a System by Graphing
- Independent, Inconsistent, and Dependent Equations
- Solving by Substitution
- Applications

We first studied linear equations in two variables in Chapter 7. In this section we will solve systems of linear equations in two variables and use systems to solve problems.

Solving a System by Graphing

Consider the two linear equations:

$$y = x + 2$$
$$x + y = 4$$

A pair of linear equations is called a **system of linear equations.** The set of points that satisfy both equations is called the **solution set of the system.** The graph of each linear equation is a straight line. If the two lines intersect at a point, the coordinates of that point satisfy the equations of both lines, and that point must be in the solution set to the system. To find the solution set of a system, we may be able to look at the graphs of the two lines and see where they intersect.

EXAMPLE 1   Solve the system by graphing.

$$y = x + 2$$
$$x + y = 4$$

Solution   We write each equation in slope-intercept form, then draw their graphs. In slope-intercept form, the system is written as follows:

$$y = x + 2$$
$$y = -x + 4$$

The graph of the system is shown in Fig. 10.1. From the graph it appears that these lines intersect at (1, 3). To be certain, we need to see if (1, 3) satisfies each equation. If $x = 1$ and $y = 3$, then

$$y = x + 2 \qquad \text{becomes} \qquad 3 = 1 + 2,$$

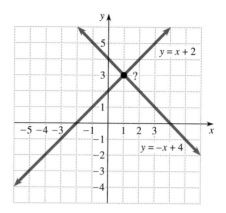

Figure 10.1

and

$$x + y = 4 \qquad \text{becomes} \qquad 1 + 3 = 4.$$

Because both of the equations are satisfied, the solution set to the system is $\{(1, 3)\}$.

◄

EXAMPLE 2 Solve the system by graphing.

$$2x - 3y = 6$$
$$3y - 2x = 3$$

Solution Write each equation in slope-intercept form.

$$y = \frac{2}{3}x - 2$$

$$y = \frac{2}{3}x + 1$$

The graph of the system is shown in Fig. 10.2. Because the two lines of Fig. 10.2 are parallel, there is no ordered pair that satisfies both equations. The solution set to the system is the empty set, $\varnothing$. ◄

Independent, Inconsistent, and Dependent Equations

Most of the systems of equations in two variables that occur in applications correspond to pairs of lines that intersect at a single point, as in Fig. 10.1. In this case, we say that the equations are **independent,** or that the system is independent. However, if the two lines are parallel, as in Fig. 10.2, they have no points in

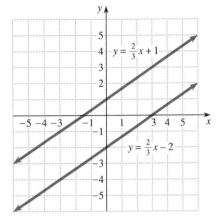

Figure 10.2

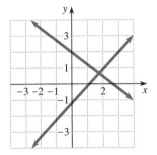

Independent

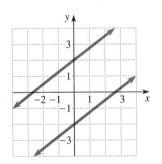

Inconsistent

common and the solution set to the system is the empty set. We say that the equations are **inconsistent,** or that the system is inconsistent.

There is also a third possibility. The equations in the system could be equations for the same line. In this case, we say that the equations are **dependent,** or that the system is dependent. Because the two equations are equivalent, the set of points that satisfies both of them is infinite. It is the same as the set of points satisfying one of them. Figure 10.3 shows what the graphs look like for the three possibilities. Each of these cases will occur in the examples given later for solving systems by substitution.

Solving by Substitution

Solving a system by graphing is certainly limited by the accuracy of the graph. If the lines intersect at a point whose coordinates are not integers, then it is difficult to identify the solution set from a graph. The method of solving a system by substitution does not depend on a graph and is totally accurate. The next example illustrates the substitution method.

EXAMPLE 3   Solve the system by substitution.

$$2x + 3y = 8$$
$$y + 2x = 6$$

Solution   We can easily solve $y + 2x = 6$ for y, to get $y = -2x + 6$. The y in the first equation must also be equal to $-2x + 6$ for a point that solves the system.

Dependent

Figure 10.3

Thus, we replace y in the first equation by the quantity $-2x + 6$ from the second equation. This technique is called **substitution.**

$$2x + 3y = 8$$
$$2x + 3(-2x + 6) = 8 \qquad \text{Substitute } -2x + 6 \text{ for } y.$$
$$2x - 6x + 18 = 8$$
$$-4x = -10$$
$$x = \frac{5}{2}$$

To find y, we use $x = 5/2$ in the equation $y = -2x + 6$.

$$y = -2\left(\frac{5}{2}\right) + 6 = -5 + 6 = 1$$

We can check by using $x = 5/2$ and $y = 1$ in each original equation. The equation

$$2x + 3y = 8 \qquad \text{becomes} \qquad 2\left(\frac{5}{2}\right) + 3(1) = 8,$$

and

$$y + 2x = 6 \qquad \text{becomes} \qquad 1 + 2\left(\frac{5}{2}\right) = 6.$$

Because both equations are true, the solution set to the system is $\{(5/2, 1)\}$. The equations of this system are independent.   ◄

EXAMPLE 4   Solve by substitution.

$$x - 2y = 3$$
$$2x - 4y = 7$$

Solution   Solve the first equation for x to get $x = 2y + 3$. Substitute $2y + 3$ for x in the second equation.

$$2x - 4y = 7$$
$$2(2y + 3) - 4y = 7$$
$$4y + 6 - 4y = 7$$
$$6 = 7$$

The equation $6 = 7$ is inconsistent. There are no values for x that will make it true, so there is no solution to this system of equations. The equations are inconsistent. To check, we write each equation in slope-intercept form.

$$x - 2y = 3 \qquad\qquad 2x - 4y = 7$$
$$-2y = -x + 3 \qquad\qquad -4y = -2x + 7$$
$$y = \frac{1}{2}x - \frac{3}{2} \qquad\qquad y = \frac{1}{2}x - \frac{7}{4}$$

The graphs of these equations are parallel lines with different y-intercepts. The solution set to the system is the empty set, $\varnothing$.   ◄

EXAMPLE 5   Solve by substitution.

$$2x + 3y = 5 + x + 4y$$
$$y = x - 5$$

Solution   If we substitute $y = x - 5$ into the first equation, we get

$$2x + 3(x - 5) = 5 + x + 4(x - 5)$$
$$2x + 3x - 15 = 5 + x + 4x - 20$$
$$2x + 3x - 15 = 5x - 15$$
$$5x - 15 = 5x - 15$$

Because the last equation is an identity, any point that satisfies $y = x - 5$ will also satisfy $2x + 3y = 5 + x + 4y$. The equations of this system are dependent. The solution set to the system is the set of all points that satisfy $y = x - 5$. We write the solution set in set notation as

$$\{(x,\ y) \mid y = x - 5\}.$$

We can verify this result by writing $2x + 3y = 5 + x + 4y$ in slope-intercept form.

$$2x + 3y = 5 + x + 4y$$
$$3y = -x + 5 + 4y$$
$$-y = -x + 5$$
$$y = x - 5$$

Because this slope-intercept form is identical to the slope-intercept form of the other equation, they are two different-looking equations for the same straight line. ◄

The strategy for solving a system by substitution can be summarized as follows.

STRATEGY
The Substitution Method

1. Solve one of the equations for one variable in terms of the others.
2. Substitute this value into the other equation to get an equation with only one variable.
*3. Solve for the remaining variable.
4. Insert this value into one of the original equations to find the value of the other variable.
5. Check your solution in both equations.

*If an identity results in step 3, the equations are dependent. The solution set is the set of all points that satisfy one of the equations.

If an inconsistent equation results, the equations are inconsistent. The solution set is empty.

Applications

Many of the problems that we solved in previous chapters involved more than one unknown quantity. To solve them, we wrote expressions for all of the unknowns in terms of one variable. Now we can solve problems involving two unknowns by using two variables, and writing a system of equations.

EXAMPLE 6   Belinda had $20,000 to invest. She invested part of it at 10% and the remainder at 12%. If her income from the two investments was $2160, how much did she invest at each rate?

Solution   Let x be the amount invested at 10%, and y be the amount invested at 12%. We can summarize all of the given information in a table.

Rate	10%	12%
Amount	x	y
Interest	$.10x$	$.12y$

We can write one equation about the amounts invested and another about the interest from the investments.

$$x + y = 20,000 \qquad \text{Total amounts invested}$$
$$.10x + .12y = 2160 \qquad \text{Total interest}$$

Solve the first equation for x to get $x = 20,000 - y$. Substitute $20,000 - y$ for x in the second equation.

$$.10x + .12y = 2160$$
$$.10(20,000 - y) + .12y = 2160$$
$$2000 - .10y + .12y = 2160$$
$$.02y = 160$$
$$y = 8,000$$
$$x = 12,000 \qquad \text{Because } x = 20,000 - y$$

To check this answer, take 10% of $12,000 and 12% of $8,000.

$$.10(12,000) = 1,200$$
$$.12(8,000) = 960$$

Because $1,200 + $960 = $2160, and $8,000 + $12,000 = $20,000, we can be certain that Belinda invested $12,000 at 10% and $8,000 at 12%.   ◄

Warm-ups

True or false?

1. The ordered pair $(1, 2)$ is in the solution set to the equation $2x + y = 4$.

2. The ordered pair $(1, 2)$ is in the solution set to the system:

$$2x + y = 4$$
$$3x - y = 6$$

3. The ordered pair $(2, 3)$ is in the solution set to the system:

$$4x - y = 5$$
$$4x - y = -5$$

4. Two distinct straight lines in the coordinate plane are either parallel or they intersect each other in exactly one point.

5. The substitution method is used to eliminate a variable.

6. The solution set to the system

$$y = 3x - 5$$
$$y = 3x + 1$$

is the empty set.

7. The equations $y = 3x - 6$ and $y = 2x + 4$ are independent.

8. The equations $y = 2x + 7$ and $y = 2x + 8$ are inconsistent.

9. The graphs of dependent equations are the same.

10. The graphs of independent equations are lines that intersect in exactly one point.

10.1 EXERCISES

Solve each system by graphing. See Examples 1 and 2.

1. $y = 2x$
   $y = -x + 3$

2. $y = x - 3$
   $y = -x + 1$

3. $y = 2x - 1$
   $2y = x - 2$

4. $y = 2x + 1$
   $x + y = -2$

5. $y = x - 3$
   $x - 2y = 4$

6. $y = -3x$
   $x + y = 2$

7. $2y - 2x = 2$
   $2y - 2x = 6$

8. $3y - 3x = 9$
   $x - y = 1$

Solve each system by the substitution method. Determine whether the equations are independent, dependent, or inconsistent. See Examples 3–5.

9. $y = x - 5$
   $2x - 5y = 1$

10. $y = x + 4$
    $3y - 5x = 6$

11. $x = 2y - 7$
    $3x + 2y = -5$

12. $x = y + 3$
    $3x - 2y = 4$

13. $x - y = 5$
    $2x = 2y + 14$

14. $2x - y = 3$
    $2y = 4x - 6$

15. $y = 2x - 5$
    $y + 1 = 2(x - 2)$

16. $3x - 6y = 5$
    $2y = 4x - 6$

17. $2x + y = 9$
    $2x - 5y = 15$

18. $3y - x = 0$
    $x - 4y = -2$

19. $x - y = 0$
    $2x + 3y = 35$

20. $2y = x + 6$
    $-3x + 2y = -2$

21. $x + y = 40$
    $.1x + .08y = 3.5$

22. $x - y = 10$
    $.2x + .05y = 7$

23. $y = 2x - 30$
    $\dfrac{1}{5}x - \dfrac{1}{2}y = -1$

24. $3x - 5y = 4$
    $y = \dfrac{3}{4}x - 2$

25. $x + y = 4$
    $x - y = 5$

26. $y = 2x - 3$
    $y = 3x - 3$

27. $2x - y = 4$
    $2x - y = 3$

28. $y = 3(x - 4)$
    $3x - y = 12$

29. $3(y - 1) = 2(x - 3)$
    $3y - 2x = -3$

30. $y = 3x$
    $y = 3x + 1$

31. $x - y = -.375$
    $1.5x - 3y = -2.25$

32. $y - 2x = 1.875$
    $2.5y - 3.5x = 11.8125$

Write a system of two equations in two unknowns for each problem. Solve each system by substitution. See Example 6.

33. Mrs. Brighton invested $30,000 and received a total of $2,300 in interest. If she invested part of the money at 10%, and the remainder at 5%, then how much did she invest at each rate?

34. Mr. Smith invested $25,000 and received a total of $2,380 in interest. If he invested part of the money at 10% and the remainder at 9%, then how much did he invest at each rate?

35. The sum of two numbers is 2, and their difference is 26. Find the numbers.

36. The sum of two numbers is -16, and their difference is 8. Find the numbers.

37. During one week, a land developer gave away Florida vacations or toasters to 100 potential customers who listened to a sales presentation. It costs the developer $6 for a toaster and $24 for a Florida vacation. If his bill for prizes that week was $708, then how many of each prize did he give away?

38. Tickets for a concert were sold to adults for $3 and students for $2. If the total receipts were $824, and twice as many adults tickets were sold, then how many of each were sold?

10.2 The Addition Method

IN THIS SECTION:

- The Addition Method
- Equations Involving Fractions or Decimals
- Applications

In Section 10.1 we solved systems of equations by using substitution. We substituted one equation into the other to eliminate a variable. The addition method of this section is another method for eliminating a variable in a system of equations.

The Addition Method

When we solve for one variable in terms of the other variable to do substitution, we may get an expression involving fractions. This expression must be substituted into the other equation. The **addition method** avoids fractions and is easier to use on certain systems.

EXAMPLE 1 Solve the system by the addition method.

$$3x - 5y = -9$$
$$4x + 5y = 23$$

Solution The addition property of equality allows us to add the same number to each side of an equation. We can also use the addition property of equality to add the two left sides, and add the two right sides.

$$
\begin{array}{rl}
3x - 5y = -9 & \\
\underline{4x + 5y = 23} & \\
7x\qquad\ = 14 & \quad \textbf{Add.} \\
x = 2 &
\end{array}
$$

The y-term was eliminated when we added the equations, because the coefficients of the y-terms were opposites of each other. Now use $x = 2$ in one of the original equations to find y. It does not matter which original equation we use. In this example we will use both equations to see that we get the same y in either case.

$$
\begin{array}{lll}
3x - 5y = -9 & & 4x + 5y = 23 \\
3(2) - 5y = -9 & \textbf{Replace } x \textbf{ by 2.} & 4(2) + 5y = 23 \\
6 - 5y = -9 & \textbf{Solve for } y. & 8 + 5y = 23 \\
-5y = -15 & & 5y = 15 \\
y = 3 & & y = 3
\end{array}
$$

Because $3(2) - 5(3) = -9$ and $4(2) + 5(3) = 23$ are both true, $(2, 3)$ satisfies both equations. The solution set is $\{(2, 3)\}$. ◄

In Example 1 the variable y was eliminated by the addition, because the coefficients of y in the two equations were opposites. If no variable will be eliminated by addition, we can use the multiplication property of equality to change the coefficients of the variables. In the next example the coefficient of y in one equation is a multiple of the coefficient of y in the other equation. We can use the multiplication property of equality to get opposite coefficients for y.

EXAMPLE 2 Solve the system by the addition method.

$$2x - 3y = -13$$
$$5x - 12y = -46$$

Solution Multiplying $-3y$ by -4 will give $12y$ in the first equation. If we multiply both sides of the first equation by -4, the coefficients of y will be 12 and -12, and y will be eliminated by addition.

$$(-4)(2x - 3y) = (-4)(-13) \quad \text{Multiply each side by } -4.$$
$$5x - 12y = -46$$

$$
\begin{aligned}
-8x + 12y &= 52 \\
\underline{5x - 12y} &= \underline{-46} \quad \text{Add the equations.} \\
-3x \quad\quad &= 6 \\
x \quad\quad &= -2
\end{aligned}
$$

Replace x by -2 in one of the original equations to find y.

$$
\begin{aligned}
2x - 3y &= -13 \\
2(-2) - 3y &= -13 \\
-4 - 3y &= -13 \\
-3y &= -9 \\
y &= 3
\end{aligned}
$$

Because $2(-2) - 3(3) = -13$ and $5(-2) - 12(3) = -46$ are both true, the solution set is $\{(-2, 3)\}$. ◀

The addition method can be used to eliminate either variable. In the next example we will eliminate the variable x by addition. In this system it is necessary to use multiplication on both equations to get opposite coefficients on one of the variables.

EXAMPLE 3 Solve the system by the addition method.

$$
\begin{aligned}
-2x + 3y &= 6 \\
3x - 5y &= -11
\end{aligned}
$$

Solution To eliminate x, we multiply the first equation by 3 and the second by 2.

$$3(-2x + 3y) = 3(6) \quad \text{Multiply each side by 3.}$$
$$2(3x - 5y) = 2(-11) \quad \text{Multiply each side by 2.}$$

$$
\begin{aligned}
-6x + 9y &= 18 \\
\underline{6x - 10y} &= \underline{-22} \quad \text{Add.} \\
-y &= -4 \quad\quad -6x + 6x = 0 \\
y &= 4
\end{aligned}
$$

Note that we could have eliminated y by multiplying by 5 and 3. Now insert $y = 4$ into one of the original equations to find x.

$$-2x + 3(4) = 6$$
$$-2x + 12 = 6$$
$$-2x = -6$$
$$x = 3$$

Check that $(3, 4)$ satisfies both equations. The solution set is $\{(3, 4)\}$. ◄

The addition method is based on the addition property of equality. We are adding equal quantities to each side of an equation. The form of the equations does not matter as long as the equal signs are lined up.

EXAMPLE 4 Solve the system. $\quad -4y = 5x + 7$
$\qquad\qquad\qquad\qquad\qquad\quad 4y = -5x + 12$

Solution If these equations are added, both variables are eliminated.

$$\begin{array}{r} -4y = 5x + 7 \\ 4y = -5x + 12 \\ \hline 0 = 19 \end{array}$$

Because this equation is inconsistent, the original equations are inconsistent. The solution set to the system is the empty set, $\varnothing$. ◄

Equations Involving Fractions or Decimals

When a system of equations involves fractions or decimals, we can use the multiplication property of equality to eliminate the fractions or decimals.

EXAMPLE 5 Solve the system. $\quad \dfrac{1}{2}x - \dfrac{2}{3}y = 7$
$\qquad\qquad\qquad\qquad\qquad\qquad \dfrac{2}{3}x - \dfrac{3}{4}y = 11$

Solution Multiplying the first equation by 6 and the second by 12 will eliminate the fractions.

$$6\left(\frac{1}{2}x - \frac{2}{3}y\right) = 6(7)$$

$$12\left(\frac{2}{3}x - \frac{3}{4}y\right) = 12(11)$$

$$3x - 4y = 42$$
$$8x - 9y = 132$$

To eliminate the variable x, multiply the first equation by -8 and the second by 3.

$$-8(3x - 4y) = -8(42)$$
$$3(8x - 9y) = 3(132)$$
$$\begin{aligned} -24x + 32y &= -336 \\ \underline{24x - 27y} &= \underline{396} \\ 5y &= 60 \\ y &= 12 \end{aligned}$$

Substitute $y = 12$ into the first of the original equations.

$$\frac{1}{2}x - \frac{2}{3}(12) = 7$$

$$\frac{1}{2}x - 8 = 7$$

$$\frac{1}{2}x = 15$$

$$x = 30$$

The solution set to the system is $\{(30, 12)\}$. Check.   ◄

Applications

Any system of two linear equations in two variables can be solved by either the addition method or substitution. We can use either method to solve the equations that result from applied problems. We will use whichever method appears to be the simplest for the problem at hand.

EXAMPLE 6   The total price for 4 nuts and 3 bolts is 48 cents, and the total price for 3 nuts and 2 bolts is 34 cents. What is the price of each?

Solution   Let x represent the price (in cents) of a nut, and y represent the price (in cents) of a bolt. We can write two equations to describe the given information.

$$4x + 3y = 48$$
$$3x + 2y = 34$$

To solve this system by addition, we multiply the first equation by -3 and the second by 4.

$$-3(4x + 3y) = -3(48) \quad \text{Multiply each side by } -3.$$
$$4(3x + 2y) = 4(34) \quad \text{Multiply each side by } 4.$$

$$
\begin{array}{l}
-12x - 9y = -144 \\
\underline{12x + 8y = 136} \quad \text{Add.} \\
\quad\quad -y = -8 \\
\quad\quad\quad y = 8
\end{array}
$$

$$4x + 3(8) = 48 \quad \text{Use } y = 8 \text{ in the first equation.}$$
$$4x + 24 = 48$$
$$4x = 24$$
$$x = 6$$

The nuts are 6 cents each and the bolts are 8 cents each. Check this solution in the original problem. ◀

Warm-ups

True or false?

1. To solve the system
$$3x - y = 9$$
$$2x + y = 7$$
by addition, we simply add the equations.

2. To solve the system
$$4x - 2y = 7$$
$$3x + y = 5$$
by addition, we can multiply the second equation by 2 and then add.

3. The system
$$x - y = 6$$
$$x - y = 7$$
has no solution.

4. Both the addition method and substitution are used to eliminate a variable from a system of two linear equations in two variables.

5. Both (2, 3) and (5, 6) are in the solution set to the system:
$$x - y = -1$$
$$x + y = 5$$

6. When we add two equations we must have the equal signs lined up.

7. To eliminate fractions in an equation, we multiply each side by the least common denominator of all fractions involved.

8. We can eliminate either variable by using the addition method.

9. The solution set to the system

$$2x - 3y = 8$$
$$4x - 6y = 16$$

is the set of all real numbers.

10. The solution set to the system

$$3x - 5y = 7$$
$$-3x + 5y = 8$$

is the empty set.

10.2 EXERCISES

Solve each system by the addition method. See Examples 1, 2, and 3.

1. $x + y = 7$
   $x - y = 9$

2. $3x - 4y = 11$
   $-3x + 2y = -7$

3. $x - y = 12$
   $2x + y = 3$

4. $x - 2y = -1$
   $-x + 5y = 4$

5. $2x - y = -5$
   $3x + 2y = 3$

6. $3x + 5y = -11$
   $x - 2y = 11$

7. $2x - 5y = 13$
   $3x + 4y = -15$

8. $3x + 4y = -5$
   $5x + 6y = -7$

9. $2x = 3y + 11$
   $7x - 4y = 6$

10. $2x = 2 - y$
    $3x + y = -1$

Solve each system by the addition method. Determine whether the equations are independent, dependent, or inconsistent. See Example 4.

11. $3x - 4y = 9$
    $-3x + 4y = 12$

12. $x - y = 3$
    $-6x + 6y = 17$

13. $5x - y = 1$
    $10x - 2y = 2$

14. $4x + 3y = 2$
    $-12x - 9y = -6$

15. $2x - y = 5$
    $2x + y = 5$

16. $-3x + 2y = 8$
    $3x + 2y = 8$

Solve each system by the addition method. See Example 5.

17. $\frac{1}{4}x + \frac{1}{3}y = 5$
    $x - y = 6$

18. $\frac{3x}{2} - \frac{2y}{3} = 10$
    $\frac{1}{2}x + \frac{1}{2}y = -1$

19. $\frac{x}{4} - \frac{y}{3} = -4$
    $\frac{x}{8} + \frac{y}{6} = 0$

20. $\dfrac{x}{3} - \dfrac{y}{2} = -\dfrac{5}{6}$

$\dfrac{x}{5} - \dfrac{y}{3} = -\dfrac{3}{5}$

21. $.05x + .10y = 1.30$

$x + y = 19$

22. $.1x + .06y = 9$

$.09x + .5y = 52.7$

23. $x + y = 1200$

$.12x + .09y = 120$

24. $x - y = 100$

$.20x + .06y = 150$

25. $1.5x - 2y = -.25$

$3x + 1.5y = 6.375$

26. $3x - 2.5y = 7.125$

$2.5x - 3y = 7.3125$

Write a system of two equations in two unknowns for each problem. Solve each system by the method of your choice. See Example 6.

27. On Monday Archie paid $2.54 for 3 doughnuts and 2 coffees. On Tuesday he paid $2.46 for 2 doughnuts and 3 coffees. On Wednesday he was tired of paying the tab and went out for coffee by himself. What was his bill for 1 doughnut and 1 coffee?

28. At Gwen's garage sale all books were one price, and all magazines were another price. Harriet bought 4 books and 3 magazines for $1.45, and June bought 2 books and 5 magazines for $1.25. What was the price of a book and what was the price of a magazine?

29. One-half of the boys and one-third of the girls of Freemont High attended the homecoming game, while one-third of the boys and one-half of the girls attended the homecoming dance. If there were 570 students at the game and 580 at the dance, then how many students are there at Freemont High?

30. There are 385 surfers in Surf City. Two-thirds of the boys are surfers, and one-twelfth of the girls are surfers. If there are two girls for every boy, then how many boys and how many girls are there in Surf City?

Photo for Exercise 30

31. Winborne has 35 coins consisting of dimes and nickels. If the value of his coins is $3.30, then how many of each type does he have?

32. Wendy has 52 coins consisting of nickels and pennies. If the value of the coins is $1.20, then how many of each type does she have?

10.3 Systems of Linear Equations in Three Variables

IN THIS SECTION:

- Definition
- Solving a System by Elimination
- Graphs of Equations in Three Variables
- Applications

The techniques that we learned in Section 10.2 can be extended to systems of equations in more than two variables. In this section we will use elimination of variables to solve systems of equations in three variables.

Definition

The equation $5x - 4y = 7$ is called a linear equation in two variables because its graph is a straight line. The equation $2x + 3y - 4z = 12$ is similar in form, and so it is called a **linear equation in three variables.** An equation in three variables is graphed in a three-dimensional coordinate system. The graph of a linear equation in three variables is a plane, not a line. We will not graph equations in three variables in this text, but we can solve systems without graphing. In general, we make the following definition.

DEFINITION
Linear Equation in
Three Variables

If A, B, C, and D are real numbers, with A, B, and C not all zero, then

$$Ax + By + Cz = D$$

is called a linear equation in three variables.

Solving a System by Elimination

A solution to an equation in three variables is an ordered triple such as $(-2, 1, 5)$, where the first coordinate is the value of x, the second coordinate is the value of y, and the third coordinate is the value of z. There are infinitely many solutions to an equation of this type.

Any collection of two or more equations is called a **system of equations.** The solution to a system of equations in three variables is the set of all ordered triples that satisfy all of the equations of the system. The techniques for solving a system of linear equations in three variables are similar to those used on systems of linear equations in two variables. We eliminate variables by either substitution or addition.

EXAMPLE 1   Solve the system.

$$
\begin{aligned}
(1) \qquad x + y - z &= -1 \\
(2) \qquad 2x - 2y + 3z &= 8 \\
(3) \qquad 2x - y + 2z &= 9
\end{aligned}
$$

Solution   We can eliminate z from Eqs. (1) and (2) by multiplying Eq. (1) by 3 and adding it to Eq. (2).

$$
\begin{array}{ll}
\quad 3x + 3y - 3z = -3 & \text{Eq. (1) multiplied by 3} \\
\underline{\quad 2x - 2y + 3z = 8} & \text{Eq. (2)} \\
(4) \quad 5x + y = 5 &
\end{array}
$$

Now we must eliminate the same variable, z, from another pair of equations. Eliminate z from Eqs. (1) and (3).

$$
\begin{array}{rl}
2x + 2y - 2z = -2 & \text{Eq. (1) multiplied by 2} \\
2x - y + 2z = 9 & \text{Eq. (3)} \\
\hline
\end{array}
$$

$$(5) \qquad 4x + y = 7$$

Equations (4) and (5) give us a system with two variables. We can eliminate y by multiplying Eq. (5) by -1 and adding the equations.

$$
\begin{array}{rl}
5x + y = 5 & \text{Eq. (4)} \\
-4x - y = -7 & \text{Eq. (5) multiplied by } -1 \\
\hline
x = -2 &
\end{array}
$$

Now that we have x, we can replace x by -2 in Eq. (5) to find y.

$$
\begin{aligned}
4x + y &= 7 \\
4(-2) + y &= 7 \\
-8 + y &= 7 \\
y &= 15
\end{aligned}
$$

Now replace x by -2 and y by 15 in Eq. (1) to find z.

$$
\begin{aligned}
x + y - z &= -1 \\
-2 + 15 - z &= -1 \\
13 - z &= -1 \\
-z &= -14 \\
z &= 14
\end{aligned}
$$

Check to see that the triple (2, 15, 14) does satisfy all three equations. The solution set is {(2, 15, 14)}. ◀

When solving a system of three equations with three variables, we can eliminate any of the three variables first. We usually look for the easiest variable to eliminate. We may eliminate that variable from the first and second, the first and third, or the second and third equations. After reducing the system to two equations in two unknowns, we solve that system using any method for two variables. After we have found the value of one variable, we again choose the simplest equations to use in finding the values of the remaining variables.

We can use either substitution or the addition technique on any system. Sometimes one is easier than the other. With three equations and three variables, we often use a combination as in the following example.

EXAMPLE 2   Solve the system.

$$
\begin{aligned}
(1) \qquad x + y \qquad\;\; &= 4 \\
(2) \qquad 2x \qquad\;\; - 3z &= 14 \\
(3) \qquad\qquad\; 2y + z &= 2
\end{aligned}
$$

Solution   From Eq. (1), we get $y = 4 - x$. If we substitute $y = 4 - x$ into Eq. (3), then Eqs. (2) and (3) will be equations involving x and z only.

$$
\begin{aligned}
(3) \qquad\qquad\quad 2y + z &= 2 \\
2(4 - x) + z &= 2 \qquad \text{Replace } y \text{ by } 4 - x. \\
8 - 2x + z &= 2 \qquad \text{Simplify.} \\
(4) \qquad\qquad -2x + z &= -6
\end{aligned}
$$

Now solve the system.

$$
\begin{aligned}
2x - 3z &= 14 \qquad \text{Eq. (2)} \\
-2x + z &= -6 \qquad \text{Eq. (4)}
\end{aligned}
$$

No multiplication is necessary, just add.

$$
\begin{aligned}
2x - \;\;\; 3z &= 14 \\
\underline{-2x + \qquad z = -6} \\
-2z &= 8 \\
z &= -4
\end{aligned}
$$

Use Eq. (3) to find y.

$$
\begin{aligned}
2y + z &= 2 \\
2y + (-4) &= 2 \qquad \text{Let } z = -4. \\
2y &= 6 \\
y &= 3
\end{aligned}
$$

Use Eq. (1) to find x.

$$
\begin{aligned}
x + y &= 4 \\
x + 3 &= 4 \qquad \text{Let } y = 3. \\
x &= 1
\end{aligned}
$$

Check that $(1, 3, -4)$ satisfies all three original equations. The solution set is $\{(1, 3, -4)\}$.   ◀

Graphs of Equations in Three Variables

The graph of any equation in three variables is drawn on a three-dimensional coordinate system. The graph of a linear equation in three variables is a plane. To solve a system of three linear equations in three variables by graphing, we would have to

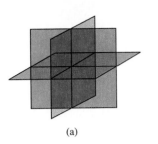
(a)

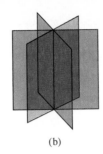
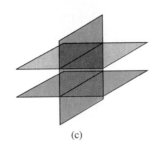
(b)

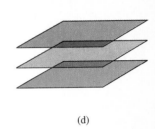
(c)

(d)

Figure 10.4

draw the three planes and then identify the points that lie on all three of them. This method would be difficult even when the points have simple coordinates. Thus, we will not attempt to solve these systems by graphing.

By considering how three planes might intersect, we can better understand the different types of solutions to a system of three equations in three variables. Figure 10.4 shows some of the possibilities for the positioning of three planes in three-dimensional space. In most of the problems we will solve, the planes intersect at a single point, as in Fig. 10.4(a). The solution set consists of one ordered triple. However, the system may include two equations whose graphs are parallel planes. Two equations with no point in common are said to be **inconsistent.** If the system has at least two inconsistent equations, then the solution set is the empty set (see Figs. 10.4c and d).

There may be infinitely many points in the intersection of all three planes. These points are either the points of a line or a plane. To get a line, we can either have three different planes intersecting along a line, as in Fig. 10.4(b), or two equations for the same plane, with the third plane intersecting that plane. If all three equations are equations of the same plane, we get that plane for the intersection. We will not solve systems corresponding to all of the possible configurations described. The following examples illustrate two of these cases.

EXAMPLE 3 Solve the system.

$$
\begin{aligned}
(1) \qquad & x + y - z = 5 \\
(2) \qquad & 3x - 2y + z = 8 \\
(3) \qquad & 2x + 2y - 2z = 7
\end{aligned}
$$

Solution We can eliminate the variable z from Eqs. (1) and (2) by adding them.

$$
\begin{aligned}
(1) \qquad & x + y - z = 5 \\
(2) \qquad & \underline{3x - 2y + z = 8} \\
& 4x - y = 13
\end{aligned}
$$

To eliminate z from Eqs. (1) and (3), multiply Eq. (1) by -2 and add the resulting equation to Eq. (3).

$$
\begin{array}{ll}
-2x - 2y + 2z = -10 & \text{Eq. (1) multiplied by } -2 \\
\underline{2x + 2y - 2z = \quad 7} & \text{Eq. (3)} \\
\qquad\qquad 0 = \;\; -3 &
\end{array}
$$

Because the last statement is false, there are two inconsistent equations in the system. Therefore, the solution set is the empty set.   ◄

EXAMPLE 4   Solve the system.

$$
\begin{array}{ll}
(1) & 2x - 3y - z = 4 \\
(2) & -6x + 9y + 3z = -12 \\
(3) & 4x - 6y - 2z = 8
\end{array}
$$

Solution   We will first eliminate x from Eqs. (1) and (2). Multiply Eq. (1) by 3 and add the resulting equation to Eq. (2).

$$
\begin{array}{ll}
6x - 9y - 3z = \quad 12 & \text{Eq. (1) multiplied by 3} \\
\underline{-6x + 9y + 3z = -12} & \text{Eq. (2)} \\
\qquad\qquad 0 = \quad 0 &
\end{array}
$$

The last statement is an identity. The identity occurred because Eq. (2) is a multiple of Eq. (1). In fact, Eq. (3) is also a multiple of Eq. (1). These equations are dependent. They are all equations for the same plane. The solution set is the set of all points on that plane,

$$
\{(x, y, z) \mid 2x - 3y - z = 4\}.   \quad ◄
$$

Applications

Problems involving three unknown quantities can often be solved using a system of three equations in three variables.

EXAMPLE 5   Theresa took in a total of $1240 last month from the rental of three apartments. She had to pay 10% of the rent from the one-bedroom apartment for taxes, 20% of the rent from the two-bedroom apartment for taxes, and 30% of the rent from the three-bedroom apartment for taxes. If the three-bedroom apartment rents for twice as much as the one-bedroom apartment and her total tax bill was $276, then what is the rent for each apartment?

Solution   Let x, y, and z represent the rent on the one-bedroom, two-bedroom, and three-bedroom apartments respectively. We can write one equation for the total rent, another equation for the total taxes, and a third equation expressing the fact that the rent for the three-bedroom apartment is twice that for the one-bedroom apartment.

$$x + y + z = 1240$$
$$.1x + .2y + .3z = 276$$
$$z = 2x$$

Substitute $z = 2x$ into both of the other equations to eliminate z.

$$x + y + 2x = 1240$$
$$.1x + .2y + .3(2x) = 276$$

$$3x + y = 1240$$
$$.7x + .2y = 276$$

$-2(3x + y) = -2(1240)$	**Multiply each side by** -2.
$10(.7x + .2y) = 10(276)$	**Multiply each side by 10.**

$$-6x - 2y = -2480$$
$$\underline{7x + 2y = 2760} \qquad \textbf{Add.}$$
$$x = 280$$

$$z = 2(280) = 560 \qquad \textbf{Because } z = 2x$$

$$280 + y + 560 = 1240 \qquad \textbf{Because } x + y + z = 1240$$
$$y = 400$$

The apartments rent for \$280, \$400, and \$560. Check these answers.   ◄

Warm-ups

True or false?

1. The point $(1, -2, 3)$ is in the solution set to the equation $x + y - z = 4$.
2. The point $(4, 1, 1)$ is the only solution to the equation $x + y - z = 4$.
3. The ordered triple $(1, -1, 2)$ is in the solution set to the system:

$$x + y + z = 2$$
$$x - y - z = 0$$
$$2x + y - z = -1$$

4. Substitution cannot be used on three equations in three variables.
5. Two distinct planes are either parallel or intersect in a single point.
6. The equations $x - y + 2z = 6$ and $x - y + 2z = 4$ are inconsistent.
7. The equations $3x + 2y - 6z = 4$ and $-6x - 4y + 12z = -8$ are dependent.

8. If you invest x, y, and z dollars in investments paying 8%, 10%, and 6% respectively, then your income from these investments is $.08x + .10y + .06z$ dollars.

9. The value of x nickels, y dimes, and z quarters is $.05x + .10y + .25z$ cents.

10. If $x = -2$, $z = 3$, and $x + y + z = 6$, then $y = 7$.

10.3 EXERCISES

Solve each system of equations. See Examples 1 and 2.

1. $x + y + z = 2$
   $x + 2y - z = 6$
   $2x + y - z = 5$

2. $2x - y + 3z = 14$
   $x + y - 2z = -5$
   $3x + y - z = 2$

3. $x - 2y + 4z = 3$
   $x + 3y - 2z = 6$
   $x - 4y + 3z = -5$

4. $2x + 3y + z = 13$
   $-3x + 2y + z = -4$
   $4x - 4y + z = 5$

5. $2x - y + z = 10$
   $3x - 2y - 2z = 7$
   $x - 3y - 2z = 10$

6. $x - 3y + 2z = -11$
   $2x - 4y + 3z = -15$
   $3x - 5y - 4z = 5$

7. $2x - 3y + z = -9$
   $-2x + y - 3z = 7$
   $x - y + 2z = -5$

8. $3x - 4y + z = 19$
   $2x + 4y + z = 0$
   $x - 2y + 5z = 17$

9. $2x - 5y + 2z = 16$
   $3x + 2y - 3z = -19$
   $4x - 3y + 4z = 18$

10. $-2x + 3y - 4z = 3$
    $3x - 5y + 2z = 4$
    $-4x + 2y - 3z = 0$

11. $x + y = 4$
    $y - z = -2$
    $x + y + z = 9$

12. $x + y - z = 0$
    $x - y = -2$
    $y + z = 10$

13. $x + y = 7$
    $y - z = -1$
    $x + 3z = 18$

14. $2x - y = -8$
    $y + 3z = 22$
    $x - z = -8$

Solve each system. See Examples 3 and 4.

15. $x - y + 2z = 3$
    $2x + y - z = 5$
    $3x - 3y + 6z = 4$

16. $2x - 4y + 6z = 12$
    $6x - 12y + 18z = 36$
    $-x + 2y - 3z = -6$

17. $3x - y + z = 5$
    $9x - 3y + 3z = 15$
    $-12x + 4y - 4z = -20$

18. $4x - 2y - 2z = 5$
    $2x - y - z = 7$
    $-4x + 2y + 2z = 6$

19. $x - y = 3$
    $y + z = 8$
    $2x + 2z = 7$

20. $2x - y = 6$
    $2y + z = -4$
    $8x + 2z = 3$

21. $.10x + .08y - .04z = 3$
    $5x + 4y - 2z = 150$
    $.3x + .24y - .12z = 9$

22. $.06x - .04y + z = 6$
    $3x - 2y + 50z = 300$
    $.03x - .02y + .5z = 3$

Use a calculator to solve each system.

23. $3x + 2y - .4z = .1$
    $3.7x - .2y + .05z = .41$
    $-2x + 3.8y - 2.1z = -3.26$

24. $3x - .4y + 9z = 1.668$
    $.3x + 5y - 8z = -.972$
    $5x - 4y - 8z = 1.8$

Solve each problem by using a system of three equations in three unknowns. See Example 5.

25. Ann invested a total of $12,000 in stocks, bonds, and a mutual fund. She received a 10% return on her stock investment, an 8% return on her bond investment, and a 12% return on her mutual fund. Her total return was $1230. If the total investment in stocks and bonds equaled her mutual-fund investment, how much did she invest in each?

26. Fearful of a bank failure, Norman split his life savings of $60,000 among three banks. He received 5%, 6%, and 7% on the three deposits, respectively. In the account earning 7% interest, he deposited twice as much as in the account earning 5% interest. If his total earnings were $3760, how much did he deposit in each account?

27. On Monday, Headley paid $1.70 for 2 cups of coffee and 1 doughnut, including the tip. On Tuesday, he paid $1.65 for 2 doughnuts and a cup of coffee, including the tip. On Wednesday, he paid $1.30 for 1 coffee and 1 doughnut, including the tip. If he always tips the same amount, then what is the amount of each item?

28. Melissa, Anna, and Barbara will not disclose their weights but agree to be weighed in pairs. Melissa and Anna together weigh 226 pounds. Anna and Barbara together weigh 210 pounds. Melissa and Barbara together weigh 200 pounds. How much does each girl weigh?

29. Salvador's Fruit Mart sells variety packs. The small pack contains 3 bananas, 2 apples, and 1 orange, for $1.80. The medium pack contains 4 bananas, 3 apples, and 3 oranges,

Photo for Exercise 29

for $3.05. The family size contains 6 bananas, 5 apples, and 4 oranges, for $4.65. What price should Salvador charge for his lunch-box special that consists of 1 banana, 1 apple, and 1 orange?

30. Edwin, his father, and his grandfather have an average age of 53. One-half of the grandfather's age, plus one-third of his father's age, plus one-fourth of Edwin's age is 65. If four years ago, Edwin's grandfather was 4 times as old as Edwin, then how old are they all now?

31. Alex is using a scale that is known to have a constant error. A can of soup and a can of tuna are placed on this scale, and it reads 24 ounces. Now 4 identical cans of soup and 3 identical cans of tuna are placed on an accurate scale, and a weight of 80 ounces is recorded. If 2 cans of tuna weigh 18 ounces on the bad scale, then what is the amount of error in the scale, and what is the correct weight of each type of can?

32. The sum of the digits of a three-digit number is 11. If the digits are reversed, the new number is 46 more than 5 times the old number. If the hundreds digit plus twice the tens digit is equal to the units digit, then what is the number?

33. To make ends meet, Mr. Farnsby works three jobs. His total income last year was $48,000. His income from teaching was just $6,000 more than his income from house painting. Royalties from his textbook sales were one-seventh the total he received from teaching and house painting. How much did he make from each source last year?

34. Harry has $2.25 in nickels, dimes, and quarters. If he has twice as many nickels, half as many dimes, and the same number of quarters, he would have $2.50. If he has 27 coins altogether, then how many of each does he have?

10.4 Nonlinear Systems of Equations

IN THIS SECTION:

- ● **Solving by Elimination**
- ● **Applications**

We have been studying systems of linear equations in this chapter. In this section we turn our attention to nonlinear systems of equations.

Solving by Elimination

Equations such as

$$y = x^2, \qquad y = \sqrt{x}, \qquad y = |x|, \qquad y = 2^x, \qquad \text{and} \qquad y = \log_2 x$$

are **nonlinear** equations, because their graphs are not straight lines. We say that a system of equations is nonlinear if there is at least one equation in the system that is nonlinear. We solve a nonlinear system just like a linear system, by elimination of variables. However, because the graphs of nonlinear equations may intersect at more than one point, there may be more than one ordered pair in the solution set to the system.

EXAMPLE 1 Solve the system of equations and draw the graph of each equation on the same coordinate system.

$$y = x^2 - 1$$
$$x + y = 1$$

Solution   We can eliminate y by substituting $y = x^2 - 1$ into $x + y = 1$.

$$x + y = 1$$
$$x + (x^2 - 1) = 1 \qquad \text{Substitute } x^2 - 1 \text{ for } y.$$
$$x^2 + x - 2 = 0$$
$$(x - 1)(x + 2) = 0$$
$$x - 1 = 0 \quad \text{or} \quad x + 2 = 0$$
$$x = 1 \quad \text{or} \qquad x = -2$$
$$y = (1)^2 - 1 \qquad y = (-2)^2 - 1 \qquad \text{Because } y = x^2 - 1$$
$$y = 0 \qquad\qquad y = 3$$

Check that the points $(1, 0)$ and $(-2, 3)$ each satisfy both of the original equations. The solution set is $\{(1, 0), (-2, 3)\}$. If we solve $x + y = 1$ for y, we get $y = -x + 1$. The line of $y = -x + 1$ has y-intercept $(0, 1)$ and slope -1.

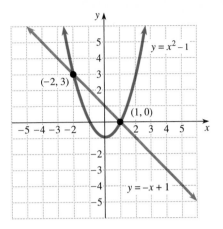

Figure 10.5

The graph of $y = x^2 - 1$ is a parabola with vertex $(0, -1)$. Of course, $(1, 0)$ and $(-2, 3)$ are on both graphs. The two graphs are shown in Fig. 10.5.   ◄

Graphing is not an accurate method for solving any system of equations. However, the graphs of the equations in a nonlinear system help us to understand how many solutions we should have for the system. It is not necessary to graph a system to solve it. Often the graphs are too difficult to sketch, but we can solve the system. In the next two examples, nonlinear systems will be solved by elimination with no graph being drawn.

EXAMPLE 2   Solve the system.

$$x^2 + y^2 + 2y = 3$$
$$x^2 - y = 5$$

Solution   If we substitute $y = x^2 - 5$ into the first equation to eliminate y, we will get a fourth-degree equation to solve. Instead, we can eliminate the variable x by solving $x^2 - y = 5$ for x^2.

$$x^2 = y + 5$$

Replace x^2 by $y + 5$ in the first equation.

$$x^2 + y^2 + 2y = 3$$
$$(y + 5) + y^2 + 2y = 3$$
$$y^2 + 3y + 5 = 3$$
$$y^2 + 3y + 2 = 0$$
$$(y + 2)(y + 1) = 0$$
$$y + 2 = 0 \quad \text{or} \quad y + 1 = 0$$
$$y = -2 \quad \text{or} \quad y = -1$$

For each y, find x.

$$x^2 = y + 5 \qquad x^2 = y + 5$$
$$= -2 + 5 \qquad\quad = -1 + 5$$
$$= 3 \qquad\qquad\;\; = 4$$
$$x = \pm\sqrt{3} \qquad\quad x = \pm 2$$

Check these values in the original equations. The solution set is $\{(\sqrt{3}, -2),$ $(-\sqrt{3}, -2), (2, -1), (-2, -1)\}$. The graphs of these two equations intersect in four points. ◀

EXAMPLE 3 Solve the system:

$$(1) \qquad \frac{2}{x} + \frac{1}{y} = \frac{1}{5}$$

$$(2) \qquad \frac{1}{x} - \frac{3}{y} = \frac{1}{3}$$

Solution When we encounter equations with rational expressions, we usually multiply by the LCD. However, that technique would make these equations more complicated. On this system we will use addition to eliminate the variable y. Multiply Eq. (1) by 3 and add the result to Eq. (2).

$$\frac{6}{x} + \frac{3}{y} = \frac{3}{5} \qquad \text{Eq. (1) multiplied by 3}$$

$$\frac{1}{x} - \frac{3}{y} = \frac{1}{3} \qquad \text{Eq. (2)}$$

$$\frac{7}{x} \qquad = \frac{14}{15} \qquad \frac{3}{5} + \frac{1}{3} = \frac{14}{15}$$

$$14x = 7 \cdot 15$$

$$x = \frac{7 \cdot 15}{14} = \frac{15}{2}$$

To find y, substitute $x = 15/2$ into Eq. (1).

$$\frac{2}{15/2} + \frac{1}{y} = \frac{1}{5}$$

$$\frac{4}{15} + \frac{1}{y} = \frac{1}{5} \qquad \frac{2}{15/2} = 2 \cdot \frac{2}{15} = \frac{4}{15}$$

$$15y \cdot \frac{4}{15} + 15y \cdot \frac{1}{y} = 15y \cdot \frac{1}{5} \qquad \text{Multiply each side by the LCD 15y.}$$

$$4y + 15 = 3y$$

$$y = -15$$

Check to see that $x = 15/2$ and $y = -15$ satisfy both original equations. The solution set is $\{(15/2, -15)\}$. ◀

Figure 10.6

Applications

The next example shows how a problem can be solved with a system of nonlinear equations.

EXAMPLE 4 A 15-foot ladder is leaning against a wall so that the distance from the bottom of the ladder to the wall is one-half the distance from the top of the ladder to the ground. Find the distance from the top of the ladder to the ground.

Solution Let x be the distance from the bottom of the ladder to the wall and y be the distance from the top of the ladder to the ground (see Fig. 10.6). We can write two equations involving x and y.

$$x^2 + y^2 = 15^2 \qquad \text{By the Pythagorean Theorem}$$
$$y = 2x$$

Solve by substitution.

$$x^2 + (2x)^2 = 225 \qquad \text{Replace } y \text{ by } 2x.$$
$$x^2 + 4x^2 = 225$$
$$5x^2 = 225$$
$$x^2 = 45$$
$$x = \pm\sqrt{45}$$
$$x = \pm 3\sqrt{5}$$

Because x represents distance, x must be positive. So $x = 3\sqrt{5}$ feet, and (because $y = 2x$) $y = 6\sqrt{5}$ feet. ◄

Warm-ups

True or false?

1. The graph of $y = x^2$ is a parabola.

2. The graph of $y = |x|$ is a straight line.

3. The point $(3, -4)$ is a solution to the system:

$$x^2 + y^2 = 25$$
$$y = \sqrt{5x + 1}$$

4. The graphs of $y = \sqrt{x}$ and $y = -x - 2$ do not intersect.

5. For nonlinear systems we can use only substitution to eliminate a variable.

6. If Judy can clean an entire house in x hours, then the fraction of the house that she does per hour is $1/x$.

7. In a triangle whose angles are 30°, 60°, and 90°, the length of the side opposite the 30°-angle is one-half the length of the hypotenuse.

> **8.** The volume of a rectangular box is $L \cdot W \cdot H$.
>
> **9.** The surface area of a rectangular box with sides of L, W, and H is $2LW + 2WH + 2LH$.
>
> **10.** If x and y represent the lengths of the legs of a right triangle, then its area is $(xy)/2$.

10.4 EXERCISES

Solve each system and graph both equations on the same set of axes. See Example 1.

1. $y = x^2$
$x + y = 6$

2. $y = x^2 - 1$
$x + y = 11$

3. $y = |x|$
$2y - x = 6$

4. $y = |x|$
$3y = x + 6$

5. $y = \sqrt{2x}$
$x - y = 4$

6. $y = \sqrt{x}$
$x - y = 6$

7. $4x - 9y = 9$
$xy = 1$

8. $2x + 2y = 3$
$xy = -1$

9. $y = -x^2 + 1$
$y = x^2$

10. $y = x^2$
$y = \sqrt{x}$

11. $y = x^2$
$y = 3x$

12. $y = x^3$
$y = 2x$

Solve each system. See Examples 2 and 3.

13. $x^2 + y^2 = 25$
$y = x^2 - 5$

14. $x^2 + y^2 = 25$
$y = x + 1$

15. $xy - 3x = 8$
$y = x + 1$

16. $xy + 2x = 9$
$x - y = 2$

17. $xy - x = 8$
$xy + 3x = -4$

18. $2xy - 3x = -1$
$xy + 5x = -7$

19. $\dfrac{1}{x} - \dfrac{1}{y} = 5$
$\dfrac{2}{x} + \dfrac{1}{y} = -3$

20. $\dfrac{2}{x} - \dfrac{3}{y} = \dfrac{1}{2}$
$\dfrac{3}{x} + \dfrac{1}{y} = \dfrac{1}{2}$

21. $x^2y = 20$
$xy + 2 = 6x$

22. $y^2x = 3$
$xy + 1 = 6x$

23. $x^2 + y^2 = 8$
$x^2 - y^2 = 2$

24. $x^2 + 2y^2 = 8$
$2x^2 - y^2 = 1$

25. $x^2 + xy - y^2 = -11$
$x + y = 7$

26. $x^2 + xy + y^2 = 3$
$y = 2x - 5$

27. $3y - 2 = x^4$
$y = x^2$

28. $y - 3 = 2x^4$
$y = 7x^2$

Solve each problem by using a system of two equations in two unknowns. See Example 4.

29. Find the lengths of the legs of a right triangle whose hypotenuse is $\sqrt{15}$ feet and whose area is 3 square feet.

30. A small television is advertised to have a picture with a diagonal measure of 5 inches and a viewing area of 12 square inches. What are the length and width of the screen?

31. Vincent has plans to build a house with seven gables. The plans call for an attic vent in the shape of an isosceles

Photo for Exercise 31

triangle in each gable. Because of the slope of the roof, the ratio of the height to the base of each triangle must be 1 to 4. If the vents are to provide a total ventilating area of 3,500 square inches, then what should be the height and base of each triangle?

32. Find the lengths of the sides of a triangle whose perimeter is 6 feet and whose angles are 30°, 60°, and 90°.

33. Pump A can either fill a tank or empty it in the same amount of time. If pump A and pump B are working together, the tank can be filled in 6 hours. When pump A was inadvertently left in the drain position while pump B was trying to fill the tank, it took 12 hours to fill the tank. How long would it take either pump working alone to fill the tank?

34. Roxanne either cleans the house or messes it up at the same rate. When she is cooperating with her mother, they can clean up a completely destroyed house in 6 hours. If she is not cooperating, it takes her mother 9 hours to clean the house, with Roxanne continually messing it up. How long would it take her mother to clean the entire house if Roxanne is sent to her grandmother's house?

35. Jan and Beth work in a seafood market that processes 200 pounds of catfish every morning. On Monday Jan started cleaning catfish at 8 A.M. and finished cleaning 100 pounds just as Beth arrived. Beth then took over and finished the job at 8:50 A.M. On Tuesday they both started at 8 A.M. and worked together to finish the job at 8:24 A.M. On Wednesday Beth was sick. If Jan is the faster worker, then how long did it take her to complete all of the catfish by herself?

36. Richard has already formed a rectangular area for a flagstone patio, but his wife Susan is unsure of the size of the patio they want. If the width is increased by 2 feet, then the area is increased by 30 square feet. If the width is increased by 1 foot and the length by 3 feet, then the area is increased by 54 square feet. What are the dimensions of the rectangle that Richard has already formed?

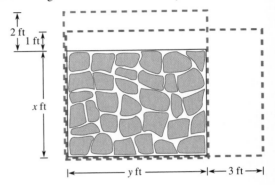

Figure for Exercise 36

37. If 34 feet of fencing are used to enclose a rectangular area of 72 square feet, then what are the dimensions of the area.

38. Find two numbers that have a sum of 8 and a product of 10.

*39. Find two complex numbers whose sum is 8 and whose product is 20.

*40. Find two complex numbers whose sum is −6 and whose product is 10.

41. Rico's sign shop has a contract to make a sign in the shape of a square with an isosceles triangle on top of it. The base of the triangle is to be the same length as the side of the square. The contract calls for a total height of 10 feet with an area of 72 square feet. How long should Rico make the side of the square, and what should be the height of the triangle?

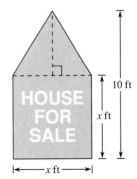

Figure for Exercise 41

42. Angelina is designing a rectangular box of 120 cubic inches that is to contain new Eaties breakfast cereal. The box must be 2 inches thick so that it is easy to hold. It must have 184 square inches of surface area to provide enough space for all of the special offers and coupons. What should be the dimensions of the box?

Figure for Exercise 42

Solve the following systems involving logarithmic and exponential functions.

43. $y = \log_2(x - 1)$
$y = 3 - \log_2(x + 1)$

44. $y = \log_3(x - 4)$
$y = 2 - \log_3(x + 4)$

45. $y = \log_2(x - 1)$
$y = 2 + \log_2(x + 2)$

46. $y = \log_4(8x)$
$y = 2 + \log_4(x - 1)$

47. $y = 2^{3x+4}$
$y = 4^{x-1}$

48. $y = 4^{3x}$
$y = (1/2)^{1-x}$

49. $y = (1/3)^x$
$y = 3^{1+x}$

50. $y = 9^{x-.5}$
$y = 27^{4x-7}$

51. $y = 2^x$
$y = 3^{x+5}$

52. $y = \log 7 - x \cdot \log 5$
$y = 2x$

The following systems require two steps to eliminate a variable. Solve each system.

*53. $x^2 + xy + y^2 = 19$
$x^2 + y^2 = 13$

*54. $x^2 - 2xy + y^2 = 25$
$x^2 + y^2 = 37$

*55. $x^2 + xy + 2y^2 = 11$
$x^2 - 2xy + 2y^2 = 5$

*56. $3x^2 - 2xy + y^2 = 18$
$3x^2 + xy + y^2 = 9$

10.5 Solution of Linear Systems in Two Variables Using Determinants

IN THIS SECTION:

- Matrices
- Determinants
- Cramer's Rule

We solved linear systems in two variables by substitution and addition in Sections 10.1 and 10.2. Those methods are performed differently on each system. In this section we will learn the method of determinants. With determinants we perform the same procedure on every system. Before we actually learn the new procedure, we need to introduce some new terminology.

Matrices

A **matrix** is a rectangular array of numbers. Some examples of *matrices* (plural of matrix) follow.

$$\begin{bmatrix} -1 & 2 \\ 5 & \sqrt{2} \\ 0 & 3 \end{bmatrix} \qquad \begin{bmatrix} 2 & 3 \\ -1 & 5 \end{bmatrix} \qquad \begin{bmatrix} 1 & 2 & 3 \\ 4 & 5 & 6 \\ -1 & 0 & 2 \end{bmatrix} \qquad \begin{bmatrix} 1 & 3 & 6 \end{bmatrix}$$

(a)   (b)   (c)   (d)

The **rows** of a matrix run horizontally, and the **columns** of a matrix run vertically. For example, matrix (a) has 3 rows and 2 columns. It is a 3 × 2 (read "3 by 2") matrix. Matrix (b) is a 2 × 2 matrix, (c) is a 3 × 3 matrix, and (d) is a 1 × 3 matrix. A **square** matrix is a matrix with the same number of rows as columns. Matrices (b) and (c) are square matrices. Each number in a matrix is called an **element** of the matrix.

Determinants

The **determinant** of a square matrix is a real number associated with the matrix. For a 2 × 2 matrix the determinant is defined as follows.

DEFINITION
Determinant of a 2 × 2 Matrix

The determinant of the matrix $\begin{bmatrix} a & b \\ c & d \end{bmatrix}$ is defined to be the real number $ad - bc$. We write

$$\begin{vmatrix} a & b \\ c & d \end{vmatrix} = ad - bc.$$

Note that the symbol for determinant is a vertical line similar to the absolute value symbol, while the symbol for a matrix uses square brackets.

EXAMPLE 1   Find the determinant of each matrix.

a) $\begin{bmatrix} 1 & 3 \\ -2 & 5 \end{bmatrix}$

b) $\begin{bmatrix} 2 & 4 \\ 6 & 12 \end{bmatrix}$

Solution

a) $\begin{vmatrix} 1 & 3 \\ -2 & 5 \end{vmatrix} = 1 \cdot 5 - 3(-2) = 5 + 6 = 11$

b) $\begin{vmatrix} 2 & 4 \\ 6 & 12 \end{vmatrix} = 2 \cdot 12 - 4 \cdot 6 = 24 - 24 = 0$ ◄

Cramer's Rule

To find a systematic method for solving a linear system we need to solve a general system of two linear equations in two variables. Consider the system

$$(1) \qquad a_1 x + b_1 y = c_1$$
$$(2) \qquad a_2 x + b_2 y = c_2$$

where $a_1, b_1, c_1, a_2, b_2,$ and c_2 represent real numbers. To eliminate y, we multiply Eq. (1) by b_2 and Eq. (2) by $-b_1$.

$$\begin{aligned}
a_1 b_2 x + b_1 b_2 y &= c_1 b_2 && \text{Eq. (1) multiplied by } b_2 \\
\underline{-a_2 b_1 x - b_1 b_2 y} &= \underline{-c_2 b_1} && \text{Eq. (2) multiplied by } -b_1 \\
a_1 b_2 x - a_2 b_1 x &= c_1 b_2 - c_2 b_1 && \text{Add.} \\
(a_1 b_2 - a_2 b_1) x &= c_1 b_2 - c_2 b_1 \\
x &= \dfrac{c_1 b_2 - c_2 b_1}{a_1 b_2 - a_2 b_1}
\end{aligned}$$

This formula for x is written in terms of determinants as

$$x = \frac{\begin{vmatrix} c_1 & b_1 \\ c_2 & b_2 \end{vmatrix}}{\begin{vmatrix} a_1 & b_1 \\ a_2 & b_2 \end{vmatrix}}.$$

If we repeat the above process to eliminate x from Eqs. (1) and (2), we get a formula for y in terms of determinants:

$$y = \frac{\begin{vmatrix} a_1 & c_1 \\ a_2 & c_2 \end{vmatrix}}{\begin{vmatrix} a_1 & b_1 \\ a_2 & b_2 \end{vmatrix}}$$

There are three determinants involved in solving for x or y. Let

$$D = \begin{vmatrix} a_1 & b_1 \\ a_2 & b_2 \end{vmatrix}, \qquad D_x = \begin{vmatrix} c_1 & b_1 \\ c_2 & b_2 \end{vmatrix}, \qquad \text{and} \qquad D_y = \begin{vmatrix} a_1 & c_1 \\ a_2 & c_2 \end{vmatrix}.$$

Note that D is the determinant made up of the original coefficients of x and y. D is used in the denominator for both x and y. D_x is obtained by replacing the first (or x) column of D by the constants c_1 and c_2. D_y is found by replacing the second (or y) column of D by the constants c_1 and c_2.

These formulas for solving a system of two linear equations in two variables are known as Cramer's rule.

Cramer's Rule

The solution to the system

$$a_1 x + b_1 y = c_1$$
$$a_2 x + b_2 y = c_2$$

is given by

$$x = \frac{\begin{vmatrix} c_1 & b_1 \\ c_2 & b_2 \end{vmatrix}}{\begin{vmatrix} a_1 & b_1 \\ a_2 & b_2 \end{vmatrix}} = \frac{D_x}{D}, \qquad y = \frac{\begin{vmatrix} a_1 & c_1 \\ a_2 & c_2 \end{vmatrix}}{\begin{vmatrix} a_1 & b_1 \\ a_2 & b_2 \end{vmatrix}} = \frac{D_y}{D},$$

provided $D \neq 0$.

EXAMPLE 2   Use Cramer's rule to solve the system:

$$3x - 2y = 4$$
$$2x + \ y = -3$$

Solution   Using Cramer's rule we get

$$x = \frac{\begin{vmatrix} 4 & -2 \\ -3 & 1 \end{vmatrix}}{\begin{vmatrix} 3 & -2 \\ 2 & 1 \end{vmatrix}} = \frac{4 - 6}{3 - (-4)} = \frac{-2}{7},$$

$$y = \frac{\begin{vmatrix} 3 & 4 \\ 2 & -3 \end{vmatrix}}{\begin{vmatrix} 3 & -2 \\ 2 & 1 \end{vmatrix}} = \frac{-9 - 8}{3 - (-4)} = \frac{-17}{7}.$$

Check these values in the original equations. The solution set is $\{(-2/7, -17/7)\}$.

◄

Recall that there are three possible types of solutions to a system of two linear equations in two variables: a single point, the empty set, and the set of all points on a line. Cramer's rule *only* works when the determinant D is *not* equal to zero. Cramer's rule only solves those systems that have a single point in their solution set. If $D = 0$, the solution set is either empty or contains all points of a line. If $D = 0$, use elimination to find the solution. The following two examples illustrate these cases.

EXAMPLE 3   Solve the system.

$$2x + 3y = 9$$
$$-2x - 3y = 5$$

Solution   Cramer's rule does not work because

$$D = \begin{vmatrix} 2 & 3 \\ -2 & -3 \end{vmatrix} = -6 - (-6) = 0.$$

If we try to eliminate a variable by addition, we get

$$\begin{array}{r} 2x + 3y = 9 \\ -2x - 3y = 5 \\ \hline 0 = 14 \end{array}.$$

Since this last statement is false, the solution set is empty. The equations are inconsistent.   ◄

Math at Work

Matrices used in this chapter for solving systems of linear equations help business managers keep track of data used for inventory control, accounting, and purchasing.

For example, suppose an inventory analyst studying the number of crates of bread and milk shipped to two convenience stores finds that store A orders 10 crates of bread and 20 crates of milk, and store B orders 30 crates of bread and 40 crates of milk. The analyst can use a 2×2 matrix to keep track of this information:

$$\begin{bmatrix} 10 & 20 \\ 30 & 40 \end{bmatrix}$$

Suppose that a crate of bread costs \$12 and a crate of milk costs \$18. This information is written as a 2×1 cost matrix:

$$\begin{bmatrix} 12 \\ 18 \end{bmatrix}$$

To find the total cost for milk and bread for each store, the inventory analyst multiplies the two matrices as follows:

$$\begin{bmatrix} 10 & 20 \\ 30 & 40 \end{bmatrix} \cdot \begin{bmatrix} 12 \\ 18 \end{bmatrix} = \begin{bmatrix} 10 \cdot 12 + 20 \cdot 18 \\ 30 \cdot 12 + 40 \cdot 18 \end{bmatrix} = \begin{bmatrix} 480 \\ 1080 \end{bmatrix}$$

Thus the cost to store A for the shipment is \$480, and the cost to store B is \$1080.

Since inventory analysts typically deal with more complicated situations than two stores with two products, much larger matrices are used for keeping track of the goods being shipped to each store and the current costs of those goods. Computers are used to store the matrices and perform the calculations. Use matrix multiplication to find the total cost to each store if the price of bread changes to \$15 and the price of milk changes to \$20.

EXAMPLE 4 Solve the system.

$$(1) \qquad 3x - 5y = 7$$
$$(2) \qquad 6x - 10y = 14$$

Solution Cramer's rule does not apply because

$$D = \begin{vmatrix} 3 & -5 \\ 6 & -10 \end{vmatrix} = -30 - (-30) = 0.$$

Multiply Eq. (1) by -2 and add the result to Eq. (2).

$$
\begin{aligned}
-6x + 10y &= -14 \qquad &\text{Eq. (1) times } -2 \\
6x - 10y &= 14 \qquad &\text{Eq. (2)} \\
\hline
0 &= 0 \qquad &\text{Add.}
\end{aligned}
$$

Because the last statement is an identity, these are dependent equations. The second equation is a multiple of the first; therefore, the solution set is described by just one of the equations. The solution set is $\{(x, y)|3x - 5y = 7\}$. ◀

To apply Cramer's rule, the equations must be in standard form.

EXAMPLE 5 Use Cramer's rule to solve the system:

$$(1) \qquad 2x - 3(y + 1) = -3$$
$$(2) \qquad 2y = 3x - 5$$

Solution First write each equation in standard form, $Ax + By = C$.

$$
\begin{array}{ll}
2x - 3(y + 1) = -3 & \qquad 2y = 3x - 5 \\
2x - 3y - 3 = -3 & \qquad -3x + 2y = -5 \\
2x - 3y = 0 &
\end{array}
$$

Now solve the system:

$$
\begin{aligned}
2x - 3y &= 0 \\
-3x + 2y &= -5
\end{aligned}
$$

$$D = \begin{vmatrix} 2 & -3 \\ -3 & 2 \end{vmatrix} = 2 \cdot 2 - (-3)(-3) = 4 - 9 = -5$$

$$D_x = \begin{vmatrix} 0 & -3 \\ -5 & 2 \end{vmatrix} = 0 \cdot 2 - (-5)(-3) = 0 - 15 = -15$$

$$D_y = \begin{vmatrix} 2 & 0 \\ -3 & -5 \end{vmatrix} = 2(-5) - (-3) \cdot 0 = -10 - 0 = -10$$

$$x = \frac{D_x}{D} = \frac{-15}{-5} = 3 \qquad y = \frac{D_y}{D} = \frac{-10}{-5} = 2$$

Check that $(3, 2)$ satisfies each of the original equations. The solution set is $\{(3, 2)\}$.
◀

Warm-ups

True or false?

1. The determinant of the matrix $\begin{bmatrix} -1 & 2 \\ 3 & -5 \end{bmatrix}$ is -1.

2. $\begin{vmatrix} 2 & 4 \\ -4 & 8 \end{vmatrix} = 0$

3. We can use Cramer's rule to solve any system of two linear equations in two variables.

4. The determinant of a 2×2 matrix is a real number.

5. If the determinant $D = 0$, then there is either no solution or infinitely many solutions to the system.

6. Cramer's rule is used to solve systems of linear equations only.

7. If a pair of linear equations have graphs that intersect in exactly one point, then this point can be found using Cramer's rule.

8. If x and y represent the digits of a two-digit number, then the value of the number is either $10x + y$ or $10y + x$.

9. If x represents the length of a side of a square and y represents the length of a side of a separate equilateral triangle, then $4x + y$ represents the total of their perimeters.

10. If a and b are the digits of a two-digit number whose value is $10a + b$, and if the digits are reversed, then the value is $b + 10a$.

10.5 EXERCISES

Find the value of each determinant. See Example 1.

1. $\begin{vmatrix} 2 & 5 \\ 3 & 7 \end{vmatrix}$
2. $\begin{vmatrix} -1 & 0 \\ 1 & 1 \end{vmatrix}$
3. $\begin{vmatrix} 0 & 3 \\ 1 & 5 \end{vmatrix}$
4. $\begin{vmatrix} 2 & 4 \\ 6 & 12 \end{vmatrix}$

5. $\begin{vmatrix} -3 & -2 \\ -4 & 2 \end{vmatrix}$
6. $\begin{vmatrix} -2 & 2 \\ -3 & -5 \end{vmatrix}$
7. $\begin{vmatrix} .05 & .06 \\ 10 & 20 \end{vmatrix}$
8. $\begin{vmatrix} .02 & -.5 \\ 30 & 50 \end{vmatrix}$

Solve each system using Cramer's rule. See Example 2.

9. $2x - y = 5$
   $3x + 2y = -3$

10. $3x + y = -1$
    $x + 2y = 8$

11. $3x - 5y = -2$
    $2x + 3y = 5$

12. $x - y = 1$
    $3x - 2y = 0$

13. $4x - 3y = 5$
    $2x + 5y = 7$

14. $2x - y = 2$
    $3x - 2y = 1$

15. $.5x + .2y = 8$
    $.4x - .6y = -5$

16. $\frac{1}{2}x + \frac{1}{4}y = 5$
    $\frac{1}{3}x - \frac{1}{2}y = -1$

17. $.6x + .5y = 18$
  $.5x - .25y = 7$

18. $\dfrac{1}{2}x + \dfrac{2}{3}y = 4$

  $\dfrac{3}{4}x + \dfrac{1}{3}y = -2$

Solve each system. Use Cramer's rule when possible. See Examples 3 and 4.

19. $2x - 3y = 5$
  $4x - 6y = 8$

20. $-x + 3y = 6$
  $3x - 9y = -18$

21. $x - y = 4$
  $x + 2y = 6$

22. $2x - y = 7$
  $3x + 2y = -7$

23. $4x - y = 6$
  $-8x + 2y = -12$

24. $-x + 2y = 3$
  $3x - 6y = 10$

25. $x^2 + y = 7$
  $2x^2 - y = 5$

26. $2x^2 - y = 3$
  $3x^2 + 2y = 22$

27. $x^2 + y^2 = 25$
  $x^2 - y = 5$

28. $x^2 - y^2 = -3$
  $2x^2 + 3y^2 = 14$

Solve each system using Cramer's rule. See Example 5.

29. $y = 3x - 12$
  $3(x + 1) - 11 = y + 4$

30. $y = x - 7$
  $y = -2x + 5$

31. $x - 6 = y + 1$
  $y = -3x + 1$

32. $1 - x = 6 + y$
  $5 - 2x = 6 - y$

33. $y = .05x$
  $x + y = 504$

34. $.05x + .10y = 6$
  $y = 98 - x$

35. $x\sqrt{2} = y\sqrt{2} - 2$
  $x + y = \sqrt{18}$

36. $x\sqrt{2} + y\sqrt{3} = 5$
  $y = x\sqrt{6} - \sqrt{3}$

Solve each problem by using two equations in two variables and Cramer's rule.

37. One serving of canned peas contains 3 grams of protein and 11 grams of carbohydrates, while one serving of canned beets contains 1 gram of protein and 8 grams of carbohydrates. How many servings of each would provide exactly 38 grams of protein and 187 grams of carbohydrates?

38. One serving of Cornies breakfast cereal contains 2 grams of protein and 25 grams of carbohydrates, while one serving of Oaties breakfast cereal contains 4 grams of protein and 20 grams of carbohydrates. How many servings of each would provide exactly 24 grams of protein and 210 grams of carbohydrates?

39. The sum of the digits of a two-digit number is 10. If the digits are reversed, the new number is 1 less than twice the original number. Find the original number.

40. The units digit of a two-digit number is twice the tens digit. If the digits are reversed, the new number is 36 more than the original number. Find the original number.

41. Althea bought a gallon of milk and a magazine for a total of $4.65, excluding tax. Including the tax, the bill was $4.95. If there is a 5% sales tax on milk and an 8% sales tax on magazines, then what was the price of each item?

42. A truck carrying 3600 cubic feet of cargo consisting of washing machines and refrigerators was hijacked. The washing machines are worth $300 each and are shipped in cartons of 36 cubic feet. The refrigerators are worth $900 each and are shipped in cartons of 45 cubic feet. If the total value of the cargo was $51,000, then how many of each were there on the truck?

43. Windy's Hamburger Palace sells singles and doubles. Toward the end of the evening, Windy himself noticed that he had on hand only 32 patties and 34 slices of tomatoes. If a single takes 1 patty and 2 slices, and a double takes 2 patties and 1 slice, then how many more singles and doubles must Windy sell to use up all of his patties and tomato slices?

44. Carman has a total of 28 wrenches, all of which are either box wrenches or open-end wrenches. For insurance purposes he values the box wrenches at $3.00 each and the open-end wrenches at $2.50 each. If the value of his wrench collection is $78, then how many of each type does he have?

45. Gary is 5 years older than Harry. Twenty-nine years ago Gary was twice as old as Harry. How old are they now?

46. One acute angle of a right triangle is 3° more than twice the other acute angle. What are the sizes of the acute angles?

47. A rope of length 80 feet is to be cut to form a square and an equilateral triangle. If the figures are to have equal perimeters, then what should be the length of a side of each?

48. For a cup of coffee and a doughnut, Thurrel spent $2.25, including a tip. Later he spent $4.00 for 2 coffees and 3 doughnuts, including a tip. If he always tips $1.00, then what is the price of a cup of coffee?

49. A 10% chlorine solution is to be mixed with a 25% chlorine solution to obtain 30 gallons of 20% solution. How many gallons of each must be used?

50. Emily and Camille started from the same city and drove in opposite directions on the freeway. After 3 hours they were 354 miles apart. If they had gone in the same direction, Emily would have been 18 miles ahead of Camille. How fast did each woman drive?

10.6 Solution of Linear Systems in Three Variables Using Determinants

IN THIS SECTION:

- Minors
- Evaluating a 3×3 Determinant
- Cramer's Rule

The solution of linear systems involving three variables using determinants is very similar to the solution of linear systems in two variables using determinants. However, we first need to learn to find the determinant of a 3×3 matrix.

Minors

To each element of a 3×3 matrix there corresponds a 2×2 matrix called the **minor** of that element. The minor for an element of a 3×3 matrix is the 2×2 matrix that remains after deleting the row and column of that element.

EXAMPLE 1 Find the minors for the elements 2, 3, and -6 of the 3×3 matrix

$$\begin{bmatrix} 2 & -1 & -8 \\ 0 & -2 & 3 \\ 4 & -6 & 7 \end{bmatrix}.$$

Solution To find the minor of 2 delete the first row and first column of the matrix.

$$\begin{bmatrix} 2 & -1 & -8 \\ 0 & -2 & 3 \\ 4 & -6 & 7 \end{bmatrix}$$ The minor of 2 is $\begin{bmatrix} -2 & 3 \\ -6 & 7 \end{bmatrix}$.

To find the minor of 3 delete the second row and third column of the matrix.

$$\begin{bmatrix} 2 & -1 & -8 \\ 0 & -2 & 3 \\ 4 & -6 & 7 \end{bmatrix}$$ The minor of 3 is $\begin{bmatrix} 2 & -1 \\ 4 & -6 \end{bmatrix}$.

To find the minor of -6 delete the third row and second column of the matrix.

$$\begin{bmatrix} 2 & -1 & -8 \\ 0 & -2 & 3 \\ 4 & -6 & 7 \end{bmatrix}$$ The minor of -6 is $\begin{bmatrix} 2 & -8 \\ 0 & 3 \end{bmatrix}$.   ◀

Evaluating a 3 × 3 Determinant

The determinant of a 3×3 matrix is defined in terms of the determinants of minors.

DEFINITION
Determinant of a
3 × 3 Matrix

If $\begin{bmatrix} a_1 & b_1 & c_1 \\ a_2 & b_2 & c_2 \\ a_3 & b_3 & c_3 \end{bmatrix}$ is a 3×3 matrix, then the determinant of this matrix

is defined to be

$$\begin{vmatrix} a_1 & b_1 & c_1 \\ a_2 & b_2 & c_2 \\ a_3 & b_3 & c_3 \end{vmatrix} = a_1 \cdot (\text{determinant of minor of } a_1)$$
$$- a_2 \cdot (\text{determinant of minor of } a_2)$$
$$+ a_3 \cdot (\text{determinant of minor of } a_3)$$
$$= a_1 \cdot \begin{vmatrix} b_2 & c_2 \\ b_3 & c_3 \end{vmatrix} - a_2 \cdot \begin{vmatrix} b_1 & c_1 \\ b_3 & c_3 \end{vmatrix} + a_3 \cdot \begin{vmatrix} b_1 & c_1 \\ b_2 & c_2 \end{vmatrix}.$$

This process of calculating a determinant in terms of minors is called **expansion by minors** about the first column. Note that we alternate the signs of the coefficients of the minors.

EXAMPLE 2 Find the determinant of the matrix by expansion by minors about the first column.

$$\begin{bmatrix} 1 & 3 & -5 \\ -2 & 4 & 6 \\ 0 & -7 & 9 \end{bmatrix}$$

Solution

$$\begin{vmatrix} 1 & 3 & -5 \\ -2 & 4 & 6 \\ 0 & -7 & 9 \end{vmatrix} = 1 \cdot \begin{vmatrix} 4 & 6 \\ -7 & 9 \end{vmatrix} - (-2) \cdot \begin{vmatrix} 3 & -5 \\ -7 & 9 \end{vmatrix} + 0 \cdot \begin{vmatrix} 3 & -5 \\ 4 & 6 \end{vmatrix}$$

$$= 1 \cdot [36 - (-42)] + 2 \cdot (27 - 35) + 0 \cdot [18 - (-20)]$$
$$= 1 \cdot 78 + 2 \cdot (-8) + 0$$
$$= 78 - 16$$
$$= 62 \qquad \blacktriangleleft$$

The value for the determinant of a 3×3 matrix is the same, if the expansion by minors is done using any row or column. When expanding about any row or column the signs of the coefficients of the minors alternate according to the **sign array** that follows.

$$\begin{bmatrix} + & - & + \\ - & + & - \\ + & - & + \end{bmatrix}$$

The sign array is easily remembered by observing that there is a "+" sign in the upper left position, followed by alternating signs for all of the remaining positions.

EXAMPLE 3   Evaluate the determinant of the matrix by expansion by minors about the second row.

$$\begin{bmatrix} 1 & 3 & -5 \\ -2 & 4 & 6 \\ 0 & -7 & 9 \end{bmatrix}$$

Solution   For expansion using the second row, we use the signs "− + −" from the second row of the sign array.

$$\begin{vmatrix} 1 & 3 & -5 \\ -2 & 4 & 6 \\ 0 & -7 & 9 \end{vmatrix} = - (-2) \cdot \begin{vmatrix} 3 & -5 \\ -7 & 9 \end{vmatrix} + 4 \cdot \begin{vmatrix} 1 & -5 \\ 0 & 9 \end{vmatrix} - 6 \cdot \begin{vmatrix} 1 & 3 \\ 0 & -7 \end{vmatrix}$$

$$= 2(27 - 35) + 4(9 - 0) - 6(-7 - 0)$$
$$= 2(-8) + 4(9) - 6(-7)$$
$$= -16 + 36 + 42$$
$$= 62$$

Note that 62 is the same value obtained for the determinant of this matrix in Example 2. $\qquad \blacktriangleleft$

Because we can use any row or column to evaluate a determinant of a 3×3 matrix, we can choose a row or column that makes the work easier. We can shorten the work considerably by picking a row or column with zeros in it.

EXAMPLE 4 Find the determinant of the matrix

$$\begin{bmatrix} 3 & -5 & 0 \\ 4 & -6 & 0 \\ 7 & 9 & 2 \end{bmatrix}.$$

Solution The easiest way to evaluate this determinant is to expand about the third column. Use the signs "+ − +" from the sign array.

$$\begin{vmatrix} 3 & -5 & 0 \\ 4 & -6 & 0 \\ 7 & 9 & 2 \end{vmatrix} = 0 \cdot \begin{vmatrix} 4 & -6 \\ 7 & 9 \end{vmatrix} - 0 \cdot \begin{vmatrix} 3 & -5 \\ 7 & 9 \end{vmatrix} + 2 \cdot \begin{vmatrix} 3 & -5 \\ 4 & -6 \end{vmatrix}$$

$$= \quad 0 \quad - \quad 0 \quad + 2[-18 - (-20)]$$

$$= \quad 0 \quad - \quad 0 \quad + 2(2)$$

$$= 4 \qquad \blacktriangleleft$$

Cramer's Rule

A system of three linear equations in three variables can be solved using determinants in a manner similar to that of the previous section. This rule is also called Cramer's rule.

Cramer's Rule

The solution to the system

$$a_1 x + b_1 y + c_1 z = d_1$$
$$a_2 x + b_2 y + c_2 z = d_2$$
$$a_3 x + b_3 y + c_3 z = d_3$$

is given by $x = \dfrac{D_x}{D}$, $y = \dfrac{D_y}{D}$, and $z = \dfrac{D_z}{D}$, where

$$D = \begin{vmatrix} a_1 & b_1 & c_1 \\ a_2 & b_2 & c_2 \\ a_3 & b_3 & c_3 \end{vmatrix}, \qquad D_x = \begin{vmatrix} d_1 & b_1 & c_1 \\ d_2 & b_2 & c_2 \\ d_3 & b_3 & c_3 \end{vmatrix},$$

$$D_y = \begin{vmatrix} a_1 & d_1 & c_1 \\ a_2 & d_2 & c_2 \\ a_3 & d_3 & c_3 \end{vmatrix}, \qquad D_z = \begin{vmatrix} a_1 & b_1 & d_1 \\ a_2 & b_2 & d_2 \\ a_3 & b_3 & d_3 \end{vmatrix},$$

provided $D \neq 0$.

Note that D_x, D_y, and D_z are obtained from D by replacing the x-, y-, or z-column with the constants d_1, d_2, and d_3.

EXAMPLE 5   Use Cramer's rule to solve the system:

$$\begin{aligned} x + y + z &= 4 \\ x - y &= -3 \\ x + 2y - z &= 0 \end{aligned}$$

Solution   We first calculate D, D_x, D_y, and D_z. To calculate D expand by minors about the third column, because the third column has a zero in it.

$$D = \begin{vmatrix} 1 & 1 & 1 \\ 1 & -1 & 0 \\ 1 & 2 & -1 \end{vmatrix} = 1 \cdot \begin{vmatrix} 1 & -1 \\ 1 & 2 \end{vmatrix} - 0 \cdot \begin{vmatrix} 1 & 1 \\ 1 & 2 \end{vmatrix} + (-1) \cdot \begin{vmatrix} 1 & 1 \\ 1 & -1 \end{vmatrix}$$

$$= 1 \cdot [2 - (-1)] - 0 + (-1)[-1 - 1]$$
$$= 3 - 0 + 2$$
$$= 5$$

For D_x expand by minors about the first column.

$$D_x = \begin{vmatrix} 4 & 1 & 1 \\ -3 & -1 & 0 \\ 0 & 2 & -1 \end{vmatrix} = 4 \cdot \begin{vmatrix} -1 & 0 \\ 2 & -1 \end{vmatrix} - (-3) \cdot \begin{vmatrix} 1 & 1 \\ 2 & -1 \end{vmatrix} + 0 \cdot \begin{vmatrix} 1 & 1 \\ -1 & 0 \end{vmatrix}$$

$$= 4 \cdot (1 - 0) + 3 \cdot (-1 - 2) + 0$$
$$= 4 - 9 + 0$$
$$= -5$$

For D_y expand by minors about the third row.

$$D_y = \begin{vmatrix} 1 & 4 & 1 \\ 1 & -3 & 0 \\ 1 & 0 & -1 \end{vmatrix} = 1 \cdot \begin{vmatrix} 4 & 1 \\ -3 & 0 \end{vmatrix} - 0 \cdot \begin{vmatrix} 1 & 1 \\ 1 & 0 \end{vmatrix} + (-1) \cdot \begin{vmatrix} 1 & 4 \\ 1 & -3 \end{vmatrix}$$

$$= 1 \cdot 3 - 0 + (-1)(-7)$$
$$= 10$$

To get D_z expand by minors about the third row.

$$D_z = \begin{vmatrix} 1 & 1 & 4 \\ 1 & -1 & -3 \\ 1 & 2 & 0 \end{vmatrix} = 1 \cdot \begin{vmatrix} 1 & 4 \\ -1 & -3 \end{vmatrix} - 2 \cdot \begin{vmatrix} 1 & 4 \\ 1 & -3 \end{vmatrix} + 0 \cdot \begin{vmatrix} 1 & 1 \\ 1 & -1 \end{vmatrix}$$

$$= 1 \cdot 1 - 2(-7) + 0$$
$$= 15$$

Now, by Cramer's rule,

$$x = \frac{D_x}{D} = \frac{-5}{5} = -1, \qquad y = \frac{D_y}{D} = \frac{10}{5} = 2,$$

and

$$z = \frac{D_z}{D} = \frac{15}{5} = 3.$$

Check that $(-1, 2, 3)$ satisfies all three original equations. The solution set is $\{(-1, 2, 3)\}$. ◄

If $D = 0$, Cramer's rule does not apply. Cramer's rule only gives us the solution to a system of three equations with three variables that has a single point in the solution set. If $D = 0$, the solution set either is empty or consists of infinitely many points. If $D = 0$, we must solve the system by elimination of variables to decide which case it is.

EXAMPLE 6 Solve the system.

$$
\begin{array}{ll}
(1) & x + y - z = 2 \\
(2) & 2x + 2y - 2z = 4 \\
(3) & -3x - 3y + 3z = -6
\end{array}
$$

Solution Calculate D by expanding about the first column:

$$D = \begin{vmatrix} 1 & 1 & -1 \\ 2 & 2 & -2 \\ -3 & -3 & 3 \end{vmatrix} = 1 \cdot \begin{vmatrix} 2 & -2 \\ -3 & 3 \end{vmatrix} - 2 \cdot \begin{vmatrix} 1 & -1 \\ -3 & 3 \end{vmatrix} + (-3) \cdot \begin{vmatrix} 1 & -1 \\ 2 & -2 \end{vmatrix}$$

$$= 1 \cdot 0 - 2 \cdot 0 + (-3) \cdot 0$$

$$= 0$$

Because $D = 0$, Cramer's rule does not apply to this system. If we multiply Eq. (1) by 2, we get Eq. (2). If we multiply Eq. (1) by -3, we get Eq. (3). Thus, all three equations are equivalent, and they are dependent. The solution set to the system is $\{(x, y, z) \mid x + y - z = 2\}$. ◄

Warm-ups

True or false?

1. A minor is a 2×2 matrix.
2. The minor for 2, in the matrix

$$\begin{bmatrix} 3 & 1 & 6 \\ 4 & 2 & 5 \\ -1 & 7 & 8 \end{bmatrix}$$

is the matrix

$$\begin{bmatrix} 2 & 5 \\ 7 & 8 \end{bmatrix}$$

3. The determinant of a 3×3 matrix is defined in terms of determinants of minors.

4. Expansion by minors is a method for stretching a 3×3 matrix into a 4×4 matrix.

5. Using Cramer's rule, we use D/D_x to get the value of x.

6. Expansion by minors about any row or any column gives the same value for the determinant of a 3×3 matrix.

7. The sign array is of no use in evaluating the determinant of a 3×3 matrix.

8. It is easier to find the determinant of a 3×3 matrix with several zero elements than one with no zero elements.

9. If $D = 0$, then x, y, and z are all zero.

10. Cramer's rule works just as well on nonlinear systems of three equations in three unknowns.

10.6 EXERCISES

Find the indicated minors using the following matrix. See Example 1.

$$\begin{bmatrix} 3 & -2 & 5 \\ 4 & -3 & 7 \\ 0 & 1 & -6 \end{bmatrix}$$

1. Minor for 3   **2.** Minor for -2 **3.** Minor for 5   **4.** Minor for -3

5. Minor for 7   **6.** Minor for 0   **7.** Minor for 1   **8.** Minor for -6

Find the determinant of each 3×3 *matrix by using expansion by minors about the first column. See Example* 2.

9. $\begin{bmatrix} 1 & 1 & 2 \\ 2 & 3 & 1 \\ 3 & 1 & 5 \end{bmatrix}$ **10.** $\begin{bmatrix} 2 & 1 & 3 \\ 1 & 1 & 2 \\ 3 & 4 & 6 \end{bmatrix}$ **11.** $\begin{bmatrix} 2 & 1 & 0 \\ 1 & 0 & 1 \\ 3 & 1 & 2 \end{bmatrix}$ **12.** $\begin{bmatrix} 1 & 0 & 2 \\ 2 & 1 & 3 \\ 4 & 3 & 0 \end{bmatrix}$

13. $\begin{bmatrix} -2 & 1 & 2 \\ -3 & 3 & 1 \\ -5 & 4 & 0 \end{bmatrix}$ **14.** $\begin{bmatrix} -2 & 1 & 3 \\ -1 & 4 & 2 \\ 2 & 1 & 1 \end{bmatrix}$ **15.** $\begin{bmatrix} 1 & 1 & 5 \\ 0 & 3 & 2 \\ 0 & 2 & 3 \end{bmatrix}$ **16.** $\begin{bmatrix} 1 & 0 & 6 \\ 0 & 1 & 4 \\ 0 & 0 & 9 \end{bmatrix}$

Evaluate the determinant of each 3×3 matrix using expansion by minors about the row or column of your choice. See Examples 3 and 4.

17. $\begin{bmatrix} 3 & 1 & 5 \\ 2 & 0 & 6 \\ 4 & 0 & 1 \end{bmatrix}$

18. $\begin{bmatrix} 2 & 1 & 2 \\ 1 & 2 & 5 \\ 3 & 0 & 0 \end{bmatrix}$

19. $\begin{bmatrix} -2 & 1 & 3 \\ 0 & 1 & -1 \\ 2 & -4 & -3 \end{bmatrix}$

20. $\begin{bmatrix} -2 & 0 & 1 \\ -3 & 2 & -5 \\ 4 & -2 & 6 \end{bmatrix}$

21. $\begin{bmatrix} -2 & -3 & 0 \\ 4 & -1 & 0 \\ 0 & 3 & 5 \end{bmatrix}$

22. $\begin{bmatrix} -2 & 6 & 3 \\ 0 & 4 & 0 \\ -1 & -4 & 5 \end{bmatrix}$

23. $\begin{bmatrix} 2 & 1 & 1 \\ 0 & 0 & 5 \\ 5 & 0 & 4 \end{bmatrix}$

24. $\begin{bmatrix} 2 & 3 & 0 \\ 6 & 4 & 1 \\ 1 & 2 & 0 \end{bmatrix}$

Use Cramer's rule to solve each system. See Example 5.

25. $\begin{aligned} x + y + z &= 6 \\ x - y + z &= 2 \\ 2x + y + z &= 7 \end{aligned}$

26. $\begin{aligned} x + y + z &= 2 \\ x - y - 2z &= -3 \\ 2x - y + z &= 7 \end{aligned}$

27. $\begin{aligned} x - 3y + 2z &= 0 \\ x + y + z &= 2 \\ x - y + z &= 0 \end{aligned}$

28. $\begin{aligned} 3x + 2y + 2z &= 0 \\ x - y + z &= 1 \\ x + y - z &= 3 \end{aligned}$

29. $\begin{aligned} x + y &= -1 \\ 2y - z &= 3 \\ x + y + z &= 0 \end{aligned}$

30. $\begin{aligned} x - y &= 8 \\ x - 2z &= 0 \\ x + y - z &= 1 \end{aligned}$

31. $\begin{aligned} x + y - z &= 0 \\ 2x + 2y + z &= 6 \\ x - 3y &= 0 \end{aligned}$

32. $\begin{aligned} x + y + z &= 1 \\ 5x - y &= 0 \\ 3x + y + 2z &= 0 \end{aligned}$

33. $\begin{aligned} x + y + z &= 0 \\ 2y + 2z &= 0 \\ 3x - y &= -1 \end{aligned}$

34. $\begin{aligned} x + z &= 0 \\ x - 3y &= 1 \\ 4y - 3z &= 3 \end{aligned}$

Solve each system. Use Cramer's rule if possible. See Example 6.

35. $\begin{aligned} 2x - y + z &= 1 \\ -6x + 3y - 3z &= -3 \\ 4x - 2y + 2z &= 2 \end{aligned}$

36. $\begin{aligned} x - y + z &= 4 \\ 2x - 2y + 2z &= 3 \\ 4x - y - z &= 1 \end{aligned}$

37. $\begin{aligned} x + y &= 1 \\ y + 2z &= 3 \\ x + 2y + 2z &= 5 \end{aligned}$

38. $\begin{aligned} x - y + z &= 5 \\ 2x - 2y + 2z &= 10 \\ 3x - 3y + 3z &= 15 \end{aligned}$

39. $\begin{aligned} x + y &= 4 \\ y + z &= -3 \\ x + z &= -5 \end{aligned}$

40. $\begin{aligned} x + y &= 0 \\ x - z &= -1 \\ y + z &= 3 \end{aligned}$

Write a system of three equations in three variables for each word problem. Use Cramer's rule to solve each system.

41. Cassandra wants to determine the weights of her two dogs, Mimi and Mitzi. However, neither dog will sit on the scale by herself. Cassandra, Mimi, and Mitzi all together weigh 175 pounds. Cassandra and Mimi together weigh 143 pounds. Cassandra and Mitzi together weigh 139 pounds. How much does each weigh individually?

42. Bernard has 41 coins consisting of nickels, dimes, and quarters, and they are worth a total of $4.00. If the number of dimes plus the number of quarters is one more than the number of nickels, then how many of each does he have?

43. If the two acute angles of a right triangle differ by 12°, then what are the measures of the three angles of this triangle?

44. The obtuse angle of a triangle is twice as large as the sum of the two acute angles. If the smallest angle is only one-eighth as large as the sum of the other two, then what is the measure of each angle?

Wrap-up

CHAPTER 10

SUMMARY

	Concepts	Examples
Solving systems of linear equations in two variables	Graphing Method: Sketch the graphs and see where they intersect.	
	Substitution Method: Solve one equation for one variable in terms of the other, then substitute into the other equation.	$y = x - 5$ $2x + 3y = 1$ Substitution: $2x + 3(x - 5) = 1$
	Addition Method: Multiply each equation as necessary in order to eliminate a variable upon addition of the equations.	$6x - 2y = 8$ $\underline{x + 2y = 1}$ $7x \quad\quad = 9$
Types of linear systems in two variables	Independent: One point in solution set Lines intersect at one point.	$y = x - 5$ $y = 2x + 3$
	Inconsistent: Empty solution set, parallel lines	$2x + y = 1$ $2x + y = 5$
	Dependent: The solution set is infinite. One equation is a multiple of the other.	$2x + 3y = 4$ $4x + 6y = 8$
Linear equation in three variables	$Ax + By + Cz = D$ In a three-dimensional coordinate system, its graph is a plane.	$2x - y + 3z = 5$
Solving linear systems in three variables	Use substitution or addition to eliminate variables. Solution set may be a single point, empty, or infinite.	
Solving nonlinear systems in two variables	Use substitution or addition to eliminate variables. Nonlinear systems often have several points in the solution set.	$y = x^2$ $x^2 + y^2 = 4$ Substitute: $y + y^2 = 4$

| 2 × 2 Determinant | $\begin{vmatrix} a_1 & b_1 \\ a_2 & b_2 \end{vmatrix} = a_1 b_2 - a_2 b_1$ | $\begin{vmatrix} 3 & 4 \\ 2 & 1 \end{vmatrix} = -5$ |

3 × 3 Determinant

$$\begin{vmatrix} a_1 & b_1 & c_1 \\ a_2 & b_2 & c_2 \\ a_3 & b_3 & c_3 \end{vmatrix} = a_1 \cdot \begin{vmatrix} b_2 & c_2 \\ b_3 & c_3 \end{vmatrix} - a_2 \cdot \begin{vmatrix} b_1 & c_1 \\ b_3 & c_3 \end{vmatrix} + a_3 \cdot \begin{vmatrix} b_1 & c_1 \\ b_2 & c_2 \end{vmatrix}$$

Expand by minors about any row or column, using signs from the sign array.

Cramer's rule (two variables)

The solution to the system

$$a_1 x + b_1 y = c_1$$
$$a_2 x + b_2 y = c_2$$

is given by

$$x = \frac{\begin{vmatrix} c_1 & b_1 \\ c_2 & b_2 \end{vmatrix}}{\begin{vmatrix} a_1 & b_1 \\ a_2 & b_2 \end{vmatrix}} = \frac{D_x}{D}, \qquad y = \frac{\begin{vmatrix} a_1 & c_1 \\ a_2 & c_2 \end{vmatrix}}{\begin{vmatrix} a_1 & b_1 \\ a_2 & b_2 \end{vmatrix}} = \frac{D_y}{D},$$

provided $D \neq 0$.

Cramer's rule (three variables)

The solution to the system

$$a_1 x + b_1 y + c_1 z = d_1$$
$$a_2 x + b_2 y + c_2 z = d_2$$
$$a_3 x + b_3 y + c_3 z = d_3$$

is given by $x = \dfrac{D_x}{D}$, $\quad y = \dfrac{D_y}{D}$, $\quad$ and $\quad z = \dfrac{D_z}{D}$, where

$$D = \begin{vmatrix} a_1 & b_1 & c_1 \\ a_2 & b_2 & c_2 \\ a_3 & b_3 & c_3 \end{vmatrix}, \qquad D_x = \begin{vmatrix} d_1 & b_1 & c_1 \\ d_2 & b_2 & c_2 \\ d_3 & b_3 & c_3 \end{vmatrix},$$

$$D_y = \begin{vmatrix} a_1 & d_1 & c_1 \\ a_2 & d_2 & c_2 \\ a_3 & d_3 & c_3 \end{vmatrix}, \qquad D_z = \begin{vmatrix} a_1 & b_1 & d_1 \\ a_2 & b_2 & d_2 \\ a_3 & b_3 & d_3 \end{vmatrix},$$

provided $D \neq 0$.

REVIEW EXERCISES

10.1 *Solve by graphing. Indicate whether each system is independent, inconsistent, or dependent.*

1. $y = 2x - 1$
   $x + y = 2$

2. $y = -x$
   $y = -x + 3$

3. $y = 3x - 4$
   $y = -2x + 1$

4. $x + y = 5$
   $x - y = -1$

Solve each system by the substitution method. Indicate whether each system is independent, inconsistent, or dependent.

5. $y = 3x + 11$
$2x + 3y = 0$

6. $x - y = 3$
$3x - 2y = 3$

7. $x = y + 5$
$2x - 2y = 12$

8. $2x - y = 3$
$6x - 9 = 3y$

10.2 *Solve each system by the addition method. Indicate whether each system is independent, inconsistent, or dependent.*

9. $5x - 3y = -20$
$3x + 2y = 7$

10. $-3x + y = 3$
$2x - 3y = 5$

11. $2(y - 5) + 4 = 3(x - 6)$
$3x - 2y = 12$

12. $3x - 4(y - 5) = x + 2$
$2y - x = 7$

10.3 *Solve each system by elimination of variables.*

13. $2x - y - z = 3$
$3x + y + 2z = 4$
$4x + 2y - z = -4$

14. $2x + 3y - 2z = -11$
$3x - 2y + 3z = 7$
$x - 4y + 4z = 14$

15. $x - 3y + z = 5$
$2x - 4y - z = 7$
$2x - 6y + 2z = 6$

16. $x - y + z = 1$
$2x - 2y + 2z = 2$
$-3x + 3y - 3z = -3$

10.4 *Graph both equations on the same set of axes, then determine the points of intersection of the graphs by solving the system.*

17. $y = x^2$
$y = -2x + 15$

18. $y = \sqrt{x}$
$y = \dfrac{1}{3}x$

19. $y = 3x$
$y = \dfrac{1}{x}$

20. $y = |x|$
$y = -3x + 5$

Solve each system.

21. $x^2 + y^2 = 4$
$y = \dfrac{1}{3}x^2$

22. $12y^2 - 4x^2 = 9$
$x = y^2$

23. $x^2 + y^2 = 34$
$y = x + 2$

24. $y = 2x + 1$
$xy - y = 5$

25. $y = \log(x - 3)$
$y = 1 - \log(x)$

26. $y = (1/2)^x$
$y = 2^{x-1}$

27. $x^4 = 2(12 - y)$
$y = x^2$

28. $x^2 - xy + 2y^2 = 7$
$x^2 + 2y^2 = 9$

10.5 *Evaluate each determinant.*

29. $\begin{vmatrix} 1 & 3 \\ 0 & 2 \end{vmatrix}$

30. $\begin{vmatrix} -1 & 2 \\ -3 & 5 \end{vmatrix}$

31. $\begin{vmatrix} .01 & .02 \\ 50 & 80 \end{vmatrix}$

32. $\begin{vmatrix} 1/2 & 1/3 \\ 1/4 & 1/5 \end{vmatrix}$

Solve each system. Use Cramer's rule if possible.

33. $2x - y = 0$
$3x + y = -5$

34. $3x - 2y = 14$
$2x + 3y = -8$

35. $y = 2x - 3$
$3x - 2y = 4$

36. $2x - 5y = -1$
$10y - 4x = 2$

37. $3x - y = -1$
$2y - 6x = 5$

38. $y = 2x - 5$
$y = 3x - 3y$

10.6 *Evaluate each determinant.*

39. $\begin{vmatrix} 2 & 3 & 1 \\ -1 & 2 & 4 \\ 6 & 1 & 1 \end{vmatrix}$

40. $\begin{vmatrix} 1 & -1 & 0 \\ -2 & 0 & 0 \\ 3 & 1 & 5 \end{vmatrix}$

41. $\begin{vmatrix} 2 & 1 & 0 \\ 2 & -1 & 0 \\ -1 & 3 & 3 \end{vmatrix}$

42. $\begin{vmatrix} 3 & -1 & 4 \\ 2 & -1 & 1 \\ -2 & 0 & 1 \end{vmatrix}$

Solve each system. Use Cramer's rule if possible.

43. $x + y = 3$
$x + y + z = 0$
$x - y - z = 2$

44. $2x - 4y + 2z = 6$
$x - y - z = 1$
$x - 2y + z = 4$

45. $2x - y + z = 0$
$4x + 6y - 2z = 0$
$x - 2y - z = -9$

46. $3x - 3y - 3z = 3$
$2x - 2y - 2z = 2$
$x - y - z = 1$

Miscellaneous

Use a system of equations in two or three variables to solve each word problem. Solve by the method of your choice.

47. The sum of the digits in a two-digit number is 15. When the digits are reversed, the new number is 9 more than the original number. What is the original number?

48. The sum of the digits in a two-digit number is 8. When the digits are reversed, the new number is 18 less than the original number. What is the original number?

49. If a rectangle has a perimeter of 16 feet and an area of 12 square feet, then find its length and width.

50. Alonzo can travel from his camp downstream to the mouth of the river in 30 minutes. If it takes him 45 minutes to come back, then how long would it take him to go that same distance in the lake with no current?

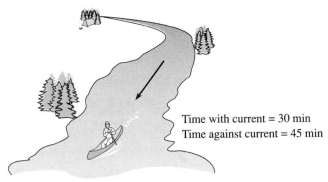

Time with current = 30 min
Time against current = 45 min

Figure for Exercise 50

51. Find the radii of two circles such that the difference in areas of the two is 10π square inches, and the difference in radii of the two is 2 inches.

52. Find two numbers that have a sum of 10 and a product of 22.

53. In four years Gasper will be old enough to drive. His father said that he must have a driver's license for 2 years before he can date. Three years ago, his age was only one-half the age necessary to date. How old must Gasper be to drive, and how old is he now?

54. A chemist has 3 solutions of acid that must be mixed to obtain 20 liters of a solution that is 38% acid. Solution A is 30% acid, solution B is 20% acid, and solution C is 60% acid. Because of another chemical in these solutions, she must keep the ratio of solution C to solution A at 2 to 1. How many liters of each should she mix together?

55. Darlene had $20,000 to invest. She invested part of it at 10% and the remainder at 8%. If her income from the two investments was $1920, how much did she invest at each rate?

56. One serving of canned beets contains 1 gram of protein and 6 grams of carbohydrates. One serving of canned red beans contains 6 grams of protein and 20 grams of carbohydrates. How many servings of each would it take to get exactly 21 grams of protein and 78 grams of carbohydrates?

CHAPTER 10 TEST

Solve the system by graphing.

1. $x + y = 4$
   $y = 2x + 1$

Solve each system by substitution.

2. $y = 2x - 8$
   $4x + 3y = 1$

3. $y = x - 5$
   $3x - 4(y - 2) = 28 - x$

Solve each system by the addition method.

4. $3x + 2y = 3$
   $4x - 3y = -13$

5. $3x - y = 5$
   $-6x + 2y = 1$

Determine whether each system is independent, inconsistent, or dependent.

6. $y = 3x - 5$
$y = 3x + 2$

7. $2x + 2y = 8$
$x + y = 4$

8. $y = 2x - 3$
$y = 5x - 14$

Solve the nonlinear system.

9. $y^2 - x^2 = 7$
$y = -x^2 + 13$

Solve the following system by elimination of variables.

10. $x + y - z = 2$
$2x - y + 3z = -5$
$x - 3y + z = 4$

Evaluate each determinant.

11. $\begin{vmatrix} 2 & 3 \\ 4 & -3 \end{vmatrix}$

12. $\begin{vmatrix} 1 & -2 & -1 \\ 2 & 3 & 1 \\ 1 & 1 & 2 \end{vmatrix}$

Solve each system by using Cramer's rule.

13. $2x - y = -4$
$3x + y = -1$

14. $x + y = 0$
$x - y + 2z = 6$
$2x + y - z = 1$

For each word problem, write a system of equations in two or three variables. Use the method of your choice to solve each system.

15. Find the length and width of a rectangular room that has an area of 108 square feet and a perimeter of 42 feet.

16. One night the manager of the Sea Breeze Motel rented 5 singles and 12 doubles for a total of $390. The next night he rented 9 singles and 10 doubles for a total of $412. What is the rental charge for each type of room?

17. Jill, Karen, and Betsy studied a total of 93 hours last week. Jill's and Karen's study time totaled only one-half as much as Betsy's. If Jill studied 3 hours more than Karen, then how many hours did each one of the girls spend studying?

Tying It All Together

CHAPTERS 1–10

Solve each equation.

1. $2(x - 3) + 5x = 8$

2. $|2x - 3| = 6$

3. $\sqrt{2x + 1} = 3$

4. $4x - 3 = 0$

5. $\dfrac{x^2}{9} = 1$

6. $4x^2 - 3 = 0$

7. $x^{2/3} = 4$

8. $x^2 + 6x + 8 = 0$

9. $x^2 + 8x - 2 = 0$

10. $\dfrac{1}{2} + \dfrac{1}{x} - \dfrac{4}{x^2} = 0$

11. $(x + 3)^2 = x^2$

12. $x^4 - 10x^2 + 9 = 0$

13. $\dfrac{x + 1}{x - 2} = \dfrac{5}{4}$

14. $2^{x-1} = 6$

15. $\ln(x) + \ln(x + 2) = 3 \cdot \ln(2)$

16. $\log_2 x = 5$

Solve each inequality and sketch the graph on the number line.

17. $3x - 5 \geq 10$

18. $3 - x < 6$ and $3x < 6$

19. $|2x - 1| < 7$

20. $x^2 + x - 6 > 0$

Sketch the graph of each equation or function.

21. $y = 3x - 4$

22. $f(x) = 3 - x$

23. $h(x) = |x| + 1$

24. $y = \sqrt{x} - 1$

25. $y = 1/x$

26. $y = 1 - x^2$

27. $y = 2^x$

28. $f(x) = \log_3 x$

29. $f(x) = 3^{-x}$

30. $g(x) = \ln(x)$

Perform the operations and simplify.

31. $(3 + \sqrt{2})^2$

32. $(3 + 2\sqrt{2})(4 - 3\sqrt{2})$

33. $6 \div \sqrt{2}$

34. $\sqrt{18} + \sqrt{50}$

35. $(5 - \sqrt{3})(5 + \sqrt{3})$

36. $\sqrt{242} - \sqrt{98}$

37. $\dfrac{3}{\sqrt{2}} + \sqrt{8}$

38. $\dfrac{\sqrt{2}}{\sqrt{3} - 2}$

39. $\dfrac{6 - \sqrt{12}}{2}$

40. $\dfrac{3}{\sqrt{2}} - \dfrac{1}{2} + \sqrt{12}$

CHAPTER 11
Conic Sections

If a punter kicks a football straight up with a velocity of 128 feet per second from a height of 6 feet above the playing field in the New Orleans Superdome, how high will it go?

The equation that relates the height of the football above the earth and the time it is in the air has a graph in the shape of a parabola. In this chapter we will study the parabola, which is one of four special curves known as the conic sections. These curves are called conic sections because they can be obtained by intersecting a cone and a plane. Will this football hit the roof of the Superdome, 273 feet above the field? We can answer these questions when we do Exercise 51 of Section 11.1. A football actually has been kicked to the roof of the Superdome. (By whom?)

11.1 The Parabola

We studied parabolas in Chapter 8, but in this section we will look at them in more detail.

Standard Equation

The parabola is one of four different curves that can be obtained by intersecting a cone and a plane. These curves, known as **conic sections,** are the parabola, circle, ellipse, and hyperbola. (They are shown in Fig. 11.1.) It can be demonstrated that any parabola that occurs as the intersection of a cone and a plane can be described by an equation of the following form.

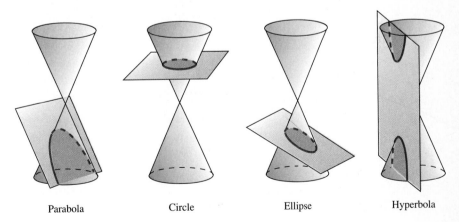

Parabola   Circle   Ellipse   Hyperbola

Figure 11.1

**Standard Equation of a
Parabola**

Suppose a, b, and c are real numbers with $a \neq 0$. The graph of any
equation of the form
$$y = ax^2 + bx + c$$
is a parabola.

We saw in Chapter 8 that a parabola given by an equation in standard form may
open upward or downward. The lowest point on a parabola that opens upward is
called the **vertex** of the parabola. Likewise, the highest point on a parabola that
opens downward also is called the **vertex**. In Chapter 8 we used $x = -b/(2a)$ to
locate the vertex. We will now learn a method for locating the vertex that will help
us understand why the x-coordinate of the vertex is $-b/(2a)$.

Locating the Vertex

Consider the equation
$$y = 3(x - 2)^2 + 1.$$

This equation could be rewritten in the standard form for the equation of a parabola;
however, in this form the vertex is easy to identify. The value of $3(x - 2)^2$ is 0 for
$x = 2$, and greater than 0 for any other value of x. Because $3(x - 2)^2$ is positive for
$x \neq 2$, adding it to 1 gives a y-coordinate greater than 1, and the parabola opens
upward. Thus the smallest y-coordinate is obtained when $x = 2$. Because $y = 1$
when $x = 2$, the vertex is $(2, 1)$. A parabola such as $y = -2(x + 5)^2 + 6$ has a
vertex at $(-5, 6)$ and opens downward. In general, the following statement may be
made.

**Opening and Vertex of a
Parabola**

The parabola $y = a(x - h)^2 + k$, $a \neq 0$, opens upward if $a > 0$ and
downward if $a < 0$. The vertex is (h, k).

EXAMPLE 1 Use completing the square to write $y = -2x^2 - 4x + 1$ in the form
$y = a(x - h)^2 + k$. Identify the vertex and determine whether the parabola opens
upward or downward.

Solution The first step is to factor out -2 from the first two terms on the right side.

$$
\begin{aligned}
y &= -2x^2 - 4x + 1 \\
&= -2(x^2 + 2x) + 1 && \text{Factor out } -2. \\
&= -2(x^2 + 2x + 1 - 1) + 1 && \text{To get an equivalent equation} \\
& && \text{we add 1 and subtract 1.} \\
&= -2(x^2 + 2x + 1) + 2 + 1 && \text{Keep just the perfect-square} \\
& && \text{trinomial in parentheses.} \\
&= -2(x + 1)^2 + 3 && \text{Note } x + 1 = x - (-1).
\end{aligned}
$$

The vertex is $(-1, 3)$, and the parabola opens downward.   ◀

We can use completing the square to rewrite the standard equation of a parabola in the form used above.

$$y = ax^2 + bx + c$$

$$y = a\left[x^2 + \frac{bx}{a}\right] + c \qquad \text{Factor } a \text{ out of the first two terms.}$$

$$y = a\left[x^2 + \frac{bx}{a} + \frac{b^2}{4a^2} - \frac{b^2}{4a^2}\right] + c \qquad \text{Complete the square.}$$

$$y = a\left[x^2 + \frac{bx}{a} + \frac{b^2}{4a^2}\right] - \frac{b^2}{4a} + c$$

$$y = a\left[x + \frac{b}{2a}\right]^2 + \frac{4ac - b^2}{4a}$$

So the x-coordinate of the vertex is $-b/(2a)$. Note that the value of a is the same for either form of the equation of the parabola. We can summarize these ideas as follows.

Opening and Vertex of a Parabola

> The parabola $y = ax^2 + bx + c$ opens upward if $a > 0$ and downward if $a < 0$. The x-coordinate of the vertex is $x = -b/(2a)$.

EXAMPLE 2 Find the vertex of the parabola $y = -3x^2 + 9x - 5$, and determine whether the parabola opens upward or downward.

Solution The x-coordinate of the vertex is

$$x = \frac{-b}{2a} = \frac{-9}{2(-3)} = \frac{-9}{-6} = \frac{3}{2}.$$

In order to find the y-coordinate of the vertex, let $x = 3/2$ in the equation $y = -3x^2 + 9x - 5$:

$$y = -3\left(\frac{3}{2}\right)^2 + 9\left(\frac{3}{2}\right) - 5 = -\frac{27}{4} + \frac{27}{2} - 5 = \frac{7}{4}$$

The vertex is $\left(\frac{3}{2}, \frac{7}{4}\right)$. Because $a = -3$, the parabola opens downward. ◄

Symmetry

Any parabola of the form $y = ax^2 + bx + c$ is symmetric with respect to a vertical line through the vertex. This line is called the **axis of symmetry** (see Fig. 11.2). Its equation is $x = -b/(2a)$. The axis of symmetry divides the parabola into two symmetric halves. If we fold the graph along this line, the two halves coincide.

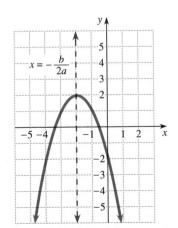

Figure 11.2

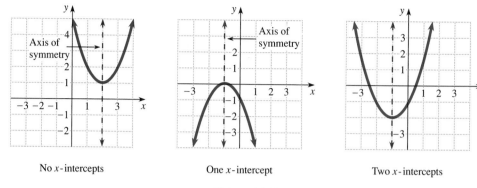

No x-intercepts   One x-intercept   Two x-intercepts

Figure 11.3

Y-intercept

The parabola $y = ax^2 + bx + c$ has exactly one y-intercept. It can be found by letting $x = 0$ in the equation of the parabola:

$$y = a(0)^2 + b(0) + c = c$$

This shows that the y-intercept is $(0, c)$.

X-intercepts

To find the x-intercepts, we let $y = 0$ in the standard equation of the parabola. Thus the x-intercepts are found by solving the quadratic equation

$$ax^2 + bx + c = 0.$$

There may be either 0, 1, or 2 x-intercepts, depending on how many solutions the equation has (see Fig. 11.3).

Graphing a Parabola

To get an accurate graph of a parabola, we should use all of the facts that we know about parabolas. These facts are summarized in the following box as a strategy for graphing parabolas.

STRATEGY
Graphing the Parabola
$y = ax^2 + bx + c$

To graph the parabola use the following facts.

1. If $a > 0$, the parabola opens upward.
   If $a < 0$, the parabola opens downward.
2. The x-coordinate of the vertex is $-b/(2a)$.

3. The axis of symmetry is the vertical line $x = -b/(2a)$.
4. The y-intercept is $(0, c)$.
5. The x-intercepts are found by solving $ax^2 + bx + c = 0$.

EXAMPLE 3 Determine whether the parabola $y = -x^2 - x + 6$ opens upward or downward, and find the vertex, axis of symmetry, x-intercepts, and y-intercept. Find several additional points on the parabola and then sketch the graph.

Solution Because $a = -1$, this parabola opens downward. Because

$$\frac{-b}{2a} = \frac{-(-1)}{2(-1)} = -\frac{1}{2},$$

the x-coordinate of the vertex is $-1/2$. And because

$$y = -\left(-\frac{1}{2}\right)^2 - \left(-\frac{1}{2}\right) + 6 = -\frac{1}{4} + \frac{1}{2} + 6 = \frac{25}{4},$$

the vertex is $(-1/2, 25/4)$. The axis of symmetry is the vertical line $x = -1/2$. To find the x-intercepts, let $y = 0$ in the original equation:

$$-x^2 - x + 6 = 0$$
$$x^2 + x - 6 = 0 \qquad \text{Multiply each side by } -1.$$
$$(x + 3)(x - 2) = 0$$
$$x + 3 = 0 \quad \text{or} \quad x - 2 = 0$$
$$x = -3 \quad \text{or} \qquad x = 2$$

The x-intercepts are $(-3, 0)$ and $(2, 0)$. The y-intercept is $(0, 6)$. Using all of this information and the additional points shown in the table in Fig. 11.4, we get the graph of $y = -x^2 - x + 6$ shown in Fig. 11.4.

x	y
1	4
-2	4
-1	6
0	6

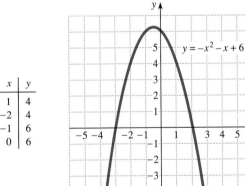

Figure 11.4

Parabolas Opening to the Left or Right

If we interchange the x and y variables in the standard form of the parabola, we get the equation

$$x = ay^2 + by + c.$$

The graph of this equation is also a parabola. By interchanging the x- and y-coordinates, the roles of the x- and y-axis are interchanged also. For example, the graph of $y = 3x^2$ opens in the direction of the positive y-axis. The graph of $x = 3y^2$ opens in the direction of the positive x-axis. All of the other facts about parabolas are modified similarly and stated in the following box.

STRATEGY

Graphing the Parabola
$x = ay^2 + by + c$

To graph the parabola use the following facts.

1. If $a > 0$, the parabola opens to the right.
   If $a < 0$, the parabola opens to the left.
2. The y-coordinate of the vertex is $-b/(2a)$.
3. The axis of symmetry is the horizontal line $y = -b/(2a)$.
4. The x-intercept is $(c, 0)$.
5. The y-intercepts are found by solving $ay^2 + by + c = 0$.

EXAMPLE 4 Determine whether the parabola $x = y^2 - 2y$ opens to the left or the right, and find the vertex, axis of symmetry, x-intercept, and y-intercepts. Find several additional points on the parabola and then sketch the graph.

Solution Because $a = 1$, the parabola opens to the right. The y-coordinate of the vertex is

$$y = \frac{-(-2)}{2(1)} = 1.$$

Because $x = (1)^2 - 2(1) = -1$, the vertex is $(-1, 1)$. The axis of symmetry is the

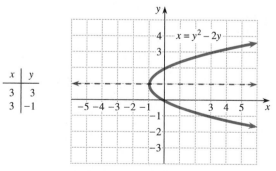

x	y
3	3
3	-1

Figure 11.5

horizontal line $y = 1$. Because $c = 0$, the x-intercept is $(0, 0)$. To find the y-intercepts, we solve

$$y^2 - 2y = 0.$$
$$y(y - 2) = 0$$
$$y = 0 \quad \text{or} \quad y = 2$$

The y-intercepts are $(0, 0)$ and $(0, 2)$. Using all of this information and the additional points shown in the table in Fig. 11.5, we get the graph shown in Fig. 11.5. ◄

Applications

On a parabola that opens upward, the minimum value that y attains is the y-coordinate of the vertex. On a parabola that opens downward, the maximum value that y attains is again the y-coordinate of the vertex. Because we know that the x-coordinate of the vertex is $-b/(2a)$, we can easily obtain the maximum or minimum value of y. Problems often involve finding the maximum or minimum value of a variable.

EXAMPLE 5 Find two numbers that have the maximum product, subject to the condition that their sum is 1.

Solution Because their sum is 1, we can let x represent one number and $1 - x$ represent the other. If we let y represent their product, we can write the equation

$$y = x(1 - x),$$
or
$$y = -x^2 + x.$$

This is the equation of a parabola that opens downward. Its highest point, and the maximum value of y, occurs when

$$x = \frac{-b}{2a} = \frac{-1}{2(-1)} = \frac{1}{2}.$$

If $x = 1/2$, then $1 - x = 1/2$. The two numbers are $1/2$ and $1/2$. Among the numbers with a sum of 1, they give the maximum product. Thus no two numbers with a sum of 1 have a product larger than $1/4$. ◄

Warm-ups

True or false?

1. The point $(1, -1)$ is on the parabola $y = -x^2 + x - 3$.

2. The graph of $y = 6x + 2x + 5$ is a parabola.

3. If x is any real number, then $-5(x - 1)^2 \geq 0$.

4. The parabola $y = 2x - x^2 + 5$ opens upward.

5. The vertex and the y-intercept are the same for the parabola $y = x^2$.

6. If the parabola $y = ax^2 + bx + c$ has its vertex at $(2, 3)$ and $a > 0$, then it has no x-intercepts.

7. The parabola $y = 3x^2 + 9x$ has no y-intercept.

8. If the axis of symmetry of a parabola is $x = 2$, and if the point $(3, 5)$ is on the graph, then the point $(1, 5)$ is also on the graph.

9. The parabola $y = (x - 2)^2$ has only one x-intercept.

10. The parabola $x = -2y^2 + 4y - 1$ opens downward.

11.1 EXERCISES

Use completing the square to write each equation in the form $y = a(x - h)^2 + k$. Identify the vertex, and determine whether the parabola opens upward or downward. See Example 1.

1. $y = x^2 - 6x + 1$

2. $y = x^2 + 4x - 7$

3. $y = 2x^2 + 12x + 5$

4. $y = 3x^2 + 6x - 7$

5. $y = -2x^2 + 16x + 1$

6. $y = -3x^2 - 6x + 7$

7. $y = 5x^2 + 40x$

8. $y = -2x^2 + 10x$

Find the vertex of each parabola (without completing the square), and determine whether the parabola opens upward or downward. See Example 2.

9. $y = x^2 - 4x + 1$

10. $y = x^2 - 6x - 7$

11. $y = -x^2 + 2x - 3$

12. $y = -x^2 + 4x + 9$

13. $y = 3x^2 - 6x + 1$

14. $y = 2x^2 + 4x - 3$

15. $y = -x^2 - 3x + 2$

16. $y = -x^2 + 3x - 1$

17. $y = 3x^2 + 5$

18. $y = -2x^2 - 6$

Determine whether each parabola opens upward or downward, and find its vertex, axis of symmetry, x-intercepts, and y-intercept. Find several additional points on the parabola and then sketch its graph. See Example 3.

19. $y = x^2 - 3x + 2$

20. $y = x^2 + 6x + 8$

21. $y = -x^2 - 2x + 8$

22. $y = -x^2 - 2x + 15$

23. $y = (x + 2)^2 + 1$

24. $y = -2(x + 1)^2 + 3$

25. $y = x^2 + 2x + 1$

26. $y = x^2 - 6x + 9$

27. $y = -4x^2 + 4x - 1$

28. $y = -4x^2 + 12x - 9$

29. $y = x^2 - 5x$

30. $y = 3x^2 - 9x$

31. $y = 3x^2 + 5$

32. $y = -2x^2 + 3$

33. $y = x^2 - 2x - 1$

34. $y = x^2 - 4x + 1$

35. $y = (x - 5)^2$

36. $y = 3(x - 1)^2 - 4$

Determine whether each parabola opens to the left or right, and find its vertex, axis of symmetry, x-intercept, and y-intercepts. Find several additional points on the parabola and then sketch its graph. See Example 4.

37. $x = y^2$

38. $x = -2y^2$

39. $x = -y^2 - 2y + 3$

40. $x = y^2 - 2y - 8$

41. $x = y^2 + 4y + 4$

42. $x = 4y^2 + 4y + 1$

43. $x = y^2 - 4$

44. $x = -2y^2 + 8$

Solve each problem. See Example 5.

45. Find two numbers that have the maximum possible product among numbers that have a sum of 8.

46. What is the maximum product that can be obtained by two numbers that have a sum of -6?

47. The length of the hypotenuse of a right triangle is 5 feet. If the perimeter is 12 feet, then what lengths for the legs would maximize the area?

48. A farmer has 160 feet of fencing that he is going to use to enclose an area in a rectangular shape. What dimensions for the length and width would maximize the area?

49. A farmer has 150 feet of fencing available to enclose a rectangular area alongside of a large barn. If the barn is used for one side and she has to fence only the other three sides of the area, then what dimensions would enclose the maximum area?

50. If the total length of the legs of a right triangle is 6 feet, then what lengths for the legs would minimize the square of the hypotenuse?

51. If a punter kicks a football straight up at a velocity of 128 feet per second from a height of 6 feet, then the football's distance above the earth after t seconds is given by the formula $s = -16t^2 + 128t + 6$. What is the maximum height reached by the ball? Could he hit the roof of the New Orleans Superdome, 273 feet above the playing field?

52. If a baseball is hit straight upward at 150 feet per second from a height of 5 feet, then its distance above the earth

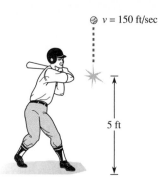

$v = 150$ ft/sec

5 ft

Figure for Exercise 52

after t seconds is given by the formula $s = -16t^2 + 150t + 5$. What is the maximum height attained by the ball?

53. It costs Acme Manufacturing C dollars per hour to operate their golf ball division. An analyst has determined that C is related to the number of golf balls produced per hour, x, by the equation $C = .009x^2 - 1.8x + 100$. What number of balls per hour should they produce in order to minimize the cost per hour of manufacturing these golf balls?

Photo for Exercise 53.

54. A chain-store manager has been told by the main office that daily profit, P, is related to the number of clerks working that day, x, according to the equation $P = -25x^2 + 300x$. What number of clerks will maximize the profit, and what is the maximum possible profit?

Graph both equations of each system on the same coordinate axes. Use elimination of variables (as done in Section 10.4) to find all points of intersection.

55. $y = -x^2 + 3$
    $y = x^2 + 1$

56. $y = x^2 - 3$
    $y = -x^2 + 5$

57. $y = x^2 - 2$
    $y = 2x - 3$

58. $y = x^2 + x - 6$
    $y = 7x - 15$

59. $y = x^2 + 3x - 4$
    $y = -x^2 - 2x + 8$

60. $y = x^2 + 2x - 8$
    $y = -x^2 - x + 12$

61. $x = y^2$
    $y = x$

62. $x = y^2 + 3y$
    $y = x + 1$

63. $x = y^2 - 2y + 1$
    $y = x + 1$

64. $y = x^2$
    $x = y^2$

65. $y = x^2 + 3x - 4$
    $y = 2x + 2$

66. $y = x^2 + 5x + 6$
    $y = x + 11$

11.2 The Circle

IN THIS SECTION:

- Equation of a Circle
- Finding the Center and Radius
- Systems of Equations

In this section we continue the study of the conic sections with the circle. A circle is obtained by cutting a cone, as shown in Fig. 11.1.

Equation of a Circle

In geometry a **circle** is defined as the set of all points in a plane that lie a fixed distance from a given point in the plane. The fixed distance is called the **radius,** and the given point is called the **center.** We can use the distance formula of Section 7.1 to write an equation for a circle with center at (h, k) and radius r. If (x, y) is a point on the circle, it must be r units from the center (see Fig. 11.6). By the distance formula, we can write

$$\sqrt{(x - h)^2 + (y - k)^2} = r.$$

We square both sides of this equation to get the standard form for the equation of a circle.

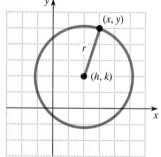

Figure 11.6

Standard Equation for a Circle

The graph of the equation

$$(x - h)^2 + (y - k)^2 = r^2,$$

with $r > 0$, is a circle with center (h, k) and radius r.

Note that an equation such as $(x - 1)^2 + (y + 3)^2 = -9$ is not satisfied by any ordered pair of real numbers, because the left side is nonnegative for any choice of x and y. The equation $(x - 1)^2 + (y + 3)^2 = 0$ is satisfied by only $(1, -3)$.

EXAMPLE 1 Write the standard equation for a circle with center at $(-2, 3)$ and radius 5.

Solution The center at $(-2, 3)$ means that $h = -2$ and $k = 3$. Thus,

$$[x - (-2)]^2 + [y - 3]^2 = 5^2.$$

Simplify this equation to get

$$(x + 2)^2 + (y - 3)^2 = 25. \qquad \blacktriangleleft$$

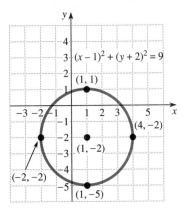

Figure 11.7

Finding the Center and Radius

EXAMPLE 2 Determine the center and radius of the circle described by the equation $x^2 + (y + 5)^2 = 2$.

Solution We can write this equation as

$$(x - 0)^2 + [y - (-5)]^2 = (\sqrt{2})^2.$$

In this form we can recognize that the center is $(0, -5)$ and the radius is $\sqrt{2}$. ◀

EXAMPLE 3 Graph the equation $(x - 1)^2 + (y + 2)^2 = 9$.

Solution The graph of this equation is a circle with center at $(1, -2)$ and radius 3 (see Fig. 11.7). ◀

If we square the binomials that appear in the standard equation of a circle, it is harder to recognize the equation as that of a circle. For example, the equation

$$(x - 1)^2 + (y + 2)^2 = 9$$

can be written as

$$x^2 - 2x + y^2 + 4y = 4.$$

The next example shows how to rewrite an equation like this one in the standard form for a circle by completing the squares for the variables x and y.

EXAMPLE 4 Find the center and radius of the circle given by the equation

$$x^2 - 6x + y^2 + 10y = -33.$$

Solution To complete the square for $x^2 - 6x$ we add 9, and for $y^2 + 10y$ we add 25. To get an equivalent equation, we must perform these operations on each side.

$$x^2 - 6x + \quad y^2 + 10y \qquad = -33$$
$$x^2 - 6x + 9 + y^2 + 10y + 25 = -33 + 9 + 25 \qquad \text{Add 9 and 25 to each side.}$$
$$(x - 3)^2 + (y + 5)^2 = 1 \qquad \text{Factor the two trinomials on the left side.}$$

Now that the equation is in standard form, we see that the center of this circle is at $(3, -5)$, and it has a radius of 1. ◀

Systems of Equations

We first solved systems of nonlinear equations in two variables in Section 10.4. We found the points of intersection of two graphs without drawing the graphs. In the exercises of Section 11.1 we solved systems of equations involving parabolas and lines. In this section we will solve systems involving circles, parabolas, and lines. In the next example we solve a system of equations that locates the points of intersection of a line and a circle.

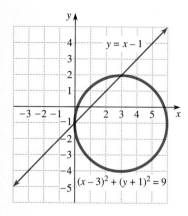

Figure 11.8

EXAMPLE 5   Graph both equations of the system

$$(x - 3)^2 + (y + 1)^2 = 9$$
$$y = x - 1$$

on the same coordinate axes, and solve the system by elimination of variables.

Solution   The graph of the first equation is a circle with center at $(3, -1)$ and radius 3. The graph of the second equation is a straight line with slope 1 and y-intercept $(0, -1)$. Both graphs are shown in Fig. 11.8. To solve the system by elimination, we substitute $y = x - 1$ into the equation of the circle.

$$(x - 3)^2 + (x - 1 + 1)^2 = 9$$
$$(x - 3)^2 + x^2 = 9$$
$$x^2 - 6x + 9 + x^2 = 9$$
$$2x^2 - 6x = 0$$
$$x^2 - 3x = 0$$
$$x(x - 3) = 0$$
$$x = 0 \quad \text{or} \quad x = 3$$
$$y = -1 \qquad y = 2 \qquad \text{Because } y = x - 1$$

The solution set to the system of equations is $\{(0, -1), (3, 2)\}$. Check these solutions in the original system.   ◀

Warm-ups

True or false?

1. The radius of a circle can be any nonzero real number.
2. The coordinates of the center must satisfy the equation of the circle.
3. The circle $x^2 + y^2 = 4$ has its center at the origin.
4. The equation of the circle centered at the origin with radius 9 is $x^2 + y^2 = 9$.
5. The equation $(x - 2)^2 + (y - 3)^2 + 4 = 0$ is the equation of a circle of radius 2.
6. The graph of $(x - 3) + (y + 5) = 9$ is a circle of radius 3 centered at $(3, -5)$.
7. If we know that a circle is centered at $(-3, -1)$ and passes through the origin, then we do not have enough information to write its equation.
8. The center of the circle $(x - 3)^2 + (y - 4)^2 = 10$ is $(-3, -4)$.
9. The center of the circle $x^2 + y^2 + 6y - 4 = 0$ is on the y-axis.
10. The radius of the circle $x^2 - 3x + y^2 = 4$ is 2.

11.2 EXERCISES

Write the standard equation for each circle with the given center and radius. See Example 1.

1. Center $(0, 3)$
   radius 5

2. Center $(2, 0)$
   radius 3

3. Center $(1, -2)$
   radius 9

4. Center $(-3, 5)$
   radius 4

5. Center $(0, 0)$
   radius $\sqrt{3}$

6. Center $(0, 0)$
   radius $\sqrt{2}$

7. Center $(-6, -3)$
   radius $1/2$

8. Center $(-3, -5)$
   radius $1/4$

9. Center $(1/2, 1/3)$
   radius .1

10. Center $(-1/2, 3)$
   radius .2

Find the center and radius for each circle. See Example 2.

11. $(x - 3)^2 + (y - 5)^2 = 2$

12. $(x + 3)^2 + (y - 7)^2 = 6$

13. $x^2 + \left(y - \dfrac{1}{2}\right)^2 = \dfrac{1}{2}$

14. $5x^2 + 5y^2 = 5$

15. $4x^2 + 4y^2 = 9$

16. $9x^2 + 9y^2 = 49$

17. $3 - y^2 = (x - 2)^2$

18. $9 - x^2 = (y + 1)^2$

Sketch the graph of each equation. See Example 3.

19. $x^2 + y^2 = 4$

20. $x^2 + y^2 = 9$

21. $x^2 + (y - 3)^2 = 9$

22. $(x - 4)^2 + y^2 = 16$

23. $(x + 1)^2 + (y - 1)^2 = 2$

24. $(x - 2)^2 + (y + 2)^2 = 8$

25. $(x - 4)^2 + (y + 3)^2 = 16$

26. $(x - 3)^2 + (y - 7)^2 = 25$

27. $\left(x - \dfrac{1}{2}\right)^2 + \left(y + \dfrac{1}{2}\right)^2 = \dfrac{1}{4}$

28. $\left(x + \dfrac{1}{3}\right)^2 + y^2 = \dfrac{1}{9}$

Rewrite each equation in the standard form for the equation of a circle and identify its center and radius. See Example 4.

29. $x^2 + 4x + y^2 + 6y = 0$

30. $x^2 - 10x + y^2 + 8y = 0$

31. $x^2 - 2x + y^2 - 4y - 3 = 0$

32. $x^2 - 6x + y^2 - 2y + 9 = 0$

33. $x^2 + y^2 = 8y + 10x - 32$

34. $x^2 + y^2 = 8x - 10y$

35. $x^2 - x + y^2 + y = 0$

36. $x^2 - 3x + y^2 = 0$

37. $x^2 - 3x + y^2 - y = 1$

38. $x^2 - 5x + y^2 + 3y = 2$

39. $x^2 - \dfrac{2}{3}x + y^2 + \dfrac{3}{2}y = 0$

40. $x^2 + \dfrac{1}{3}x + y^2 - \dfrac{2}{3}y = \dfrac{1}{9}$

Graph both equations of each system on the same coordinate axes. Solve the system by elimination of variables to find all points of intersection of the graphs. See Example 5.

41. $x^2 + y^2 = 10$
   $y = 3x$

42. $x^2 + y^2 = 4$
   $y = x - 2$

43. $x^2 + y^2 = 9$
   $y = x^2 - 3$

44. $x^2 + y^2 = 4$
   $y = x^2 - 2$

45. $(x - 2)^2 + (y + 3)^2 = 4$
   $y = x - 3$

46. $(x + 1)^2 + (y - 4)^2 = 17$
   $y = x + 2$

Solve each problem.

47. Determine all points of intersection of the circle $(x - 1)^2 + (y - 2)^2 = 4$ with the y-axis.

48. Determine the points of intersection of the circle $x^2 + (y - 3)^2 = 25$ with the x-axis.

49. Find the radius of the circle that has center $(2, -5)$ and passes through the origin.

50. Find the radius of the circle that has center $(-2, 3)$ and passes through $(3, -1)$.

51. Determine the equation of the circle that is centered at $(2, 3)$ and passes through the point $(-2, -1)$.

52. Determine the equation of the circle that is centered at $(3, 4)$ and passes through the origin.

53. Find all points of intersection of the circles $x^2 + y^2 = 9$ and $(x - 5)^2 + y^2 = 9$.

54. Suppose lighthouse A is located at the origin and lighthouse B is located at coordinates $(0, 6)$. The captain of a ship has determined that his distance from A is 2 and his distance from B is 5. What are the possible coordinates for the location of the ship?

55. What does the graph of $x^2 + y^2 + 9 = 0$ look like?

56. A donkey is tied at the point $(2, -3)$ on a rope of length 12 feet. Turnips are growing at the point $(6, 7)$. Can he reach them?

57. Graph the equation $x^2 + y^2 = 0$.

58. Graph the equation $x^2 - y^2 = 0$.

*59. Graph the equation $y = \sqrt{1 - x^2}$.

*60. Graph the equation $y = -\sqrt{1 - x^2}$.

11.3   The Ellipse and Hyperbola

IN THIS SECTION:

- **The Ellipse**
- **The Hyperbola**

In this section we study the remaining two conic sections, the ellipse and the hyperbola.

The Ellipse

An ellipse can be obtained by intersecting a plane and a cone as shown in Fig. 11.1. We can also describe an **ellipse** as the set of points in a plane with the property that the sum of their distances from two fixed points is a constant. Each fixed point is called a focus (plural: foci). A common way to draw an ellipse is illustrated in Fig. 11.9. A string is attached at two fixed points, and a pencil is used to take up the slack. As the pencil is moved around the paper, the sum of the distances of the pencil point from the two fixed points remains constant. Of course, the length of the string is that constant. You may wish to try this.

The ellipse also has interesting reflecting properties. All light or sound waves emitted from one focus are reflected off the ellipse to concentrate at the other focus (see Fig. 11.10). This property is used in light fixtures where a concentration of light at a point is desired or in a whispering gallery like the rotunda of the United States Capitol Building.

The orbits of the planets around the sun and satellites around the earth are elliptical. For a satellite, the earth is at one focus and a point in space is at the other focus. The sun is at one focus for the orbit of the earth.

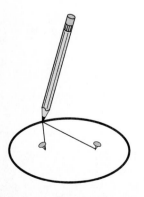

Figure 11.9

The whispering gallery of the U.S. Capitol Building

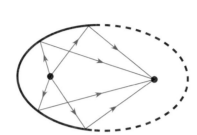

Figure 11.10

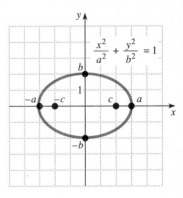

Figure 11.11

Figure 11.11 shows an ellipse with foci $(c, 0)$ and $(-c, 0)$. The origin is the **center** of this ellipse. It has x-intercepts at $(a, 0)$ and $(-a, 0)$ and y-intercepts at $(0, b)$ and $(0, -b)$. The distance formula can be used to write the following equation for this ellipse. The derivation of this formula is usually done in a college algebra course.

Equation of an Ellipse Centered at (0, 0)

For $a > 0$ and $b > 0$, the graph of the equation

$$\frac{x^2}{a^2} + \frac{y^2}{b^2} = 1$$

is an ellipse centered at the origin having x-intercepts $(a, 0)$ and $(-a, 0)$ and y-intercepts $(0, b)$ and $(0, -b)$.

EXAMPLE 1 Sketch the graph of $\dfrac{x^2}{4} + \dfrac{y^2}{9} = 1$.

Solution To sketch this ellipse, we find the x-intercepts and the y-intercepts. If $x = 0$, we get

$$\frac{0}{4} + \frac{y^2}{9} = 1$$

$$\frac{y^2}{9} = 1$$

$$y^2 = 9$$

$$y = \pm 3$$

If $y = 0$, we get $x = \pm 2$. For a rough sketch of an ellipse we plot only the intercepts and draw an ellipse through them, as shown in Fig. 11.12. ◄

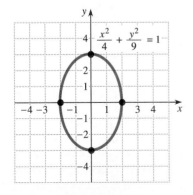

Figure 11.12

Ellipses, like circles, may be centered at any point in the plane. In the next example, we graph an ellipse that is not centered at the origin.

EXAMPLE 2 Sketch the graph of $\dfrac{(x-2)^2}{25} + \dfrac{(y+4)^2}{9} = 1$.

Solution The center of an ellipse is determined the same way the center for a circle is obtained. The center for this ellipse is $(2, -4)$. The denominator 25 is used to determine that the ellipse passes through points that are 5 units to the left and 5 units to the right of the center, $(7, -4)$ and $(-3, -4)$ (see Fig. 11.13). The denominator 9 is used to determine that the ellipse passes through points that are 3 units above and 3 units below the center, $(2, -1)$ and $(2, -7)$.

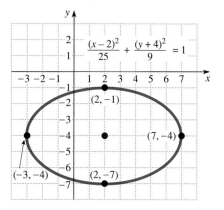

Figure 11.13 ◀

The Hyperbola

A hyperbola is the curve that occurs at the intersection of a cone and a plane as shown in Fig. 11.1. A **hyperbola** can also be described as the set of all points in the plane such that the difference between their distances from two fixed points (foci) is constant (see Fig. 11.14). The definitions of a hyperbola and an ellipse are similar, and so are their equations. Their graphs, however, are very different. A hyperbola has two parts called **branches.** These branches look like parabolas, *but they are not parabolas*.

The hyperbola shown in Fig. 11.15 can be described by the equation

$$\frac{x^2}{36} - \frac{y^2}{9} = 1.$$

The dashed lines are not part of the hyperbola, but are used as an aid in sketching the hyperbola. We say that this hyperbola opens to the left and right. The diagonal

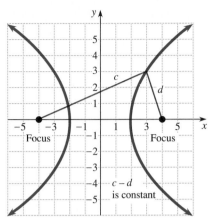

Figure 11.14

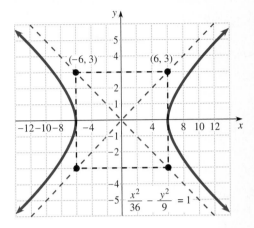

Figure 11.15

lines are called **asymptotes.** The asymptotes are the extended diagonals of the dashed rectangle of Fig. 11.15. The rectangle is called the **fundamental rectangle.** The hyperbola approaches these asymptotes but does not intersect them.

Note that if $x = 0$ in the equation of this hyperbola, we get

$$\frac{0^2}{36} - \frac{y^2}{9} = 1$$

$$-\frac{y^2}{9} = 1$$

$$y^2 = -9.$$

This equation has no solution. There are no y-intercepts. If $y = 0$ in the equation of this hyperbola, we get

$$\frac{x^2}{36} = 1$$

$$x = \pm 6.$$

Thus, the x-intercepts are $(6, 0)$ and $(-6, 0)$.

The key to drawing a hyperbola is getting the fundamental rectangle and extending its diagonals to get the asymptotes. Notice that the rectangle of Fig. 11.15 extends to the x-intercepts along the x-axis, and extends 3 units above and below the origin along the y-axis. The 3 comes from $\sqrt{9}$, and the 9 is found under y^2 in the equation. The facts necessary for graphing a hyperbola centered at the origin and opening to the left and to the right are summarized as follows.

STRATEGY

Graphing a Hyperbola Centered at the Origin, Opening Left and Right

Suppose $a > 0$ and $b > 0$. The graph of

$$\frac{x^2}{a^2} - \frac{y^2}{b^2} = 1$$

is a hyperbola that opens left and right.

1. Its x-intercepts are $(a, 0)$ and $(-a, 0)$.
2. Its asymptotes are found by extending the diagonals of the rectangle that extends to the intercepts along the x-axis and to $(0, b)$ and $(0, -b)$ along the y-axis.

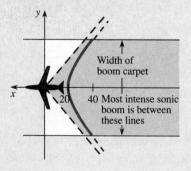

Math at Work

The Concorde aircraft can fly from London to New York in less than three hours. Why are there no plans to use supersonic jets to shorten the trip from New York to Los Angeles?

Supersonic jets travel great distances in short periods of time because they fly faster than the speed of sound. However, for this reason jets such as the Concorde pose unique noise problems. Traveling at approximately 770 mph, the jet creates a cone-shaped wave through the air. If the plane is flying parallel to the ground, the ground and cone intersect along one branch of a hyperbola, which moves along the ground with the aircraft. The noise from the aircraft creates a momentary change in air pressure along the hyperbola, and a thunderlike sound called a sonic boom is heard at any point on the hyperbola. Sonic booms have been blamed for physical destruction such as broken windows and cracked plaster, as well as for psychological disturbances caused by the loud noise. For this reason, the Federal Aviation Association restricts supersonic operations over land areas in the United States.

The area where the sonic boom is most noticeable is called the "boom carpet." A pilot knows that the boom carpet is roughly 5 miles wide for each 1 mile of altitude of the aircraft. The diagram at the left shows a bird's-eye-view of a jet and the boom carpet. In this case the sonic boom is heard along the hyperbola given by the equation

$$\frac{x^2}{400} - \frac{y^2}{100} = 1,$$

where the units are miles. Assume that the width of the boom carpet is measured 40 miles behind the aircraft and find the altitude of the aircraft.

A hyperbola may open up and down. In this case the graph intersects only the y-axis. The facts necessary for graphing a hyperbola that opens up and down are summarized as follows.

STRATEGY
Graphing a Hyperbola
Centered at the Origin,
Opening Up and Down

Suppose $a > 0$ and $b > 0$. The graph of

$$\frac{y^2}{b^2} - \frac{x^2}{a^2} = 1$$

is a hyperbola that opens up and down.

1. Its y-intercepts are $(0, b)$ and $(0, -b)$.
2. Its asymptotes are found by extending the diagonals of the rectangle that extends to the intercepts along the y-axis, and to $(a, 0)$ and $(-a, 0)$ along the x-axis.

EXAMPLE 3 Graph the hyperbola

$$\frac{y^2}{9} - \frac{x^2}{4} = 1.$$

Solution This graph has no x-intercepts, because if $y = 0$, we get

$$-\frac{x^2}{4} = 1,$$

which has no solution. If $x = 0$, we get

$$\frac{y^2}{9} = 1$$

$$y = \pm 3.$$

The y-intercepts are $(0, 3)$ and $(0, -3)$. The hyperbola opens up and down. Because $a^2 = 4$, $a = 2$. The rectangle extends to the intercepts $(0, 3)$ and $(0, -3)$ on the y-axis, and to the points $(2, 0)$ and $(-2, 0)$ along the x-axis. We extend the diagonals of the rectangle and draw the graph of the hyperbola as shown in Fig. 11.16. ◄

EXAMPLE 4 Sketch the graph of the hyperbola $4x^2 - y^2 = 4$.

Solution We first write the equation in standard form. Divide each side by 4 to get the equation

$$x^2 - \frac{y^2}{4} = 1.$$

There are no y-intercepts. If $y = 0$, then $x = \pm 1$. The hyperbola opens left and right, with x-intercepts at $(1, 0)$ and $(-1, 0)$. The dashed rectangle extends to the intercepts along the x-axis, and to the points $(0, 2)$ and $(0, -2)$ along the y-axis. We

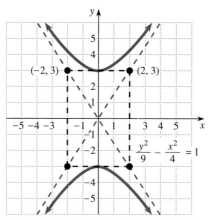

Figure 11.16

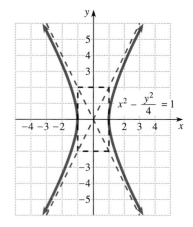

Figure 11.17

extend the diagonals of the rectangle for the asymptotes and draw the graph as shown in Fig. 11.17.   ◀

Warm-ups

True or false?

1. The x-intercepts of the ellipse $\dfrac{x^2}{36} + \dfrac{y^2}{25} = 1$ are $(5, 0)$ and $(-5, 0)$.

2. The graph of $\dfrac{x^2}{9} + \dfrac{y}{4} = 1$ is an ellipse.

3. If the foci of an ellipse coincide, then the ellipse is a circle.

4. The graph of $2x^2 + y^2 = 2$ is an ellipse centered at the origin.

5. The y-intercepts of $x^2 + \dfrac{y^2}{3} = 1$ are $(0, \sqrt{3})$ and $(0, -\sqrt{3})$.

6. The graph of $\dfrac{x^2}{9} - \dfrac{y}{4} = 1$ is a hyperbola.

7. The graph of $\dfrac{x^2}{25} - \dfrac{y^2}{16} = 1$ has y-intercepts at $(0, 4)$ and $(0, -4)$.

8. The hyperbola $\dfrac{y^2}{9} - x^2 = 1$ opens up and down.

9. The graph of $4x^2 - y^2 = 4$ is a hyperbola.

10. The asymptotes of a hyperbola are the extended diagonals of a rectangle.

11.3 EXERCISES

Sketch the graph of each ellipse. See Example 1.

1. $\dfrac{x^2}{9} + \dfrac{y^2}{4} = 1$

2. $\dfrac{x^2}{9} + \dfrac{y^2}{16} = 1$

3. $\dfrac{x^2}{9} + y^2 = 1$

4. $x^2 + \dfrac{y^2}{4} = 1$

5. $\dfrac{x^2}{36} + \dfrac{y^2}{25} = 1$

6. $\dfrac{x^2}{25} + \dfrac{y^2}{49} = 1$

7. $\dfrac{x^2}{24} + \dfrac{y^2}{5} = 1$

8. $\dfrac{x^2}{6} + \dfrac{y^2}{17} = 1$

9. $9x^2 + 16y^2 = 144$

10. $9x^2 + 25y^2 = 225$

11. $25x^2 + y^2 = 25$

12. $x^2 + 16y^2 = 16$

13. $4x^2 + 9y^2 = 1$

14. $25x^2 + 16y^2 = 1$

Sketch the graph of each ellipse. See Example 2.

15. $\dfrac{(x-3)^2}{4} + \dfrac{(y-1)^2}{9} = 1$

16. $\dfrac{(x+5)^2}{49} + \dfrac{(y-2)^2}{25} = 1$

17. $\dfrac{(x+1)^2}{16} + \dfrac{(y-2)^2}{25} = 1$

18. $\dfrac{(x-3)^2}{36} + \dfrac{(y+4)^2}{64} = 1$

19. $(x-2)^2 + \dfrac{(y+1)^2}{36} = 1$

20. $\dfrac{(x+3)^2}{9} + (y+1)^2 = 1$

Sketch the graph of each hyperbola. See Examples 3 and 4.

21. $\dfrac{x^2}{4} - \dfrac{y^2}{9} = 1$

22. $\dfrac{x^2}{16} - \dfrac{y^2}{9} = 1$

23. $\dfrac{y^2}{4} - \dfrac{x^2}{25} = 1$

24. $\dfrac{y^2}{9} - \dfrac{x^2}{16} = 1$

25. $\dfrac{x^2}{25} - y^2 = 1$

26. $x^2 - \dfrac{y^2}{9} = 1$

27. $x^2 - \dfrac{y^2}{25} = 1$

28. $\dfrac{x^2}{9} - y^2 = 1$

29. $9x^2 - 16y^2 = 144$

30. $9x^2 - 25y^2 = 225$

31. $x^2 - y^2 = 1$

32. $y^2 - x^2 = 1$

33. $\dfrac{(x+1)^2}{4} - \dfrac{(y-2)^2}{9} = 1$

34. $\dfrac{(x+3)^2}{16} - \dfrac{(y+2)^2}{25} = 1$

Graph both equations of each system on the same coordinate axes. Use elimination of variables (as in Section 10.4) to find all points of intersection.

35. $\dfrac{x^2}{4} + \dfrac{y^2}{9} = 1$

$x^2 - \dfrac{y^2}{9} = 1$

36. $x^2 - \dfrac{y^2}{4} = 1$

$\dfrac{x^2}{9} + \dfrac{y^2}{4} = 1$

37. $\dfrac{x^2}{4} + \dfrac{y^2}{16} = 1$

$x^2 + y^2 = 1$

38. $x^2 + \dfrac{y^2}{9} = 1$

$x^2 + y^2 = 4$

39. $x^2 + y^2 = 4$

$x^2 - y^2 = 1$

40. $x^2 + y^2 = 16$

$x^2 - y^2 = 4$

41. $x^2 + 9y^2 = 9$

$x^2 + y^2 = 4$

42. $x^2 + y^2 = 25$

$x^2 + 25y^2 = 25$

43. $x^2 + 9y^2 = 9$

$y = x^2 - 1$

44. $4x^2 + y^2 = 4$

$y = 2x^2 - 2$

45. $9x^2 - 4y^2 = 36$

$2y = x - 2$

46. $25y^2 - 9x^2 = 225$

$y = 3x + 3$

11.4 Second-degree Inequalities

IN THIS SECTION:

- Graphing a Second-degree Inequality
- Systems of Inequalities

In this section we will graph second-degree inequalities and systems of inequalities involving second-degree inequalities.

Graphing a Second-degree Inequality

A second-degree inequality is an inequality involving squares of at least one of the variables. Changing the equal sign to an inequality symbol for any of the equations of the conic sections gives us a second-degree inequality. Second-degree inequalities are graphed in the same manner as linear inequalities.

EXAMPLE 1   Graph the inequality $y < x^2 + 2x - 3$.

Solution   We first graph $y = x^2 + 2x - 3$. This parabola has x-intercepts at $(1, 0)$ and $(-3, 0)$, y-intercept of $(0, -3)$, and vertex $(-1, -4)$. The graph of the parabola is drawn with a dashed line as shown in Fig. 11.18. The graph of the parabola divides the plane into two regions. Every point on one side of the parabola satisfies the inequality $y < x^2 + 2x - 3$, and every point on the other side satisfies the inequality $y > x^2 + 2x - 3$. To determine which side is which, we test a point not on

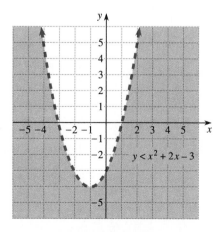

Figure 11.18

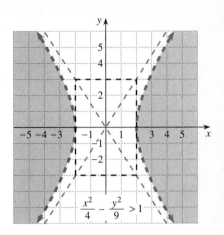

Figure 11.19   Figure 11.20

the parabola, say $(0, 0)$. Because $0 < 0^2 + 2 \cdot 0 - 3$ is false, the region not containing the origin is shaded as in Fig. 11.18.   ◄

EXAMPLE 2   Graph the inequality $x^2 + y^2 < 9$.

Solution   The graph of $x^2 + y^2 = 9$ is a circle of radius 3, centered at the origin. The circle divides the plane into two regions. Each point in one region satisfies $x^2 + y^2 < 9$, and each point in the other region satisfies $x^2 + y^2 > 9$. To identify the regions, we pick some point and test it. Select $(0, 0)$. The inequality

$$0^2 + 0^2 < 9$$

is true. Because $(0, 0)$ is inside the circle, all points inside the circle satisfy the inequality $x^2 + y^2 < 9$. (The graph is shown in Fig. 11.19.) The points outside the circle satisfy the inequality $x^2 + y^2 > 9$.   ◄

EXAMPLE 3   Graph the inequality $\dfrac{x^2}{4} - \dfrac{y^2}{9} > 1$.

Solution   First graph the hyperbola $\dfrac{x^2}{4} - \dfrac{y^2}{9} = 1$. Because the hyperbola shown in Fig. 11.20 divides the plane into three regions, we select a test point in each region and check to see if it satisfies the inequality. Testing the points $(-3, 0)$, $(0, 0)$, and $(3, 0)$ gives us the inequalities

$$\frac{(-3)^2}{4} - \frac{0^2}{9} > 1, \qquad \frac{0^2}{4} - \frac{0^2}{9} > 1, \qquad \text{and} \qquad \frac{3^2}{4} - \frac{0^2}{9} > 1.$$

Because only the first and third inequalities are correct, we shade only the regions containing $(-3, 0)$ and $(3, 0)$, as shown in Fig. 11.20.   ◄

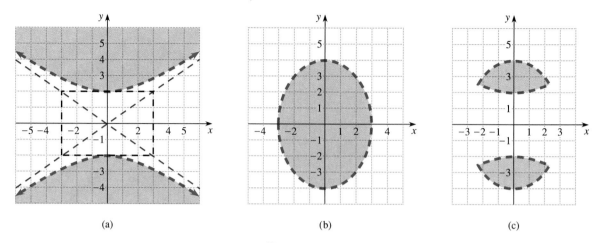

(a)   (b)   (c)

Figure 11.21

Systems of Inequalities

A point is in the solution set to a system of inequalities if it satisfies all inequalities of the system. We graph a system of inequalities by first determining the graph of each inequality and then finding the intersection of the graphs.

EXAMPLE 4   Graph the system of inequalities.

$$\frac{y^2}{4} - \frac{x^2}{9} > 1$$

$$\frac{x^2}{9} + \frac{y^2}{16} < 1$$

Solution   Figure 11.21(a) shows the graph of the first inequality. Figure 11.21(b) shows the graph of the second inequality. In Fig. 11.21(c) we have shaded only the points that satisfy both inequalities. Figure 11.21(c) is the graph of the system. ◄

Warm-ups

True or false?

1. The graph of $x^2 + y = 4$ is a circle of radius 2.

2. The graph of $x^2 + 9y^2 = 9$ is an ellipse.

3. The graph of $y^2 = x^2 + 1$ is a hyperbola.

4. The point $(0, 0)$ satisfies the inequality $2x^2 - y < 3$.

5. The graph of the inequality $y > x^2 - 3x + 2$ contains the origin.

6. The origin should be used as a test point to graph $x^2 > y$.

7. The region containing the origin should be shaded when graphing the inequality $x^2 + 3x + y^2 + 8y + 3 < 0$.

8. The graph of $x^2 + y^2 < 4$ is the region enclosed by a circle of radius 2.

9. The point $(0, 4)$ is in the solution set to the system:

$$x^2 - y^2 < 1$$
$$y > x^2 - 2x + 3$$

10. The point $(0, 0)$ is in the solution set to the system:

$$x^2 + y^2 < 1$$
$$y < x^2 + 1$$

11.4 EXERCISES

Graph each inequality. See Examples 1–3.

1. $y > x^2$

2. $y \le x^2 + 1$

3. $y < x^2 - x$

4. $y > x^2 + x$

5. $y > x^2 - x - 2$

6. $y < x^2 + x - 6$

7. $x^2 + y^2 \le 9$

8. $x^2 + y^2 > 16$

9. $x^2 + 4y^2 > 4$

10. $4x^2 + y^2 \le 4$

11. $4x^2 - 9y^2 < 36$

12. $25x^2 - 4y^2 > 100$

13. $(x - 2)^2 + (y - 3)^2 < 4$

14. $(x + 1)^2 + (y - 2)^2 > 1$

15. $x^2 + y^2 > 1$

16. $x^2 + y^2 < 25$

17. $4x^2 - y^2 > 4$

18. $x^2 - 9y^2 \le 9$

19. $y^2 - x^2 \le 1$

20. $x^2 - y^2 > 1$

21. $x \le y^2 - 3y$

22. $x - 1 \ge y^2 + 2y$

23. $x > y$

24. $x < 2y - 1$

Graph the solution set to each system of inequalities. See Example 4.

25. $x^2 + y^2 < 9$
    $y > x$

26. $x^2 + y^2 > 1$
    $x > y$

27. $x^2 - y^2 > 1$
    $x^2 + y^2 < 4$

28. $y^2 - x^2 < 1$
    $x^2 + y^2 > 9$

29. $y > x^2 + x$
    $y < 5$

30. $y > x^2 + x - 6$
    $y < x + 3$

31. $y \ge x + 2$
    $y \le 2 - x$

32. $y \ge 2x - 3$
    $y \le 3 - 2x$

33. $4x^2 - y^2 < 4$
    $x^2 + 4y^2 > 4$

34. $x^2 - 4y^2 < 4$
    $x^2 + 4y^2 > 4$

35. $x - y < 0$
    $y + x^2 < 1$

36. $y + 1 > x^2$
    $x + y < 2$

37. $y < 5x - x^2$
    $x^2 + y^2 < 9$

38. $y < x^2 + 4x$
    $x^2 + y^2 < 16$

39. $y \ge 3$
    $x \le 1$

40. $x > -3$
    $y < 2$

41. $4y^2 - 9x^2 < 36$
    $x^2 + y^2 < 16$

42. $25y^2 - 16x^2 < 400$
    $x^2 + y^2 > 4$

43. $y < x^2$
    $x^2 + y^2 < 1$

44. $y > x^2$
    $4x^2 + y^2 < 4$

Wrap-up

CHAPTER 11

SUMMARY

Parabola

$y = a(x - h)^2 + k$

Opens up if $a > 0$
Opens down if $a < 0$
Vertex at (h, k)

For $y = ax^2 + bx + c$, the
x-coordinate of vertex is $-b/(2a)$.

Examples

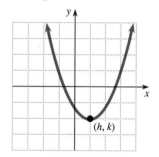

$x = a(y - h)^2 + k$

Opens to right if $a > 0$
Opens to left if $a < 0$
Vertex at (k, h)

For $x = ay^2 + by + c$, the
y-coordinate of vertex is $-b/(2a)$.

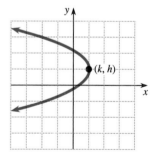

Circle

$(x - h)^2 + (y - k)^2 = r^2$

Center (h, k)
Radius r

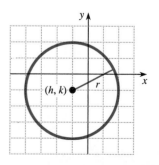

Ellipse

$$\frac{x^2}{a^2} + \frac{y^2}{b^2} = 1$$

Center: (0, 0)
x-intercepts: $(a, 0)$ and $(-a, 0)$
y-intercepts: $(0, b)$ and $(0, -b)$

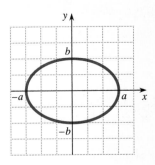

Hyperbola

$$\frac{x^2}{a^2} - \frac{y^2}{b^2} = 1$$

Centered at (0, 0)
x-intercepts: $(a, 0)$ and $(-a, 0)$
y-intercepts: none
Opens left and right

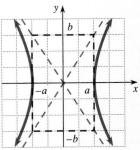

$$\frac{y^2}{b^2} - \frac{x^2}{a^2} = 1$$

Centered at (0, 0)
x-intercepts: none
y-intercepts: $(0, b)$ and $(0, -b)$
Opens up and down

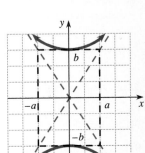

Inequalities

Use test points to determine which
regions satisfy the inequality.

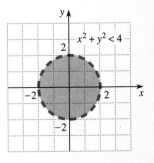

Systems of inequalities	Shade only points that satisfy all inequalities of the system.	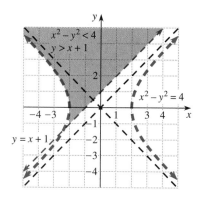

REVIEW EXERCISES

11.1 *Sketch the graph of each parabola. Determine the x- and y-intercepts, the vertex, and the axis of symmetry for each.*

1. $y = x^2 + 3x - 18$ **2.** $y = x - x^2$ **3.** $y = x^2 + 3x + 2$

4. $y = -x^2 - 3x + 4$ **5.** $x = -y^2 + 10y - 25$ **6.** $x = y^2 - 4y - 6$

Solve each problem.

7. Find two numbers that have the maximum possible product among numbers that have a sum of 6.

8. Find two numbers that have the smallest possible product among numbers that have a difference of 10.

9. Otto has 90 feet of fencing that he plans to use to enclose a rectangular area. What dimensions will maximize the area of the rectangle?

10. Suppose Otto already has a 6-foot-wide gate in place, and he wants to use the gate and an additional 90 feet of fencing to enclose a rectangular area. What dimensions will maximize the area of the rectangle?

11.2 *Determine the center and radius of each circle and sketch its graph.*

11. $x^2 + y^2 = 100$ **12.** $x^2 + y^2 = 20$ **13.** $(x - 2)^2 + (y + 3)^2 = 81$

14. $x^2 + 2x + y^2 = 8$ **15.** $9y^2 + 9x^2 = 4$ **16.** $x^2 + 4x + y^2 - 6y - 3 = 0$

Write the standard equation for each circle with the given center and radius.

17. Center $(0, -3)$
radius 6

18. Center $(0, 0)$
radius $\sqrt{6}$

19. Center $(2, -7)$
radius 5

20. Center $(1/2, -3)$
radius $1/2$

11.3 *Sketch the graph of each ellipse.*

21. $\dfrac{x^2}{36} + \dfrac{y^2}{49} = 1$ **22.** $\dfrac{x^2}{25} + y^2 = 1$ **23.** $25x^2 + 4y^2 = 100$ **24.** $6x^2 + 4y^2 = 24$

Sketch the graph of each hyperbola.

25. $\dfrac{x^2}{49} - \dfrac{y^2}{36} = 1$ **26.** $\dfrac{y^2}{25} - \dfrac{x^2}{49} = 1$ **27.** $4x^2 - 25y^2 = 100$ **28.** $6y^2 - 16x^2 = 96$

11.4 *Graph each inequality.*

29. $4x - 2y > 3$

30. $y < x^2 - 3x$

31. $y^2 < x^2 - 1$

32. $y^2 < 1 - x^2$

33. $4x^2 + 9y^2 > 36$

34. $x^2 + y > 2x - 1$

Graph the solution set to each system of inequalities.

35. $y < 3x - x^2$
$x^2 + y^2 < 9$

36. $x^2 - y^2 < 1$
$y < 1$

37. $4x^2 + 9y^2 > 36$
$x^2 + y^2 < 9$

38. $y^2 - x^2 > 4$
$y^2 + 16x^2 < 16$

Miscellaneous

Identify each equation as the equation of a straight line, parabola, circle, hyperbola, or ellipse. Do these mentally, without rewriting the equations.

39. $x^2 = y^2 + 1$

40. $x = y + 1$

41. $x^2 = 1 - y^2$

42. $x^2 = y + 1$

43. $x^2 + x = 1 - y^2$

44. $(x - 3)^2 + (y + 2)^2 = 7$

45. $x^2 + 4x = 6y - y^2$

46. $4x + 6y = 1$

47. $\dfrac{x^2}{3} - \dfrac{y^2}{5} = 1$

48. $x^2 + \dfrac{y^2}{3} = 1$

49. $4y^2 - x^2 = 8$

50. $9x^2 + y = 9$

Sketch the graph of each equation.

51. $x^2 = 4 - y^2$

52. $x^2 = 4y^2 + 4$

53. $x^2 = 4y + 4$

54. $x = 4y + 4$

55. $x^2 = 4 - 4y^2$

56. $x^2 = 4y - y^2$

57. $x^2 = 4 - (y - 4)^2$

58. $(x - 2)^2 + (y - 4)^2 = 4$

Determine the equation of each of the circles described below.

59. Centered at the origin and passing through $(3, 4)$

60. Centered at $(2, -3)$ and passing through $(-1, 4)$

61. Centered at $(-1, 5)$ with radius 6

62. Centered at $(0, -3)$ and passing through the origin

Solve each system of equations.

63. $x^2 + y^2 = 25$
$y = -x + 1$

64. $x^2 - y^2 = 1$
$x^2 + y^2 = 7$

65. $4x^2 + y^2 = 4$
$x^2 - y^2 = 21$

66. $y = x^2 + x$
$y = -x^2 + 3x + 12$

CHAPTER 11 TEST

Sketch the graph of each equation.

1. $x^2 + y^2 = 25$

2. $\dfrac{x^2}{16} - \dfrac{y^2}{25} = 1$

3. $y^2 + 4x^2 = 4$

4. $y = x^2 + 4x + 4$

5. $y^2 - 4x^2 = 4$

6. $y = -x^2 - 2x + 3$

Sketch the graph of each inequality.

7. $x^2 - y^2 < 9$

8. $x^2 + y^2 > 9$

9. $y > x^2 - 9$

10. $y > x - 9$

11. $36x^2 + 9y^2 < 324$

12. $\dfrac{y^2}{4} - \dfrac{x^2}{16} > 1$

Graph the solution set to each system of inequalities.

13. $x^2 + y^2 < 9$
$x^2 - y^2 > 1$

14. $y < -x^2 + x$
$y < x - 4$

Solve each system of equations.

15. $y = x^2 - 2x - 8$
$y = 7 - 4x$

16. $x^2 + y^2 = 12$
$y = x^2$

17. Find the center and radius of the circle
$$x^2 + 2x + y^2 + 10y = 10.$$

18. Determine the vertex of the parabola $y = x^2 + x + 3$.

19. If a ball is thrown straight up with a velocity of 64 feet per second from a height of 20 feet, then its height above the ground at any time t is given by $s = -16t^2 + 64t + 20$. What is its maximum height?

20. Find the equation of a circle with center $(-1, 3)$ that passes through $(2, 5)$.

Tying It All Together

CHAPTERS 1–11

Sketch the graph of each equation.

1. $y = 9x - x^2$
2. $y = 9x$
3. $y = (x - 9)^2$
4. $y^2 = 9 - x^2$
5. $y = 9x^2$
6. $y = |9x|$
7. $4x^2 + 9y^2 = 36$
8. $4x^2 - 9y^2 = 36$
9. $y = 9 - x$
10. $y = 9^x$

Find the following products.

11. $(x + 2y)^2$
12. $(x + y)(x^2 + 2xy + y^2)$
13. $(a + b)^3$
14. $(a - 3b)^2$
15. $(2a + 1)(3a - 5)$
16. $(x - y)(x^2 + xy + y^2)$

Solve each system of equations.

17. $2x - 3y = -4$
$x + 2y = 5$

18. $x^2 + y^2 = 25$
$x + y = 7$

19. $2x - y + z = 7$
$x - 2y - z = 2$
$x + y + z = 2$

20. $y = x^2$
$y - 2x = 3$

Solve each formula for the specified variable.

21. $ax + b = 0$, for x

22. $wx^2 + dx + m = 0$, for x

23. $A = \dfrac{1}{2}h(B + b)$, for B

24. $\dfrac{1}{x} + \dfrac{1}{y} = \dfrac{1}{2}$, for x

25. $L = m + mxt$, for m

26. $y = 3a\sqrt{t}$, for t

27. Write the equation of the line in slope-intercept form that goes through the points $(2, -3)$ and $(-4, 1)$.

28. Write the equation of the line in slope-intercept form that contains the origin and is perpendicular to the line $2x - 4y = 5$.

29. Write the equation of the circle that has center $(2, 5)$ and passes through the point $(-1, -1)$.

30. Find the center and radius of the circle
$$x^2 + 3x + y^2 - 6y = 0.$$

Perform the computations with complex numbers.

31. $2i(3 + 5i)$

32. i^6

33. $(2i - 3) + (6 - 7i)$

34. $(3 + i\sqrt{2})^2$

35. $(2 - 3i)(5 - 6i)$

36. $(3 - i) + (-6 + 4i)$

37. $(5 - 2i)(5 + 2i)$

38. $(2 - 3i) \div (2i)$

39. $(4 + 5i) \div (1 - i)$

40. $\dfrac{4 - \sqrt{-8}}{2}$

Sequences and Series

Retirement might seem to be far in the future, but banks and investment companies are continually stressing the importance of long-term investment for retirement. Why should we start saving for something that might be as much as 45 years away? Periodically saving a modest sum can amount to a substantial "nest egg" in 45 years.

Suppose that you invest $2000 each year for 45 years in an investment paying 9% compounded annually. At the end of 45 years, the first $2000 you invested will be worth $2000(1.09)^{45}$. The second $2000 will receive compound interest for 44 years, and will be worth $2000(1.09)^{44}$. The sum of 45 such amounts is what you will have at the end of 45 years. This sum is a geometric series. In this chapter we will learn an easy way to calculate the sum of a geometric series.

At $2000 per year for 45 years, you will invest only $90,000. What will this investment be worth after 45 years? We will calculate the actual amount of this investment in Exercise 51 of Section 12.4. It turns out to be well over one million dollars!

12.1 Sequences

IN THIS SECTION:

- **Definition**
- **Finding a Formula for the *n*th Term**

The word ''sequence'' is a familiar word. We may speak of a sequence of events, or say that something is out of sequence. In this section we will give the mathematical definition of a sequence.

Definition

In mathematics we think of a **sequence** as a list of numbers. Each number in the sequence is called a **term** of the sequence. There is a first term, a second term, a third term, and so on. For example, the daily high-temperature readings in Minot, North Dakota, for the first 10 days in January, can be thought of as a finite sequence with 10 terms.

$$-9, -2, 8, -11, 0, 6, 14, 1, -5, -11$$

The set of all positive even integers,

$$2, 4, 6, 8, 10, 12, 14, \ldots,$$

can be thought of as an infinite sequence.

Actually, a sequence is not defined as a list of numbers; it is defined as a function. The list of numbers is the range of the function. This may not sound like what we have just described, but it is more precise than saying that a sequence is some kind of list.

DEFINITION
Sequence

> A **finite sequence** is a function whose domain is the set of positive integers less than or equal to some fixed positive integer. An **infinite sequence** is a function whose domain is the set of all positive integers.

When the domain is apparent, we will refer to either a finite sequence or an infinite sequence as a sequence. For the independent variable of the function, we will usually use n (for natural number), rather than x. For the dependent variable, we write a_n (read ''a sub n''), rather than y. We call a_n the **nth term,** or the **general term,** of the sequence. Rather than use the $f(x)$ notation for functions, we will define sequences with formulas. For example, the function

$$a_n = n^2 \qquad \text{for } 1 \le n \le 5$$

is a finite sequence with five terms. When we use n for the variable, we assume that n is a natural number. Note that

$$a_1 = 1^2 = 1, \quad \text{the first term,}$$
$$a_2 = 2^2 = 4, \quad \text{the second term,}$$
$$a_3 = 3^2 = 9, \quad \text{the third term,}$$
$$a_4 = 4^2 = 16, \quad \text{the fourth term,}$$
and
$$a_5 = 5^2 = 25, \quad \text{the fifth term.}$$

The five terms of this sequence are

$$1, 4, 9, 16, 25.$$

We often refer to the listing of the terms of the sequence as the sequence.

EXAMPLE 1   List all of the terms of the sequence

$$a_n = \frac{1}{n + 2} \qquad \text{for } 1 \le n \le 6.$$

Solution   Letting n equal each integer from 1 through 6, we get the following.

$$a_1 = \frac{1}{1 + 2} = \frac{1}{3}, \qquad a_2 = \frac{1}{2 + 2} = \frac{1}{4}, \qquad a_3 = \frac{1}{5},$$

$$a_4 = \frac{1}{6}, \qquad a_5 = \frac{1}{7}, \qquad a_6 = \frac{1}{8}$$

The 6 terms of the sequence are

$$\frac{1}{3}, \frac{1}{4}, \frac{1}{5}, \frac{1}{6}, \frac{1}{7}, \frac{1}{8}. \qquad \blacktriangleleft$$

EXAMPLE 2   List the first four terms of the infinite sequence whose nth term is

$$a_n = \frac{(-1)^n}{2^{n+1}}.$$

Solution   We find the first four terms by letting n take the values 1 through 4 in the formula for the nth term.

$$a_1 = \frac{(-1)^1}{2^{1+1}} = -\frac{1}{4} \qquad a_2 = \frac{(-1)^2}{2^{2+1}} = \frac{1}{8}$$

$$a_3 = \frac{(-1)^3}{2^{3+1}} = -\frac{1}{16} \qquad a_4 = \frac{(-1)^4}{2^{4+1}} = \frac{1}{32}$$

Using the first four terms just obtained, we may write the terms of this sequence as follows:

$$-\frac{1}{4}, \frac{1}{8}, -\frac{1}{16}, \frac{1}{32}, \ldots$$ ◀

Finding a Formula for the nth Term

We often know the terms of a sequence and want to write a formula that will produce the known terms. There is no definite procedure for writing the formula, and there may be more than one formula. We must examine the terms and discover what kind of pattern there is to them. We will assume the most obvious pattern. Each term is a function of the term number. The first term is a function of 1, the second term is a function of 2, and so on.

EXAMPLE 3 Write the general term for the infinite sequence whose terms are

$$3, 5, 7, 9, 11, \ldots$$

Solution The even numbers are all multiples of two, and can be represented as $2n$. Because each odd number is one more than an even number, a formula for the nth term might be

$$a_n = 2n + 1.$$

To be sure, we write out a few terms using the formula:

$$a_1 = 2(1) + 1 = 3$$
$$a_2 = 2(2) + 1 = 5$$
$$a_3 = 2(3) + 1 = 7$$

Thus, the general term is $a_n = 2n + 1$. ◀

EXAMPLE 4 Write the general term for the infinite sequence whose terms are

$$1, -\frac{1}{4}, \frac{1}{9}, -\frac{1}{16}, \ldots$$

Solution To obtain the alternating signs, we use a power of -1, $(-1)^{n+1}$. Note that any even power of -1 is positive, and any odd power of -1 is negative. The denominators are the squares of the positive integers. Thus, the nth term of this infinite sequence is given by the formula

$$a_n = \frac{(-1)^{n+1}}{n^2}.$$ ◀

The next example illustrates how a sequence can occur in a physical situation.

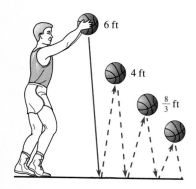

6 ft

4 ft

$\frac{8}{3}$ ft

Figure 12.1

EXAMPLE 5   Suppose that a ball always rebounds 2/3 of the height from which it falls, and the ball is dropped from a height of 6 feet. Write a sequence whose terms are the heights from which the ball falls. What is a formula for the nth term of this sequence?

Solution   On the first fall the ball travels 6 feet (see Fig. 12.1). On the second fall it travels $(2/3)(6) = 4$ feet. On the third fall it travels $(2/3)(4) = 8/3$ feet, and so on. We can write the sequence as follows:

$$6, 4, \frac{8}{3}, \frac{16}{9}, \frac{32}{27}, \ldots$$

The nth term can be written using powers of 2/3.

$$a_n = 6\left(\frac{2}{3}\right)^{n-1}$$

◀

Warm-ups

True or false?

1. The nth term of the sequence 2, 4, 6, 8, 10, . . . is $a_n = 2n$.
2. The nth term of the sequence 1, 3, 5, 7, 9, . . . is $a_n = 2n - 1$.
3. A sequence is a function.
4. The domain of a finite sequence is the set of positive integers.
5. The nth term of the sequence -1, 4, -9, 16, -25, . . . is $a_n = (-1)^{n+1}n^2$.
6. For the infinite sequence $b_n = 1/n$, the independent variable is $1/n$.
7. For the sequence $c_n = n^3$, the dependent variable is c_n.
8. The sixth term of the sequence $a_n = (-1)^{n+1}2^n$ is -64.
9. The symbol a_n is used for the dependent variable of a sequence.
10. The tenth term of the sequence 1, 1, 2, 3, 5, 8, 13, . . . (called a Fibonacci sequence) is 55.

12.1 EXERCISES

List all terms of each finite sequence. See Examples 1 and 2.

1. $a_n = n^2$ $1 \leq n \leq 8$

2. $a_n = -n^2$ $1 \leq n \leq 4$

3. $b_n = \dfrac{(-1)^n}{n}$ $1 \leq n \leq 10$

4. $b_n = \dfrac{(-1)^{n+1}}{n}$ $1 \leq n \leq 6$

5. $c_n = (-2)^{n-1}$ $1 \leq n \leq 5$

6. $c_n = (-3)^{n-2}$ $1 \leq n \leq 5$

7. $a_n = 2^{-n}$ $1 \leq n \leq 6$

8. $a_n = 2^{-n+2}$ $1 \leq n \leq 5$

9. $b_n = 2n - 3 \quad 1 \le n \le 7$

10. $b_n = 2n + 6 \quad 1 \le n \le 7$

11. $c_n = n^{-1/2} \quad 1 \le n \le 5$

12. $c_n = n^{1/2}2^{-n} \quad 1 \le n \le 4$

Write the first four terms of the infinite sequence whose nth term is given. See Examples 1 and 2.

13. $a_n = \dfrac{1}{n^2 + n}$

14. $b_n = \dfrac{1}{(n + 1)(n + 2)}$

15. $b_n = \dfrac{1}{2n - 5}$

16. $a_n = \dfrac{4}{2n + 5}$

17. $c_n = (-1)^n(n - 2)^2$

18. $c_n = (-1)^n(2n - 1)^2$

19. $a_n = \dfrac{(-1)^{2n}}{n^2}$

20. $a_n = (-1)^{2n+1}2^{n-1}$

Write a formula for the general term of each infinite sequence. See Examples 3 and 4.

21. $1, 3, 5, 7, 9, \ldots$

22. $5, 7, 9, 11, 13, \ldots$

23. $1, -1, 1, -1, \ldots$

24. $-1, 1, -1, 1, \ldots$

25. $0, 2, 4, 6, 8, \ldots$

26. $4, 6, 8, 10, 12, \ldots$

27. $3, 6, 9, 12, \ldots$

28. $4, 8, 12, 16, \ldots$

29. $4, 7, 10, 13, \ldots$

30. $3, 7, 11, 15, \ldots$

31. $-1, 2, -4, 8, -16, \ldots$

32. $1, -3, 9, -27, \ldots$

33. $0, 1, 4, 9, 16, \ldots$

34. $0, 1, 8, 27, 64, \ldots$

Solve each problem. See Example 5.

35. A football is on the 8-yard line, and five penalties in a row are given that move the ball "half the distance to the (closest) goal." Write a sequence of five terms that specify the location of the ball after each penalty.

36. Leon planted nine acres of soybeans, but by the end of each week insects had destroyed 1/3 of the acreage that was healthy at the beginning of the week. How many acres does he have left after six weeks?

37. Each year the price of a new car increases 5% over its price the previous year. For a car that sells for $20,000 now, find its price to the nearest dollar each year for the next five years.

38. Each year a manufacturer adds $1000 to the price of a new car. If the car sells for $20,000 now, find its price each year for the next five years.

39. The sequence whose nth term is $a_n = (.999)^n$ is a sequence of numbers that get closer and closer to 0. Calculate a_{100}, a_{1000}, and a_{10000}.

40. The sequence whose nth term is

$$a_n = (1 + 1/n)^n$$

is a sequence of numbers that get closer and closer to the number e, the base of the natural logarithm. Find a_{100}, a_{1000}, and $a_{1000000}$, and compare them to the value of e.

41. Everyone has 2 (biological) parents, 4 grandparents, 8 great-grandparents, 16 great-great-grandparents, and so on. If we put the word "great" in front of the word "grandparents" 35 times, then how many of this type of relative do you have? Is this more or less than the present population of the earth?

Photo for Exercise 41

42. If you deposit one cent into your piggy bank on September 1 and each day thereafter deposit twice as much as on the previous day, then how much will you be depositing on September 30?

12.2 Series

IN THIS SECTION:

- Summation Notation
- Series
- Changing the Index

We are frequently interested in the total value of the terms of a sequence. Of course, if we have only a few terms of a sequence and we know what they are, we can add them up. However, we are often interested in the sum of many terms or even the sum of all the terms of an infinite sequence. In this section we will learn a notation for expressing the sum of the terms of a sequence.

Summation Notation

To describe the sum of the terms of a sequence, we adopt a new notation called **summation notation.** We use the Greek letter Σ (sigma) in summation notation. For example, the sum of the first five terms of the sequence whose general term is $a_n = n^2$ is written as

$$\sum_{i=1}^{5} i^2.$$

The letter i is used in place of n to represent the independent variable. To actually add this up, we let i take the values 1 through 5 in the expression i^2. Thus,

$$\sum_{i=1}^{5} i^2 = 1^2 + 2^2 + 3^2 + 4^2 + 5^2 = 1 + 4 + 9 + 16 + 25 = 55.$$

The letter i used here is called the **index of summation.** Other letters may also be used. The expressions

$$\sum_{i=1}^{5} i^2, \quad \sum_{j=1}^{5} j^2, \quad \text{and} \quad \sum_{n=1}^{5} n^2$$

have the same value.

EXAMPLE 1 Find the value of the expression

$$\sum_{i=1}^{3} (-1)^i (2i + 1).$$

Solution Let i take the values from 1 through 3 and add the results.

$$\sum_{i=1}^{3}(-1)^i(2i + 1) = (-1)[2(1) + 1] + (-1)^2[2(2) + 1] + (-1)^3[2(3) + 1]$$

$$= -3 + 5 - 7$$
$$= -5 \qquad \blacktriangleleft$$

Series

When an expression is written that indicates the sum of the terms of a sequence, the expression is called a **series.** For example, the expression

$$1 + 4 + 9 + 16 + 25 + 36$$

is a series. The formal definition of series follows.

DEFINITION
Series

> The indicated sum of the terms of a sequence is called a **series.**

Just as a sequence may be finite or infinite, a series may be finite or infinite. In this section we will discuss finite series only. In Section 12.4 we will discuss one type of infinite series.

It is often convenient to write a series using the compact summation notation.

EXAMPLE 2 Write the series

$$2 + 4 + 6 + 8 + 10 + 12 + 14$$

in summation notation.

Solution The general term for the sequence of positive even integers is $2n$, but for summation notation, we will use $2i$. If we let i take the values from 1 through 7, we will get these even integers. Thus,

$$2 + 4 + 6 + 8 + 10 + 12 + 14 = \sum_{i=1}^{7}2i. \qquad \blacktriangleleft$$

EXAMPLE 3 Write the series

$$\frac{1}{2} - \frac{1}{3} + \frac{1}{4} - \frac{1}{5} + \frac{1}{6} - \frac{1}{7} + \cdots + \frac{1}{50}$$

in summation notation.

Solution It is not necessary to start the index with the number 1. In this case we will let i be 2 through 50. The alternating signs are produced by a power of -1. The series is written as

$$\sum_{i=2}^{50} \frac{(-1)^i}{i}.$$ ◀

Changing the Index

In Example 3 we saw the index go from 2 through 50, but this is arbitrary. A series can be written with the index starting at any given number.

EXAMPLE 4   Rewrite the series

$$\sum_{i=1}^{6} \frac{(-1)^i}{i^2}$$

with an index j, where j starts at 0.

Solution   If j starts at 0, then j takes the values 0 through 5. This will give us six terms. Because $i = j + 1$, we replace i by $j + 1$ in the summation notation.

$$\sum_{j=0}^{5} \frac{(-1)^{j+1}}{(j+1)^2}$$

Check that these two series have exactly the same six terms.   ◀

Warm-ups

True or false?

1. A series is the indicated sum of the terms of a sequence.

2. There are eight terms in the series

$$\sum_{i=2}^{10} i^3.$$

3. The series $\displaystyle\sum_{i=1}^{9}(-1)^i i^2$ and $\displaystyle\sum_{j=0}^{8}(-1)^j(j+1)^2$ have the same sum.

4. $\displaystyle\sum_{i=1}^{2}(-1)^i 2^i = 2$

5. The ninth term of the series $\displaystyle\sum_{i=1}^{100} \frac{(-1)^i}{(i+1)(i+2)}$ is $1/110$.

6. The sum of a series can never be negative.

$$7.\ \sum_{i=1}^{5} 3i = 3\sum_{i=1}^{5} i \qquad\qquad 8.\ \sum_{i=1}^{5} 2i + \sum_{i=1}^{5} 7i = \sum_{i=1}^{5} 9i$$

$$9.\ \sum_{i=1}^{5} 4 = 20 \qquad\qquad 10.\ \sum_{i=1}^{3} (2i+1) = \left(\sum_{i=1}^{3} 2i\right) + 1$$

12.2 EXERCISES

Find the sum of each series. See Example 1.

1. $\sum_{i=1}^{4} i^2$
2. $\sum_{j=0}^{3} (j+1)^2$
3. $\sum_{j=0}^{5} (2j-1)$
4. $\sum_{i=1}^{6} (2i-3)$

5. $\sum_{i=1}^{5} 2^{-i}$
6. $\sum_{i=1}^{5} (-2)^{-i}$
7. $\sum_{i=1}^{10} 5i^0$
8. $\sum_{j=1}^{20} 3$

9. $\sum_{i=1}^{3} (i-3)(i+1)$
10. $\sum_{i=0}^{5} i(i-1)(i-2)(i-3)$
11. $\sum_{j=1}^{10} (-1)^j$
12. $\sum_{j=1}^{11} (-1)^j$

Write each series in summation notation. Use the index i, and let i begin at 1 in each summation. See Examples 2 and 3.

13. $1+2+3+4+5+6$
14. $2+4+6+8+10$
15. $-1+3-5+7-9+11$

16. $1-3+5-7+9$
17. $1+4+9+16+25+36$
18. $1+8+27+64+125$

19. $\frac{1}{3}+\frac{1}{4}+\frac{1}{5}+\frac{1}{6}$
20. $1-\frac{1}{2}+\frac{1}{3}-\frac{1}{4}+\frac{1}{5}-\frac{1}{6}$

21. $\ln(2)+\ln(3)+\ln(4)$
22. $e^1+e^2+e^3+e^4$
23. $a_1+a_2+a_3+a_4$

24. $a^2+a^3+a^4+a^5$
25. $x_1+x_2+\cdots+x_{50}$
26. $y_1+y_2+y_3+\cdots+y_{30}$

27. $w_1+w_2+w_3+\cdots+w_n$
28. $m_1+m_2+m_3+\cdots+m_k$

Complete the rewriting of each series using the new index as indicated. See Example 4.

29. $\sum_{i=1}^{5} i^2 = \sum_{j=0}$
30. $\sum_{i=1}^{6} i^3 = \sum_{j=0}$
31. $\sum_{i=0}^{5} (2i-1) = \sum_{j=1}$
32. $\sum_{i=1}^{3} (3i+2) = \sum_{j=0}$

33. $\sum_{i=4}^{8} \frac{1}{i} = \sum_{j=1}$
34. $\sum_{i=5}^{10} 2^{-i} = \sum_{j=1}$
35. $\sum_{i=1}^{5} x^{2i+3} = \sum_{j=0}$
36. $\sum_{i=0}^{2} x^{3-2i} = \sum_{j=1}$

37. $\sum_{i=1}^{n} x^i = \sum_{j=0}$
38. $\sum_{i=0}^{n} x^{-i} = \sum_{j=1}$

Write out the terms of each series.

39. $\sum_{i=1}^{6} x^i$
40. $\sum_{i=1}^{5} (-1)^i x^{i-1}$
41. $\sum_{j=0}^{3} (-1)^j x_j$

42. $\displaystyle\sum_{j=1}^{5} \frac{1}{x_j}$ **43.** $\displaystyle\sum_{i=1}^{3} i x^i$ **44.** $\displaystyle\sum_{i=1}^{5} \frac{x}{i}$

A series can be used to describe the situation in each of the following problems.

45. A frog with a vision problem is one yard away from a dead cricket. He spots the cricket and jumps halfway to it. After he realizes that he has not reached it, he again jumps halfway to it. Write a series in summation notation to describe how far he has moved after nine such jumps.

46. Wilma deposited $1000 at the beginning of each year for five years in an account paying 10% interest compounded annually. Write a series using summation notation to describe how much she has in the account at the end of the fifth year. Note that the first $1000 will receive interest for five years, the second $1000 will receive interest for four years, and so on.

12.3 Arithmetic Sequences and Series

IN THIS SECTION:

- **Arithmetic Sequences**
- **Arithmetic Series**

We defined sequences and series in Sections 12.1 and 12.2. In this section we will study a special type of sequence known as an arithmetic sequence. We will also study the corresponding series made from this sequence.

Arithmetic Sequences

Consider the sequence

$$5, 9, 13, 17, 21, \ldots$$

This sequence is called an arithmetic sequence because of the pattern for the terms. Each term is four more than the previous term.

DEFINITION
Arithmetic Sequence

> A sequence where each term after the first is obtained by adding a fixed amount to the previous term is called an **arithmetic sequence.**

The fixed amount is called the **common difference** and is denoted by the letter d. If a_1 is the first term, then the second term must be $a_1 + d$. The third term must be $a_1 + 2d$, and the fourth must be $a_1 + 3d$, and so on.

Formula for the *n*th Term of an Arithmetic Sequence

> The *n*th term, a_n, of an arithmetic sequence with first term a_1 and common difference d is
> $$a_n = a_1 + (n - 1)d.$$

EXAMPLE 1 Write a formula for the *n*th term of the sequence

$$5, 9, 13, 17, 21, \ldots$$

Solution In this sequence each term after the first is four more than the previous term. Because the common difference is four and the first term is five, we can write the *n*th term as

$$a_n = 5 + (n - 1)4.$$

We can simplify this to

$$a_n = 4n + 1. \qquad \blacktriangleleft$$

In the next example the common difference is negative.

EXAMPLE 2 Write a formula for the *n*th term of the sequence

$$4, 1, -2, -5, -8, \ldots$$

Solution Each term is three less than the previous term, so $d = -3$. Because $a_1 = 4$, we can write the *n*th term as

$$a_n = 4 + (n - 1)(-3),$$

or

$$a_n = -3n + 7. \qquad \blacktriangleleft$$

In the next example we will write some terms of an arithmetic sequence from a formula for the *n*th term.

EXAMPLE 3 Write the first five terms of the arithmetic sequence whose *n*th term is $a_n = 3 + (n - 1)6$.

Solution Let n take the values from 1 through 5 and find a_n:

$$a_1 = 3 + (1 - 1)6 = 3$$
$$a_2 = 3 + (2 - 1)6 = 9$$
$$a_3 = 3 + (3 - 1)6 = 15$$
$$a_4 = 3 + (4 - 1)6 = 21$$
$$a_5 = 3 + (5 - 1)6 = 27$$

The first five terms of the sequence are 3, 9, 15, 21, 27. $\qquad \blacktriangleleft$

The formula $a_n = a_1 + (n - 1)d$ involves four variables: a_1, a_n, n, and d. If we know the values of any three of these variables, we can find the fourth.

EXAMPLE 4 Find the twelfth term of the arithmetic sequence whose first term is 2 and whose fifth term is 14.

Solution The fifth term, a_5, is given by

$$a_5 = a_1 + (5 - 1)d.$$

Because we know that $a_1 = 2$, and $a_5 = 14$, we can find d:

$$14 = 2 + (5 - 1)d$$
$$14 = 2 + 4d$$
$$12 = 4d$$
$$3 = d$$

Now we use the same formula to find a_{12}.

$$a_{12} = 2 + (12 - 1)3$$
$$a_{12} = 35$$
◀

Arithmetic Series

Consider the series

$$2 + 4 + 6 + 8 + 10 + \cdots + 54.$$

The terms of this series are the terms of an arithmetic sequence. The indicated sum of an arithmetic sequence is called an **arithmetic series.**

To find the actual sum of the above arithmetic series we use a useful trick. Write the series in increasing order, and below that write the series in decreasing order. We then add the corresponding terms.

$$
\begin{array}{rcccccccc}
S = & 2 + & 4 + & 6 + & 8 + \cdots + & 52 + 54 \\
S = & 54 + & 52 + & 50 + & 48 + \cdots + & 4 + 2 \\
\hline
2S = & 56 + & 56 + & 56 + & 56 + \cdots + & 56 + 56
\end{array}
$$

All that remains is to find how many times 56 appears in the sum on the right. Because

$$2 + 4 + 6 + \cdots + 54 = 2 \cdot 1 + 2 \cdot 2 + 2 \cdot 3 + \cdots + 2 \cdot 27,$$

there are 27 terms in this sum. Therefore, 56 appears 27 times in the sum. Thus,

$$2S = 27 \cdot 56$$
$$S = \frac{27 \cdot 56}{2} = 27 \cdot 28 = 756.$$

We can apply the same idea as above to develop a formula for the sum of the first n terms of any arithmetic series. Let $S_n = a_1 + a_2 + a_3 + \cdots + a_n$. We can also write S_n as follows.

$$S_n = a_1 \qquad\;\; + (a_1 + d) + (a_1 + 2d) + \cdots + a_n$$
$$\underline{S_n = a_n \qquad\;\; + (a_n - d) + (a_n - 2d) + \cdots + a_1} \qquad \text{Add.}$$
$$2S_n = (a_1 + a_n) + (a_1 + a_n) + (a_1 + a_n) + \cdots + (a_1 + a_n)$$

On the right there are n terms of the type $(a_1 + a_n)$. Thus,

$$2S_n = n(a_1 + a_n)$$

$$S_n = \frac{n}{2}(a_1 + a_n).$$

Sum of Arithmetic Series

The sum, S_n, of the first n terms of an arithmetic series with first term a_1 and nth term a_n is given by the formula $S_n = \dfrac{n}{2}(a_1 + a_n)$.

EXAMPLE 5   Find the sum of the positive integers from 1 to 100 inclusive.

Solution   The series $1 + 2 + 3 + \cdots + 100$ is a series with 100 terms; the first is 1, and the last is 100. Thus,

$$S_{100} = \frac{100}{2}(1 + 100)$$

$$S_{100} = 50(101) = 5050. \qquad \blacktriangleleft$$

EXAMPLE 6   Find the sum of the series

$$12 + 16 + 20 + \cdots + 84.$$

Solution   This series is an arithmetic series with $a_n = 84$, $a_1 = 12$, and $d = 4$. To get the number of terms, n, we can use the formula

$$a_n = a_1 + (n - 1)d.$$

Thus,

$$84 = 12 + (n - 1)4$$
$$84 = 8 + 4n$$
$$76 = 4n$$
$$19 = n.$$

We can now find the sum of these 19 terms.

$$S_{19} = \frac{19}{2}(12 + 84) = 912 \qquad \blacktriangleleft$$

Warm-ups

True or false?

1. The common difference in the arithmetic sequence 3, 1, -1, -3, -5, . . . is 2.

2. The sequence 2, 5, 9, 14, 20, 27, . . . is an arithmetic sequence.

3. The sequence 2, 4, 2, 0, 2, 4, 2, 0, . . . is an arithmetic sequence.

4. The nth term of an arithmetic sequence with first term a_1 and common difference d is given by the formula $a_n = a_1 + nd$.

5. If the first term of an arithmetic sequence is 5 and the third term is 10, then the fourth term is 15.

6. If the first term of an arithmetic sequence is 6 and the third term is 2, then the second term is 10.

7. An arithmetic series is the indicated sum of an arithmetic sequence.

8. The series $\sum_{i=1}^{5}(3 + 2i)$ is an arithmetic series.

9. The sum of the first n counting numbers is $n(n + 1)/2$.

10. The sum of the even integers from 8 through 28 inclusive is $\dfrac{10}{2}(8 + 28)$.

12.3 EXERCISES

Write a formula for the nth term of each arithmetic sequence. See Examples 1 and 2.

1. 0, 6, 12, 18, 24, . . .
2. 0, 5, 10, 15, 20, . . .
3. 7, 12, 17, 22, 27, . . .
4. 4, 15, 26, 37, 48, . . .
5. 5, 1, -3, -7, -11, . . .
6. 8, 5, 2, -1, -4, . . .
7. -2, -9, -16, -23, . . .
8. -5, -7, -9, -11, -13, . . .
9. -3, -2.5, -2, -1.5, -1, . . .
10. -2, -1.25, $-.5$, $.25$, . . .
11. -6, -6.5, -7, -7.5, -8, . . .
12. 1, $.5$, 0, $-.5$, -1, . . .

Write the first five terms of the arithmetic sequence whose nth term is given. See Example 3.

13. $a_n = 9 + (n - 1)4$
14. $a_n = 13 + (n - 1)6$
15. $a_n = 7 + (n - 1)(-2)$
16. $a_n = 6 + (n - 1)(-3)$
17. $a_n = -4 + (n - 1)3$
18. $a_n = -19 + (n - 1)12$
19. $a_n = -2 + (n - 1)(-3)$
20. $a_n = -1 + (n - 1)(-2)$
21. $a_n = -4n - 3$
22. $a_n = -3n + 1$
23. $a_n = .5n + 4$
24. $a_n = .3n + 1$
25. $a_n = 20n + 1000$
26. $a_n = -600n + 4000$

Find the indicated part of each arithmetic sequence. See Example 4.

27. Find the eighth term of the sequence that has a first term of 9 and a common difference of 6.

28. Find the twelfth term of the sequence that has a first term of −2 and a common difference of −3.

29. Find the sixth term of the sequence that has a fifth term of 13 and a first term of −3.

30. Find the eighth term of the sequence that has a sixth term of −42 and a first term of 3.

31. Find the common difference if the first term is 6 and the twentieth term is 82.

32. Find the common difference if the first term is −8 and the ninth term is −64.

33. If the common difference is −2 and the seventh term is 14, then what is the first term?

34. If the common difference is five and the twelfth term is −7, then what is the first term?

Find the sum of each given series. See Examples 5 and 6.

35. $1 + 2 + 3 + \cdots + 48$

37. $8 + 10 + 12 + \cdots + 36$

39. $-1 + (-7) + (-13) + \cdots + (-73)$

41. $-6 + (-1) + 4 + 9 + \cdots + 64$

43. $20 + 12 + 4 + (-4) + \cdots + (-92)$

45. $\sum_{i=1}^{12} (3i - 7)$ **46.** $\sum_{i=1}^{7} (-4i + 6)$

36. $1 + 2 + 3 + \cdots + 12$

38. $9 + 12 + 15 + \cdots + 72$

40. $-7 + (-12) + (-17) + \cdots + (-72)$

42. $-9 + (-1) + 7 + \cdots + 103$

44. $19 + 1 + (-17) + \cdots + (-125)$

47. $\sum_{i=1}^{11} (-5i + 2)$ **48.** $\sum_{i=1}^{19} (3i - 5)$

Solve each problem using the ideas of arithmetic sequences and series.

49. If a teacher has a salary of $22,000 her first year and is due to get a $500 raise each year, then what will her salary be in her seventh year?

50. What is the total salary for seven years of teaching for the teacher of Exercise 49?

51. On the first day of October an English teacher suggests to her students that they read five pages of a novel, and every day thereafter increase their daily reading by two pages. If her students follow her suggestion, then how many pages will they read during October?

52. If an air-conditioning system is not completed by the agreed-upon date, the contractor pays a penalty of $500 for the first day that he is overdue, $600 for the second day, $700 for the third day, and so on. If the system is completed 10 days late, then what is the total amount of the penalties that he must pay?

12.4 Geometric Sequences and Series

IN THIS SECTION:

- **Geometric Sequences**
- **Geometric Series**
- **Infinite Geometric Series**

In Section 12.3 we studied the arithmetic sequences and series. In this section we will study sequences in which each term is a multiple of the term preceding it. We will also learn how to find the sum of the corresponding series.

Geometric Sequences

Consider the sequence

$$3, 6, 12, 24, 48, \ldots$$

Unlike an arithmetic sequence, these terms do not have a common difference, but there is a simple pattern to the terms. Each term after the first is twice the term preceding it. Such a sequence is called a geometric sequence.

DEFINITION
Geometric Sequence

> A sequence in which each term after the first is obtained by multiplying the preceding term by a constant is called a **geometric sequence.**

The constant is denoted by the letter r and is called the **common ratio.** If a_1 is the first term, then the second term is $a_1 r$. The third term is $a_1 r^2$, the fourth term is $a_1 r^3$, and so on. We can write a formula for the nth term of a geometric sequence by following this pattern.

Formula for the nth Term of a Geometric Sequence

> The nth term, a_n, of a geometric sequence with first term a_1 and common ratio r, is
>
> $$a_n = a_1 r^{n-1}.$$

The first term and the common ratio determine all of the terms of a geometric sequence.

EXAMPLE 1 Write a formula for the nth term of the geometric sequence

$$6, 2, \frac{2}{3}, \frac{2}{9}, \ldots$$

Solution We can obtain the common ratio by dividing any term after the first by the term preceding it. Thus,

$$r = 2 \div 6 = \frac{1}{3}.$$

Because each term after the first is $1/3$ of the term preceding it, the nth term is given by

$$a_n = 6\left(\frac{1}{3}\right)^{n-1}. \qquad \blacktriangleleft$$

EXAMPLE 2 Find a formula for the nth term of the geometric sequence

$$2, -1, \frac{1}{2}, -\frac{1}{4}, \ldots$$

Solution   We obtain the ratio by dividing a term by the term preceding it.

$$r = -1 \div 2 = -\frac{1}{2}$$

Each term after the first is obtained by multiplying the preceding term by $-1/2$. The formula for the nth term is

$$a_n = 2\left(-\frac{1}{2}\right)^{n-1}.$$ ◀

In the next example we use the formula for the nth term to write some terms of a geometric sequence.

EXAMPLE 3   Write the first five terms of the geometric sequence whose nth term is

$$a_n = 3(-2)^{n-1}.$$

Solution   Let n take the values 1 through 5 in the formula for the nth term.

$$a_1 = 3(-2)^{1-1} = 3$$
$$a_2 = 3(-2)^{2-1} = -6$$
$$a_3 = 3(-2)^{3-1} = 12$$
$$a_4 = 3(-2)^{4-1} = -24$$
$$a_5 = 3(-2)^{5-1} = 48$$

Thus, the first five terms are 3, −6, 12, −24, 48. ◀

The formula for the nth term involves four variables: a_n, a_1, r, and n. If we know the value of any three of them, we can find the value of the fourth.

EXAMPLE 4   Find the first term of a geometric sequence whose fourth term is 8 and whose common ratio is $1/2$.

Solution   If we let $a_4 = 8$, $r = 1/2$, and $n = 4$ in the formula for the nth term, we get

$$8 = a_1\left(\frac{1}{2}\right)^{4-1}$$

$$8 = a_1\left(\frac{1}{2}\right)^{3}$$

$$8 = a_1 \cdot \frac{1}{8}$$

$$64 = a_1.$$ ◀

Geometric Series

Consider the series

$$1 + 2 + 4 + 8 + 16 + \cdots + 512.$$

The terms of this series are the terms of a geometric sequence. The indicated sum of a geometric sequence is called a **geometric series.**

If we want the actual sum of the geometric series above, we can get it with a trick similar to the one used for the sum of an arithmetic series. Let S represent the sum of the above series with common ratio 2.

$$S = 1 + 2 + 4 + 8 + \cdots + 256 + 512$$

Math at Work

It is hard to make ends meet when you have a steady income but even harder after retiring, since your income stops but your expenses do not. Financial planners help clients plan for retirement by setting up an annuity where the client makes monthly payments while working in order to receive a monthly payment after retiring.

For example, suppose a client decides to pay $200 per month for 360 months (30 years) into an account paying 9% interest compounded monthly. The formula for calculating the total amount saved comes from the sum of a geometric series of 360 terms:

$$S_{360} = \$200\left(\frac{(1 + i)^{360} - 1}{i}\right),$$

where i is the monthly interest rate (annual rate/12). For the 9% annual interest rate, $i = .0075$. Thus after 30 years, the account has a balance of $366,148.70.

To find out how much the client can withdraw from this account after retiring, the financial planner again uses a formula that comes from the sum of a geometric series. For example, if the client decides to withdraw monthly payments for 20 years (240 months), and the account continues to earn 9% annual interest, the formula for finding the monthly payment is

$$R = \$366,148.70\left(\frac{i}{1 - (1 + i)^{-240}}\right),$$

where i is the monthly interest rate and R is the monthly payment the client receives. Thus the client withdraws 240 monthly payments of $3294.33 each.

If a client deposits $300 per month for 35 years into an account earning 8% interest compounded monthly, how much could be withdrawn monthly after retirement over 15 years if the account continued to earn 8% compounded monthly?

Now multiply each side by -2, the opposite of the common ratio.

$$-2S = -2 - 4 - 8 - \cdots - 512 - 1024$$

If we add these two equations, all but two of the terms on the right are eliminated.

$$
\begin{aligned}
S &= 1 + 2 + 4 + 8 + \cdots + 256 + 512 \\
-2S &= \quad\; -2 - 4 - 8 - \cdots \qquad\qquad - 512 - 1024 \\
\hline
-S &= 1 \qquad\qquad\qquad\qquad\qquad\qquad\quad - 1024 \\
-S &= -1023 \\
S &= 1023
\end{aligned}
$$

Add.

We can apply the same idea to get a general formula for the sum of the first n terms of any geometric series. We denote the sum of the first n terms by S_n:

$$S_n = a_1 + a_1 r + a_1 r^2 + \cdots + a_1 r^{n-1}.$$

Multiply each side of this equation by $-r$.

$$-rS_n = -a_1 r - a_1 r^2 - a_1 r^3 - \cdots - a_1 r^n$$

If we add these two equations, all but two of the terms on the right are eliminated.

$$
\begin{aligned}
S_n &= a_1 + a_1 r + a_1 r^2 + \cdots + a_1 r^{n-1} \\
-rS_n &= \quad\; -a_1 r - a_1 r^2 - a_1 r^3 - \cdots - a_1 r^n \\
\hline
S_n - rS_n &= a_1 \qquad\qquad\qquad\qquad\qquad\qquad - a_1 r^n \\
(1 - r)S_n &= a_1(1 - r^n) \\
S_n &= \frac{a_1(1 - r^n)}{1 - r}
\end{aligned}
$$

Add.

Sum of n Terms of a Geometric Series

> If S_n represents the sum of the first n terms of a geometric series with first term a_1 and common ratio r, then
>
> $$S_n = \frac{a_1(1 - r^n)}{1 - r}.$$

EXAMPLE 5 Find the sum of the series

$$\frac{1}{3} + \frac{1}{9} + \frac{1}{27} + \cdots + \frac{1}{729}.$$

Solution The first term is $1/3$, and the common ratio is $1/3$. The nth term can be written as

$$a_n = \left(\frac{1}{3}\right)\left(\frac{1}{3}\right)^{n-1}.$$

We can find n from the formula for the nth term.

$$\frac{1}{729} = \left(\frac{1}{3}\right)\left(\frac{1}{3}\right)^{n-1}$$

$$\frac{1}{729} = \left(\frac{1}{3}\right)^{n}$$

$$n = 6 \qquad \text{Because } 3^6 = 729$$

We now use the formula for the sum of six terms of this geometric series.

$$S_6 = \frac{\left(\frac{1}{3}\right)\left[1 - \left(\frac{1}{3}\right)^6\right]}{1 - \left(\frac{1}{3}\right)} = \frac{\left(\frac{1}{3}\right)\left(1 - \frac{1}{729}\right)}{\frac{2}{3}}$$

$$= \frac{1}{3} \cdot \frac{728}{729} \cdot \frac{3}{2} = \frac{364}{729} \qquad \blacktriangleleft$$

EXAMPLE 6   Find the sum of the series

$$\sum_{i=1}^{12} 3(-2)^{i-1}.$$

Solution   This is a geometric series with first term 3, ratio -2, and $n = 12$. We use the formula for the sum of the first 12 terms of a geometric series.

$$S_{12} = \frac{3[1 - (-2)^{12}]}{1 - (-2)}$$

$$= \frac{3[1 - 4096]}{3}$$

$$= -4095 \qquad \blacktriangleleft$$

Infinite Geometric Series

In the formula for the sum of a geometric series, n appears as an exponent in the term r^n. Suppose $r = 2/3$. Using a calculator, we can calculate some powers of $2/3$.

$$\left(\frac{2}{3}\right)^{10} = .0173415$$

$$\left(\frac{2}{3}\right)^{20} = .0003007$$

$$\left(\frac{2}{3}\right)^{100} = 2.459 \times 10^{-18}$$

Notice that the powers of 2/3 get very small as n gets large. This happens only when $|r| < 1$. Thus, for a very large number of terms of a geometric series, the term r^n is approximately 0, and the expression $1 - r^n$ is approximately equal to 1. In the expression for S_n we omit $1 - r^n$ and write

$$S_n \approx \frac{a_1}{1 - r} \quad \text{for large values of } n.$$

S_n gets closer and closer to this expression as n gets larger and larger.

In light of the previous discussion, it is now possible to give meaning to the idea of the sum of all of the terms of an infinite geometric series.

Sum of an Infinite Geometric Series

If $a_1 + a_1 r + a_1 r^2 + \cdots$ is an infinite geometric series, with $|r| < 1$, then the sum S of all of the terms of this series is given by

$$S = \frac{a_1}{1 - r}.$$

We can use the symbol ∞ to represent infinity, and write the sum of all of the terms of an infinite geometric series as

$$a_1 + a_1 r + a_1 r^2 + \cdots = \sum_{i=1}^{\infty} a_1 r^{i-1}.$$

EXAMPLE 7 Find the value of the sum

$$\sum_{i=1}^{\infty} 8\left(\frac{3}{4}\right)^{i-1}.$$

Solution This is an infinite geometric series with first term 8 and ratio 3/4. The sum S is therefore

$$S = \frac{8}{1 - \dfrac{3}{4}} = \frac{8}{\dfrac{1}{4}} = 32.$$

The reader should check this by using a calculator to find

$$\sum_{i=1}^{50} 8\left(\frac{3}{4}\right)^{i-1}.$$ ◀

EXAMPLE 8 Find the value of the infinite sum

$$\frac{1}{2} + \frac{1}{4} + \frac{1}{8} + \frac{1}{16} + \cdots$$

Solution This is an infinite geometric series with $a_1 = 1/2$ and $r = 1/2$. Thus,

$$S = \frac{\frac{1}{2}}{1 - \frac{1}{2}} = 1.$$ ◄

EXAMPLE 9 Suppose that a ball always rebounds 2/3 of the height from which it falls, and the ball is dropped from a height of 6 feet. Find the total distance that the ball travels.

Solution The ball falls 6 feet and rebounds 4 feet, then falls 4 feet and rebounds 4/3 feet. The sum of the distances that the ball falls is given by the series

$$6 + 4 + \frac{4}{3} + \frac{8}{9} + \cdots$$

The sum of the distances that the ball rebounds is given by the series

$$4 + \frac{4}{3} + \frac{8}{9} + \cdots$$

Each of these series is a geometric series with a ratio of 2/3. The sum for the first series is

$$\frac{6}{1 - \frac{2}{3}} = \frac{6}{\frac{1}{3}} = 18.$$

The sum for the second series is

$$\frac{4}{1 - \frac{2}{3}} = \frac{4}{\frac{1}{3}} = 12.$$

The total distance traveled by the ball is 30 feet. ◄

Warm-ups

True or false?

1. The sequence 2, 6, 24, 120, . . . is a geometric sequence.
2. In a geometric sequence there is a common difference between adjacent terms.
3. If the nth term of a geometric sequence is $a_n = 3(.5)^{n-1}$, then the common ratio is .5.
4. If the nth term of a geometric sequence is $a_n = 3(2)^{-n+3}$, then the first term is 12.

5. If the nth term of a geometric sequence is $a_n = 3(2)^{-n+3}$, then the value of r is $1/2$.

6. The terms of a geometric series are the terms of a geometric sequence.

7. There is no way to find the value of $\sum_{i=1}^{10}(2)^i$ without adding up all of the terms.

8. $\sum_{i=1}^{5}6(3/4)^{i-1} = \dfrac{6[1-(3/4)^5]}{1-3/4}$ **9.** $10+5+\dfrac{5}{2}+\cdots = \dfrac{10}{1-1/2}$

10. $2+4+8+16+\cdots = \dfrac{2}{1-2}$

12.4 EXERCISES

Write a formula for the nth term of each geometric sequence. See Examples 1 and 2.

1. $\dfrac{1}{3}$, 1, 3, 9, . . . **2.** $\dfrac{1}{4}$, 2, 16, . . . **3.** 64, 8, 1, . . .

4. 100, 10, 1, . . . **5.** 8, −4, 2, −1, . . . **6.** −9, 3, −1, . . .

7. 2, −4, 8, −16, . . . **8.** $-\dfrac{1}{2}$, 2, −8, 32, . . .

9. $-\dfrac{1}{3}$, $-\dfrac{1}{4}$, $-\dfrac{3}{16}$, . . . **10.** $-\dfrac{1}{4}$, $-\dfrac{1}{5}$, $-\dfrac{4}{25}$, . . .

Write the first five terms of the geometric sequence with the given nth term. See Example 3.

11. $a_n = 2\left(\dfrac{1}{3}\right)^{n-1}$ **12.** $a_n = -5\left(\dfrac{1}{2}\right)^{n-1}$ **13.** $a_n = (-2)^{n-1}$

14. $a_n = \left(-\dfrac{1}{3}\right)^{n-1}$ **15.** $a_n = 2^{-n}$ **16.** $a_n = 3^{-n}$

17. $a_n = (.78)^n$ **18.** $a_n = (-.23)^n$

Find the required part of each geometric sequence. See Example 4.

19. Find the first term of the geometric sequence that has a fourth term of 40 and a common ratio of 2.

20. Find the first term of a geometric sequence that has a fifth term of 4 and a common ratio of 1/2.

21. Find the common ratio for a geometric sequence that has a first term of 6 and a fourth term of 2/9.

22. Find the common ratio for a geometric sequence that has a first term of 1 and a fourth term of −27.

23. Find the fourth term of the geometric sequence that has a first term of −3 and common ratio of 1/3.

24. Find the fifth term of the geometric sequence that has a first term of −2/3 and a common ratio of −2/3.

Find the sum of each series. See Examples 5 and 6.

25. $\dfrac{1}{2} + \dfrac{1}{4} + \dfrac{1}{8} + \cdots + \dfrac{1}{512}$

26. $1 + \dfrac{1}{3} + \dfrac{1}{9} + \cdots + \dfrac{1}{81}$

27. $\dfrac{1}{2} - \dfrac{1}{4} + \dfrac{1}{8} - \dfrac{1}{16} + \dfrac{1}{32}$

28. $3 - 1 + \dfrac{1}{3} - \dfrac{1}{9} + \dfrac{1}{27} - \dfrac{1}{81}$

29. $\displaystyle\sum_{i=1}^{10} 5(2)^{i-1}$

30. $\displaystyle\sum_{i=1}^{7} (10000)(.1)^{i-1}$

31. $\displaystyle\sum_{i=1}^{6} (.1)^{i}$

32. $\displaystyle\sum_{i=1}^{5} (.2)^{i}$

Find the sum of each infinite geometric series. See Examples 7 and 8.

33. $\dfrac{1}{8} + \dfrac{1}{16} + \dfrac{1}{32} + \cdots$

34. $\dfrac{1}{9} + \dfrac{1}{27} + \dfrac{1}{81} + \cdots$

35. $3 + 2 + \dfrac{4}{3} + \cdots$

36. $2 + 1 + \dfrac{1}{2} + \cdots$

37. $4 - 2 + 1 - \dfrac{1}{2} + \cdots$

38. $16 - 12 + 9 - \dfrac{27}{4} + \cdots$

39. $\displaystyle\sum_{i=1}^{\infty} (.3)^{i}$

40. $\displaystyle\sum_{i=1}^{\infty} (.2)^{i}$

41. $\displaystyle\sum_{i=1}^{\infty} 3(.5)^{i-1}$

42. $\displaystyle\sum_{i=1}^{\infty} 7(.4)^{i-1}$

43. $\displaystyle\sum_{i=1}^{\infty} 3(.1)^{i}$

44. $\displaystyle\sum_{i=1}^{\infty} 6(.1)^{i}$

45. $\displaystyle\sum_{i=1}^{\infty} 12(.01)^{i}$

46. $\displaystyle\sum_{i=1}^{\infty} 72(.01)^{i}$

Use the ideas of geometric series to solve each problem. See Example 9.

47. The repeating decimal number .44444. . . can be written as the infinite geometric series

$$\frac{4}{10} + \frac{4}{100} + \frac{4}{1000} + \cdots$$

Find the sum of this geometric series.

48. The repeating decimal number .24242424. . . can be written as the infinite geometric series

$$\frac{24}{100} + \frac{24}{10,000} + \frac{24}{1,000,000} + \cdots$$

Find the sum of this geometric series.

49. Suppose you deposit one cent into your piggy bank on the first day of December, and on each day of December after that, deposit twice as much as on the previous day. How much will you have in the bank after the last deposit?

50. Consider your relatives, including yourself, your parents, your grandparents, your great-grandparents, your great-great-grandparents, and so on, down to your grandparents with the word "great" used in front 40 times. What is the total number of people you are considering?

51. If you deposit $2000 in an account paying 9% compounded annually, then after n years, this $2000 amounts to $2000(1.09)^n$. Suppose that you deposit $2000 at the beginning of each year for 45 years in an account paying 9% compounded annually. How much will you have in the account at the end of the 45th year?

52. Suppose that you deposit $2000 at the beginning of each year for 45 years into an account paying 12% compounded annually. How much will you have in the account at the end of the 45th year?

12.5 Binomial Expansions

In Chapter 4 we learned how to square a binomial. In this section we will learn how to write the terms of higher powers of a binomial.

Some Examples

In Chapter 4 we learned that $(x + y)^2 = x^2 + 2xy + y^2$. To find $(x + y)^3$, we multiply $x^2 + 2xy + y^2$ by $x + y$:

$$\begin{array}{r} x^2 + 2xy + y^2 \\ x + y \\ \hline x^2y + 2xy^2 + y^3 \\ x^3 + 2x^2y + xy^2 \\ \hline x^3 + 3x^2y + 3xy^2 + y^3 \end{array}$$

Thus,

$$(x + y)^3 = x^3 + 3x^2y + 3xy^2 + y^3.$$

This equation is called the **binomial expansion** of $(x + y)^3$. If we again multiply by $x + y$, we will get the binomial expansion of $(x + y)^4$. This method is rather tedious. If we examine these expansions, we can find a pattern. From the pattern, we can find binomial expansions without multiplying.

Consider the binomial expansions shown below.

$$(x + y)^1 = x + y$$
$$(x + y)^2 = x^2 + 2xy + y^2$$
$$(x + y)^3 = x^3 + 3x^2y + 3xy^2 + y^3$$
$$(x + y)^4 = x^4 + 4x^3y + 6x^2y^2 + 4xy^3 + y^4$$
$$(x + y)^5 = x^5 + 5x^4y + 10x^3y^2 + 10x^2y^3 + 5xy^4 + y^5$$

Observe that the exponents on the variable x are decreasing while the exponents on the variable y are increasing as we read from left to right. Also notice that the sum of the exponents in each term is the same for that entire line. For instance, in the fourth expansion the terms x^4, x^3y, x^2y^2, xy^3, and y^4 all have exponents with a sum of four. If we continue the pattern, the expansion of $(x + y)^6$ will have seven terms whose variable parts are x^6, x^5y, x^4y^2, x^3y^3, x^2y^4, xy^5, and y^6. Now we must find the pattern for the coefficients of these terms.

Obtaining the Coefficients

If we write out only the coefficients of the expansions that we already have, we can easily see a pattern. This triangular array of coefficients for the binomial expansions is called Pascal's triangle.

$$
\begin{array}{ccccccccc}
& & & & 1 & \ \ 1 & & & \\
& & & 1 & 2 & 1 & & & \\
& & 1 & 3 & & 3 & 1 & & \\
& 1 & 4 & & 6 & & 4 & 1 & \\
1 & 5 & & 10 & & 10 & & 5 & 1
\end{array}
$$

Coefficients of $(x + y)^1$
Coefficients of $(x + y)^2$
Coefficients of $(x + y)^3$
Coefficients of $(x + y)^4$
Coefficients of $(x + y)^5$

Notice that each line starts and ends with a 1, and that each entry of a line is the sum of the two entries above it in the previous line. For instance, $4 = 3 + 1$, and $10 = 6 + 4$. Following this pattern, the sixth and seventh lines of coefficients are:

$$
\begin{array}{ccccccc}
1 & 6 & 15 & 20 & 15 & 6 & 1 \\
1 & 7 & 21 & 35 & 35 & 21 & 7 \quad 1
\end{array}
$$

Pascal's triangle gives us an easy way to get the coefficients for the binomial expansion with small powers, but it is impractical for larger powers. For larger powers of $(x + y)$, the coefficients can be obtained with a formula involving a new notation called **factorial notation.**

DEFINITION
n! (*n* factorial)

> If n is a positive integer, $n!$ (read "*n factorial*") is defined to be the product of all of the positive integers from 1 through n.

For example, $3! = 3 \cdot 2 \cdot 1 = 6$, and $5! = 5 \cdot 4 \cdot 3 \cdot 2 \cdot 1 = 120$. We also define 0! to be 1. The pattern for the coefficients in Pascal's triangle can be described using factorials. Consider the following expressions.

$$
\frac{4!}{4!0!} = \frac{4 \cdot 3 \cdot 2 \cdot 1}{4 \cdot 3 \cdot 2 \cdot 1 \cdot 1} = 1
$$

$$
\frac{4!}{3!1!} = \frac{4 \cdot 3 \cdot 2 \cdot 1}{3 \cdot 2 \cdot 1 \cdot 1} = 4
$$

$$
\frac{4!}{2!2!} = \frac{4 \cdot 3 \cdot 2 \cdot 1}{2 \cdot 1 \cdot 2 \cdot 1} = 6
$$

$$
\frac{4!}{1!3!} = \frac{4 \cdot 3 \cdot 2 \cdot 1}{1 \cdot 3 \cdot 2 \cdot 1} = 4
$$

$$
\frac{4!}{0!4!} = \frac{4 \cdot 3 \cdot 2 \cdot 1}{1 \cdot 4 \cdot 3 \cdot 2 \cdot 1} = 1
$$

Observe that these numbers are the coefficients appearing in the fourth line of Pascal's triangle. For example, the coefficient of x^3y is $\dfrac{4!}{3!1!}$. Note that each expression involving factorials has 4! in the numerator, and the factorials in the denominator match the exponents on x and y. If we think of the term x^4 as x^4y^0, then this pattern holds for that term as well. The coefficient of x^4y^0 is $\dfrac{4!}{4!0!}$.

The Binomial Theorem

We now summarize these ideas in the binomial theorem.

The Binomial Theorem

In the expansion of $(x + y)^n$, for a positive integer n, there are $n + 1$ terms, given by the following formula:

$$(x + y)^n = \frac{n!}{n!0!}x^n + \frac{n!}{(n-1)!1!}x^{n-1}y + \frac{n!}{(n-2)!2!}x^{n-2}y^2 + \cdots + y^n.$$

We use the binomial theorem to write the terms in a binomial expansion.

EXAMPLE 1 Write out the first three terms of $(x + y)^9$.

Solution

$$(x + y)^9 = \frac{9!}{9!0!}x^9 + \frac{9!}{8!1!}x^8y + \frac{9!}{7!2!}x^7y^2 + \cdots$$
$$= x^9 + 9x^8y + 36x^7y^2 + \cdots \qquad \blacktriangleleft$$

EXAMPLE 2 Write the binomial expansion for $(x^2 - 2a)^5$.

Solution We expand a difference by writing it as a sum and using the binomial theorem.

$$(x^2 - 2a)^5 = (x^2 + (-2a))^5$$
$$= \frac{5!}{5!0!}(x^2)^5 + \frac{5!}{4!1!}(x^2)^4(-2a)^1 + \frac{5!}{3!2!}(x^2)^3(-2a)^2$$
$$+ \frac{5!}{2!3!}(x^2)^2(-2a)^3 + \frac{5!}{1!4!}(x^2)(-2a)^4 + \frac{5!}{0!5!}(-2a)^5$$
$$= x^{10} - 10x^8a + 40x^6a^2 - 80x^4a^3 + 80x^2a^4 - 32a^5 \qquad \blacktriangleleft$$

EXAMPLE 3 Find the fourth term of the expansion of $(a + b)^{12}$.

Solution Notice that the power of b in the first term is 0, and that b in the second term has a power of 1. Thus, b in the fourth term has a power of 3 and a has a power of 9. The fourth term is

$$\frac{12!}{9!3!}a^9b^3 = 220a^9b^3.$$ ◄

Using the ideas of Example 3, we can write a formula for any term of a binomial expansion.

Formula for the *k*th Term of $(x + y)^n$

For k ranging from 1 to $n + 1$, the kth term of the expansion of $(a + b)^n$ is given by the formula

$$\frac{n!}{(n - k + 1)!(k - 1)!}x^{n-k+1}y^{k-1}.$$

EXAMPLE 4 Find the sixth term of the expansion of $(a^2 - 2b)^7$.

Solution For the sixth term, use the general formula with $k = 6$ and $n = 7$.

$$\frac{7!}{(7 - 6 + 1)!(6 - 1)!}(a^2)^2(-2b)^5 = 21a^4(-32b^5) = -672a^4b^5$$ ◄

We can think of the binomial expansion as a finite series. Using summation notation, we can write the binomial theorem as follows.

The Binomial Theorem (using summation notation)

For any positive integer n,

$$(x + y)^n = \sum_{i=0}^{n}\frac{n!}{(n - i)!i!}x^{n-i}y^i.$$

EXAMPLE 5 Write $(a + b)^5$ using summation notation.

Solution Use $n = 5$ in the binomial theorem.

$$(a + b)^5 = \sum_{i=0}^{5}\frac{5!}{(5 - i)!i!}a^{5-i}b^i$$ ◄

Warm-ups

True or false?
1. There are 12 terms in the expansion of $(a + b)^{12}$.
2. The seventh term of $(a + b)^{12}$ has variable part a^5b^7.
3. For all values of x, $(x + 2)^5 = x^5 + 32$.

4. In the expansion of $(x - 5)^8$, the signs of the terms alternate.

5. The eighth line of Pascal's triangle is

$$1 \quad 8 \quad 28 \quad 56 \quad 70 \quad 56 \quad 28 \quad 8 \quad 1.$$

6. The binomial coefficients in the fourth row of Pascal's triangle have a sum of 2^4.

7. $(a + b)^3 = \displaystyle\sum_{i=0}^{3} \frac{3!}{(3 - i)!i!} a^{3-i} b^i$

8. The binomial coefficients in the nth row of Pascal's triangle have a sum of 2^n.

9. $0! = 1!$

10. $\dfrac{7!}{5!2!} = 21$

12.5 EXERCISES

Use the binomial theorem to expand each of the following. See Examples 1 and 2.

1. $(r + t)^5$

2. $(r + t)^6$

3. $(m - n)^3$

4. $(m - n)^4$

5. $(x + 2a)^3$

6. $(a + 3b)^4$

7. $(x^2 - 2)^4$

8. $(x^2 - a^2)^5$

9. $(x - 1)^7$

10. $(x + 1)^6$

For each of the following expressions, write out the first four terms. See Examples 1 and 2.

11. $(a - 3b)^{12}$

12. $(x - 2y)^{10}$

13. $(x^2 + 5)^9$

14. $(x^2 + 1)^{20}$

15. $(x - 1)^{22}$

16. $(2x - 1)^{15}$

17. $\left(\dfrac{x}{2} + \dfrac{y}{3}\right)^{10}$

18. $\left(\dfrac{a}{2} + \dfrac{b}{5}\right)^{8}$

Find the indicated term of the binomial expansion. See Examples 3 and 4.

19. $(a + w)^{13}$,   6th term

20. $(m + n)^{12}$,   7th term

21. $(m - n)^{16}$,   8th term

22. $(a - b)^{14}$,   6th term

23. $(2a^2 - b)^{20}$,   7th term

24. $(a^2 - w^2)^{12}$,   5th term

Write the following using summation notation. See Example 5.

25. $(a + m)^8$

26. $(z + w)^{13}$

27. $(a - 2x)^5$

28. $(w - 3m)^7$

Wrap-up

CHAPTER 12

SUMMARY

	Concepts	Examples
Sequences	Finite—A function whose domain is the set of positive integers less than or equal to a fixed positive integer	3, 5, 7, 9, 11 $a(n) = 2n + 1$ $1 \leq n \leq 5$
	Infinite—A function whose domain is the set of positive integers	2, 4, 6, 8, . . . $a(n) = 2n$
Series	The indicated sum of a sequence	$2 + 4 + 6 + \ldots + 50$
Summation notation	$\displaystyle\sum_{i=1}^{n} f(i)$	$\displaystyle\sum_{i=1}^{25} 2i$
Arithmetic sequence	A sequence where each term after the first is obtained by adding a fixed amount to the previous term	6, 11, 16, 21, . . .
General term	The nth term of an arithmetic sequence is $a_n = a_1 + (n - 1)d.$	$a_1 = 6, \quad d = 5$ $a_n = 6 + (n - 1)5$
Arithmetic series	The sum of an arithmetic sequence	$6 + 11 + 16 + 21 + \cdots$
Sum of first n terms	$S_n = \dfrac{n}{2}(a_1 + a_n)$ We do not consider the sum of an infinite arithmetic series, because it would always be infinite.	$S_4 = \dfrac{4}{2}(6 + 21) = 54$
Geometric sequence	Each term after the first is obtained by multiplying the preceding term by a constant.	2, 6, 18, 54, . . .
nth Term	The nth term of a geometric sequence is $a_n = a_1 r^{n-1}.$	$a_1 = 2, \quad r = 3$ $a_n = 2 \cdot 3^{n-1}$
Geometric series	The indicated sum of a geometric sequence	$2 + 6 + 18 + 54 + 162$
Sum of first n terms	$S_n = \dfrac{a_1(1 - r^n)}{1 - r}$	$a_1 = 2, \quad r = 3, \quad n = 5$ $S_5 = \dfrac{2(1 - 3^5)}{1 - 3} = 242$

Infinite geometric series	$a_1 + a_1 r + a_1 r^2 + a_1 r^3 + \cdots$	$8 + 4 + 2 + 1 + \dfrac{1}{2} + \cdots$
Sum of all terms	$S = \dfrac{a_1}{1 - r}$, provided $|r| < 1$	$a_1 = 8, \quad r = 1/2$ $S = \dfrac{8}{1 - 1/2} = 16$
Definition of $n!$	Product of positive integers from 1 through n	$5! = 5 \cdot 4 \cdot 3 \cdot 2 \cdot 1 = 120$

Binomial theorem

$$(x + y)^n = \frac{n!}{n!0!}x^n + \frac{n!}{(n - 1)!1!}x^{n-1}y$$

$$+ \frac{n!}{(n - 2)!2!}x^{n-2}y^2 + \cdots + y^n$$

Using summation notation:

$$(x + y)^n = \sum_{i=0}^{n} \frac{n!}{(n - i)!i!}x^{n-i}y^i$$

$$(x + y)^3 = x^3 + 3x^2 y + 3xy^2 + y^3$$

kth term of $(x + y)^n$

$$\frac{n!}{(n - k + 1)!(k - 1)!}x^{n-k+1}y^{k-1} \qquad 1 \le k \le n + 1$$

3rd term of $(a + b)^{10}$

$$\frac{10!}{8!2!}a^8 b^2 = 45a^8 b^2$$

REVIEW EXERCISES

12.1 *List all terms of each finite sequence.*

1. $a_n = n^3 \qquad 1 \le n \le 5$

2. $b_n = (n - 1)^4 \qquad 1 \le n \le 4$

3. $c_n = (-1)^n (2n - 3) \qquad 1 \le n \le 6$

4. $d_n = (-1)^{n-1}(3 - n) \qquad 1 \le n \le 7$

List the first three terms of the infinite sequence whose nth term is given.

5. $a_n = -\dfrac{1}{n}$

6. $b_n = \dfrac{(-1)^n}{n^2}$

7. $b_n = \dfrac{(-1)^{2n}}{2n + 1}$

8. $a_n = \dfrac{-1}{(2n - 3)}$

9. $c_n = \log_2(2^{n+3})$

10. $c_n = \ln(e^{2n})$

12.2 *Find the sum of each series.*

11. $\displaystyle\sum_{i=1}^{3} i^3$

12. $\displaystyle\sum_{i=0}^{4} 6$

13. $\displaystyle\sum_{i=1}^{5} i(i - 1)$

14. $\displaystyle\sum_{j=0}^{3} (-2)^j$

Write each series in summation notation. Use the index i, and let i begin at 1.

15. $\dfrac{1}{4} + \dfrac{1}{6} + \dfrac{1}{8} + \cdots$

16. $\dfrac{1}{3} + \dfrac{1}{4} + \dfrac{1}{5} + \cdots$

17. $0 + 1 + 4 + 9 + 16 + \cdots$

18. $-1 + 2 - 3 + 4 - 5 + 6 - \cdots$ **19.** $x_1 - x_2 + x_3 - x_4 + \cdots$ **20.** $-x^2 + x^3 - x^4 + x^5 - \cdots$

12.3 *Write a formula for the nth term of each arithmetic sequence.*

21. $\dfrac{1}{3}, \dfrac{2}{3}, 1, \dfrac{4}{3}, \ldots$ **22.** $10, 6, 2, -2, \ldots$

23. $2, 4, 6, 8, \ldots$ **24.** $20, 10, 0, -10, \ldots$

Find the sum of each arithmetic series.

25. $1 + 2 + 3 + \cdots + 24$ **26.** $-5 + (-2) + 1 + 4 + \cdots + 34$ **27.** $\dfrac{1}{6} + \dfrac{1}{2} + \dfrac{5}{6} + \dfrac{7}{6} + \cdots + \dfrac{11}{2}$

28. $-3 - 6 - 9 - 12 - \cdots - 36$ **29.** $\displaystyle\sum_{i=1}^{7} (2i - 3)$ **30.** $\displaystyle\sum_{i=1}^{6} 12 + (i - 1)5$

12.4 *Write the first four terms of the geometric sequence with the given nth term.*

31. $a_n = 3\left(\dfrac{1}{2}\right)^{n-1}$ **32.** $a_n = 6\left(-\dfrac{1}{3}\right)^{n}$ **33.** $a_n = 2^{1-n}$

34. $a_n = 5(10)^{n-1}$ **35.** $a_n = 23(10)^{-2n}$ **36.** $a_n = 4(10)^{-n}$

Write a formula for the nth term of each geometric sequence.

37. $\dfrac{1}{2}, 3, 18, \ldots$ **38.** $-6, 2, -\dfrac{2}{3}, \dfrac{2}{9}, \ldots$

39. $\dfrac{7}{10}, \dfrac{7}{100}, \dfrac{7}{1000}, \ldots$ **40.** $2, 2x, 2x^2, 2x^3, \ldots$

Find the sum of each geometric series.

41. $\dfrac{1}{3} + \dfrac{1}{9} + \dfrac{1}{27} + \dfrac{1}{81}$ **42.** $2 + 4 + 8 + 16 + \ldots + 512$ **43.** $\displaystyle\sum_{i=1}^{10} 3(10)^{-i}$

44. $\displaystyle\sum_{i=1}^{5} (.1)^i$ **45.** $\dfrac{1}{4} + \dfrac{1}{12} + \dfrac{1}{36} + \dfrac{1}{108} + \ldots$ **46.** $12 + (-6) + 3 + \left(-\dfrac{3}{2}\right) + \cdots$

47. $\displaystyle\sum_{i=1}^{\infty} 18\left(\dfrac{2}{3}\right)^{i-1}$ **48.** $\displaystyle\sum_{i=1}^{\infty} 9(.1)^i$

12.5 *Use the binomial theorem to expand the following.*

49. $(m + n)^5$ **50.** $(2m - y)^4$

51. $(a^2 - 3b)^3$ **52.** $\left(\dfrac{x}{2} + 2a\right)^5$

Find the indicated term of the binomial expansion.

53. $(x + y)^{12}$,   5th term   **54.** $(x - 2y)^9$,   5th term

Write the following in summation notation.

55. $(a + w)^7$ **56.** $(m - 3y)^9$

Miscellaneous

Identify each sequence as an arithmetic sequence, a geometric sequence, or neither.

57. 1, 3, 6, 10, 15, . . .

58. 9, 12, 16, $\dfrac{64}{3}$, . . .

59. 9, 12, 15, 18, . . .

60. 2, 4, 8, 16, . . .

61. 0, 2, 4, 6, 8, . . .

62. 0, 3, 9, 27, 81, . . .

Solve each problem.

63. Find the common ratio for the geometric sequence that has a first term of 6 and a fourth term of $\dfrac{1}{30}$.

64. Find the common difference for an arithmetic sequence that has a first term of 6 and a fourth term of 36.

65. Write out all of the terms of the series:

$$\sum_{i=1}^{5} \frac{(-1)^i}{i!}$$

66. Write out the first eight rows of Pascal's triangle.

67. Write out all of the terms of the series

$$\sum_{i=0}^{5} \frac{5!}{(5-i)!i!} a^{5-i} b^i.$$

68. Write out all of the terms of the series

$$\sum_{i=0}^{8} \frac{8!}{(8-i)!i!} x^{8-i} y^i.$$

69. How many terms are there in the expansion of $(a+b)^{25}$?

70. Calculate

$$\frac{12!}{8!4!}.$$

71. If \$3000 is deposited in an account paying 10% compounded annually, then how much will be in the account at the end of 15 years?

72. If \$3000 is deposited in an account paying 10% compounded annually at the beginning of each year for 15 years, then how much will be in the account at the end of the 15th year?

CHAPTER 12 TEST

List the first four terms of the sequence whose nth term is given.

1. $a_n = -10 + (n-1)6$

2. $a_n = 5(.1)^{n-1}$

3. $a_n = \dfrac{(-1)^n}{n!}$

4. $a_n = \dfrac{2n-1}{n^2}$

Write a formula for the general term of each sequence.

5. 7, 4, 1, -2, . . .

6. -25, 5, -1, $\dfrac{1}{5}$, . . .

7. 2, -4, 6, -8, 10, -12, . . .

8. 1, 4, 9, 16, 25, . . .

Write out all of the terms of each series.

9. $\displaystyle\sum_{i=1}^{5} 2i + 3$

10. $\displaystyle\sum_{i=1}^{6} 5(2)^{i-1}$

11. $\displaystyle\sum_{i=0}^{4} \frac{4!}{(4-i)!i!} m^{4-i} q^i$

12. $\displaystyle\sum_{i=1}^{10} i^2$

Find the sum of each series.

13. $\displaystyle\sum_{i=1}^{20} 6 + 3i$

14. $\displaystyle\sum_{i=1}^{5} 10\left(\frac{1}{2}\right)^{i-1}$

15. $\displaystyle\sum_{i=1}^{200} 2i$

16. $\displaystyle\sum_{i=1}^{\infty} .35(.93)^{i-1}$

17. Find the common ratio for a geometric sequence that has a first term of 3 and a fifth term of 48.

18. Find the common difference for an arithmetic sequence that has a first term of 1 and a twelfth term of 122.

19. Find the fifth term in the expansion of $(r - t)^{15}$.

20. Find the fourth term in the expansion of $(a^2 - 2b)^8$.

Tying It All Together

CHAPTERS 1–12

Let $f(x) = x^2 - 3$ and $g(x) = 2x - 1$. Find the following.

1. $f(3)$

2. $f(n)$ $n^2 -$

3. $f(x + h)$

4. $f(x) - g(x)$

5. $g(f(3))$

6. $(f \circ g)(2)$

7. $(g \circ f)(x)$

8. $f(x) \cdot g(x)$

9. $g(3)$

10. $g^{-1}(5)$

11. $g^{-1}(x)$

12. $(g \circ g^{-1})(x)$

13. If y varies directly as x, and $y = -6$ when $x = 4$, find y when $x = 9$.

14. If a varies inversely as b, and $a = 2$ when $b = -4$, find a when $b = 3$.

15. If y varies directly as w and inversely as t, and $y = 16$ when $w = 3$ and $t = -4$, find y when $w = 2$ and $t = 3$.

16. If y varies jointly as h and the square of r, and $y = 12$ when $h = 2$ and $r = 3$, find y when $h = 6$ and $r = 2$.

Sketch the graph of each inequality.

17. $x > 3$ and $x + y < 0$

18. $|x - y| \geq 2$

19. $y < -2x + 3$ and $y > 2x - 6$

20. $|y + 2x| < 1$

21. $x^2 + y^2 < 4$

22. $x^2 - y^2 < 1$

23. $x + 2y < 4$

24. $x^2 + 2y < 4$

25. $\dfrac{x^2}{4} + \dfrac{y^2}{9} < 1$ and $y > x^2$

26. $|x^2 + y^2 - 10| < 6$

Perform the indicated operation and simplify. Write answers with positive exponents.

27. $\dfrac{a}{b} + \dfrac{b}{a}$

28. $1 - \dfrac{3}{y}$

29. $\dfrac{x - 2}{x^2 - 9} - \dfrac{x - 4}{x^2 - 2x - 3}$

30. $\dfrac{x^2 - 16}{2x + 8} \cdot \dfrac{4x^2 + 16x + 64}{x^3 - 16}$

31. $\dfrac{(a^2b)^3}{(ab^2)^4} \cdot \dfrac{ab^3}{a^{-4}b^2}$

32. $\dfrac{x^2 y}{(xy)^3} \div \dfrac{xy^2}{x^2 y^4}$

Simplify.

33. $8^{2/3}$

34. $16^{-5/4}$

35. $-4^{1/2}$

36. $27^{-2/3}$

37. -2^{-3}

38. $2^{-3/5} \cdot 2^{-7/5}$

39. $5^{-2/3} \div 5^{1/3}$

40. $(9^{1/2} + 4^{1/2})^2$

APPENDIX A

Table of Squares and Square Roots

n	n^2	$\sqrt{n}$	n	n^2	$\sqrt{n}$	n	n^2	$\sqrt{n}$
1	1	1.0000	41	1681	6.4031	81	6561	9.0000
2	4	1.4142	42	1764	6.4807	82	6724	9.0554
3	9	1.7321	43	1849	6.5574	83	6889	9.1104
4	16	2.0000	44	1936	6.6332	84	7056	9.1652
5	25	2.2361	45	2025	6.7082	85	7225	9.2195
6	36	2.4495	46	2116	6.7823	86	7396	9.2736
7	49	2.6458	47	2209	6.8557	87	7569	9.3274
8	64	2.8284	48	2304	6.9282	88	7744	9.3808
9	81	3.0000	49	2401	7.0000	89	7921	9.4340
10	100	3.1623	50	2500	7.0711	90	8100	9.4868
11	121	3.3166	51	2601	7.1414	91	8281	9.5394
12	144	3.4641	52	2704	7.2111	92	8464	9.5917
13	169	3.6056	53	2809	7.2801	93	8649	9.6437
14	196	3.7417	54	2916	7.3485	94	8836	9.6954
15	225	3.8730	55	3025	7.4162	95	9025	9.7468
16	256	4.0000	56	3136	7.4833	96	9216	9.7980
17	289	4.1231	57	3249	7.5498	97	9409	9.8489
18	324	4.2426	58	3364	7.6158	98	9604	9.8995
19	361	4.3589	59	3481	7.6811	99	9801	9.9499
20	400	4.4721	60	3600	7.7460	100	10000	10.0000
21	441	4.5826	61	3721	7.8102	101	10201	10.0499
22	484	4.6904	62	3844	7.8740	102	10404	10.0995
23	529	4.7958	63	3969	7.9373	103	10609	10.1489
24	576	4.8990	64	4096	8.0000	104	10816	10.1980
25	625	5.0000	65	4225	8.0623	105	11025	10.2470
26	676	5.0990	66	4356	8.1240	106	11236	10.2956
27	729	5.1962	67	4489	8.1854	107	11449	10.3441
28	784	5.2915	68	4624	8.2462	108	11664	10.3923
29	841	5.3852	69	4761	8.3066	109	11881	10.4403
30	900	5.4772	70	4900	8.3666	110	12100	10.4881
31	961	5.5678	71	5041	8.4261	111	12321	10.5357
32	1024	5.6569	72	5184	8.4853	112	12544	10.5830
33	1089	5.7446	73	5329	8.5440	113	12769	10.6301
34	1156	5.8310	74	5476	8.6023	114	12996	10.6771
35	1225	5.9161	75	5625	8.6603	115	13225	10.7238
36	1296	6.0000	76	5776	8.7178	116	13456	10.7703
37	1369	6.0828	77	5929	8.7750	117	13689	10.8167
38	1444	6.1644	78	6084	8.8318	118	13924	10.8628
39	1521	6.2450	79	6241	8.8882	119	14161	10.9087
40	1600	6.3246	80	6400	8.9443	120	14400	10.9545

Common Logarithms

This table gives the common logarithms for numbers between 1 and 10. The common logarithms for other numbers can be found by using scientific notation and the properties of logarithms. For example, to find log(1230) we write

$$\log(1230) = \log(1.23 \times 10^3) = \log(1.23) + \log(10^3)$$
$$= .0899 + 3 = 3.0899$$

n	0	1	2	3	4	5	6	7	8	9
1.0	.0000	.0043	.0086	.0128	.0170	.0212	.0253	.0294	.0334	.0374
1.1	.0414	.0453	.0492	.0531	.0569	.0607	.0645	.0682	.0719	.0755
1.2	.0792	.0828	.0864	.0899	.0934	.0969	.1004	.1038	.1072	.1106
1.3	.1139	.1173	.1206	.1239	.1271	.1303	.1335	.1367	.1399	.1430
1.4	.1461	.1492	.1523	.1553	.1584	.1614	.1644	.1673	.1703	.1732
1.5	.1761	.1790	.1818	.1847	.1875	.1903	.1931	.1959	.1987	.2014
1.6	.2041	.2068	.2095	.2122	.2148	.2175	.2201	.2227	.2253	.2279
1.7	.2304	.2330	.2355	.2380	.2405	.2430	.2455	.2480	.2504	.2529
1.8	.2553	.2577	.2601	.2625	.2648	.2672	.2695	.2718	.2742	.2765
1.9	.2788	.2810	.2833	.2856	.2878	.2900	.2923	.2945	.2967	.2989
2.0	.3010	.3032	.3054	.3075	.3096	.3118	.3139	.3160	.3181	.3201
2.1	.3222	.3243	.3263	.3284	.3304	.3324	.3345	.3365	.3385	.3404
2.2	.3424	.3444	.3464	.3483	.3502	.3522	.3541	.3560	.3579	.3598
2.3	.3617	.3636	.3655	.3674	.3692	.3711	.3729	.3747	.3766	.3784
2.4	.3802	.3820	.3838	.3856	.3874	.3892	.3909	.3927	.3945	.3962
2.5	.3979	.3997	.4014	.4031	.4048	.4065	.4082	.4099	.4116	.4133
2.6	.4150	.4166	.4183	.4200	.4216	.4232	.4249	.4265	.4281	.4298
2.7	.4314	.4330	.4346	.4362	.4378	.4393	.4409	.4425	.4440	.4456
2.8	.4472	.4487	.4502	.4518	.4533	.4548	.4564	.4579	.4594	.4609
2.9	.4624	.4639	.4654	.4669	.4683	.4698	.4713	.4728	.4742	.4757
3.0	.4771	.4786	.4800	.4814	.4829	.4843	.4857	.4871	.4886	.4900
3.1	.4914	.4928	.4942	.4955	.4969	.4983	.4997	.5011	.5024	.5038
3.2	.5051	.5065	.5079	.5092	.5105	.5119	.5132	.5145	.5159	.5172
3.3	.5185	.5198	.5211	.5224	.5237	.5250	.5263	.5276	.5289	.5302
3.4	.5315	.5328	.5340	.5353	.5366	.5378	.5391	.5403	.5416	.5428
3.5	.5441	.5453	.5465	.5478	.5490	.5502	.5514	.5527	.5539	.5551
3.6	.5563	.5575	.5587	.5599	.5611	.5623	.5635	.5647	.5658	.5670
3.7	.5682	.5694	.5705	.5717	.5729	.5740	.5752	.5763	.5775	.5786
3.8	.5798	.5809	.5821	.5832	.5843	.5855	.5866	.5877	.5888	.5899
3.9	.5911	.5922	.5933	.5944	.5955	.5966	.5977	.5988	.5999	.6010
4.0	.6021	.6031	.6042	.6053	.6064	.6075	.6085	.6096	.6107	.6117
4.1	.6128	.6138	.6149	.6160	.6170	.6180	.6191	.6201	.6212	.6222
4.2	.6232	.6243	.6253	.6263	.6274	.6284	.6294	.6304	.6314	.6325
4.3	.6335	.6345	.6355	.6365	.6375	.6385	.6395	.6405	.6415	.6425
4.4	.6435	.6444	.6454	.6464	.6474	.6484	.6493	.6503	.6513	.6522
n	0	1	2	3	4	5	6	7	8	9

n	0	1	2	3	4	5	6	7	8	9
4.5	.6532	.6542	.6551	.6561	.6571	.6580	.6590	.6599	.6609	.6618
4.6	.6628	.6637	.6646	.6656	.6665	.6675	.6684	.6693	.6702	.6712
4.7	.6721	.6730	.6739	.6749	.6758	.6767	.6776	.6785	.6794	.6803
4.8	.6812	.6821	.6830	.6839	.6848	.6857	.6866	.6875	.6884	.6893
4.9	.6902	.6911	.6920	.6928	.6937	.6946	.6955	.6964	.6972	.6981
5.0	.6990	.6998	.7007	.7016	.7024	.7033	.7042	.7050	.7059	.7067
5.1	.7076	.7084	.7093	.7101	.7110	.7118	.7126	.7135	.7143	.7152
5.2	.7160	.7168	.7177	.7185	.7193	.7202	.7210	.7218	.7226	.7235
5.3	.7243	.7251	.7259	.7267	.7275	.7284	.7292	.7300	.7308	.7316
5.4	.7324	.7332	.7340	.7348	.7356	.7364	.7372	.7380	.7388	.7396
5.5	.7404	.7412	.7419	.7427	.7435	.7443	.7451	.7459	.7466	.7474
5.6	.7482	.7490	.7497	.7505	.7513	.7520	.7528	.7536	.7543	.7551
5.7	.7559	.7566	.7574	.7582	.7589	.7597	.7604	.7612	.7619	.7627
5.8	.7634	.7642	.7649	.7657	.7664	.7672	.7679	.7686	.7694	.7701
5.9	.7709	.7716	.7723	.7731	.7738	.7745	.7752	.7760	.7767	.7774
6.0	.7782	.7789	.7796	.7803	.7810	.7818	.7825	.7832	.7839	.7846
6.1	.7853	.7860	.7868	.7875	.7882	.7889	.7896	.7903	.7910	.7917
6.2	.7924	.7931	.7938	.7945	.7952	.7959	.7966	.7973	.7980	.7987
6.3	.7993	.8000	.8007	.8014	.8021	.8028	.8035	.8041	.8048	.8055
6.4	.8062	.8069	.8075	.8082	.8089	.8096	.8102	.8109	.8116	.8122
6.5	.8129	.8136	.8142	.8149	.8156	.8162	.8169	.8176	.8182	.8189
6.6	.8195	.8202	.8209	.8215	.8222	.8228	.8235	.8241	.8248	.8254
6.7	.8261	.8267	.8274	.8280	.8287	.8293	.8299	.8306	.8312	.8319
6.8	.8325	.8331	.8338	.8344	.8351	.8357	.8363	.8370	.8376	.8382
6.9	.8388	.8395	.8401	.8407	.8414	.8420	.8426	.8432	.8439	.8445
7.0	.8451	.8457	.8463	.8470	.8476	.8482	.8488	.8494	.8500	.8506
7.1	.8513	.8519	.8525	.8531	.8537	.8543	.8549	.8555	.8561	.8567
7.2	.8573	.8579	.8585	.8591	.8597	.8603	.8609	.8615	.8621	.8627
7.3	.8633	.8639	.8645	.8651	.8657	.8663	.8669	.8675	.8681	.8686
7.4	.8692	.8698	.8704	.8710	.8716	.8722	.8727	.8733	.8739	.8745
7.5	.8751	.8756	.8762	.8768	.8774	.8779	.8785	.8791	.8797	.8802
7.6	.8808	.8814	.8820	.8825	.8831	.8837	.8842	.8848	.8854	.8859
7.7	.8865	.8871	.8876	.8882	.8887	.8893	.8899	.8904	.8910	.8915
7.8	.8921	.8927	.8932	.8938	.8943	.8949	.8954	.8960	.8965	.8971
7.9	.8976	.8982	.8987	.8993	.8998	.9004	.9009	.9015	.9020	.9025
8.0	.9031	.9036	.9042	.9047	.9053	.9058	.9063	.9069	.9074	.9079
8.1	.9085	.9090	.9096	.9101	.9106	.9112	.9117	.9122	.9128	.9133
8.2	.9138	.9143	.9149	.9154	.9159	.9165	.9170	.9175	.9180	.9186
8.3	.9191	.9196	.9201	.9206	.9212	.9217	.9222	.9227	.9232	.9238
8.4	.9243	.9248	.9253	.9258	.9263	.9269	.9274	.9279	.9284	.9289
8.5	.9294	.9299	.9304	.9309	.9315	.9320	.9325	.9330	.9335	.9340
8.6	.9345	.9350	.9355	.9360	.9365	.9370	.9375	.9380	.9385	.9390
8.7	.9395	.9400	.9405	.9410	.9415	.9420	.9425	.9430	.9435	.9440
8.8	.9445	.9450	.9455	.9460	.9465	.9469	.9474	.9479	.9484	.9489
8.9	.9494	.9499	.9504	.9509	.9513	.9518	.9523	.9528	.9533	.9538
9.0	.9542	.9547	.9552	.9557	.9562	.9566	.9571	.9576	.9581	.9586
9.1	.9590	.9595	.9600	.9605	.9609	.9614	.9619	.9624	.9628	.9633
9.2	.9638	.9643	.9647	.9652	.9657	.9661	.9666	.9671	.9675	.9680
9.3	.9685	.9689	.9694	.9699	.9703	.9708	.9713	.9717	.9722	.9727
9.4	.9731	.9736	.9741	.9745	.9750	.9754	.9759	.9763	.9768	.9773
9.5	.9777	.9782	.9786	.9791	.9795	.9800	.9805	.9809	.9814	.9818
9.6	.9823	.9827	.9832	.9836	.9841	.9845	.9850	.9854	.9859	.9863
9.7	.9868	.9872	.9877	.9881	.9886	.9890	.9894	.9899	.9903	.9908
9.8	.9912	.9917	.9921	.9926	.9930	.9934	.9939	.9943	.9948	.9952
9.9	.9956	.9961	.9965	.9969	.9974	.9978	.9983	.9987	.9991	.9996
n	0	1	2	3	4	5	6	7	8	9

Answers to Selected Exercises

CHAPTER 1

Section 1.1 Warm-ups FFFFTTTFTT
1. F **3.** F **5.** T **7.** T **9.** {1, 2, 3, 4, 5, 6, 7, 8, 9}
11. {1, 3, 5} **13.** {1, 2, 3, 4, 5, 6, 8}
15. {1, 3, 5, 7, 9} **17.** ∅ **19.** = **21.** ∪ **23.** ∩
25. ∉ **27.** ∈ **29.** T **31.** T **33.** T **35.** T
37. F **39.** T **41.** {2, 3, 4, 5, 6, 7, 8} **43.** {3, 5}
45. {1, 2, 3, 4, 5, 6, 8} **47.** {2, 3, 4, 5}
49. {2, 3, 4, 5, 7} **51.** {2, 3, 4, 5} **53.** {2, 3, 4, 5, 7}
55. ⊆ **57.** ∈ **59.** ∩ **61.** ⊆ **63.** ∩ **65.** ∪

Section 1.2 Warm-ups FTFFTFTTFT
1. T **3.** F **5.** T **7.** T
9. {0, 1, 2, 3, 4, 5}
11. {−4, −3, −2, −1, 0, 1, ...}
13. {1, 2, 3, 4}
15. {−2, −1, 0, 1, 2, 3, 4}
17. All of them **19.** {0, 8/2}
21. {−3, −5/2, −.025, 0, 3½, 8/2} **23.** T **25.** F

27. T **29.** F **31.** F **33.** F **35.** T **37.** ⊆
39. ⊄ **41.** ⊆ **43.** ⊆ **45.** ⊆ **47.** ∈ **49.** ∈
51. ∈ **53.** ∉ **55.** ⊆ **57.** ⊆

Section 1.3 Warm-ups TTFTTFFFFT
1. 34 **3.** 34 **5.** 0 **7.** −9 **9.** 4 **11.** 4 **13.** −7
15. −2 **17.** −10 **19.** −26 **21.** −2 **23.** −5
25. 0 **27.** $\frac{1}{10}$ **29.** $-\frac{1}{6}$ **31.** −14.98 **33.** −2.71
35. 2.803 **37.** −.2649 **39.** −3 **41.** −11 **43.** 13
45. −6 **47.** −9 **49.** 23 **51.** 1 **53.** $-\frac{1}{2}$ **55.** 1.97
57. 7.3 **59.** −50.73 **61.** −9.05 **63.** −75 **65.** 24
67. −2 **69.** −6 **71.** −.09 **73.** .2 **75.** −8
77. .25 **79.** −18 **81.** $\frac{3}{5}$ **83.** −91.25 **85.** 17000
87. −4 **89.** −1 **91.** 5 **93.** −3 **95.** $\frac{3}{4}$ **97.** 3
99. .8 **101.** −.33 **103.** −4 **105.** 10 **107.** 0
109. −4 **111.** $-\frac{1}{6}$ **113.** 2 **115.** −1 **117.** $-\frac{3}{2}$
119. 1.99 **121.** −2.3 **123.** −.6 **125.** 2

Section 1.4 Warm-ups FTTFFFFTTFF
1. 32 **3.** 1 **5.** 16 **7.** $-\frac{27}{64}$ **9.** .001 **11.** −8
13. 17 **15.** 2 **17.** −40 **19.** −25 **21.** −200
23. 40 **25.** −25 **27.** −2 **29.** −1 **31.** −6 **33.** 2
35. −46 **37.** 3 **39.** −8 **41.** −7 **43.** $-\frac{4}{3}$
45. −8 **47.** 25 **49.** $-\frac{5}{2}$ **51.** 4 **53.** $\frac{5}{8}$ **55.** $\frac{3}{4}$

57. 0   **59.** -2.67 **61.** 16   **63.** 49   **65.** -44
67. 29   **69.** -8 **71.** 60   **73.** 14.65   **75.** 37.12
77. 41   **79.** 27   **81.** 11   **83.** 125   **85.** 0   **87.** 24
89. 33   **91.** 144   **93.** 1   **95.** 9   **97.** -1 **99.** 26
101. $-\frac{3}{2}$ **103.** -17 **105.** -2 **107.** 17

Section 1.5   Warm-ups TFFFFFTTFT

1. 1   **3.** -14 **5.** -24 **7.** -1.7 **9.** -19.8
11. $4x - 24$ **13.** $2(m + 5)$ **15.** $3a + at$
17. $-2w + 10$ **19.** $-6 + 2y$ **21.** $5(x - 1)$
23. $-x + 3$ **25.** $-w - 5$ **27.** $3(y - 5)$ **29.** $3(a + 3)$
31. $2x + 4$ **33.** $-x + 2$ **35.** 2   **37.** $-\frac{1}{3}$ **39.** $\frac{1}{6}$
41. 4   **43.** $\frac{10}{7}$ **45.** $\frac{2}{3}$ **47.** 1   **49.** Commutative
51. Distributive   **53.** Associative   **55.** Inverse
57. Commutative   **59.** Identity   **61.** Distributive
63. Identity   **65.** Multiplication property of zero
67. Distributive   **69.** $w + 5$ **71.** $(5x)y$ **73.** $\frac{1}{2}(x - 1)$
75. $3(2x + 3)$ **77.** 1   **79.** 0   **81.** 4

Section 1.6   Warm-ups TFTTFTFFFT

1. 9000   **3.** 1   **5.** 527   **7.** 470   **9.** 38   **11.** 48,000
13. 0   **15.** 398   **17.** 374   **19.** 1   **21.** 55   **23.** 68
25. 30   **27.** 0   **29.** $2n$ **31.** $7w$ **33.** $-11mw^2$
35. $-3x$ **37.** $-4 - 7z$ **39.** $9t^2$ **41.** $-4a + 3a^2$
43. $8n$ **45.** $-2y$ **47.** $-2k$ **49.** $28t$ **51.** $10x^2$
53. $-21ab$ **55.** $-5a - 5ab$ **57.** h^2 **59.** $-28w$
61. $-x + x^2$ **63.** $25k^2$ **65.** y **67.** $2y$ **69.** $3x$
71. $x + 5$ **73.** $-x + 2$ **75.** $\frac{1}{2}x - 5$ **77.** $-3a + 1$
79. $10 - x$ **81.** $3m + 1$ **83.** $-12 + a$ **85.** $2t - 3$
87. $y + z$ **89.** $9x + 8$ **91.** $2x - 2$ **93.** $-2a + 2$
95. $7t - 9$ **97.** $m - 12$ **99.** $-7k - 17$
101. $.96x - 2$ **103.** $.06x - 1.5$ **105.** $-4k + 16$
107. $-4.5x + 17.83$ **109.** $-x + 8$ **111.** $15w^2$
113. $-2w$ **115.** $-2x + 9$ **117.** $25m^2$ **119.** $2x$
121. $4t - 1$

Chapter 1 Review Exercises

1. T   **3.** F   **5.** T   **7.** F   **9.** T   **11.** T   **13.** F
15. T   **17.** F   **19.** T   **21.** $\{0, 1, 31\}$
23. $\{-1, 0, 1, 31\}$ **25.** $\{-\sqrt{2}, \sqrt{3}, \pi\}$ **27.** F   **29.** F
31. F   **33.** T   **35.** F   **37.** 5   **39.** -12 **41.** -24
43. 2   **45.** $-\frac{1}{6}$ **47.** 10   **49.** 9.96   **51.** -4 **53.** 0
55. -4 **57.** 0   **59.** -3 **61.** 39   **63.** 121   **65.** 23
67. 2   **69.** 17   **71.** -96 **73.** 5   **75.** -1 **77.** .76
79. 1   **81.** -3 **83.** 4   **85.** 1   **87.** -8 **89.** 1
91. -35 **93.** 2   **95.** 5   **97.** -1 **99.** Commutative
101. Distributive   **103.** Associative   **105.** Identity
107. Inverse   **109.** Identity   **111.** Inverse
113. Multiplication property of zero   **115.** $3(x - a)$
117. $3w + 3$ **119.** $7(x + 1)$ **121.** $5x - 25$

123. $-6x + 15$ **125.** $p(1 - t)$ **127.** $7a + 2$
129. $5a - 20$ **131.** $-t - 2$ **133.** $-2a + 4$
135. $4x + 33$ **137.** $-.8x - .48$ **139.** $-.05x - 1.85$
141. $\frac{1}{4}x + 4$ **143.** 0, Inverse, Multiplication property of 0
145. 7680, Distributive   **147.** 48, Associative, Inverse
149. 0, Distributive, Inverse   **151.** 47, Associative
153. -24, Commutative, Associative
155. 0, Inverse, Multiplication property of 0

Chapter 1 Test

1. $\{2, 3, 4, 5, 6, 7, 8, 10\}$ **2.** $\{6, 7\}$ **3.** $\{4, 6, 8, 10\}$
4. $\{0, 8\}$ **5.** $\{-4, 0, 8\}$ **6.** $\{-4, -\frac{1}{2}, 0, 1.65, 8\}$
7. $\{-\sqrt{3}, \sqrt{5}, \pi\}$ **8.** -9 **9.** 44   **10.** -2 **11.** -4
12. -11 **13.** -1.98 **14.** 16   **15.** 598   **16.** 7
17. -3 **18.** 0   **19.** 4780   **20.** Distributive
21. Commutative   **22.** Associative   **23.** Inverse
24. Commutative   **25.** Multiplication property of 0
26. $5x - 17$ **27.** $.95x + 2.9$ **28.** $11m - 3$
29. $4t - 11$ **30.** $\frac{3}{4}x - \frac{5}{4}$ **31.** $5(x - 8)$ **32.** $8(t - 1)$

CHAPTER 2

Section 2.1   Warm-ups FTFFTTFTTT

1. Yes   **3.** No   **5.** No   **7.** Yes   **9.** $\{\frac{3}{2}\}$ **11.** $\{-1\}$
13. $\{-12\}$ **15.** $\{8\}$ **17.** $\{-6\}$ **19.** $\{34\}$ **21.** $\{\frac{9}{4}\}$
23. $\{-6\}$ **25.** $\{12\}$ **27.** $\{-\frac{28}{3}\}$ **29.** $\{-\frac{28}{5}\}$ **31.** $\{-7\}$
33. $\{2\}$ **35.** $\{\frac{3}{2}\}$ **37.** $\{\frac{7}{2}\}$ **39.** $\{12\}$ **41.** $\{6\}$
43. $\{90\}$ **45.** $\{1000\}$ **47.** $\{800\}$ **49.** $\varnothing$, inconsistent
51. R, identity   **53.** $\{1\}$, conditional   **55.** $\varnothing$, inconsistent
57. R, identity   **59.** $\{-4\}$, conditional   **61.** R, identity
63. R, identity   **65.** -3 **67.** -6 **69.** -2 **71.** 0
73. $\{-2\}$ **75.** $\{53, 191.49\}$ **77.** $\{2\}$ **79.** $\{.3\}$
81. $\{4.7\}$ **83.** $\{-47.102\}$

Section 2.2   Warm-ups FTTTTTFFTF

1. $\{3\}$ **3.** $\{\frac{8}{3}\}$ **5.** $\{16\}$ **7.** $\{-5\}$ **9.** $\{-\frac{1}{2}\}$ **11.** $\{6\}$
13. $\{-11\}$ **15.** $\{\frac{9}{2}\}$ **17.** $\{23\}$ **19.** $\{6\}$ **21.** $\{1\}$
23. $\{27\}$ **25.** $\{500\}$ **27.** $\{3\}$ **29.** $\{\frac{4}{3}\}$ **31.** $\{-7\}$
33. $\{-2\}$ **35.** $\{7\}$ **37.** $\{5\}$ **39.** $\{18\}$ **41.** $\{20\}$
43. $\{-\frac{5}{3}\}$ **45.** $\{28\}$ **47.** $\{\frac{4}{3}\}$ **49.** $\{\frac{5}{2}\}$ **51.** $\{3\}$
53. $\{\frac{11}{4}\}$ **55.** $\{-6\}$ **57.** $\{8\}$ **59.** $\{3\}$ **61.** $\{\frac{15}{2}\}$
63. $\{11\}$ **65.** $\{-11\}$ **67.** $\{8\}$ **69.** $\{-5\}$ **71.** R
73. $\varnothing$ **75.** $\{0\}$ **77.** $\{50\}$ **79.** $\{200\}$ **81.** $\{400\}$
83. $\{2\}$ **85.** $\{2\}$

Section 2.3   Warm-ups FFFTFTFTTF

1. $t = \dfrac{I}{Pr}$ **3.** $C = \frac{5}{9}(F - 32)$ **5.** $W = \dfrac{A}{L}$

7. $b_1 = 2A - b_2$ **9.** $t = \dfrac{S - P}{Pr}$ **11.** $a = \dfrac{1}{b + 1}$

13. $y = \dfrac{12}{1-x}$ **15.** $x = \dfrac{6}{w^2 - y^2 - z^2}$

17. $R_1 = \dfrac{RR_2}{R_2 - R}$ **19.** $y = -\frac{2}{3}x + 3$ **21.** $y = x - 4$

23. $y = 2x - 3$ **25.** $y = -\frac{1}{3}x - 2$ **27.** $y = \frac{1}{2}x - \frac{5}{2}$

29. $y = 2x - 6$ **31.** $y = \frac{3}{4}x - 6$ **33.** $y = \frac{1}{2}x + \frac{1}{2}$

35. $y = 4x - 29$ **37.** $y = -3x + 984$ **39.** $\frac{1}{3}$ **41.** $\frac{13}{2}$

43. -1 **45.** 4   **47.** -4.4507 **49.** $\frac{5}{8}$ **51.** $-\frac{7}{2}$

53. -10 **55.** $-\frac{8}{3}$ **57.** 4   **59.** $\frac{23}{4}$ yards   **61.** 7.2 feet

63. 5 feet   **65.** 15%   **67.** 15 feet   **69.** 14 inches

71. 168 feet   **73.** 3979 miles   **75.** 1.5 meters

77. 9.55 inches   **79.** 150 feet   **81.** $1200

Section 2.4   Warm-ups FTFFTTFTFF

1. $x, x + 2$ **3.** $x, 10 - x$ **5.** $.85x$ **7.** $3x$

9. $4x + 10$ **11.** 27, 28, 29   **13.** 82, 84, 86

15. 63, 65   **17.** 46 meters, 93 meters

19. 161 feet, 312 feet, 211 feet

21. W = 11.25 feet, L = 27.5 feet

23. $1500 at 6%, $2500 at 10%   **25.** $80,000

27. $\frac{40}{3}$ gallons   **29.** 1.36 ounces   **31.** 40 mph

33. 15 mph   **35.** $86,957   **37.** $8450   **39.** $8.75

41. 30 pounds   **43.** $\frac{40}{7}$ quarts

45. Brian, $14,400; Daniel, $7200; Raymond, $3800

47. 22, 23   **49.** 20 meters by 20 meters

51. $500 at 8%, $2500 at 10%

53. 2 gallons of 5%, 3 gallons of 10%

55. Todd 46, Darla 32

Section 2.5   Warm-ups FFTFFFTTTT

1. F   **3.** T   **5.** T   **7.** T

9. (number line: $-5\ -4\ -3\ -2\ -1\ 0\ 1$)

11. (number line: 18 19 20 21 22 23 24)

13. (number line: 1 2 3 4 5 6 7)

15. (number line: 2.3; 1 2 3 4 5 6 7)

17. $\{x \mid x > 5\}$ (number line: 3 4 5 6 7 8 9)

19. $\{x \mid x \geq -3\}$ (number line: $-5\ -4\ -3\ -2\ -1\ 0\ 1$)

21. $\{x \mid x > 13\}$ (number line: 11 12 13 14 15 16 17)

23. $\{x \mid x \geq -1\}$ (number line: $-3\ -2\ -1\ 0\ 1\ 2\ 3$)

25. $\{x \mid x > \frac{8}{3}\}$ (number line: $\frac{8}{3}$; 1 2 3 4 5 6 7)

27. $\{x \mid x > \frac{4}{3}\}$ (number line: $\frac{4}{3}$; 0 1 2 3 4 5 6)

29. $\{x \mid x \leq 12\}$ (number line: 8 9 10 11 12 13 14)

31. $\{x \mid x \leq 4\}$ (number line: 0 1 2 3 4 5 6)

33. $\{x \mid x > \frac{2}{3}\}$ (number line: $\frac{2}{3}$; $-1\ 0\ 1\ 2\ 3\ 4$)   **35.** $x < \$9100$

37. $x \geq \$9100.92$ **39.** $x \geq 77$ **41.** $x < \$16.67$

43. $\{x \mid x \leq 2.397\}$ (number line: 2.397; $-1\ 0\ 1\ 2\ 3\ 4$)

45. $\{x \mid x < -17\}$ (number line: $-21\ -20\ -19\ -18\ -17\ -16\ -15$)

47. $\{x \mid x \leq \frac{49}{30}\}$ (number line: $\frac{49}{30}$; $-3\ -2\ -1\ 0\ 1\ 2\ 3$)

49. $\{x \mid x > -\frac{148}{33}\}$ (number line: $-\frac{148}{33}$; $-6\ -5\ -4\ -3\ -2\ -1$)   **51.** $(1, +\infty)$

53. $(-\infty, -3]$ **55.** $(-\infty, 5)$ **57.** $[-4, +\infty)$

Section 2.6   Warm-ups TTFTTTFTFT

1. No   **3.** Yes   **5.** Yes

7. (number line: $-1\ 0\ 1\ 2\ 3\ 4$)   **9.** (number line: 0 1 2 3 4 5 6)

11. (number line: $-2\ -1\ 0\ 1\ 2\ 3\ 4\ 5\ 6$)   **13.** $\varnothing$

15. (number line: 4 5 6 7 8 9 10 11)   **17.** $\varnothing$

19. $\{x \mid x > 10 \text{ or } x < 1\}$ (number line: $-1\ 0\ 1\ 2\ 3\ 4\ 5\ 6\ 7\ 8\ 9\ 10\ 11$)

21. $\{x \mid x > 9\}$ (number line: 7 8 9 10 11 12 13)

23. $\{x \mid x > -6\}$ (number line: $-8\ -7\ -6\ -5\ -4\ -3\ -2$)

25. $\{x \mid 1 < x \leq 4\}$ (number line: 0 1 2 3 4 5)

27. R (number line: $-3\ -2\ -1\ 0\ 1\ 2\ 3$)   **29.** $\varnothing$ **31.** $\varnothing$

33. R (number line: $-3\ -2\ -1\ 0\ 1\ 2\ 3$)

35. $\{x \mid 4 < x < 7\}$ (number line: 3 4 5 6 7 8)

37. $\{x \mid -3 \leq x < 2\}$ (number line: $-3\ -2\ -1\ 0\ 1\ 2$)

39. $\{m \mid -\frac{7}{3} < m \leq 3\}$ (number line: $-\frac{7}{3}$; $-3\ -2\ -1\ 0\ 1\ 2\ 3$)

41. $\{x \mid -1 < x < 5\}$

43. $\{x \mid 2 \le x \le 3\}$

45. $\$11{,}033 \le x \le \$13{,}811$ **47.** $82 \le x \le 109$
49. $4 \le x \le 18$ **51.** $x > 2$ **53.** $x < 3$
55. $x > 2$ or $x \le -1$ **57.** $-2 \le x < 3$ **59.** $x \ge -3$
61. $\{x \mid 5 < x < 7\}$

63. $\{x \mid -1 < x \le 1$ or $x > 10\}$

65. $(-\infty, 0) \cup (3, +\infty)$ **67.** $[-9, 2]$ **69.** $(-3, 8)$
71. $[8, 9)$

Section 2.7   Warm-ups TFFTFTFTTF

1. $\{-5, 5\}$ **3.** $\{2, 4\}$ **5.** $\{-3, 9\}$ **7.** $\{-\frac{8}{3}, \frac{16}{3}\}$
9. $\{\frac{9}{5}\}$ **11.** $\{6\}$ **13.** $\varnothing$ **15.** $\{0, 5\}$ **17.** $|x| < 2$
19. $|x| > 3$ **21.** $|x| \le 1$ **23.** $|x| \ge 2$ **25.** No
27. Yes   **29.** No   **31.** Yes   **33.** No
35. $\{x \mid x > 6$ or $x < -6\}$

37. $\{a \mid -3 < a < 3\}$

39. $\{x \mid x \le -1$ or $x \ge 5\}$

41. $\{x \mid -\frac{1}{2} < x < \frac{9}{2}\}$

43. $\{x \mid -2 \le x \le 12\}$

45. $\{x \mid x \le -\frac{9}{2}$ or $x \ge \frac{15}{2}\}$

47. $\{w \mid w < -5$ or $w > -2\}$

49. $\{x \mid x \ne 2\}$

51. R **53.** $\varnothing$ **55.** R **57.** 15, 21
59. $121 < x < 133$ **61.** $\{-2, 2\}$ **63.** $\{-11, 5\}$
65. $\{.143, 1.298\}$ **67.** $\{-6, \frac{4}{3}\}$ **69.** $\{1, 3\}$ **71.** R
73. $\{0\}$ **75.** $\{\frac{5}{2}\}$ **77.** $\{1\}$ **79.** $\{\frac{3}{2}\}$
81. $\{x \mid -2 < x < 2\}$

83. $\{x \mid x < -3$ or $x > -1\}$

85. $\{x \mid -4 < x < 4\}$

87. $\{x \mid -1 < x < 1\}$

89. $\{x \mid .255 < x < .847\}$

91. $\{x \mid \frac{1}{2} < x < 2\}$

93. $\{x \mid 4 < x < 5\}$

95. $\{x \mid x < 0\}$

97. $\{x \mid x > \frac{4}{3}\}$

Chapter 2 Review Exercises

1. $\{8\}$ **3.** $\varnothing$ **5.** $\{-\frac{3}{2}\}$ **7.** R **9.** $\{20\}$ **11.** $\{5\}$
13. $\{5\}$ **15.** $\{3\}$ **17.** $\{11\}$ **19.** $\{12\}$ **21.** $\{-6\}$
23. $\{\frac{5}{2}\}$ **25.** $\{-2\}$ **27.** $\{2\}$ **29.** $\{70\}$ **31.** $\{-\frac{5}{3}\}$
33. $\{0\}$ **35.** R **37.** $\{1200\}$ **39.** $\varnothing$ **41.** $\{\frac{4}{3}\}$
43. $x = -\dfrac{b}{a}$ **45.** $x = \dfrac{2}{c - a}$ **47.** $x = \dfrac{P}{mw}$
49. $x = \dfrac{2}{2w - 1}$ **51.** $y = \dfrac{3}{2}x + 3$ **53.** $y = -\dfrac{1}{3}x + 4$
55. $y = 2x - 20$ **57.** $L = 8.5$ inches, $W = 14$ inches
59. $\$27{,}000$, $\$35{,}000$ **61.** $\$9500$
63. 11 nickels, 4 dimes   **65.** 15 miles
67. $\{x \mid x > -3\}$

69. $\{x \mid x > 0\}$

71. $\{x \mid x \le -8\}$

73. $\{x \mid x < \frac{11}{2}\}$

75. $\{x \mid x \ge 48\}$

77. $\{x \mid x < -4$ or $x > 1\}$

79. $\{x \mid 0 < x < 9\}$

81. $\{x \mid x > 0\}$

83. $\{x \mid x < 4\}$ **85.** $\varnothing$ **87.** R

89. $\{x \mid -\frac{17}{2} \le x \le \frac{13}{2}\}$

91. $\{x \mid x \le -4$ or $x \ge 4\}$

93. $\varnothing$
95. $\{x \mid x < -4$ or $x > 14\}$

97. $\varnothing$

99. $\{-4, 4\}$ **101.** R
-4-3-2-1 0 1 2 3 4

103. $\{x \mid x < 1 \text{ or } x > 3\}$
-1 0 1 2 3 4 5

105. $\$3 \le x \le \$5$ **107.** 81 or 91   **109.** $x > 1$
111. $x = 2$ **113.** $|x| = 3$ **115.** $x \le -1$ **117.** $|x| \le 2$
119. $x \le 2$ or $x \ge 7$ **121.** $|x| > 3$ **123.** $5 < x < 7$
125. $|x| > 0$

Chapter 2 Test

1. $\{-4\}$ **2.** R **3.** $\{-6, 6\}$ **4.** $\{2, 5\}$ **5.** $y = \frac{2}{5}x - 4$

6. $a = \dfrac{S}{1 - xt}$

7. $\{w \mid w < -4 \text{ or } w > 4\}$
-6 -4 -2 0 2 4 6

8. $\{m \mid 4 \le m \le 8\}$
3 4 5 6 7 8 9

9. $\{w \mid w > 5\}$
3 4 5 6 7 8 9

10. $\{x \mid -8 < x < -\frac{1}{2}\}$
-8-7-6-5-4-3-2-1 0

11. $\{x \mid -5 \le x < 3\}$
-5-4-3-2-1 0 1 2 3

12. $\{y \mid y < 15\}$
11 12 13 14 15 16 17

13. $\varnothing$ **14.** R **15.** $\varnothing$ **16.** $\{\frac{5}{2}\}$ **17.** $\varnothing$ **18.** R
19. $\varnothing$ **20.** R **21.** $\{\frac{7}{5}\}$ **22.** 13 meters   **23.** 14 inches
24. $300   **25.** 30 liters

Tying It All Together   Chapters 1–2

1. $11x$ **2.** $30x^2$ **3.** $3x + 1$ **4.** $4x - 3$ **5.** 899
6. 961   **7.** 841   **8.** 25   **9.** 13   **10.** -25 **11.** 5
12. -4 **13.** $-2x + 13$ **14.** 60   **15.** 72   **16.** -9
17. $-3x^3$ **18.** 1   **19.** $\{0\}$ **20.** R **21.** $\{0\}$ **22.** $\{1\}$
23. $\{-\frac{4}{3}\}$ **24.** $\{1\}$ **25.** R **26.** $\{1000\}$

CHAPTER 3

Section 3.1   Warm-ups TFTTTFTTFT

1. 16   **3.** $\frac{1}{16}$ **5.** $\frac{1}{16}$ **7.** $\frac{1}{5}$ **9.** $\frac{16}{9}$ **11.** 1   **13.** 2
15. 5   **17.** 1   **19.** $-\frac{1}{9}$ **21.** 49   **23.** 2^{17} **25.** $2y^{12}$

27. $30x$ **29.** 2   **31.** 1   **33.** x^9y^5 **35.** $\dfrac{6}{x^5}$ **37.** $\dfrac{1}{a^{17}}$

39. 32   **41.** $\dfrac{100}{3}$ **43.** $\dfrac{4y^2}{5x^3}$ **45.** $\dfrac{1}{x^2}$ **47.** x^2 **49.** x^8

51. $\dfrac{1}{a^3}$ **53.** $\dfrac{1}{w^8}$ **55.** $\dfrac{4}{x^5}$ **57.** x^2 **59.** 3^8w^5 **61.** $\dfrac{y^4}{x}$

63. $\dfrac{3}{x^{13}}$ **65.** $\dfrac{y}{2x}$ **67.** -8 **69.** 16   **71.** $\dfrac{1}{4}$ **73.** 27

75. -16 **77.** -8 **79.** $\frac{9}{4}$ **81.** $\frac{1}{8}$ **83.** 72   **85.** -1
87. $\dfrac{5}{6}$ **89.** $\dfrac{1}{5}$ **91.** $15a^4$ **93.** $8a^2$ **95.** $\dfrac{a^2}{2}$ **97.** $\dfrac{x}{3y^4}$
99. $4x^6$ **101.** $4a^4$ **103.** 850.56   **105.** 1.533

Section 3.2   Warm-ups FTTFFFTTTT

1. y^{10} **3.** 64   **5.** $\dfrac{1}{x^{14}}$ **7.** a^9 **9.** 1   **11.** $\dfrac{1}{x^2}$

13. $81y^2$ **15.** $25w^2$ **17.** $\dfrac{x^9}{y^6}$ **19.** $9a^2b^2$ **21.** $\dfrac{6x^2}{y}$

23. $\dfrac{1}{a^3b^4}$ **25.** $\dfrac{w^3}{8}$ **27.** $\dfrac{27}{64}$ **29.** $\dfrac{4x^2}{y^2}$ **31.** $-\dfrac{x^3}{8y^6}$

33. $\frac{25}{4}$ **35.** -8 **37.** $-\dfrac{27}{8x^3}$ **39.** $\dfrac{27y^3}{8x^6}$

41. $3885.09   **43.** $2958.64   **45.** $6x^9$ **47.** $-8x^6$

49. $\dfrac{3z}{x^2y}$ **51.** $-\frac{3}{2}$ **53.** $\dfrac{4x^6}{9}$ **55.** $-\dfrac{x^2}{2}$ **57.** $\dfrac{4}{x^2}$

59. $\dfrac{1}{x^6}$ **61.** $\dfrac{y^3}{x^3}$ **63.** $\dfrac{b^{12}}{a^6}$ **65.** $\dfrac{x^4}{y}$ **67.** ac^8

69. $\dfrac{a^{28}}{b^{38}}$ **71.** $\dfrac{x^{11}z^{11}}{81y^7}$ **73.** 2^{5m+7}

Section 3.3   Warm-ups FFFTFTFTTF

1. Yes   **3.** No   **5.** Yes   **7.** No   **9.** Binomial, 4, -8
11. Trinomial, 6, 1   **13.** Monomial, 7, 0   **15.** 80
17. -29 **19.** 0   **21.** $75   **23.** $3a + 2$ **25.** $5xy + 25$
27. $11x^2 - 2x - 9$ **29.** $x + 5$ **31.** $-6x^2 + 5x + 2$
33. $2x$ **35.** $-10x^2 + 15x$ **37.** $2x^2 + 9x - 18$
39. $a^2 - b^2$ **41.** $a^3 + b^3$ **43.** $x^2 + 2x - 9$
45. $2x - 5$ **47.** $-15x^6$ **49.** $xy + 2$ **51.** $x^2 - 4$
53. $-3x + 2$ **55.** $-xt - 9y$ **57.** $4a^2 - 11a - 4$
59. $4w^2 - 8w - 7$ **61.** $2x^3 - x^2 + x - 6$ **63.** $x^3 - 8$
65. $15x^3 - 20x^2$ **67.** $xz - wz + 2xw - 2w^2$
69. $a^2z^2 + 2awz + w^2$ **71.** $6x^2 + 7x - 5$ **73.** $\frac{3}{4}x + \frac{3}{2}$
75. $-\frac{1}{2}x^2 + x$ **77.** $\frac{1}{4}x^2 + x - 3$
79. $15.369x^2 + 1.88x - .866$
81. $14.4375x^2 - 29.592x - 9.72$ **83.** $3x^2 - 21x + 1$
85. $-4x^2 - 62x - 168$ **87.** $x^4 - m^2 - 4m - 4$
89. $-6x^3 - 53x^2 - 8x$

Section 3.4   Warm-ups TTFTFTTTFF

1. $x^2 + 2x - 8$ **3.** $a^2 + 5a + 6$ **5.** $2x^2 + 5x - 3$
7. $2a^2 + 7a - 15$ **9.** $w^2 + 5wz - 6z^2$
11. $8m^2 - 6m - 9$ **13.** $t^2 + 3t + 2$ **15.** $3x^4 - x^2 - 4$
17. $xy - 3y + xw - 3w$ **19.** $s^6 - 5s^3t - 6t^2$
21. $ac + ad + bc + bd$ **23.** $m^2 + 6m + 9$
25. $a^2 - 8a + 16$ **27.** $4w^2 + 4w + 1$
29. $9t^2 - 30tu + 25u^2$ **31.** $x^2 - 2x + 1$

33. $a^2 - 6ay + 9y^2$ **35.** $9x^2 - 6x + 1$ **37.** $a^2 - \frac{2}{3}a + \frac{1}{9}$
39. $g^2t^2 + 4gt + 4$ **41.** $64y^4 - 48y^2 + 9$
43. $x^2 - 8x + 16$ **45.** $x^2y^2 - 4xy + 4$ **47.** $w^2 - 81$
49. $w^2 - y^2$ **51.** $4x^2 - 49$ **53.** $9x^4 - 4$ **55.** $a^4 - 1$
57. $x^2 + 3x - 54$ **59.** $25 - x^2$ **61.** $6x^2 - 23ax + 20a^2$
63. $2t^2 + 2tw - 3t - 3w$ **65.** $9x^4 + 12x^2y^3 + 4y^6$
67. $6y^2 - 4y - 10$ **69.** 2496   **71.** $49x^2 + 42x + 9$
73. $25y^2 - 30y + 9$ **75.** $x^2 + \frac{5}{6}x + \frac{1}{6}$ **77.** $9y^2 + 3y + \frac{1}{4}$
79. $x^2 + 4x + 3$ **81.** $x^{3m} + 2x^{2m} + 3x^m + 6$
83. $a^{3n+1} + a^{2n+1} - 3a^{n+1}$ **85.** $a^{2m} + 2a^{m+n} + a^{2n}$
87. $15y^{3m} + 24y^{2m}z^k + 20y^mz^{3-k} + 32z^3$
89. $16.32x^2 - 10.47x - 17.55$
91. $12.96y^2 + 31.68y + 19.36$

Section 3.5   Warm-ups TTFTTTFFFT

1. 12   **3.** 3   **5.** $6xy$ **7.** $x(x^2 - 5)$ **9.** $12w(4x + 3y)$
11. $(x - 6)(a + b)$ **13.** $(y - 1)^2(y + z)$
15. $2x(x^2 - 2x + 3)$ **17.** $2(x - y), -2(-x + y)$
19. $3x(-1 + 2x), -3x(1 - 2x)$
21. $w^2(-w + 3), -w^2(w - 3)$ **23.** $a(a + 1), -a(-a - 1)$
25. $1(2x - 7), -1(-2x + 7)$ **27.** $(x - 10)(x + 10)$
29. $(2y - 7)(2y + 7)$ **31.** $(3x - 5a)(3x + 5a)$
33. $(12wz - 1)(12wz + 1)$ **35.** $(x - 10)^2$ **37.** $(w - t)^2$
39. $(3t + 5)^2$ **41.** $2(x + 2)(x - 2)$ **43.** $x(x + 5)^2$
45. $(2x + 1)^2$ **47.** $(x + 3)(x + 7)$ **49.** $3y(2y + 1)$
51. $(2x - 5)^2$ **53.** $(2x - 3)(x - 2)$
55. $a(3a + w)(3a - w)$ **57.** $-5(a - 3)^2$
59. $-3y(y + 3)^2$ **61.** $-7(ab + 1)(ab - 1)$
63. $(x + 2)(x - 3)(x + 3)$

Section 3.6   Warm-ups TFFFTFFFTT

1. $(x + 1)(x + 3)$ **3.** $(a + 10)(a + 5)$
5. $(x - 2)(x - 4)$ **7.** $(y - 7)(y + 2)$ **9.** $(x - 6)(x + 3)$
11. $(a + 10)(a - 3)$ **13.** $(2x + 1)(x - 3)$
15. $(3w + 1)(2w + 1)$ **17.** $(3m + 4)(2m - 3)$
19. $(4x + 3)(x + 2)$ **21.** $(3y + 1)(4y - 1)$
23. $(6a - 5)(a + 1)$ **25.** $2(x + 5)^2$
27. $a(a - 6)(a + 6)$ **29.** $5(a + 6)(a - 1)$
31. $2(x + 8y)(x - 8y)$ **33.** $-3(x + 3)(x - 4)$
35. $m^3(m + 10)^2$ **37.** $(3x + 4)(2x + 5)$ **39.** $(y - 6)^2$
41. $(3m - 5n)(3m + 5n)$ **43.** $5(a + 6)(a - 2)$
45. $-2(w - 10)(w + 1)$ **47.** $x^2(w + 10)(w - 10)$
49. $(3x + 1)(3x - 1)$ **51.** $(4x + 5)(2x - 3)$
53. $(2x - 5)^2$ **55.** $(3a + 10)^2$
57. $10(3y - 10)(3y + 10)$ **59.** $(5x - 1)^2$
61. $(5x - 4)(2x + 3)$ **63.** $4(a + 2)(a + 4)$
65. $(x^a + 5)(x^a - 3)$ **67.** $(x^a - y^b)(x^a + y^b)$
69. $x^a(x - 1)(x + 1)$ **71.** $(x^a + 3)^2$

Section 3.7   Warm-ups FTTTTTTFFF

1. $12x^4$ **3.** -2 **5.** $2b - 3$ **7.** $x + 2$
9. $-5x^2 + 4x - 3$ **11.** $\frac{7}{2}x^2 - 2x$ **13.** $x + 5, -2$
15. $x - 4, 8$ **17.** $3, 3$ **19.** $x^2 - 2x + 4, 0$
21. $a^2 + 2a + 8, 11$ **23.** $x^2 - 2x + 3, -6$
25. $x^3 + 3x^2 + 6x + 11, 21$ **27.** $-3x^2 - 1, x - 4$
29. $\frac{1}{2}x - \frac{5}{4}, -\frac{7}{4}$ **31.** $\frac{2}{3}x + \frac{1}{9}, \frac{56}{9}$ **33.** $2 + \dfrac{10}{x - 5}$
35. $x - 1 + \dfrac{1}{x + 1}$ **37.** $x + 1 + \dfrac{2}{x - 1}$
39. $-3 + \dfrac{6}{x + 2}$ **41.** $x^2 - 2x + 4 + \dfrac{-8}{x + 2}$ **43.** $x + \dfrac{2}{x}$
45. $\frac{1}{2}x - \frac{3}{4} + \dfrac{9}{8x + 12}$ **47.** $\frac{1}{2}x^2 + \frac{1}{4}x - \frac{19}{8} - \dfrac{19}{16x - 8}$
49. Yes   **51.** No   **53.** Yes   **55.** Yes   **57.** No
59. Yes   **61.** $(a - 1)(a^2 + a + 1)$
63. $(w + 3)(w^2 - 3w + 9)$ **65.** $(2x - 1)(4x^2 + 2x + 1)$
67. $(a + 2)(a^2 - 2a + 4)$ **69.** $(m - n)(m^2 + mn + n^2)$
71. $(2 - 3x)(4 + 6x + 9x^2)$
73. $(wx^2 - 5y^3)(w^2x^4 + 5wx^2y^3 + 25y^6)$
75. $(5a^3 + 2cb^2)(25a^6 - 10a^3cb^2 + 4c^2b^4)$
77. $(.1x - .2y)(.01x^2 + .02xy + .04y^2)$ **79.** $-3y^9, 0$
81. $z + 3, 0$ **83.** $m - 3, 0$ **85.** $w^2 + 2w + 4, 16$
87. $x - 3, 18$

Section 3.8   Warm-ups FTTFFTFTTF

1. Prime   **3.** Not prime   **5.** Prime   **7.** Prime
9. Prime   **11.** $(a^2 - 5)^2$ **13.** $(x^2 - 2)(x - 2)(x + 2)$
15. $3(x - 2)(3x - 4)$ **17.** $(m + 5)(m + 1)$
19. $4w(w - 4)$ **21.** $2(4a^2 + 3)(4a^2 - 3)$
23. $(x + 3)(x - 2)(x^2 - x + 6)$ **25.** $(3y - 7)(y + 4)$
27. $(a + b)(x + y)$ **29.** $(x - 3)(x + 3)(x + 1)$
31. $(w - 3)(a - b)$ **33.** $(a + 3)^2(a^2 - 3a + 9)$
35. $(y - 5)(y + 2)(y^2 - 2y + 4)$ **37.** $(d - w)(ay + 1)$
39. $(a + y)(x - 1)(x + 1)$ **41.** $(y + b)(y + 1)(y^2 - y + 1)$
43. $(3x - 4)^2$ **45.** $(3x - 1)(4x - 3)$
47. $3a(a + 3)(a^2 - 3a + 9)$ **49.** $2(x^2 + 16)$ **51.** Prime
53. $(x + y - 1)(x + y + 1)$ **55.** $ab(a - b)(a + b)$
57. $(x + 2)(x - 2)(x^2 + 2x + 4)$ **59.** $n(m + n)^2$
61. $(2 + w)(m + n)$ **63.** $(2w + 3)(2w - 1)$
65. $(t^2 + 7)(t^2 - 3)$ **67.** $a(a + 10)(a - 3)$
69. $(y + 2)(y + 6)$ **71.** $2(w - 5)(w + 5)(w^2 + 25)$
73. $8a(a^2 + 1)$ **75.** $(w + 2)(w + 8)$ **77.** $a(2w - 3)^2$
79. $(x - 3)^2$ **81.** $3x^2(x - 5)(x + 5)$
83. $n(m - 1)(m^2 + m + 1)$ **85.** $2(3x + 5)(2x - 3)$
87. $2(a^3 - 16)$ **89.** $(a^m - 1)(a^{2m} + a^m + 1)$

91. $(a^w - b^{2n})(a^{2w} + a^w b^{2n} + b^{4n})$
93. $(t^{2n} - 2)(t^{2n} + 2)(t^{4n} + 4)$ **95.** $a(a^n - 5)(a^n + 3)$
97. $(a^n - 3)(a^n + b)$

Section 3.9   Warm-ups FTTFTFTTFF
1. $\{-4, 5\}$ **3.** $\{\frac{5}{2}, -\frac{4}{3}\}$ **5.** $\{-7, 2\}$ **7.** $\{0, 7\}$
9. $\{-4, 5\}$ **11.** $\{-3, 4\}$ **13.** $\{-4, \frac{5}{2}\}$ **15.** $\{-2, 0, 2\}$
17. $\{-5, -4, 5\}$ **19.** $\{-1, 1, 2\}$ **21.** $\{-3, -1, 1, 3\}$
23. $\{-8, -6, 4, 6\}$ **25.** $\{-4, 0, -2\}$ **27.** $\{-7, -3, 1\}$
29. $W = 4$ inches, $L = 6$ inches   **31.** 4 and 9
33. $W = 5$ feet, $L = 12$ feet   **35.** $W = 5$ feet, $L = 12$ feet
37. 3, 4 or -4, -3 **39.** 4 seconds
41. 6 feet by 8 feet   **43.** $\{-\frac{3}{2}, 2\}$ **45.** $\{-6, -3, -2, 1\}$
47. $\{-6, 1\}$ **49.** $\{1\}$ **51.** $\{-5, -4, 0\}$ **53.** $\{-3, 0, 3\}$
55. $\{0, -b\}$ **57.** $\left\{-\frac{b}{a}, \frac{b}{a}\right\}$ **59.** $\left\{-\frac{b}{2}\right\}$ **61.** $\left\{-\frac{3}{a}, 1\right\}$

Chapter 3 Review Exercises
1. 2   **3.** 36   **5.** $-\dfrac{1}{27}$ **7.** 1   **9.** $\dfrac{8}{x^3}$ **11.** $\dfrac{1}{y^2}$

13. a^7 **15.** $\dfrac{2}{x^4}$ **17.** $\dfrac{4}{9}$ **19.** $\dfrac{4}{3}$ **21.** 36   **23.** 4

25. x^5 **27.** $\dfrac{x^8}{81y^4}$ **29.** $\dfrac{4y^6}{9x^2}$ **31.** $\dfrac{1}{m^{16}n^6}$ **33.** $8w + 2$

35. $-6x + 3$ **37.** $x^3 - 4x^2 + 8x - 8$ **39.** $-4xy + 22z$
41. $5m^5 - m^3 + 2m^2$ **43.** $3x^3 - 15x^2 + 17x$
45. $x^2 + 4x - 21$ **47.** $z^2 - 25y^2$ **49.** $m^2 + 16m + 64$
51. $w^2 - 10xw + 24x^2$ **53.** $n^2 - 10n + 25$
55. $2w^2 + 9w - 18$ **57.** $4m^2 + 4m + 1$
59. $64 - 16a + a^2$ **61.** $x - 2$ **63.** $-a + 5$
65. $w - 3$ **67.** 5   **69.** $b - a$ **71.** $-a - b$
73. $(y - 9)(y + 9)$ **75.** $(2x + 7)^2$ **77.** $(t - 9)^2$
79. $(x - 10)(x + 3)$ **81.** $(w - 7)(w + 4)$
83. $(2m + 7)(m - 1)$ **85.** $x^2 + 3x - 5$, 0
87. $m^3 - m^2 + m - 1$, 0   **89.** $m^2 + 2m - 6$, 0
91. -1, 0   **93.** $6x^2 - 4x + 3$, 0   **95.** $x + 1 + \dfrac{-4}{x - 1}$
97. $3 + \dfrac{6}{x - 2}$ **99.** $-1 + \dfrac{1}{1 - x}$ **101.** $x + \dfrac{-x}{x^2 + 1}$
103. $-2 + \dfrac{10}{x + 5}$ **105.** Yes   **107.** Yes   **109.** No
111. Yes   **113.** $5(x + 2)(x^2 - 2x + 4)$
115. $(3x + 2)(3x + 1)$ **117.** $(x - 1)(x + 1)^2$
119. $y(x - 4)(x + 4)$ **121.** $ab^2(a - 1)^2$
123. $(x - 1)(x^2 + 9)$ **125.** $(x - 2)(x + 2)(x^2 + 3)$
127. $a^3(a - 1)(a^2 + a + 1)$ **129.** $2(2m + 3)^2$

131. $(2x - 7)(2x + 1)$ **133.** $\{0, 5\}$ **135.** $\{0, 5\}$
137. $\{-\frac{1}{2}, 5\}$ **139.** $\{-5, -1, 1\}$ **141.** $\{-3, -1, 1, 3\}$
143. 6 feet

Chapter 3 Test
1. $\frac{1}{9}$ **2.** 36   **3.** 8   **4.** $12x^7$ **5.** $4y^{12}$ **6.** $64a^6b^3$
7. $\dfrac{27}{x^6}$ **8.** $\dfrac{2a^3}{b^3}$ **9.** $3x^3 + 3x^2 - 2x + 3$
10. $-2x^2 - 8x - 3$ **11.** $x^3 - 5x^2 + 13x - 14$
12. $x^2 + 4x - 5$ **13.** $x^3 - 6x^2 + 12x - 8$ **14.** -1
15. $x^2 - 4x - 21$ **16.** $x^2 - 12x + 36$ **17.** $y^2 - 25$
18. $4x^2 + 20x + 25$ **19.** $5 + \dfrac{-15}{x + 3}$
20. $x + 5 + \dfrac{4}{x - 2}$ **21.** $(a - 6)(a + 4)$ **22.** $(2x + 7)^2$
23. $3(m - 2)(m^2 + 2m + 4)$ **24.** $2y(x - 4)(x + 4)$
25. $(a - 5)(2x + 3)$ **26.** $(x - 1)(x + 1)(x^2 + 4)$
27. $\{-5, \frac{3}{2}\}$ **28.** $\{-2, 0, 2\}$ **29.** $\{-4, -3, 2, 3\}$
30. $W = 8$ inches, $H = 6$ inches

Tying It All Together   Chapters 1–3
1. 16   **2.** -8 **3.** $\frac{1}{16}$ **4.** 2   **5.** 1   **6.** $\frac{1}{6}$ **7.** $\frac{1}{12}$
8. 49   **9.** 64   **10.** 8   **11.** -29 **12.** $\frac{11}{30}$ **13.** $\{200\}$
14. $\{\frac{9}{5}\}$ **15.** $\{-9, \frac{3}{2}\}$ **16.** $\{-\frac{15}{2}, 0\}$ **17.** $\{2, \frac{8}{5}\}$
18. $\{\frac{9}{5}\}$ **19.** $\varnothing$ **20.** $\{-3, -2, 1, 2\}$ **21.** $\{-5, \frac{1}{2}\}$
22. $\{-\frac{2}{3}, \frac{4}{3}\}$ **23.** $\{\frac{22}{15}\}$ **24.** $\{-\frac{6}{5}\}$

CHAPTER 4

Section 4.1   Warm-ups TTFTFFFTTF
1. $\{x \mid x \neq 1\}$ **3.** $\{x \mid x \neq -2 \text{ and } x \neq 2\}$ **5.** $\{x \mid x \neq \frac{5}{2}\}$
7. $\{x \mid x \neq -2 \text{ and } x \neq -3\}$ **9.** $\{x \mid x \neq -4 \text{ and } x \neq 0\}$
11. $\{x \mid x \neq 0\}$ **13.** $\dfrac{2}{19}$ **15.** $\dfrac{1}{5}$ **17.** $\dfrac{x + 1}{2}$ **19.** $-\dfrac{3}{5}$
21. $\dfrac{b}{a^2}$ **23.** $-\frac{1}{2}$ **25.** $\dfrac{x + 2}{x}$ **27.** $\dfrac{x - 2}{x + 2}$ **29.** $\dfrac{x - 2}{2x - 6}$
31. $\dfrac{w}{x^2 y}$ **33.** $\dfrac{9x^2 - 3xy + y^2}{2}$ **35.** $\dfrac{(x + 2)(x^2 + 4)}{2}$
37. $\dfrac{10}{50}$ **39.** $\dfrac{3x}{3x^2}$ **41.** $\dfrac{5x - 5}{x^2 - 2x + 1}$ **43.** $\dfrac{-3}{-6x - 6}$
45. $\dfrac{x^2 + x - 2}{x^2 + 2x - 3}$ **47.** $\dfrac{-7}{1 - x}$ **49.** $\dfrac{5a}{a}$ **51.** $\dfrac{3x - 3}{x^2 - 1}$
53. $\dfrac{7x - x^2}{7 - x}$ **55.** 7   **57.** 2   **59.** -6 **61.** 10
63. $3a$ **65.** -2 **67.** $2x + 2$ **69.** $3 - w$ **71.** $x + 2$
73. $x - 4$ **75.** $a + 1$

Section 4.2   Warm-ups FFTFTTTTTF

1. $\dfrac{5}{11}$ **3.** $\dfrac{ab}{4}$ **5.** $\dfrac{x+1}{x^2+1}$ **7.** $\dfrac{1}{a-b}$ **9.** $\dfrac{a+2}{2}$

11. $\dfrac{3}{2}$ **13.** $\dfrac{2}{x+3}$ **15.** $\dfrac{-a^2+6a-9}{8}$ **17.** $2x+10$

19. $\dfrac{a-b}{6}$ **21.** $3a-3b$ **23.** $\dfrac{5x}{6}$ **25.** 3 **27.** 14

29. $\dfrac{5}{3}$ **31.** $\dfrac{1}{18}$ **33.** 3 **35.** $\dfrac{1}{12}$ **37.** $\dfrac{b}{2a}$ **39.** $\dfrac{2x}{3}$

41. -1 **43.** -2 **45.** $\dfrac{a+b}{a}$ **47.** $5x^2y$ **49.** $2a+2b$

51. $\dfrac{3x}{5y}$ **53.** $\dfrac{3a}{10b}$ **55.** $\dfrac{b}{2a}$ **57.** $\dfrac{x^5}{6}$ **59.** -2

61. $\dfrac{2x+4}{x-2}$ **63.** $\dfrac{7x^2}{3x+2}$ **65.** $-\dfrac{a^6b^2}{8c^2}$ **67.** $\dfrac{2m^8n^2}{3}$

69. $\dfrac{2x-3}{8x-4}$ **71.** $\dfrac{k+m}{m-k}$ **73.** $\dfrac{3}{2}$

Section 4.3   Warm-ups FFTFFFTTTF

1. 120 **3.** a^3b^4 **5.** $(x-1)(x+1)^2$

7. $x(x+2)(x-2)$ **9.** $x(x-4)(x+4)(x+2)$ **11.** $\frac{17}{140}$

13. $\dfrac{1}{40}$ **15.** $\dfrac{5}{2}$ **17.** $\dfrac{2}{x-3}$ **19.** $\dfrac{11x}{10a}$ **21.** $\dfrac{6x-1}{3}$

23. $\dfrac{1+4a}{4}$ **25.** $\dfrac{-x-3}{x(x+1)}$ **27.** $\dfrac{8x-4}{(x+2)(x-2)}$

29. $\dfrac{2b-3a}{a^2b^2}$ **31.** 0 **33.** $\dfrac{11}{2x-4}$ **35.** $\dfrac{4x+9}{(x+3)(x-3)}$

37. $\dfrac{11x+27}{(x-1)(x+2)(x+3)}$ **39.** $\dfrac{x^2+7x-18}{(x+1)(x+3)(x-3)}$

41. $\dfrac{6x^2+2x-2}{x(x-1)(x+2)}$ **43.** $\dfrac{5}{6}$ **45.** $-\dfrac{19}{40}$ **47.** $\dfrac{2x+3}{6}$

49. $\dfrac{3a-2b}{3b}$ **51.** $\dfrac{3a+2}{3}$ **53.** $\dfrac{a+3}{a}$ **55.** $\dfrac{5-2x}{x}$

57. $\dfrac{3}{x}$ **59.** $\dfrac{ms-nr}{ns}$ **61.** $\dfrac{8x+3}{12x}$ **63.** $\dfrac{3x+7}{(x+2)(x+3)}$

65. $\dfrac{2a}{(a-1)(a+1)}$ **67.** $\dfrac{5x^2+x+4}{(x-1)(x^2+x+1)}$

69. $\dfrac{-w^2+2w+8}{(w+3)(w^2-3w+9)}$

71. $\dfrac{-w^3-2w^2-5w+6}{(w-2)(w+2)(w^2+2w+4)}$

73. $\dfrac{x^3+x^2+2x+1}{(x-1)(x+1)(x^2+x+1)}$

Section 4.4   Warm-ups FTFTFFFFFTT

1. $\dfrac{10}{3}$ **3.** -8 **5.** $\dfrac{a^2+ab}{a-b}$ **7.** $\dfrac{a^2b+3a}{a+b^2}$

9. $\dfrac{x-3y}{x+y}$ **11.** $\dfrac{60m-3m^2}{8m+36}$ **13.** $\dfrac{a^2-ab}{b^2}$ **15.** $\dfrac{x+2}{x-2}$

17. $\dfrac{y^2-y-2}{(y-1)(3y+4)}$ **19.** $\dfrac{4x-10}{x-4}$ **21.** $\dfrac{6w-3}{2w^2+w-4}$

23. $\dfrac{2b-a}{a-3b}$ **25.** $\dfrac{3a-7}{5a-2}$ **27.** $\dfrac{3m^2-12m+12}{(m-3)(2m-1)}$

29. $\dfrac{2x^2+4x+5}{4x^2-2x-6}$ **31.** $\dfrac{x}{x+1}$ **33.** $\dfrac{a^2+b^2}{ab^3}$

35. $\dfrac{yz+wz}{wy+wz}$ **37.** $x-2$ **39.** $\dfrac{1}{x^2-x+1}$

41. $2m-3$ **43.** $\dfrac{a-1}{a}$ **45.** $\dfrac{xy}{x+y}$ **47.** $\dfrac{a^2b^3-a^3b^2}{a+b}$

49. $-\dfrac{1}{ab}$ **51.** 47.4%

Section 4.5   Warm-ups TFFTFTFFFT

1. $\{-12\}$ **3.** $\{\frac{22}{15}\}$ **5.** $\{5\}$ **7.** $\{1, 6\}$ **9.** $\varnothing$
11. $\{20, 25\}$ **13.** $\{\frac{8}{3}\}$ **15.** $\{-\frac{3}{4}\}$ **17.** $\{-\frac{14}{5}\}$ **19.** $\{20\}$
21. $\{-6\}$ **23.** $\{\frac{11}{2}\}$ **25.** $\{-3, 3\}$ **27.** $\{-5, 6\}$
29. $\{-6, 6\}$ **31.** $\{8\}$ **33.** $\{2, 4\}$ **35.** $\{-1\}$ **37.** $\{5\}$
39. $\{-3, 3\}$ **41.** $\{8\}$ **43.** $\{\frac{25}{2}\}$ **45.** $\{4\}$ **47.** $\{-5, 2\}$

Section 4.6   Warm-ups FTFTTTFTFF

1. $y=5x-7$ **3.** $y=-\frac{1}{3}x+1$ **5.** $y=mx-bm+a$

7. $y=-\dfrac{7}{3}x-\dfrac{29}{3}$ **9.** $f=\dfrac{F}{M}$ **11.** $D^2=\dfrac{4A}{\pi}$

13. $r^2=\dfrac{km_1m_2}{F}$ **15.** $q=\dfrac{pf}{p-f}$ **17.** $a^2=\dfrac{b^2}{1-e^2}$

19. $T_1=\dfrac{P_1V_1T_2}{P_2V_2}$ **21.** $h=\dfrac{3V}{4\pi r^2}$ **23.** $\dfrac{1}{2}$ **25.** 24

27. $\frac{32}{3}$ **29.** -6.517 **31.** 19.899 **33.** 1.910
35. 4 mph
37. Patrick, 24 minutes; Guy, 36 minutes; 100 mph
39. 10 mph   **41.** 6 hours   **43.** 60 minutes   **45.** 6 hours
47. 10 pounds apples, 12 pounds oranges
49. 25.255 days

Chapter 4 Review Exercises

1. $\dfrac{c^2}{a^2b}$ **3.** $\dfrac{4x^2}{3y}$ **5.** $-a$ **7.** $\dfrac{3}{2}$ **9.** $6x(x-2)$

11. $12a^5b^3$ **13.** $\dfrac{3x+11}{2(x-3)(x+3)}$ **15.** $\dfrac{aw-5b}{a^2b^2}$

17. $\dfrac{3}{4x - 24}$ **19.** $\dfrac{7 - 3y}{4y - 3}$ **21.** $\dfrac{b^3 - a^2}{ab}$ **23.** $\left\{-\dfrac{16}{3}\right\}$

25. $\{10\}$ **27.** $y = mx + b$ **29.** $m = \dfrac{Fr}{v^2}$ **31.** $r = \dfrac{3A}{2\pi h}$

33. $y = 2x - 17$ **35.** 10 hours   **37.** 400 hours

39. $\dfrac{4}{3x}$ **41.** $\dfrac{10 + 7y}{6xy}$ **43.** $\dfrac{8a + 10}{(a - 5)(a + 5)}$

45. $\{-8, 8\}$ **47.** $-\dfrac{3}{2}$ **49.** $\{-9, 9\}$ **51.** $\dfrac{1}{x - m}$

53. $\dfrac{8a + 20}{(a - 5)(a + 5)(a + 1)}$ **55.** $\dfrac{-15a + 10}{2(a + 2)(a + 3)(a - 3)}$

57. $\{2\}$ **59.** $\left\{-\dfrac{5}{2}\right\}$ **61.** $-\dfrac{1}{3}$ **63.** $\dfrac{x - 2}{3x + 3}$

65. $\dfrac{3a^2 + 7a + 16}{a^3 - 8}$ **67.** $\dfrac{3 - x}{(x + 1)(x + 3)}$ **69.** $\dfrac{a + w}{a + 4}$

71. $\{-6, 8\}$ **73.** 18   **75.** -3 **77.** $4x$ **79.** $10x$

81. $\dfrac{1}{3}$ **83.** $a + 3$ **85.** $\dfrac{3}{10}$ **87.** $\dfrac{5a}{6}$ **89.** $\dfrac{b - a}{ab}$

91. $\dfrac{a - 3}{3}$ **93.** $\dfrac{2a + 1}{a}$ **95.** 1   **97.** $1 - x$ **99.** -1

101. $\dfrac{b}{6a}$ **103.** -2 **105.** -2

Chapter 4 Test

1. $\{x \mid x \neq 3 \text{ and } x \neq -3\}$ **2.** $\{x \mid x \neq \frac{4}{3}\}$

3. $\{a \mid a \neq 0 \text{ and } a \neq -4\}$ **4.** $-\dfrac{1}{36}$ **5.** $\dfrac{7y + 3}{y}$

6. $\dfrac{5}{a - 9}$ **7.** $\dfrac{-4x + 2}{(x + 2)(x - 2)(x - 5)}$ **8.** $\dfrac{m + 1}{3}$

9. $-\dfrac{3}{a + b}$ **10.** $\dfrac{a^3}{30b^2}$ **11.** $\left\{\dfrac{12}{7}\right\}$ **12.** $\{4, 10\}$

13. $\{-2, 2\}$ **14.** $t = \dfrac{a^2}{W}$ **15.** $b = \dfrac{2a}{a - 2}$

16. 24 minutes   **17.** 30 miles

18. $\dfrac{16}{9 - 6x}$ **19.** $w - m$ **20.** $\dfrac{1 + a}{a^2}$ **21.** $\dfrac{3a^2}{2}$

Tying It All Together   Chapters 1–4

1. $\{\frac{15}{4}\}$ **2.** $\{-4, 4\}$ **3.** $\{\frac{12}{5}\}$ **4.** $\{-6, 3\}$ **5.** $\{\frac{1}{4}\}$
6. $\{6\}$ **7.** $\{\frac{1}{2}\}$ **8.** $\{\frac{3}{2}\}$ **9.** $\{-3, 3\}$ **10.** $\{-8, 3\}$
11. $\{\frac{1}{3}, \frac{2}{3}\}$ **12.** $\{-3, 9\}$ **13.** $\{-\frac{3}{2}, \frac{1}{2}\}$ **14.** $\varnothing$

15. $y = \dfrac{C - Ax}{B}$ **16.** $y = -\dfrac{1}{3}x + \dfrac{4}{3}$ **17.** $y = \dfrac{C}{A - B}$

18. $y = -A$ or $y = A$ **19.** $y = 2A - 2B$

20. $y = \dfrac{2AC}{2B + C}$ **21.** $y = \dfrac{3}{4}x - \dfrac{3}{2}$ **22.** $y = A$ or $y = 2$

23. $y = \dfrac{2A - BC}{B}$ **24.** $y = B$ or $y = -C$ **25.** $12x^{13}$

26. $3x^5 + 15x^8$ **27.** $25x^{12}$ **28.** $27a^9b^6$ **29.** $-4a^6b^6$

30. $\dfrac{1}{32x^{10}}$ **31.** $\dfrac{27x^{12}y^{15}}{8}$ **32.** $\dfrac{a^2}{4b^6c^2}$

CHAPTER 5

Section 5.1   Warm-ups TFTFTTTFFF
1. 10   **3.** -3 **5.** Undefined   **7.** 10   **9.** -4
11. -2 **13.** 8   **15.** -8 **17.** $\frac{1}{8}$ **19.** $-\frac{1}{8}$ **21.** 81
23. 32   **25.** $\frac{1}{2}$ **27.** $\frac{1}{8}$ **29.** $\frac{1}{16}$ **31.** Undefined   **33.** 4
35. 81   **37.** $\frac{1}{4}$ **39.** 1   **41.** $2^{5/6}$ **43.** 6   **45.** $2^{1/4}$
47. 3   **49.** 9   **51.** 4^8 **53.** 2   **55.** $\frac{1}{2}$ **57.** $\frac{1}{2}$ **59.** 2

61. $\dfrac{4}{9}$ **63.** Undefined   **65.** $x^{3/4}$ **67.** $\dfrac{y}{x^{1/2}}$ **69.** $\dfrac{y^{3/2}}{x^{1/4}}$

71. $y^{1/2}$ **73.** $\dfrac{1}{a^{3/2}}$ **75.** $\dfrac{a^3}{b^2}$ **77.** x^8 **79.** $\dfrac{1}{a^{1/2}b^{7/3}}$

81. $\dfrac{x^{1/4}}{y^{1/2}z^{1/4}}$ **83.** 1   **85.** $a^{5/4}b$ **87.** $k^{9/2}m^4$ **89.** t^2

91. a^2 **93.** x^2 **95.** $\dfrac{w^6}{x^5}$ **97.** $\dfrac{a^{3/4}}{w^{17}}$ **99.** 13 inches

101. 1.2599   **103.** -1.4142 **105.** 2   **107.** 1.9862
109. 2.5   **111.** 2.1392

Section 5.2   Warm-ups TFTFTFTFTT
1. $x^{5/2}$ **3.** $a^{-12/3}$ **5.** $\sqrt{3}$ **7.** $-\sqrt[3]{5}$ **9.** $\sqrt[5]{2^2}$
11. $\sqrt[3]{x^{-2}}$ **13.** 11   **15.** -10 **17.** Undefined   **19.** 27
21. $\frac{1}{10}$ **23.** x^3 **25.** x^5 **27.** 9   **29.** a^4 **31.** 4^8
33. a^4 **35.** $2\sqrt{5}$ **37.** $3\sqrt{5w}$ **39.** $12\sqrt{2}$
41. $3 \cdot \sqrt[3]{2}$ **43.** $2 \cdot \sqrt[4]{2a}$ **45.** $\dfrac{5\sqrt{2}}{3}$ **47.** $\dfrac{2 \cdot \sqrt[3]{2x}}{3}$

49. $\dfrac{2\sqrt{2x}}{7}$ **51.** $\dfrac{2 \cdot \sqrt[4]{2a}}{3}$ **53.** $\dfrac{\sqrt{2}}{2}$ **55.** $\dfrac{\sqrt{6}}{4}$

57. $\dfrac{2\sqrt{5}}{5}$ **59.** $\dfrac{\sqrt{15}}{6}$ **61.** $\dfrac{1}{2}$ **63.** $\dfrac{\sqrt[3]{2}}{2}$ **65.** $\dfrac{\sqrt[3]{150}}{5}$

67. $2x^4 \cdot \sqrt{3}$ **69.** $2a^4 \cdot \sqrt{15a}$ **71.** $\dfrac{\sqrt{xy}}{y}$ **73.** $\dfrac{a\sqrt{ab}}{b^4}$

75. $\dfrac{\sqrt{b}}{a}$ **77.** $x^4 \cdot \sqrt[3]{x}$ **79.** $x^2y \cdot \sqrt[4]{xy^2}$ **81.** $2x^4 \cdot \sqrt[5]{2x^2}$

83. $\dfrac{\sqrt[3]{ab^2}}{b}$ **85.** $\dfrac{\sqrt[3]{18}}{6b^2}$ **87.** $\sqrt[5]{3240m^3n}$ **89.** 2.887

91. .693   **93.** 1.310   **95.** 1

Section 5.3　Warm-ups FTFFTFTFFT

1. $-\sqrt{3}$　**3.** $9\sqrt{7x}$　**5.** $5 \cdot \sqrt[3]{2}$　**7.** $4\sqrt{3} - 2\sqrt{5}$

9. $5 \cdot \sqrt[3]{x}$　**11.** $2 \cdot \sqrt[3]{x} - \sqrt{2x}$　**13.** $3\sqrt{2}$　**15.** $\dfrac{3\sqrt{2}}{2}$

17. $2\sqrt{2} + 2\sqrt{7}$　**19.** $\sqrt{5} + 2\sqrt{2}$　**21.** $\dfrac{21\sqrt{5}}{5}$

23. $\sqrt[3]{2t}$　**25.** $7 \cdot \sqrt[3]{3}$　**27.** $-\sqrt[4]{3}$　**29.** $\sqrt[3]{15}$
31. $30\sqrt{2}$　**33.** $50\sqrt{c}$　**35.** $3 \cdot \sqrt[4]{3}$　**37.** 12
39. $2x \cdot \sqrt[3]{x}$　**41.** $6\sqrt{2} + 18$　**43.** $5\sqrt{2} - 2\sqrt{5}$
45. $3t\sqrt{2} + 3\sqrt{3t}$　**47.** $-7 - 3\sqrt{3}$　**49.** 8
51. $-8 - 6\sqrt{5}$　**53.** $-6 + 9\sqrt{2}$　**55.** $11 + 6\sqrt{2}$
57. -1　**59.** 3　**61.** 19　**63.** 13　**65.** -29
67. $11\sqrt{3}$　**69.** $10\sqrt{30}$　**71.** $14 - 4\sqrt{7}$

73. $28 + \sqrt{10}$　**75.** $\dfrac{8\sqrt{2}}{15}$　**77.** 17　**79.** $9 + 6\sqrt{x} + x$

81. $25x - 30\sqrt{x} + 9$　**83.** $x + 3 + 2\sqrt{x + 2}$
85. $x + 1 + 4\sqrt{x - 3}$　**87.** $5\sqrt{a}$　**89.** $-\sqrt{w}$
91. $7\sqrt{2a}$　**93.** $3x^2\sqrt{x}$　**95.** $4a - a\sqrt{a}$
97. $x + 13 + 8\sqrt{x - 3}$　**99.** $2x + 8 + 6\sqrt{2x - 1}$
101. $w - 4$　**103.** $x - 10\sqrt{x} + 25$　**105.** $15m$
107. $3x^2 \cdot \sqrt{2x}$

Section 5.4　Warm-ups TTFTFTFTTT

1. $\dfrac{\sqrt{15}}{5}$　**3.** $\sqrt[3]{10}$　**5.** $\dfrac{3\sqrt{2}}{10}$　**7.** $\dfrac{\sqrt{2}}{3}$　**9.** x^2

11. $2 + \sqrt{5}$　**13.** $1 - \sqrt{3}$　**15.** $\dfrac{2 - \sqrt{2}}{5}$

17. $\dfrac{1 + \sqrt{3}}{2}$　**19.** $\dfrac{\sqrt{6} + \sqrt{2}}{2}$　**21.** $\dfrac{2\sqrt{3} - \sqrt{6}}{3}$

23. $\dfrac{6\sqrt{6} + 2\sqrt{15}}{13}$　**25.** $\dfrac{18\sqrt{5} + 3\sqrt{10} - 2\sqrt{6} - 12\sqrt{3}}{66}$

27. $4\sqrt{2}$　**29.** $24\sqrt{3}$　**31.** 24　**33.** $4 \cdot \sqrt[3]{25}$

35. $x^2 \cdot \sqrt{x}$　**37.** x^4　**39.** $\dfrac{\sqrt{6} + 2\sqrt{2}}{2}$　**41.** $\dfrac{\sqrt{2}}{2}$

43. $\sqrt{7}$　**45.** $\tfrac{2}{3}$　**47.** $a - 3\sqrt{a}$　**49.** $4a\sqrt{a} + 4a$
51. $12m$　**53.** $4xy^2z$　**55.** $m - m^2$　**57.** $5x \cdot \sqrt[3]{x}$

59. $8m^4 \cdot \sqrt[4]{8m^2}$　**61.** $2 + 8\sqrt{2}$　**63.** $\dfrac{3\sqrt{2} + 2\sqrt{3}}{6}$

65. $7\sqrt{2} - 1$　**67.** $\dfrac{4x + 4\sqrt{x}}{x - 4}$　**69.** $\dfrac{x + \sqrt{x}}{x - x^2}$　**71.** 3

73. 1　**75.** $\sqrt{2}$　**77.** $\sqrt{15} + \sqrt{3}$　**79.** $\sqrt{x} + 1$
81. $3\sqrt{2} - 3x$　**83.** $4\sqrt{x - 1}$　**85.** 3.968　**87.** 12.124
89. $.725$　**91.** 8.873　**93.** $-.419$　**95.** 14.697

Section 5.5　Warm-ups FTFFTFFTTT

1. $\{-10\}$　**3.** $\{\tfrac{1}{2}\}$　**5.** $\{1\}$　**7.** $\{-2\}$　**9.** $\{-5, 5\}$
11. $\{-2\sqrt{5}, 2\sqrt{5}\}$　**13.** $\varnothing$　**15.** $\{-1, 7\}$
17. $\{-1 - 2\sqrt{2}, -1 + 2\sqrt{2}\}$　**19.** $\{-\sqrt{10}, \sqrt{10}\}$

21. $\{-\tfrac{3}{2}, -\tfrac{1}{2}\}$　**23.** $\{3\}$　**25.** $\{-2, 2\}$　**27.** $\{-\tfrac{2}{3}, \tfrac{2}{3}\}$
29. $\{52\}$　**31.** $\{\tfrac{9}{4}\}$　**33.** $\{9\}$　**35.** $\varnothing$　**37.** $\{3\}$
39. $\{-5, 3\}$　**41.** $\{1\}$　**43.** $\varnothing$　**45.** $\{-1\}$　**47.** $\{6\}$
49. $\{7\}$　**51.** $\{-5\}$　**53.** $\{-8, 8\}$　**55.** $\{-\tfrac{1}{27}, \tfrac{1}{27}\}$
57. $\{512\}$　**59.** $\{\tfrac{1}{81}\}$　**61.** $\{8\}$　**63.** $\{0, \tfrac{2}{3}\}$

65. $\left\{\dfrac{4 - \sqrt{2}}{4}, \dfrac{4 + \sqrt{2}}{4}\right\}$　**67.** $4\sqrt{2}$ feet　**69.** $2\sqrt{2}$ feet

71. $5\sqrt{2}$ feet　**73.** $2 \cdot \sqrt[3]{10}$ feet　**75.** 50 feet
77. $\sqrt[6]{32}$ meters　**79.** $\sqrt{73}$ km　**81.** $\{-\sqrt{2}, \sqrt{2}\}$
83. $\{-5\}$　**85.** $\{-\tfrac{1}{3}, \tfrac{1}{3}\}$　**87.** $\varnothing$　**89.** $\{-9\}$　**91.** $\{\tfrac{5}{4}\}$
93. $\{-9, 4\}$　**95.** $\{-\tfrac{2}{3}, 2\}$
97. $\{-2 - 2 \cdot \sqrt[4]{2}, -2 + 2 \cdot \sqrt[4]{2}\}$　**99.** $\{0\}$　**101.** $\{\tfrac{1}{2}\}$

103. $\left\{-\dfrac{\sqrt{3}}{3}, \dfrac{\sqrt{3}}{3}\right\}$　**105.** $\{0, 8\}$　**107.** $\{-1.8, 1.8\}$

109. $\{4.993\}$　**111.** $\{-26.372, 26.372\}$

Section 5.6　Warm-ups FFTTTTFTTF

1. $486{,}000{,}000$　**3.** $-.00162$　**5.** $4{,}132$　**7.** $1{,}000{,}000$
9. $496{,}000$　**11.** 3.2×10^5　**13.** 7.1×10^{-7}
15. -3.58×10^8　**17.** 7×10^{-5}　**19.** 2.35×10^7
21. 1.8×10^{15}　**23.** 3×10^{22}　**25.** 5×10^{15}
27. 4×10^{-9}　**29.** 4×10^5　**31.** 4.5×10^4
33. 6.4×10^{22}　**35.** 4×10^6　**37.** 2×10^5　**39.** $.75$
41. -5×10^{17}　**43.** 5×10^{-11}　**45.** 1.578×10^5
47. 9.187×10^{-5}　**49.** 5.961×10^{157}　**51.** 1.1×10^{100}
53. 3.828×10^{30}　**55.** -5.523×10^{-11}
57. 4.910×10^{11} feet　**59.** 3.83×10^7 seconds
61. 3.333×10^{-4} centimeters

Chapter 5 Review Exercises

1. $\tfrac{1}{9}$　**3.** 4　**5.** $\tfrac{1}{1000}$　**7.** $27x^{1/2}$　**9.** $a^{7/2}b^{7/2}$
11. $x^{3/4}y^{5/4}$　**13.** $6x^2 \cdot \sqrt{2x}$　**15.** $2x \cdot \sqrt[3]{9x^2}$　**17.** 8

19. $\dfrac{\sqrt{6}}{6}$　**21.** $\dfrac{\sqrt[3]{3}}{3}$　**23.** $\dfrac{3\sqrt{2}}{2}$　**25.** $\dfrac{\sqrt{15}}{3}$

27. $\dfrac{3\sqrt{2a}}{2a}$　**29.** $\dfrac{5 \cdot \sqrt[3]{9x}}{3x}$　**31.** $2xy^3 \cdot \sqrt[3]{3x}$

33. $3\sqrt{5} - 2\sqrt{3}$　**35.** $\dfrac{10\sqrt{3}}{3}$　**37.** $30 - 21\sqrt{6}$

39. $6 - 3\sqrt{3} + 2\sqrt{2} - \sqrt{6}$　**41.** $\sqrt[3]{5}$　**43.** 13

45. $\dfrac{5\sqrt{2}}{2}$　**47.** 9　**49.** $1 - \sqrt{2}$　**51.** $\dfrac{-\sqrt{6} - 3\sqrt{2}}{2}$

53. $\dfrac{3\sqrt{2} + 2}{7}$　**55.** $\{-4, 4\}$　**57.** $\{3, 7\}$

59. $\{-1 - \sqrt{5}, -1 + \sqrt{5}\}$　**61.** $\varnothing$　**63.** $\{10\}$　**65.** $\{9\}$
67. $\{-8, 8\}$　**69.** $\{124\}$　**71.** $\{7\}$　**73.** $\{2, 3\}$　**75.** $\{9\}$
77. $\{4\}$　**79.** $5\sqrt{30}$ seconds　**81.** $10\sqrt{7}$ feet
83. $200\sqrt{2}$ feet　**85.** $8{,}360{,}000$　**87.** $-.00057$
89. 8.07×10^6　**91.** -7.09×10^{-4}　**93.** -1.5×10^{-30}

95. 1.25×10^{38} **97.** F   **99.** T   **101.** T   **103.** F
105. F   **107.** F   **109.** F   **111.** T   **113.** F   **115.** T
117. F   **119.** F   **121.** T

Chapter 5 Test

1. 4   **2.** $\frac{1}{8}$ **3.** $\sqrt{3}$ **4.** 30   **5.** $3\sqrt{5}$ **6.** $\frac{6\sqrt{5}}{5}$

7. 2   **8.** $6\sqrt{2}$ **9.** $\frac{\sqrt{15}}{6}$ **10.** $\frac{2+\sqrt{2}}{2}$

11. $4 - 3\sqrt{3}$ **12.** $3 \cdot \sqrt[3]{2}$ **13.** $2 \cdot \sqrt[4]{2}$ **14.** $\sqrt[5]{2}$

15. $\frac{\sqrt[3]{4}}{2}$ **16.** $a^4 \cdot \sqrt{a}$ **17.** x^3 **18.** $2m\sqrt{5m}$ **19.** $x^{3/4}$

20. $x^{1/4}$ **21.** $2x^2 \cdot \sqrt[3]{5x}$ **22.** $19 + 8\sqrt{3}$ **23.** $\frac{5+\sqrt{3}}{11}$

24. $\frac{6\sqrt{2}-\sqrt{3}}{23}$ **25.** $\{-5, 9\}$ **26.** $\left\{-\frac{7}{4}\right\}$ **27.** $\{-8, 8\}$

28. $\{5\}$ **29.** $\frac{3\sqrt{2}}{2}$ feet   **30.** 25 and 36

31. 4 feet and 6 feet   **32.** 3.24×10^5 **33.** 8.67×10^{-5}
34. 2.4×10^{-5} **35.** 2×10^{-13}

Tying It All Together   Chapters 1–5

1. $\left\{-\frac{4}{7}\right\}$ **2.** $\left\{\frac{3}{2}\right\}$
3. $\{x \mid x < -3 \text{ or } x > -2\}$ **4.** $\left\{\frac{3}{2}\right\}$

5. $\{x \mid x < 1\}$

6. $\varnothing$ **7.** $\{9\}$ **8.** $\varnothing$ **9.** $\{-12, -2\}$ **10.** $\left\{\frac{1}{16}\right\}$
11. $\{x \mid x > -6\}$

12. $\left\{-\frac{1}{64}, \frac{1}{64}\right\}$ **13.** $\left\{-\frac{\sqrt{3}}{3}, \frac{\sqrt{3}}{3}\right\}$ **14.** R

15. $\{x \mid -\frac{1}{3} < x < 3\}$

16. $\left\{\frac{1}{3}\right\}$ **17.** $\{82\}$ **18.** $\left\{\frac{6}{5}, \frac{12}{5}\right\}$ **19.** $\{100\}$ **20.** R
21. $\{4\sqrt{30}\}$ **22.** $\{400\}$ **23.** $\left\{\frac{13+9\sqrt{2}}{3}\right\}$
24. $\{-3\sqrt{2}, 3\sqrt{2}\}$ **25.** $\{5\}$ **26.** $\{7 + 3\sqrt{6}\}$
27. $\{-2, 3\}$ **28.** $\{-5, 2\}$ **29.** 3   **30.** -2 **31.** $\frac{1}{2}$
32. $\frac{1}{3}$

CHAPTER 6

Section 6.1   Warm-ups FFFFTFTTFF
1. $\{-2, 3\}$ **3.** $\{-5, 3\}$ **5.** $\{-1, \frac{3}{2}\}$ **7.** $\{-3, 4\}$
9. $\{-7\}$ **11.** $\{-4, 4\}$ **13.** $\{-9, 9\}$ **15.** $\{-\frac{4}{3}, \frac{4}{3}\}$
17. $\{-1, 7\}$ **19.** $\{-\frac{5}{2}, \frac{5}{2}\}$ **21.** $\{-1 - \sqrt{5}, -1 + \sqrt{5}\}$

23. $\left\{\frac{3-\sqrt{7}}{2}, \frac{3+\sqrt{7}}{2}\right\}$ **25.** $\{-2, 1\}$

27. $\left\{\frac{1-3\sqrt{2}}{2}, \frac{1+3\sqrt{2}}{2}\right\}$

29. $\left\{\frac{-2-4\sqrt{2}}{3}, \frac{-2+4\sqrt{2}}{3}\right\}$ **31.** $x^2 + 2x + 1$

33. $x^2 - 6x + 9$ **35.** $w^2 - 3w + \frac{9}{4}$ **37.** $m^2 + 12m + 36$

39. $x^2 + \frac{x}{2} + \frac{1}{16}$ **41.** $x^2 + \frac{2}{3}x + \frac{1}{9}$ **43.** $\{-3, 5\}$

45. $\{-7, 1\}$ **47.** $\{-10, 2\}$ **49.** $\{-5, 7\}$ **51.** $\{-4, 5\}$
53. $\{-7, 2\}$ **55.** $\{-1, \frac{3}{2}\}$ **57.** $\{-2, \frac{1}{3}\}$
59. $\{-2 - \sqrt{10}, -2 + \sqrt{10}\}$
61. $\{-4 - 2\sqrt{5}, -4 + 2\sqrt{5}\}$

63. $\left\{\frac{-3-\sqrt{41}}{4}, \frac{-3+\sqrt{41}}{4}\right\}$

65. $\left\{\frac{-2-\sqrt{19}}{5}, \frac{-2+\sqrt{19}}{5}\right\}$ **67.** $\{-2, 3\}$

69. $\left\{\frac{1+\sqrt{17}}{8}\right\}$ **71.** $\left\{\frac{3}{4}, \frac{9}{4}\right\}$ **73.** $\{1, 6\}$

75. $\{-2 - \sqrt{2}, -2 + \sqrt{2}\}$ **77.** $\left\{\frac{9-\sqrt{65}}{2}, \frac{9+\sqrt{65}}{2}\right\}$

79. $\{-3 - \sqrt{9-k}, -3 + \sqrt{9-k}\}$

Section 6.2   Warm-ups TFTFTFTTFT
1. $\{-3, -2\}$ **3.** $\{-3, 2\}$ **5.** $\{-\frac{1}{3}, \frac{3}{2}\}$ **7.** $\{\frac{1}{2}\}$ **9.** $\varnothing$

11. $\left\{\frac{-5 \pm \sqrt{29}}{2}\right\}$ **13.** $\{-4 \pm \sqrt{10}\}$ **15.** $\left\{\frac{3 \pm \sqrt{7}}{2}\right\}$

17. $\{-2, \frac{1}{2}\}$ **19.** 28, 2   **21.** $-23, 0$ **23.** 0, 1
25. $-\frac{3}{4}, 0$ **27.** 97, 2   **29.** 0, 1

31. $\frac{1+\sqrt{65}}{2}, \frac{-1+\sqrt{65}}{2}$ **33.** $3 + \sqrt{5}, 3 - \sqrt{5}$

35. $L = \frac{1+\sqrt{5}}{2}$ feet, $W = \frac{-1+\sqrt{5}}{2}$ feet

37. $L = 2 + \sqrt{14}$ feet, $W = -2 + \sqrt{14}$ feet
39. 3 seconds   **41.** 7.02 seconds   **43.** $\{-4.474, 1.274\}$
45. $\varnothing$ **47.** $\{3.7\}$ **49.** $\{-2.979, -.653\}$
51. $\{-4792.983, -.017\}$ **53.** $\{-.079, .078\}$

Section 6.3   Warm-ups TFFTFFFTFF
1. $\{\pm\sqrt{5}, \pm3\}$ **3.** $\{-2, 1\}$ **5.** $\{-1 \pm \sqrt{5}, -3, 1\}$

7. $\{-\frac{3}{2}, \frac{3}{2}\}$ **9.** $\left\{\frac{-3 \pm \sqrt{5}}{2}\right\}$ **11.** $\{16, 81\}$ **13.** $\{9\}$

15. $\{64\}$ **17.** $\left\{\frac{2}{3}, \frac{3}{2}\right\}$ **19.** $\left\{\pm\frac{\sqrt{14}}{2}, \pm\frac{\sqrt{38}}{2}\right\}$ **21.** 2 P.M.

23. $-5 + \sqrt{265} \approx 11.3$ mph before,
$-9 + \sqrt{265} \approx 7.3$ mph after

25. $\dfrac{13 + \sqrt{265}}{2} \approx 14.6$ hours for Andrew,

$\dfrac{19 + \sqrt{265}}{2} \approx 17.6$ hours for John

27. $L = 5 + 5\sqrt{41} \approx 37.02$ feet,
$W = -5 + 5\sqrt{41} \approx 27.02$ feet

Section 6.4   Warm-ups FFFFTTTTTF

1. $\{x \mid x < -2 \text{ or } x > 2\}$

3. $\{x \mid -3 \le x \le 3\}$

5. $\{x \mid -3 < x < 2\}$

7. $\{x \mid x \le -4 \text{ or } x \ge \frac{3}{2}\}$

9. $\{x \mid -4 < x < 4\}$

11. $\{x \mid x \le 0 \text{ or } x \ge 4\}$

13. $\{x \mid x < 0 \text{ or } x > \frac{1}{2}\}$

15. $\{x \mid x < -1 \text{ or } x > 0\}$

17. $\{x \mid -2 < x < 4\}$

19. $\{x \mid -\frac{3}{2} < x < \frac{5}{3}\}$

21. $\{x \mid x \le -2 \text{ or } x \ge 6\}$

23. $\varnothing$

25. $\{x \mid x < -3 \text{ or } x > 5\}$

27. $\{x \mid -4 \le x \le 6\}$

29. $\{x \mid -1 \le x \le 2 \text{ or } x \ge 5\}$

31. $\{x \mid x < -3 \text{ or } -1 < x < 1\}$

33. $\{x \mid x < 0 \text{ or } x > 3\}$

35. $\{x \mid -2 \le x < 0\}$

37. $\{x \mid x < -6 \text{ or } x > 3\}$

39. $\{x \mid x < -2 \text{ or } x > -1\}$

41. $\{x \mid -2 < x < 1 \text{ or } x > 7\}$

43. $\{x \mid x < -5 \text{ or } 1 < x < 3 \text{ or } x > 5\}$

45. $\{x \mid -6 \le x < 3 \text{ or } 4 \le x < 6\}$

47. $\{x \mid x < 1 - \sqrt{5} \text{ or } x > 1 + \sqrt{5}\}$

49. $\left\{x \mid x \le \dfrac{3 - \sqrt{3}}{2} \text{ or } x \ge \dfrac{3 + \sqrt{3}}{2}\right\}$

51. $\left\{x \mid \dfrac{3 - 3\sqrt{5}}{2} \le x \le \dfrac{3 + 3\sqrt{5}}{2}\right\}$

53. $\{x \mid -27.58 < x < -.68\}$

55. $\{x \mid x < -4.4 \text{ or } -3 < x < .4 \text{ or } x > 2\}$

57. 6, 7, 8, …   **59.** 4 seconds
61. $t < .43$ seconds or $t > 1.44$ seconds

Section 6.5   Warm-ups TFFTTTTFTT

1. $-2 + 8i$ **3.** $-4 + 4i$ **5.** -2 **7.** $-8 - 2i$
9. $6 + 15i$ **11.** $-2 - 10i$ **13.** $-4 - 12i$
15. $-7 + 22i$ **17.** $-1 + 3i$ **19.** $-5i$ **21.** 29
23. $-16 - 30i$ **25.** 20   **27.** -9 **29.** 2   **31.** 5
33. 5   **35.** 52   **37.** 7   **39.** $1 + 3i$ **41.** $\frac{12}{17} - \frac{3}{17}i$
43. $\frac{4}{13} + \frac{7}{13}i$ **45.** $2 + 5i$ **47.** $-\frac{3}{2} + 2i$ **49.** $2 + 2i$
51. $5 + 6i$ **53.** $7 - i\sqrt{6}$ **55.** $5i\sqrt{2}$ **57.** $4 + i$
59. $1 + i\sqrt{3}$ **61.** $-1 - \frac{1}{2}i\sqrt{6}$ **63.** $\{\pm 3i\}$
65. $\{\pm 2i\sqrt{3}\}$ **67.** $\left\{\pm\dfrac{i\sqrt{10}}{2}\right\}$ **69.** $\{\pm i\sqrt{2}\}$ **71.** $\{\pm i\}$
73. $\left\{\dfrac{-5 \pm i\sqrt{11}}{6}\right\}$ **75.** $\left\{\dfrac{2}{5} \pm \dfrac{1}{5}i\right\}$ **77.** $\{-3 \pm i\}$
79. $\left\{\dfrac{-1 \pm 3i\sqrt{11}}{10}\right\}$ **81.** $18 - i$ **83.** $-\dfrac{6}{25} - \dfrac{17}{25}i$
85. 13   **87.** $\sqrt{3} + i\sqrt{3}$

Chapter 6 Review Exercises

1. $\{-3, 5\}$ **3.** $\{-3, \frac{5}{2}\}$ **5.** $\{-5, 5\}$ **7.** $\{\frac{3}{2}\}$
9. $\{\pm 2\sqrt{3}\}$ **11.** $\{-2, 4\}$ **13.** $\left\{\dfrac{4 \pm \sqrt{3}}{2}\right\}$ **15.** $\left\{\pm\dfrac{3}{2}\right\}$
17. $\{2, 4\}$ **19.** $\{2, 3\}$ **21.** $\{\frac{1}{2}, 3\}$ **23.** $\{-2 \pm \sqrt{3}\}$
25. $\{-2, 5\}$ **27.** $\{-\frac{1}{3}, \frac{3}{2}\}$ **29.** $\{-2 \pm \sqrt{2}\}$
31. $\left\{\dfrac{5 \pm \sqrt{13}}{6}\right\}$ **33.** $\{-2, 1\}$ **35.** $\{\pm 2, \pm 3\}$

37. $\{-6, -5, 2, 3\}$ **39.** $\{-2, 8\}$ **41.** $\{-\frac{1}{9}, \frac{1}{4}\}$
43. $\{16, 81\}$ **45.** $-2 + 2\sqrt{2} \approx .83$ and $2 + 2\sqrt{2} \approx 4.83$
47. $W = \dfrac{4 + \sqrt{706}}{2} \approx 15.3$ inches,   **49.** 2 inches
$H = \dfrac{-4 + \sqrt{706}}{2} \approx 11.3$ inches
51. $\{a \mid a < -3 \text{ or } a > 2\}$
53. $\{x \mid -4 \le x \le 5\}$
55. $\{w \mid 0 < w < 1\}$
57. $\{x \mid x < -2 \text{ or } x \ge 4\}$
59. $\{x \mid x > -3\}$
61. $\{x \mid -2 < x < -1 \text{ or } x > -\frac{1}{2}\}$
63. $5 + 25i$ **65.** $-1 + 2i$ **67.** $12 + i$ **69.** $2 + i$
71. $2 - i\sqrt{3}$ **73.** $\dfrac{5}{17} - \dfrac{14}{17}i$ **75.** $\left\{\dfrac{2 \pm i\sqrt{2}}{2}\right\}$
77. $\left\{\dfrac{3 \pm i\sqrt{15}}{4}\right\}$ **79.** $\left\{\dfrac{-1 \pm i\sqrt{5}}{3}\right\}$ **81.** $\{-3 \pm i\sqrt{7}\}$

Chapter 6 Test

1. $-7, 0$ **2.** $13, 2$ **3.** $0, 1$ **4.** $\{-3, \frac{1}{2}\}$
5. $\{-3 \pm \sqrt{3}\}$ **6.** $\{-5\}$ **7.** $\{-2, \frac{3}{2}\}$ **8.** $\{-4, 3\}$
9. $\{\pm 1, \pm 2\}$ **10.** $\{11, 27\}$
11. $\{x \mid -6 < x < 3\}$
12. $\{x \mid -1 < x < 2 \text{ or } x > 8\}$
13. $W = -1 + \sqrt{17}, L = 1 + \sqrt{17}$ **14.** $9 - 3i$
15. $-1 - 4i$ **16.** $-4 + 7i$ **17.** 13 **18.** $\frac{1}{10} - \frac{7}{10}i$
19. $\{-3 \pm i\}$ **20.** $\left\{\dfrac{1 \pm i\sqrt{11}}{6}\right\}$

Tying It All Together   Chapters 1–6

1. $\left\{\dfrac{15}{2}\right\}$ **2.** $\left\{\pm\dfrac{\sqrt{30}}{2}\right\}$ **3.** $\left\{-3, \dfrac{5}{2}\right\}$ **4.** $\left\{\dfrac{-2 \pm \sqrt{34}}{2}\right\}$
5. $\left\{-\dfrac{7}{2}, -2\right\}$ **6.** $\left\{-3, \dfrac{1}{4}, \dfrac{-11 \pm \sqrt{73}}{8}\right\}$ **7.** $\{9\}$
8. $\{-\frac{3}{2}, \frac{13}{2}\}$ **9.** $y = \frac{2}{3}x - 3$ **10.** $y = -\frac{1}{2}x + 2$
11. $y = \dfrac{-c \pm \sqrt{c^2 - 12d}}{6}$ **12.** $y = \dfrac{n \pm \sqrt{n^2 + 4mw}}{2m}$
13. $y = \frac{5}{6}x - \frac{25}{12}$ **14.** $y = -\frac{2}{3}x + \frac{17}{3}$ **15.** $\frac{4}{3}$ **16.** $-\frac{11}{7}$
17. 0 **18.** -2 **19.** $\frac{5}{4}$ **20.** $\frac{58}{5}$

CHAPTER 7

Section 7.1   Warm-ups FFFTTTFFTT

1. I
3. III
5. y-axis
7. IV
9. II
11. x-axis
13. y-axis
15. x-axis and y-axis

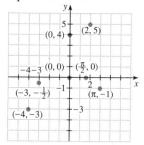

17.

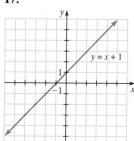

19.

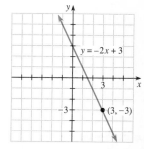

21.

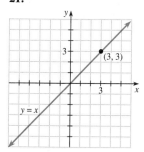

23.

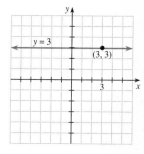

25.

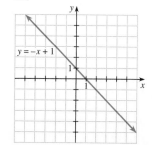

27.

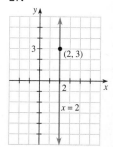

29.

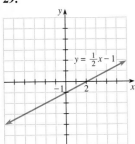

31.

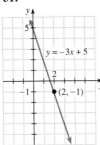

45.

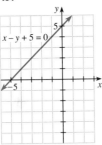

47.

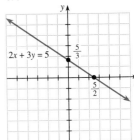

49.

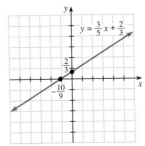

33.

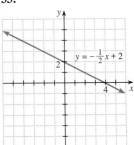

35.

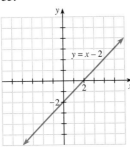

51. $\sqrt{13}$ **53.** $2\sqrt{17}$ **55.** $\sqrt{65}$ **57.** 7   **59.** 5

61. 10   **63.** $2\sqrt{5}$ **65.** 11.046   **67.** 1.862   **69.** $\dfrac{\sqrt{13}}{12}$

71. Yes   **73.** No   **75.** Yes   **77.** Yes   **79.** Yes **81.** Yes

83. $4.45   **85.** $146

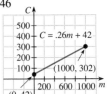

37.

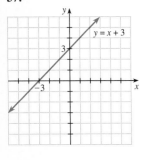

39.

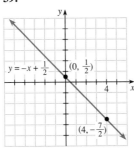

Section 7.2   Warm-ups TFFTFFFTFF

1. $\frac{2}{3}$ **3.** Undefined   **5.** -1 **7.** $\frac{3}{2}$ **9.** -1 **11.** $-\frac{5}{3}$

13. $\frac{4}{7}$ **15.** 5   **17.** -1 **19.** $-\frac{3}{5}$ **21.** $-\frac{2}{5}$ **23.** -3

25. 0   **27.** Undefined   **29.** .169   **31.** -2.828 **33.** -1.273

35. $-\frac{5}{4}$ **37.** $\frac{8}{3}$

41.

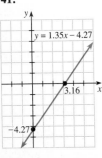

43.

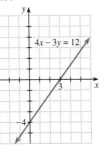

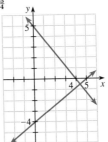

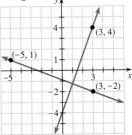

39. $\frac{3}{7}$

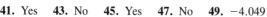

41. Yes   **43.** No   **45.** Yes   **47.** No   **49.** -4.049
51. Since the slopes are $\frac{1}{2}$ and -2, the diagonals are perpendicular.

Section 7.3   Warm-ups TFFTTFTFFT

1. $y = -2x + 1$ **3.** $y = -\frac{1}{2}x + 2$ **5.** $y = -\frac{3}{7}x + \frac{27}{7}$
7. $y = 4x - 8$ **9.** $y = -\frac{3}{4}x$ **11.** $y = \frac{1}{2}x + 2$
13. $x = 2$ **15.** $y = -x$ **17.** $y = \frac{3}{2}x - 3$
19. $y = -x + 2$ **21.** $x - 3y = 6$ **23.** $x - 2y = -13$
25. $2x - 6y = 11$ **27.** $5x + 6y = 890$
29. $x + 50y = -2150$ **31.** $y = -\frac{2}{5}x + \frac{1}{5}$, $-\frac{2}{5}$, $(0, \frac{1}{5})$
33. $y = 3x - 2$, 3, $(0, -2)$ **35.** $y = \frac{3}{2}x - 2$, $\frac{3}{2}$, $(0, -2)$
37. $y = 3x - 1$, 3, $(0, -1)$ **39.** $y = \frac{1}{2}x - 6$, $\frac{1}{2}$, $(0, -6)$
41. $y = \frac{3}{2}x + 11$, $\frac{3}{2}$, $(0, 11)$ **43.** $y = \frac{1}{3}x + \frac{7}{12}$, $\frac{1}{3}$, $(0, \frac{7}{12})$
45. $y = .01x + 6057$, $.01$, $(0, 6057)$ **47.** $y = 2$, 0, $(0, 2)$
49. $y = \frac{3}{2}x - 50$, $\frac{3}{2}$, $(0, -50)$ **51.** $x - 2y = -10$
53. $2x + y = 4$ **55.** $2x - y = -5$ **57.** $x + 2y = 7$
59. $2x + y = 3$ **61.** $3x - y = -9$ **63.** $y = 5$
65. $x = -3$

67.

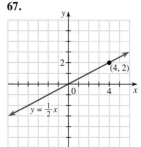

69.

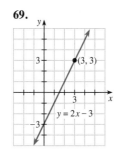

71.

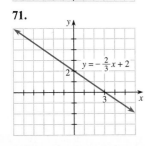

73.

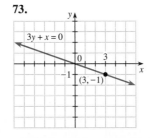

75.

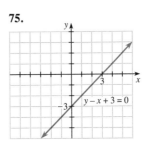

77.

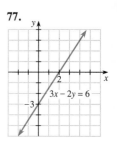

79.

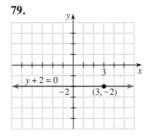

81.

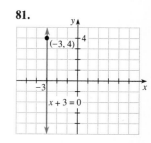

83. $t = \frac{7}{6}s + 60$, $95°F$ **85.** $P = 5t - 9696$, 304

Section 7.4   Warm-ups TFTFTTFTFT

1.

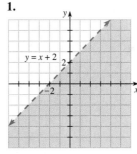

3.

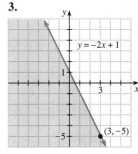

5.

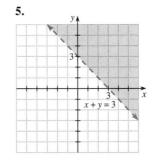

7.

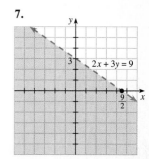

9.

11.

25.

27.

13.

15.

29.

31.

17.

19.

33.

35.

21.

23.

37.

39.

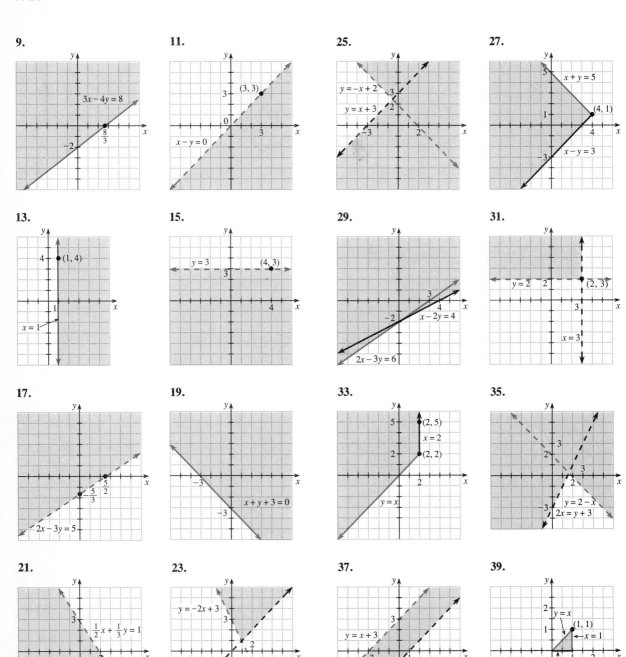

41.

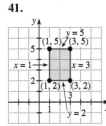

43.

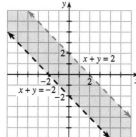

57.

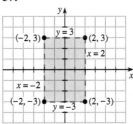

59.

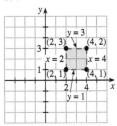

45.

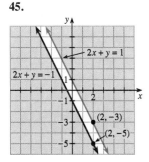

47.

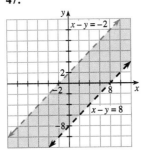

Chapter 7 Review Exercises

1. III   **3.** x-axis   **5.** y-axis   **7.** IV   **9.** $\sqrt{5}$ **11.** 10
13. $(0, 2)$, $(\frac{2}{3}, 0)$, $(4, -10)$, $(\frac{5}{3}, -3)$ **15.** 1   **17.** $\frac{3}{7}$
19. -3, $(0, 4)$ **21.** $\frac{2}{3}$, $(0, \frac{7}{3})$ **23.** $2x - 3y = 12$
25. $x - 2y = -5$ **27.** $x - 2y = 7$ **29.** $3x + 4y = 18$
31. $y = 5$

33.

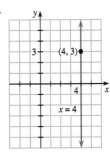

35.

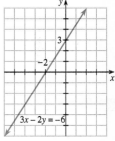

37.

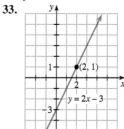

39.

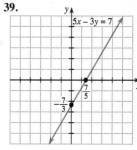

49.

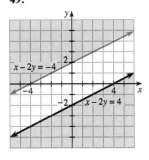

51.

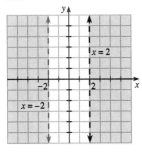

41.

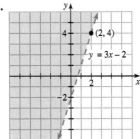

43.

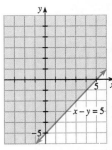

53.

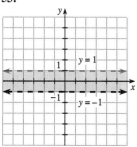

55.

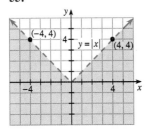

45.

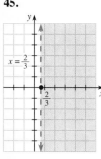

47.

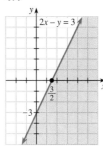

61. $3x - y = 6$ **63.** $3x - y = -6$ **65.** $y = 5$
67. $x + 2y = 7$ **69.** $x = 2$ **71.** $y = 0$ **73.** $2x - y = -6$
75. Because the slopes of the sides are $\frac{1}{3}$, $\frac{1}{3}$, 2, and 2, it is a parallelogram.
77. Because two sides have length $2\sqrt{17}$, it is an isosceles triangle.
79. Because all sides have length 4, it is an equilateral triangle.
81. $4\sqrt{17} + 2\sqrt{34}$ **83.** 62 days

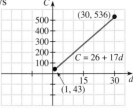

49.

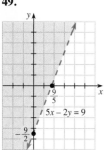

51.

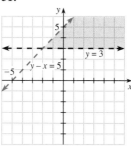

Chapter 7 Test

1. $(0, 5)$, $(\frac{5}{2}, 0)$, $(4, -3)$, $(\frac{13}{2}, -8)$ **2.** $2\sqrt{5}$ **3.** 5
4. $-\frac{6}{5}$ **5.** 2   **6.** $-\frac{1}{2}$, $(0, -2)$ **7.** $\frac{8}{5}$, $(0, 2)$
8. $x + 2y = 6$ **9.** $4x + y = -7$ **10.** $x - y = 3$
11. $5x + 3y = 19$ **12.** $y = 6$ **13.** $x - 2y = -4$
14. Because two sides have slopes -3 and $\frac{1}{3}$, it is a right triangle.
15. Because opposite sides have equal slopes, it is a parallelogram.

53.

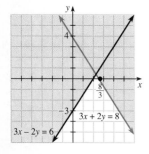

55.

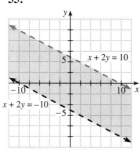

16.

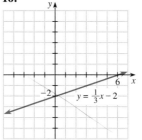

17.

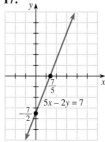

57.

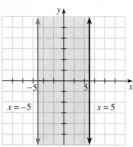

59.

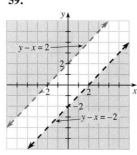

18.

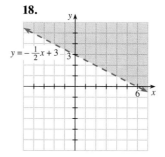

19.

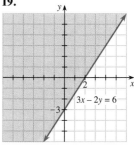

20.

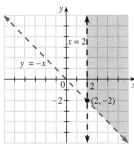

21.

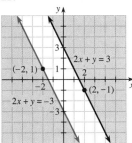

22. $V = -2000a + 22,000$

Tying It All Together Chapters 1–7
1. 125 **2.** 5 **3.** 18 **4.** 7 **5.** 3 **6.** 4 **7.** $2x + h$
8. $6x + 3h - 2$

9.

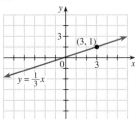

10.

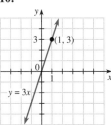

11.

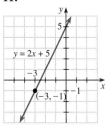

12.

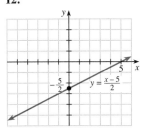

13. $\{7\}$ **14.** $\{-1\}$ **15.** $\{\frac{3}{2}\}$ **16.** $\{4\}$ **17.** $\varnothing$ **18.** $\{2, 6\}$

19.

20.

21.

22.

23.

24.

25. $\{x \mid x \neq -3\}$ **26.** $\{x \mid x \geq 0\}$ **27.** R
28. $\{x \mid x \neq 1\}$ **29.** $\{x \mid x \neq 1 \text{ and } x \neq 9\}$ **30.** R

CHAPTER 8

Section 8.1   Warm-ups FTTFFFTTFT
1. Yes **3.** Yes **5.** No **7.** Yes **9.** Yes **11.** No
13. Yes **15.** Yes **17.** Yes **19.** Yes **21.** Yes
23. No **25.** $\{2\}, \{3, 5, 7\}$ **27.** $R, \{y \mid y \geq 0\}$
29. $\{x \mid x \geq 0\}, \{y \mid y \geq 0\}$ **31.** $\{x \mid x \geq 8\}, \{y \mid y \geq 2\}$
33. 10 **35.** 12 **37.** 1 **39.** $\frac{7}{3}$ **41.** 1 or 2
43. -5 or 1 **45.** $3a - 2$ **47.** $b^2 + b$
49. $x^2 + 2xh + h^2 - 3x - 3h + 2$ **51.** $A = s^2$
53. $C = 3.98y$ **55.** $C = .50n + 14.95$ **57.** $h = 4t + 70$
59. 2.337 **61.** 66.75 **63.** 29.36 **65.** .279 or 2.387
67. 6 **69.** 3 **71.** $2a + 1$ **73.** $2x + h - 1$

Section 8.2   Warm-ups TTTFFFTTFT
1. Domain = R
   Range = R
3. Domain = R
   Range = R

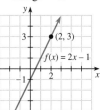

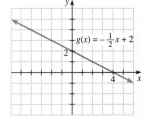

5. Domain = R
   Range = $\{3\}$
7. Domain = R
   Range = R

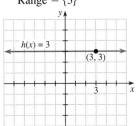

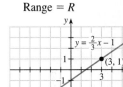

9. Domain = R
   Range = R
11. Domain = R
   Range $\{y \mid y \geq 0\}$

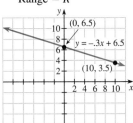

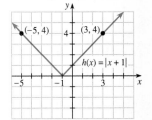

13. Domain = R
Range $\{y \mid y \geq 0\}$

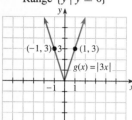

15. Domain = R
Range $\{y \mid y \geq 0\}$

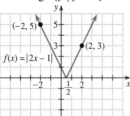

29. Domain = R
Range $\{y \mid y \leq 1\}$

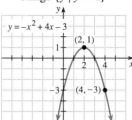

31. Domain $\{x \mid x \geq 1\}$
Range $\{y \mid y \geq 0\}$

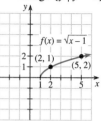

17. Domain = R
Range $\{y \mid y \geq 1\}$

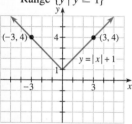

19. Domain = R
Range $\{y \mid y \geq 2\}$

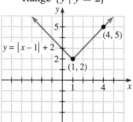

33. Domain $\{x \mid x \geq 0\}$
Range $\{y \mid y \geq 0\}$

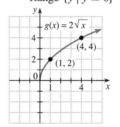

35. Domain $\{x \mid x \geq 0\}$
Range $\{y \mid y \leq 0\}$

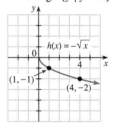

21. Domain = R
Range $\{y \mid y \geq 2\}$

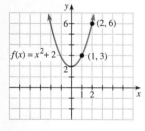

23. Domain = R
Range $\{y \mid y \geq -\frac{9}{4}\}$

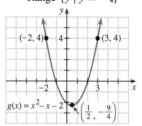

37. Domain $\{x \mid x \geq 0\}$
Range $\{y \mid y \leq 2\}$

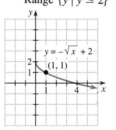

39. Domain $\{x \mid x \geq 0\}$
Range = R

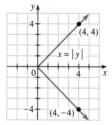

25. Domain = R
Range $\{y \mid y \geq -9\}$

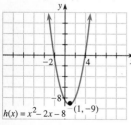

27. Domain = R
Range $\{y \mid y \leq 1\}$

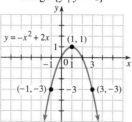

41. Domain $\{x \mid x \leq 0\}$
Range = R

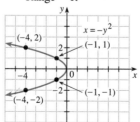

43. Domain $\{x \mid x \geq -9\}$
Range = R

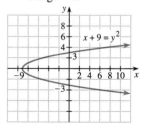

45. Domain $\{x \mid x \geq 0\}$
Range $\{y \mid y \geq 0\}$

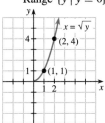

47. Domain $\{x \mid x \geq 0\}$
Range $= R$

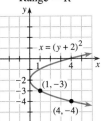

65. Domain $= R$
Range $\{y \mid y \geq 0\}$

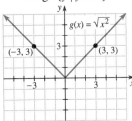

67. Domain $= R$
Range $= R$

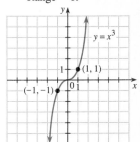

49. Domain $\{5\}$
Range $= R$

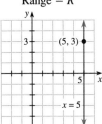

51. no   **53.** Yes   **55.** No

69. Domain $\{x \mid x \neq 0\}$
Range $\{y \mid y \neq 0\}$

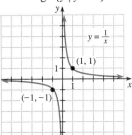

71. Domain $= R$
Range $= R$

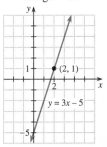

57. 64 feet

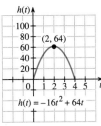

59. $62.50

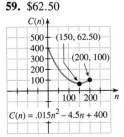

Section 8.3   Warm-ups TTFFFFTTTT

1. $x^2 + 2x - 3$ **3.** $4x^3 - 11x^2 + 6x$ **5.** 12   **7.** -30
9. -21 **11.** $\frac{13}{8}$ **13.** $y = 6x - 20$ **15.** $y = x + 2$
17. $y = x^2 + 2x$ **19.** $y = x$ **21.** -2 **23.** 5   **25.** 7
27. 5   **29.** 5   **31.** 4   **33.** 22.2992   **35.** $4x^2 - 6x$
37. $2x^2 + 6x - 3$ **39.** $4x - 9$ **41.** x **43.** $F = f \circ h$
45. $G = g \circ h$ **47.** $H = h \circ g$ **49.** $J = h \circ h$ **51.** $K = g \circ g$
61. $P(x) = -x^2 + 20x - 170$ **63.** $J = .025I$

Section 8.4   Warm-ups FFFTTFTTTF

1. No   **3.** $\{(4, 16), (3, 9), (0, 0)\}$ **5.** No
7. $\{(0, 0), (2, 2), (9, 9)\}$ **9.** Yes   **11.** No   **13.** Yes

15. Yes   **17.** $f^{-1}(x) = \frac{x}{5}$ **19.** $g^{-1}(x) = x + 9$

21. $k^{-1}(x) = \frac{x + 9}{5}$ **23.** $m^{-1}(x) = \frac{2}{x}$ **25.** $n^{-1}(x) = x$

27. $p^{-1}(x) = x^4$ for $x \geq 0$ **29.** $u^{-1}(x) = -x + 1$

31. $f^{-1}(x) = \frac{x + 9}{3}$ **33.** $f^{-1}(x) = x^3 + 4$

35. $f^{-1}(x) = \frac{x^3 - 7}{3}$ **37.** $f^{-1}(x) = \frac{2x + 1}{x - 1}$

61. Domain $= R$
Range $\{y \mid y \leq 1\}$

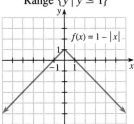

63. Domain $= R$
Range $\{y \mid y \geq 0\}$

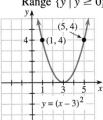

39. $f^{-1}(x) = \dfrac{6}{x}$ **41.** $f^{-1}(x) = \dfrac{3}{x} + 4$

43. $f^{-1}(x) = \sqrt{x-3}$ **45.** $f^{-1}(x) = x^2 - 2,\ x \ge 0$

47. $f^{-1}(x) = \dfrac{x-3}{2}$ **49.** $f^{-1}(x) = \sqrt{x+1}$

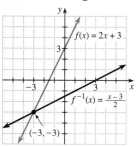 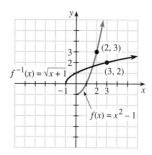

51. $f^{-1}(x) = \dfrac{x}{5}$ **53.** $f^{-1}(x) = \sqrt[3]{x}$

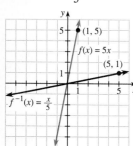 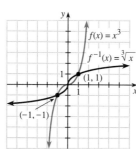

55. $f^{-1}(x) = x^2 + 2,\ x \ge 0$

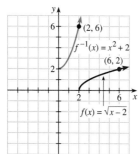

57. $(f^{-1}\!\circ\! f)(x) = x$ **59.** $(f^{-1}\!\circ\! f)(x) = x$ **61.** $(f^{-1}\!\circ\! f)(x) = x$

63. $(f^{-1}\!\circ\! f)(x) = x$

Section 8.5   Warm-ups TFFTFTFTFT

1. $a = km$ **3.** $d = \dfrac{k}{e}$ **5.** $I = krt$ **7.** $m = kp^2$

9. $B = k \cdot \sqrt[3]{w}$ **11.** $t = \dfrac{k}{x^2}$ **13.** $v = \dfrac{km}{n}$ **15.** $y = \dfrac{3}{2}x$

17. $A = \dfrac{30}{B}$ **19.** $m = \dfrac{36}{\sqrt{p}}$ **21.** $A = \dfrac{2}{5}tu$

23. $y = \dfrac{1.067x}{z}$ **25.** $y = -\dfrac{21}{5}$ **27.** $-\dfrac{3}{2}$ **29.** $\dfrac{1}{2}$

31. $D = 3.293$ **33.** \$420   **35.** 9 cm^3 **37.** \$360

39. \$18.90   **41.** 4 feet   **43.** 400 pounds

Section 8.6   Warm-ups FFTTTTTTTT

1. $x^2 - 3x,\ -3$ **3.** $2x - 6,\ 11$

5. $3x^3 + 9x^2 + 12x + 43,\ 120$

7. $x^4 + x^3 + x^2 + x + 1,\ 0$ **9.** $x^2 - 2x - 1,\ 8$

11. $2.3x + .596,\ .79072$ **13.** $x - 6 + \dfrac{21}{x+2}$

15. $2 + \dfrac{-15}{x+6}$ **17.** $3x^2 - x - 1 + \dfrac{6}{x-1}$ **19.** Yes

21. Yes   **23.** No   **25.** No   **27.** No   **29.** Yes

31. Yes   **33.** Yes, $(x - 3)(x^2 + 3x + 3)$

35. Yes, $(x + 5)(x + 3)(x + 1)$ **37.** No

39. Yes, $(x + 1)(x - 2)(x^2 + 2x + 4)$

41. Yes, $(2x - 1)(x - 3)(x + 2)$ **43.** $\{-4, 1, 3\}$

45. $\{-\frac{1}{2}, 2, 3\}$ **47.** $\{-4, -\frac{1}{2}, 6\}$ **49.** $\{-3, 1\}$

51. $\{-1, 1, 2\}$

Chapter 8 Review Exercises

1. No   **3.** Yes   **5.** No   **7.** $\{1, 2\}, \{3\}$ **9.** R, R

11. $R, \{y \mid y \ge 0\}$ **13.** -5 **15.** -6 **17.** $-\frac{21}{4}$

19. $2a - 5$ **21.** $a^2 - a - 6$ **23.** 3

25. R, R **27.** $R, \{y \mid y \ge -2\}$

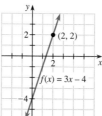 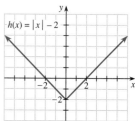

29. $R, \{y \mid y \ge 0\}$ **31.** $\{x \mid x \ge 0\}, \{y \mid y \ge 2\}$

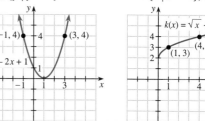

33. $\{x \mid x \neq 0\}, \{y \mid y \neq 0\}$ **35.** $\{2\}, R$

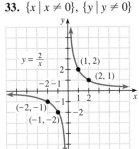

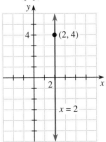

37. $\{x \mid x \geq 1\}, R$ **39.** -4 **41.** $\sqrt{2}$ **43.** 99 **45.** 17

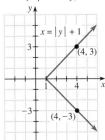

47. $3x^3 - x^2 - 10x$ **49.** 20 **51.** $F = f \circ g$ **53.** $H = g \circ h$

55. $I = g \circ g$ **57.** No   **59.** $f^{-1}(x) = \dfrac{x}{8}$

61. $h^{-1}(x) = x^3 + 6$ **63.** $j^{-1}(x) = \dfrac{x+1}{x-1}$ **65.** No

67. $f^{-1}(x) = \dfrac{x+1}{3}$ **69.** $f^{-1}(x) = \sqrt[3]{2x}$

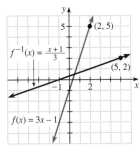

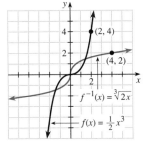

71. 24 **73.** -16 **75.** $x^2 - 2x - 2, -18$
77. $x^3 - x^2 + x - 5, 7$ **79.** $2, 5, -3$ **81.** $-2, 3, 4$
83. $C = \frac{5}{2}n + 800$ **85.** $a = 15w - 16$ **87.** $(\frac{3}{2}, -\frac{1}{4})$
89. $6x - 4 + 3h$

Chapter 8 Test
1. $R, \{y \mid y \geq 0\}$ **2.** $\{x \mid x \geq 0\}, \{y \mid y \geq 0\}$ **3.** 11

4. $a^2 + 2a + 5$ **5.** $4x^2 - 20x + 29$ **6.** $f^{-1}(x) = \dfrac{x-5}{-2}$

7. 9 **8.** 15 **9.** -1782 **10.** $\pm\sqrt{5}$
11.   **12.**

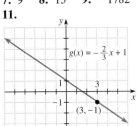

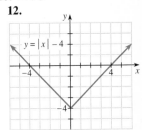

13.   **14.**

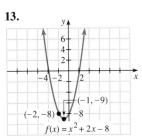

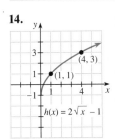

15. $f^{-1}(x) = \dfrac{x+1}{x-2}$ **16.** $\dfrac{32\pi}{3}$ cubic feet

17. $S = .50n + 3$ **18.** $\frac{36}{7}$

19.

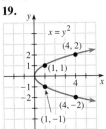

20. $\frac{5}{4}$ seconds   **21.** $2x^3 - 7x^2 + 10x - 4, -8$ **22.** $\{3, 4, 5\}$

Tying It All Together   Chapters 1–8
1. $\frac{1}{25}$ **2.** $\frac{3}{2}$ **3.** $\sqrt{2}$ **4.** x^8 **5.** 2 **6.** x^9 **7.** $\{\pm 3\}$
8. $\{2\}$ **9.** $\{\frac{1}{3}\}$ **10.** $\{81\}$ **11.** $\varnothing$ **12.** $\{2\}$ **13.** $\{\frac{1}{3}\}$
14. $\{\pm 8\}$ **15.** $\{\frac{1}{4}\}$ **16.** $\{\pm 2\sqrt{2}\}$ **17.** $\{2 \pm \sqrt{10}\}$
18. $\{-3, \frac{7}{2}\}$ **19.** $\{-\frac{17}{5}, 5\}$ **20.** $\{42\}$
21.   **22.**

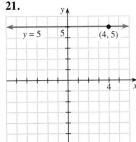

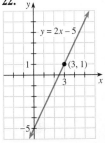

23.

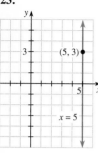

24.

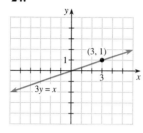

25.

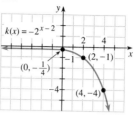

27.

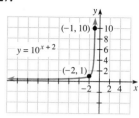

29.

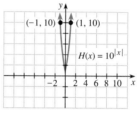

31.

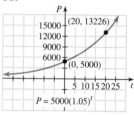

25.

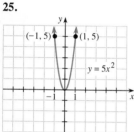

26.

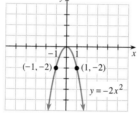

33. 2   **35.** $\frac{4}{3}$ **37.** -2 **39.** 0   **41.** $\frac{3}{2}$ **43.** $\frac{1}{2}$
45. $\{6\}$ **47.** $\{-2\}$ **49.** $\{-\frac{1}{2}\}$ **51.** $\{-2\}$ **53.** $\{\frac{9}{2}\}$
55. $\{-3, 3\}$ **57.** $\{-2\}$ **59.** \$9861.72
61. 300 grams, 1.7×10^{-55} grams   **63.** \$45, \$12.92
65. \$616.84   **67.** \$6230.73   **69.** 2884.032   **71.** 25.457
73. 7.389   **75.** 1

CHAPTER 9

Section 9.1   Warm-ups FTFTFTTFTF
1. 16   **3.** 2   **5.** $\frac{1}{16}$ **7.** 8   **9.** 3   **11.** $\frac{1}{3}$ **13.** 1
15. 100

17.

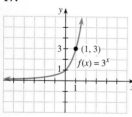

19.

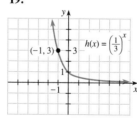

21.

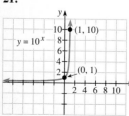

23.

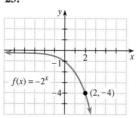

Section 9.2   Warm-ups FTFTFTTTFF
1. $2^3 = 8$ **3.** $\log(100) = 2$ **5.** $5^y = x$ **7.** $\log_2 b = a$
9. $3^{10} = x$ **11.** $\ln(x) = 3$ **13.** 2   **15.** 4   **17.** 6
19. 3   **21.** -2 **23.** 2   **25.** -2 **27.** 1   **29.** -3
31. 3   **33.** 2   **35.** .69897   **37.** 1.8307
39.

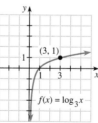

41.

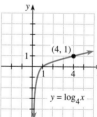

43.

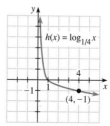

45. $f^{-1}(x) = \log_6 x$ **47.** $f^{-1}(x) = e^x$ **49.** $f^{-1}(x) = (\tfrac{1}{2})^x$
51. $\{4\}$ **53.** $\{\tfrac{1}{2}\}$ **55.** $\{.001\}$ **57.** $\{6\}$ **59.** $\{\tfrac{1}{5}\}$
61. $\{\pm 3\}$ **63.** $\{.4771\}$ **65.** $\{-.3010\}$ **67.** 5.776 years
69. 1.9269 years   **71.** 7   **73.** 11.50

Section 9.3   Warm-ups TFTTFTFFFT

1. $\log(21)$ **3.** $\log_3 5x$ **5.** $\log(x^5)$ **7.** $\ln(30)$
9. $\log(4)$ **11.** $\log_2(x^4)$ **13.** $\log(\tfrac{7}{9})$ **15.** $3\log(3)$
17. $\tfrac{1}{2}\log(3)$ **19.** $x\log(3)$ **21.** $\log(3) + \log(5)$
23. $\log(5) - \log(3)$ **25.** $2\log(5)$ **27.** $2\log(5) + \log(3)$
29. $-\log(3)$ **31.** $-\log(5)$ **33.** $\log(x) + \log(y) + \log(z)$
35. $3 + \log_2 x$ **37.** $\ln(x) - \ln(y)$ **39.** $1 + 2\log(x)$
41. $2\log_5(x - 3) - \tfrac{1}{2}\log_5 w$
43. $\ln(y) + \ln(z) + \dfrac{1}{2}\ln(x) - \ln(w)$ **45.** $\ln\!\left(\dfrac{xz}{w}\right)$
47. $\ln\!\left(\dfrac{x^2 y^3}{w}\right)$ **49.** $\log\!\left(\dfrac{(x - 3)^{1/2}}{(x + 1)^{2/3}}\right)$
51. $\log_2\!\left(\dfrac{(x - 1)^{2/3}}{(x + 2)^{1/4}}\right)$ **53.** F   **55.** T   **57.** T   **59.** F
61. T   **63.** T   **65.** T   **67.** F

Section 9.4   Warm-ups TTTFTFTFTT

1. $\{7\}$ **3.** $\{2\}$ **5.** $\{3\}$ **7.** $\varnothing$ **9.** $\{6\}$ **11.** $\{3\}$
13. $\{2\}$ **15.** $\{4\}$ **17.** $\{\log_3 7\}$ **19.** $\{-6\}$ **21.** $\{-\tfrac{1}{2}\}$
23. $\dfrac{\ln 7}{\ln 2} \approx 2.807$ **25.** $\dfrac{1}{3 - \ln 2} \approx .433$
27. $\dfrac{\ln 5}{\ln 3} \approx 1.465$ **29.** $\dfrac{\ln 9}{\ln 2} + 1 \approx 4.170$
31. $\dfrac{5 \ln 3}{\ln 2 - \ln 3} \approx -13.548$ **33.** $\dfrac{4 + 2\log 5}{1 - \log 5} \approx 17.932$
35. $\dfrac{\ln 9}{\ln 9 - \ln 8} \approx 18.655$ **37.** 1.585   **39.** $-.6309$
41. -2.202 **43.** 1.523   **45.** 41 months   **47.** 7.1 years
49. 16.8 years   **51.** 6.31×10^{-5} **53.** $\{\log_3 20\}$ **55.** $\{\tfrac{3}{5}\}$
57. $\{\tfrac{1}{2}\}$

Chapter 9 Review Exercises

1. $\tfrac{1}{25}$ **3.** 125   **5.** 1   **7.** $\tfrac{1}{10}$ **9.** 4   **11.** $\tfrac{1}{2}$ **13.** 2
15. 4   **17.** $-\tfrac{5}{2}$ **19.** 8.642   **21.** 177.828   **23.** .020
25. .141
27.

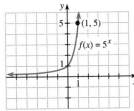

29.

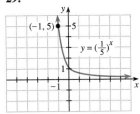

31.

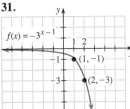

33.

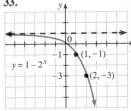

35. -3 **37.** -1 **39.** 2   **41.** 0   **43.** 256
45. 6.267   **47.** -5.083 **49.** 5.560
51. $f^{-1}(x) = \log x$ **53.** $f^{-1}(x) = \ln x$

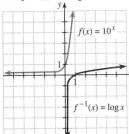

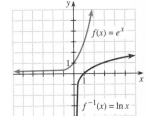

55. $2 \cdot \log x + \log y$ **57.** $4 \cdot \ln 2$ **59.** $-\log_5 x$
61. $\log\!\left(\dfrac{\sqrt{x + 2}}{(x - 1)^2}\right)$ **63.** $\{256\}$ **65.** $\{3\}$ **67.** $\{2\}$
69. $\{3\}$ **71.** $\left\{\dfrac{\ln(7)}{\ln(3) - 1}\right\}$ **73.** $\left\{\dfrac{\ln(5)}{\ln(5) - \ln(3)}\right\}$
75. $\{\tfrac{1}{3}\}$ **77.** $\{22\}$ **79.** $\{\tfrac{200}{99}\}$ **81.** $\{1.387\}$ **83.** $\{.465\}$
85. \$51,182.68   **87.** 161.5 grams   **89.** 5 years

Chapter 9 Test

1. 25   **2.** $\tfrac{1}{5}$ **3.** 1   **4.** 3   **5.** 0   **6.** -1
7.

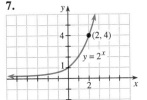

8.

9.

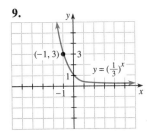

10.

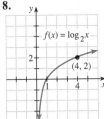

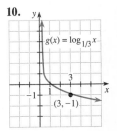

11. 10   **12.** 8   **13.** $\frac{3}{2}$ **14.** 15   **15.** -4

16. $\{\log_3 12\}$ **17.** $\{\sqrt{3}\}$ **18.** $\left\{\dfrac{\ln 8}{\ln 8 - \ln 5}\right\}$ **19.** $\{5\}$

20. $\{3\}$ **21.** $\{.537\}$ **22.** $\{20.516\}$ **23.** 10 and 147,648
24. 1.733 hours

Tying It All Together   Chapters 1–9

1. $\{3 \pm 2\sqrt{2}\}$ **2.** $\{259\}$ **3.** $\{6\}$ **4.** $\left\{\dfrac{11}{2}\right\}$ **5.** $\{-5, 11\}$

6. $\{67\}$ **7.** $\{6\}$ **8.** $\{4\}$ **9.** $\{-\frac{52}{15}\}$ **10.** $\left\{\dfrac{3 \pm \sqrt{3}}{3}\right\}$

11. $f^{-1}(x) = 3x$ **12.** $g^{-1}(x) = 3^x$ **13.** $f^{-1}(x) = \dfrac{x + 4}{2}$

14. $h^{-1}(x) = x^2$ for $x \geq 0$ **15.** $j^{-1}(x) = \dfrac{1}{x}$

16. $k^{-1}(x) = \log_5 x$ **17.** $m^{-1}(x) = 1 + \ln x$ **18.** $n^{-1}(x) = e^x$
19.

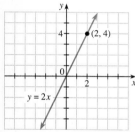

20.

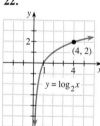

21.

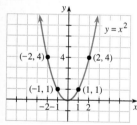

22.

23.

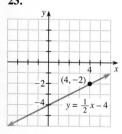

24.

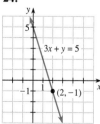

25.

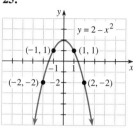

26.

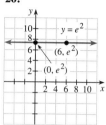

CHAPTER 10

Section 10.1   Warm-ups TFFTTTTTTT
1. $\{(1, 2)\}$ **3.** $\{(0, -1)\}$ **5.** $\{(2, -1)\}$ **7.** $\varnothing$
9. $\{(8, 3)\}$, independent   **11.** $\{(-3, 2)\}$, independent
13. $\varnothing$, inconsistent   **15.** $\{(x, y) \mid y = 2x - 5\}$, dependent
17. $\{(5, -1)\}$, independent   **19.** $\{(7, 7)\}$, independent
21. $\{(15, 25)\}$, independent   **23.** $\{(20, 10)\}$, independent
25. $\{(\frac{9}{2}, -\frac{1}{2})\}$, independent   **27.** $\varnothing$, inconsistent
29. $\{(x, y) \mid 3y - 2x = -3\}$, dependent
31. $\{(.75, 1.125)\}$, independent
33. \$14,000 at 5%, \$16,000 at 10%   **35.** -12 and 14
37. 94 toasters, 6 vacations

Section 10.2   Warm-ups TTTTFTTTFT
1. $\{(8, -1)\}$ **3.** $\{(5, -7)\}$ **5.** $\{(-1, 3)\}$
7. $\{(-1, -3)\}$ **9.** $\{(-2, -5)\}$ **11.** $\varnothing$, inconsistent
13. $\{(x, y) \mid 5x - y = 1\}$, dependent
15. $\{(\frac{5}{2}, 0)\}$, independent   **17.** $\{(12, 6)\}$ **19.** $\{(-8, 6)\}$
21. $\{(12, 7)\}$ **23.** $\{(400, 800)\}$ **25.** $\{(1.5, 1.25)\}$
27. \$1.00   **29.** 660 boys, 720 girls
31. 31 dimes, 4 nickels

Section 10.3   Warm-ups FFTFFTTTFF
1. $\{(1, 2, -1)\}$ **3.** $\{(1, 3, 2)\}$ **5.** $\{(1, -5, 3)\}$
7. $\{(-1, 2, -1)\}$ **9.** $\{(-1, -2, 4)\}$ **11.** $\{(1, 3, 5)\}$
13. $\{(3, 4, 5)\}$ **15.** $\varnothing$ **17.** $\{(x, y, z) \mid 3x - y + z = 5\}$
19. $\varnothing$ **21.** $\{(x, y, z) \mid 5x + 4y - 2z = 150\}$
23. $\{(.1, .3, 2)\}$
25. \$1500 stocks, \$4500 bonds, \$6000 mutual fund
27. \$.40 coffee, \$.35 doughnut, \$.55 tip
29. \$.95
31. 14 ounces soup, 8 ounces tuna, 2 ounces error
33. \$24,000 teaching, \$18,000 painting, \$6,000 royalties

Section 10.4 Warm-ups TFFTFTTTTT

1. $\{(2, 4), (-3, 9)\}$

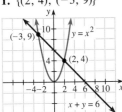

3. $\{(-2, 2), (6, 6)\}$

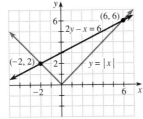

5. $\{(8, 4)\}$

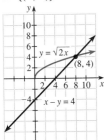

7. $\{(-\frac{3}{4}, -\frac{4}{3}), (3, \frac{1}{3})\}$

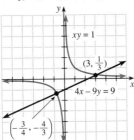

9. $\left\{\left(\frac{\sqrt{2}}{2}, \frac{1}{2}\right), \left(-\frac{\sqrt{2}}{2}, \frac{1}{2}\right)\right\}$

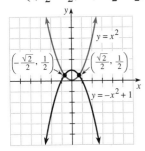

11. $\{(0, 0), (3, 9)\}$

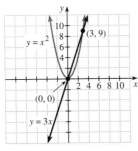

13. $\{(0, -5), (3, 4), (-3, 4)\}$ **15.** $\{(4, 5), (-2, -1)\}$
17. $\{(-3, -\frac{5}{3})\}$ **19.** $\{(\frac{2}{3}, -\frac{3}{3})\}$ **21.** $\{(-\frac{5}{3}, \frac{36}{5}), (2, 5)\}$
23. $\{(\sqrt{5}, \sqrt{3}), (\sqrt{5}, -\sqrt{3}), (-\sqrt{5}, \sqrt{3}), (-\sqrt{5}, -\sqrt{3})\}$
25. $\{(2, 5), (19, -12)\}$
27. $\{(\sqrt{2}, 2), (-\sqrt{2}, 2), (1, 1), (-1, 1)\}$
29. $\sqrt{3}$ feet and $2\sqrt{3}$ feet
31. $20\sqrt{10}$ inches base, $5\sqrt{10}$ inches height
33. 24 hours for A alone, 8 hours for B alone
35. 40 minutes **37.** 8 feet by 9 feet **39.** $4 - 2i$ and $4 + 2i$
41. 8 feet side of square, 2 feet height of triangle
43. $\{(3, 1)\}$ **45.** $\varnothing$ **47.** $\{(-6, 4^{-7})\}$ **49.** $\{(-\frac{1}{2}, \sqrt{3})\}$
51. $\{(-13.5, .00008)\}$

53. $\{(2, 3), (-2, -3), (3, 2), (-3, -2)\}$
55. $\left\{\left(2\sqrt{2}, \frac{\sqrt{2}}{2}\right), \left(-2\sqrt{2}, -\frac{\sqrt{2}}{2}\right), (1, 2), (-1, -2)\right\}$

Section 10.5 Warm-ups TFFTTTTTFF

1. -1 **3.** -3 **5.** -14 **7.** $.4$ **9.** $\{(1, -3)\}$
11. $\{(1, 1)\}$ **13.** $\{(\frac{23}{13}, \frac{9}{13})\}$ **15.** $\{(10, 15)\}$
17. $\{(20, 12)\}$ **19.** $\varnothing$ **21.** $\{(\frac{14}{3}, \frac{2}{3})\}$
23. $\{(x, y) \mid 4x - y = 6\}$ **25.** $\{(2, 3), (-2, 3)\}$
27. $\{(0, -5), (3, 4), (-3, 4)\}$ **29.** $\{(x, y) \mid y = 3x - 12\}$
31. $\{(2, -5)\}$ **33.** $\{(480, 24)\}$ **35.** $\{(\sqrt{2}, 2\sqrt{2})\}$
37. 9 servings peas, 11 servings beets **39.** 37
41. \$2.40 milk, \$2.25 magazine
43. 12 singles, 10 doubles **45.** Gary 39, Harry 34
47. 10 feet side of square, $\frac{40}{3}$ feet side of triangle
49. 10 gallons 10% solution, 20 gallons 25% solution

Section 10.6 Warm-ups TFTFFTFTFF

1. $\begin{bmatrix} -3 & 7 \\ 1 & -6 \end{bmatrix}$ **3.** $\begin{bmatrix} 4 & -3 \\ 0 & 1 \end{bmatrix}$ **5.** $\begin{bmatrix} 3 & -2 \\ 0 & 1 \end{bmatrix}$

7. $\begin{bmatrix} 3 & 5 \\ 4 & 7 \end{bmatrix}$ **9.** -7 **11.** -1 **13.** 9 **15.** 5 **17.** 22

19. 6 **21.** 70 **23.** 25 **25.** $\{(1, 2, 3)\}$
27. $\{(-1, 1, 2)\}$ **29.** $\{(-3, 2, 1)\}$ **31.** $\{(\frac{3}{2}, \frac{1}{2}, 2)\}$
33. $\{(0, 1, -1)\}$ **35.** $\{(x, y, z) \mid 2x - y + z = 1\}$ **37.** $\varnothing$
39. $\{(1, 3, -6)\}$
41. 36 pounds Mimi, 32 pounds Mitzi, 107 pounds
Cassandra
43. $39°, 51°, 90°$

Chapter 10 Review Exercises

1. $\{(1, 1)\}$, independent **3.** $\{(1, -1)\}$, independent
5. $\{(-3, 2)\}$, independent **7.** $\varnothing$, inconsistent
9. $\{(-1, 5)\}$, independent
11. $\{(x, y) \mid 3x - 2y = 12\}$, dependent **13.** $\{(1, -3, 2)\}$
15. $\varnothing$ **17.** $\{(3, 9), (-5, 25)\}$

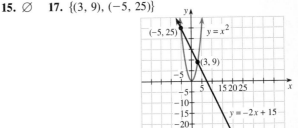

19. $\left\{\left(\dfrac{\sqrt{3}}{3}, \sqrt{3}\right), \left(-\dfrac{\sqrt{3}}{3}, -\sqrt{3}\right)\right\}$

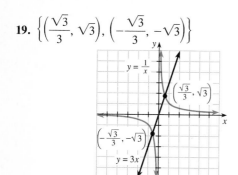

21. $\{(\sqrt{3}, 1), (-\sqrt{3}, 1)\}$ **23.** $\{(-5, -3), (3, 5)\}$
25. $\{(5, \log 2)\}$ **27.** $\{(2, 4), (-2, 4)\}$ **29.** 2   **31.** $-.2$
33. $\{(-1, -2)\}$ **35.** $\{(2, 1)\}$ **37.** $\varnothing$ **39.** 58
41. -12 **43.** $\{(1, 2, -3)\}$ **45.** $\{(-1, 2, 4)\}$ **47.** 78
49. $L = 6$ feet, $W = 2$ feet   **51.** $\frac{7}{2}$ inches and $\frac{3}{2}$ inches
53. 16, 12   **55.** \$16,000 at 10%, \$4,000 at 8%

Chapter 10 Test

1. $\{(1, 3)\}$ **2.** $\{(\frac{5}{2}, -3)\}$ **3.** $\{(x, y) \mid y = x - 5\}$
4. $\{(-1, 3)\}$ **5.** $\varnothing$ **6.** Inconsistent   **7.** Dependent
8. Independent
9. $\{(3, 4), (-3, 4), (3\sqrt{2}, -5), (-3\sqrt{2}, -5)\}$
10. $\{(1, -2, -3)\}$ **11.** -18 **12.** 12   **13.** $\{(-1, 2)\}$
14. $\{(2, -2, 1)\}$ **15.** $L = 12$ feet, $W = 9$ feet
16. Singles \$18, doubles \$25
17. Jill 17 hours, Karen 14 hours, Betsy 62 hours

Tying It All Together   Chapters 1–10

1. $\{2\}$ **2.** $\{-\frac{3}{2}, \frac{9}{2}\}$ **3.** $\{4\}$ **4.** $\{\frac{3}{4}\}$ **5.** $\{-3, 3\}$
6. $\left\{\pm\dfrac{\sqrt{3}}{2}\right\}$ **7.** $\{-8, 8\}$ **8.** $\{-4, -2\}$
9. $\{-4 \pm 3\sqrt{2}\}$ **10.** $\{-4, 2\}$ **11.** $\{-\frac{3}{2}\}$
12. $\{\pm3, \pm1\}$ **13.** $\{14\}$ **14.** $\{1 + \log_2 6\}$ **15.** $\{2\}$
16. $\{32\}$
17.

18.

19.

20.

21.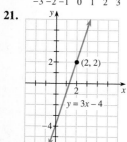

22.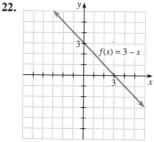

23.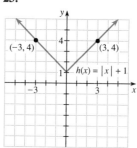

24.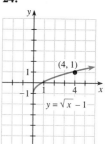

25.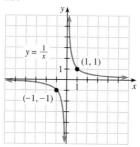

26.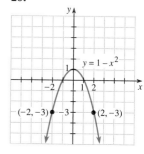

27.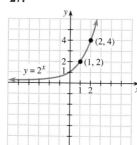

28.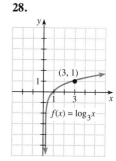

29.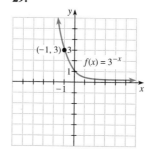

30.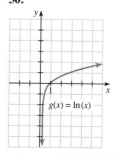

31. $11 + 6\sqrt{2}$ **32.** $-\sqrt{2}$ **33.** $3\sqrt{2}$ **34.** $8\sqrt{2}$

35. 22 **36.** $4\sqrt{2}$ **37.** $\dfrac{7\sqrt{2}}{2}$ **38.** $-\sqrt{6} - 2\sqrt{2}$

39. $3 - \sqrt{3}$ **40.** $\dfrac{3\sqrt{2} + 4\sqrt{3} - 1}{2}$

CHAPTER 11

Section 11.1   Warm-ups FFFFTTFTTF

1. $y = (x - 3)^2 - 8$, vertex $(3, -8)$, upward
3. $y = 2(x + 3)^2 - 13$, vertex $(-3, -13)$, upward
5. $y = -2(x - 4)^2 + 33$, vertex $(4, 33)$, downward
7. $y = 5(x + 4)^2 - 80$, vertex $(-4, -80)$, upward
9. $(2, -3)$, upward   **11.** $(1, -2)$, downward
13. $(1, -2)$, upward   **15.** $(-\frac{3}{2}, \frac{17}{4})$, downward
17. $(0, 5)$, upward   **19.** Vertex $(\frac{3}{2}, -\frac{1}{4})$

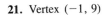
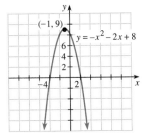

21. Vertex $(-1, 9)$ **23.** Vertex $(-2, 1)$

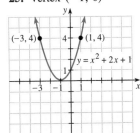

25. Vertex $(-1, 0)$ **27.** Vertex $(\frac{1}{2}, 0)$

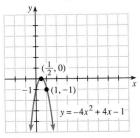

29. Vertex $(\frac{5}{2}, -\frac{25}{4})$ **31.** Vertex $(0, 5)$

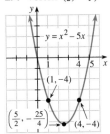
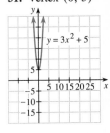

33. Vertex $(1, -2)$ **35.** Vertex $(5, 0)$

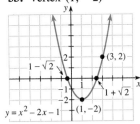
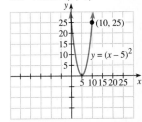

37. Vertex $(0, 0)$ **39.** Vertex $(4, -1)$

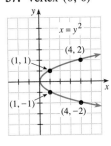
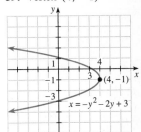

41. Vertex $(0, -2)$ **43.** Vertex $(-4, 0)$

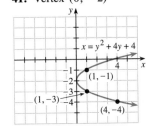
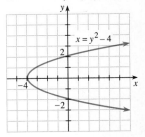

45. 4 and 4   **47.** 3 feet and 4 feet
49. 2 sides 37.5 feet, 1 side 75 feet   **51.** 262 feet, no
53. 100

55. $\{(-1, 2), (1, 2)\}$

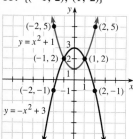

57. $\{(1, -1)\}$

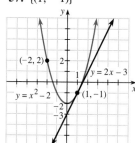

19.

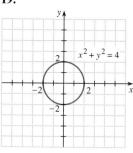

21.

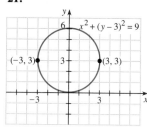

59. $\{(\frac{3}{2}, \frac{11}{4}), (-4, 0)\}$

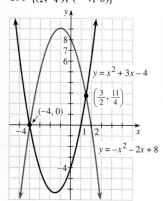

61. $\{(0, 0), (1, 1)\}$

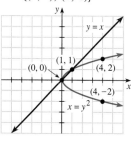

23.

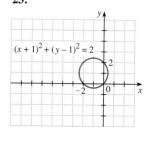

25.

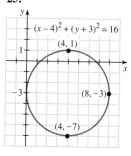

27.

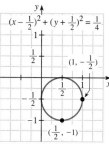

63. $\{(0, 1), (1, 2)\}$

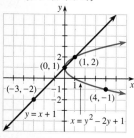

65. $\{(-3, -4), (2, 6)\}$

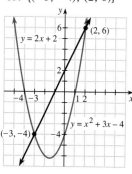

29. $(x + 2)^2 + (y + 3)^2 = 13$ **31.** $(x - 1)^2 + (y - 2)^2 = 8$
33. $(x - 5)^2 + (y - 4)^2 = 9$ **35.** $(x - \frac{1}{2})^2 + (y + \frac{1}{2})^2 = \frac{1}{2}$
37. $(x - \frac{3}{2})^2 + (y - \frac{1}{2})^2 = \frac{7}{2}$ **39.** $(x - \frac{1}{3})^2 + (y + \frac{3}{4})^2 = \frac{97}{144}$
41. $\{(1, 3), (-1, -3)\}$ **43.** $\{(0, -3), (\sqrt{5}, 2), (-\sqrt{5}, 2)\}$

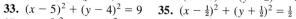
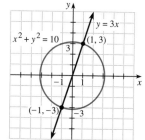

Section 11.2   Warm-ups FFTFFFFFTF

1. $x^2 + (y - 3)^2 = 25$ **3.** $(x - 1)^2 + (y + 2)^2 = 81$
5. $x^2 + y^2 = 3$ **7.** $(x + 6)^2 + (y + 3)^2 = \frac{1}{4}$
9. $(x - \frac{1}{2})^2 + (y - \frac{1}{3})^2 = .01$ **11.** $(3, 5), \sqrt{2}$
13. $\left(0, \frac{1}{2}\right), \frac{\sqrt{2}}{2}$ **15.** $(0, 0), \frac{3}{2}$ **17.** $(2, 0), \sqrt{3}$

45. $\{(0, -3), (2, -1)\}$

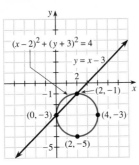

47. $(0, 2 + \sqrt{3}), (0, 2 - \sqrt{3})$ **49.** $\sqrt{29}$

51. $(x - 2)^2 + (y - 3)^2 = 32$

53. $\left(\dfrac{5}{2}, -\dfrac{\sqrt{11}}{2}\right), \left(\dfrac{5}{2}, \dfrac{\sqrt{11}}{2}\right)$ **55.** No graph

57. Graph consists of $(0, 0)$ only.

59.

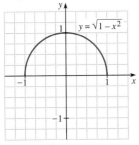

Section 11.3   Warm-ups FFTTTFFTTT

1.

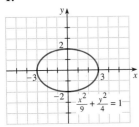

3.

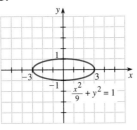

5.

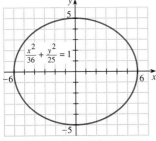

7.

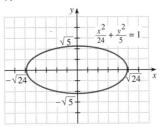

9.

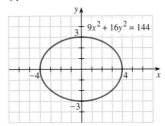

11.

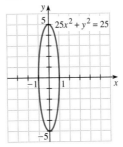

13.

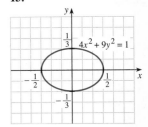

15.

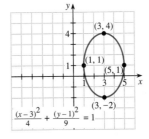

17.

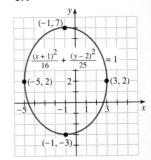

19.

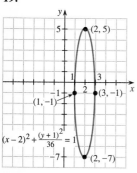

21.

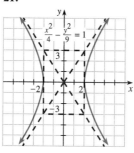

35. $\left\{\left(\dfrac{2\sqrt{10}}{5}, \dfrac{3\sqrt{15}}{5}\right), \left(\dfrac{2\sqrt{10}}{5}, -\dfrac{3\sqrt{15}}{5}\right),\right.$
$\left.\left(-\dfrac{2\sqrt{10}}{5}, \dfrac{3\sqrt{15}}{5}\right), \left(-\dfrac{2\sqrt{10}}{5}, -\dfrac{3\sqrt{15}}{5}\right)\right\}$

37. $\varnothing$

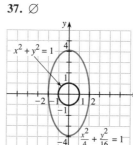

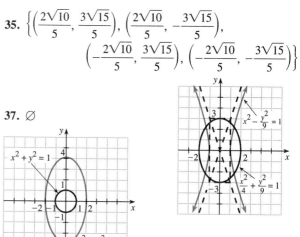

23.

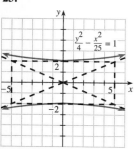

25.

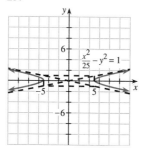

39. $\left\{\left(\dfrac{\sqrt{10}}{2}, \dfrac{\sqrt{6}}{2}\right), \left(\dfrac{\sqrt{10}}{2}, -\dfrac{\sqrt{6}}{2}\right), \left(-\dfrac{\sqrt{10}}{2}, \dfrac{\sqrt{6}}{2}\right),\right.$
$\left.\left(-\dfrac{\sqrt{10}}{2}, -\dfrac{\sqrt{6}}{2}\right)\right\}$

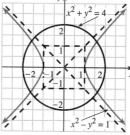

27.

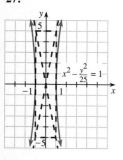

29.

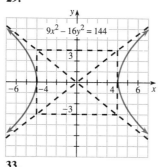

41. $\left\{\left(\dfrac{3\sqrt{6}}{4}, \dfrac{\sqrt{10}}{4}\right), \left(\dfrac{3\sqrt{6}}{4}, -\dfrac{\sqrt{10}}{4}\right), \left(-\dfrac{3\sqrt{6}}{4}, \dfrac{\sqrt{10}}{4}\right),\right.$
$\left.\left(-\dfrac{3\sqrt{6}}{4}, -\dfrac{\sqrt{10}}{4}\right)\right\}$

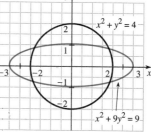

31.

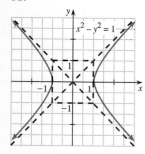

33.

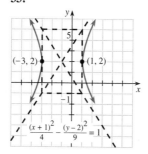

43. $\left\{ \left(\frac{\sqrt{17}}{3}, \frac{8}{9} \right), \left(-\frac{\sqrt{17}}{3}, \frac{8}{9} \right), (0, -1) \right\}$

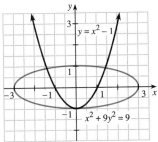

45. $\{(2, 0), (-\frac{5}{2}, -\frac{9}{4})\}$

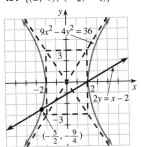

Section 11.4   Warm-ups FTTTFFFTTT

1.

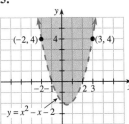

3.

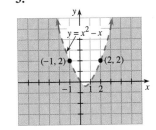

5.

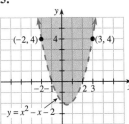

7.

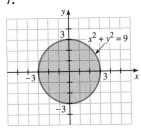

9.

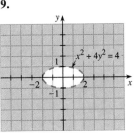

11.

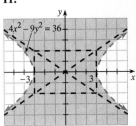

13.

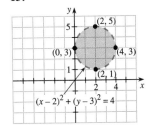

15.

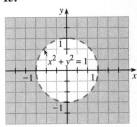

17.

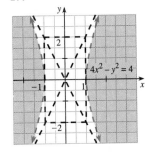

19.

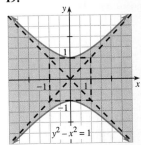

21.

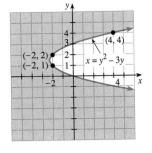

23.

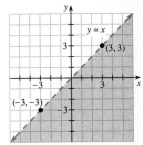

25.

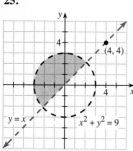

27.

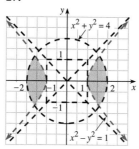

41.

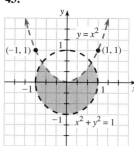

43.

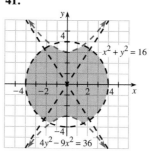

29.

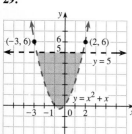

31.

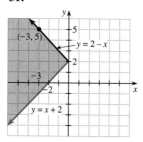

Chapter 11 Review Exercises

1. Vertex $\left(-\frac{3}{2}, -\frac{81}{4}\right)$

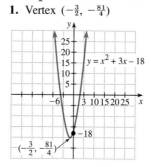

3. Vertex $\left(-\frac{3}{2}, -\frac{1}{4}\right)$

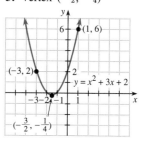

33.

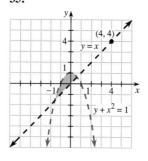

35.

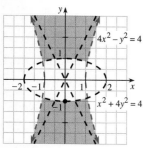

5. Vertex $(0, 5)$

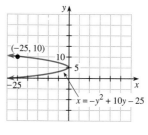

7. 3 and 3   **9.** 22.5 feet by 22.5 feet
11. $(0, 0)$, 10   **13.** $(2, -3)$, 9

37.

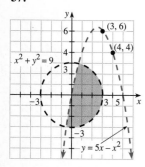

39.

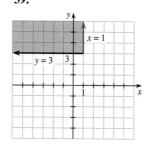

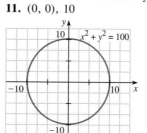

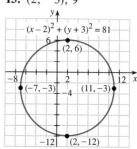

15. $(0, 0), \frac{2}{3}$

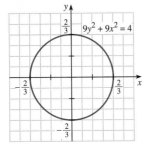

17. $x^2 + (y + 3)^2 = 36$ **19.** $(x - 2)^2 + (y + 7)^2 = 25$

21.

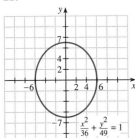

23.

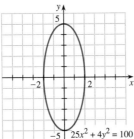

25.

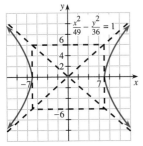

27.

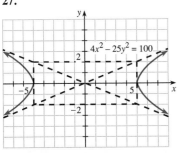

29.

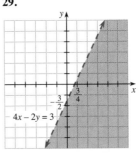

31.

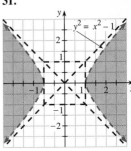

33.

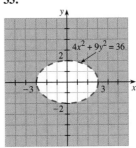

35.

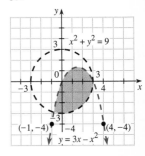

37.

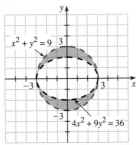

39. Hyperbola   **41.** Circle   **43.** Circle   **45.** Circle
47. Hyperbola   **49.** Hyperbola

51.

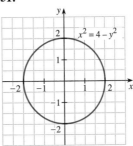

53.

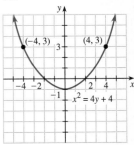

55.

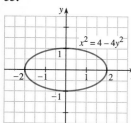

57.

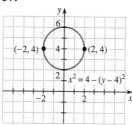

7.

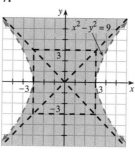

8.

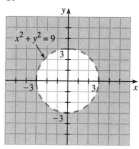

59. $x^2 + y^2 = 25$ **61.** $(x + 1)^2 + (y - 5)^2 = 36$
63. $\{(4, -3), (-3, 4)\}$ **65.** $\varnothing$

Chapter 11 Test

9.

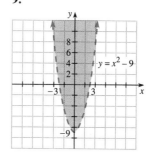

10.

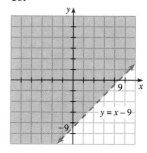

1.

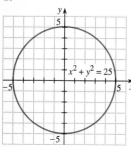

2.

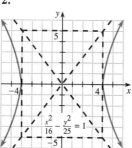

11.

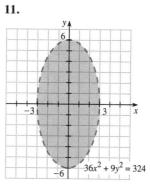

12.

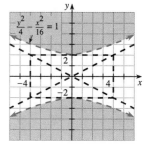

3.

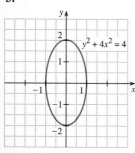

4.

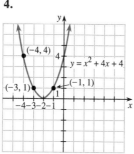

5.

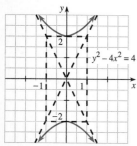

6.

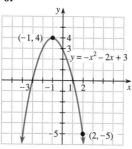

13.

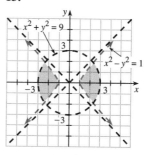

14.

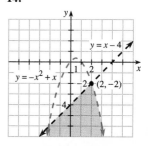

15. $\{(-5, 27), (3, -5)\}$ **16.** $\{(\sqrt{3}, 3), (-\sqrt{3}, 3)\}$
17. $(-1, -5)$, 6   **18.** $(-\frac{1}{2}, \frac{11}{4})$ **19.** 84 feet
20. $(x + 1)^2 + (y - 3)^2 = 13$

Tying It All Together   Chapters 1–11

1.

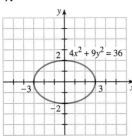

2.

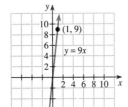

3.

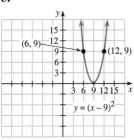

4.

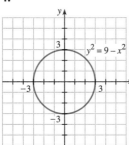

5.

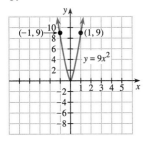

6.

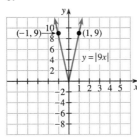

7.

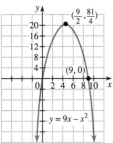

8.

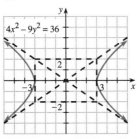

9.

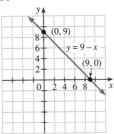

10.

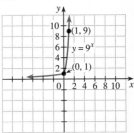

11. $x^2 + 4xy + 4y^2$ **12.** $x^3 + 3x^2y + 3xy^2 + y^3$
13. $a^3 + 3a^2b + 3ab^2 + b^3$ **14.** $a^2 - 6ab + 9b^2$
15. $6a^2 - 7a - 5$ **16.** $x^3 - y^3$ **17.** $\{(1, 2)\}$
18. $\{(3, 4), (4, 3)\}$ **19.** $\{(1, -2, 3)\}$
20. $\{(-1, 1), (3, 9)\}$ **21.** $x = -\dfrac{b}{a}$

22. $x = \dfrac{-d \pm \sqrt{d^2 - 4wm}}{2w}$ **23.** $B = \dfrac{2A - bh}{h}$

24. $x = \dfrac{2y}{y - 2}$ **25.** $m = \dfrac{L}{1 + xt}$ **26.** $t = \dfrac{y^2}{9a^2}$

27. $y = -\frac{2}{3}x - \frac{5}{3}$ **28.** $y = -2x$

29. $(x - 2)^2 + (y - 5)^2 = 45$ **30.** $(-\frac{3}{2}, 3), \dfrac{3\sqrt{5}}{2}$

31. $-10 + 6i$ **32.** -1 **33.** $3 - 5i$ **34.** $7 + 6i\sqrt{2}$
35. $-8 - 27i$ **36.** $-3 + 3i$ **37.** 29   **38.** $-\frac{3}{2} - i$
39. $-\frac{1}{2} - \frac{9}{2}i$ **40.** $2 - i\sqrt{2}$

CHAPTER 12

Section 12.1   Warm-ups TTTFFFTTTT
1. 1, 4, 9, 16, 25, 36, 49, 64
3. $-1, \frac{1}{2}, -\frac{1}{3}, \frac{1}{4}, -\frac{1}{5}, \frac{1}{6}, -\frac{1}{7}, \frac{1}{8}, -\frac{1}{9}, \frac{1}{10}$
5. $1, -2, 4, -8, 16$ **7.** $\frac{1}{2}, \frac{1}{4}, \frac{1}{8}, \frac{1}{16}, \frac{1}{32}, \frac{1}{64}$
9. $-1, 1, 3, 5, 7, 9, 11$ **11.** $1, \dfrac{\sqrt{2}}{2}, \dfrac{\sqrt{3}}{3}, \dfrac{1}{2}, \dfrac{\sqrt{5}}{5}$
13. $\frac{1}{2}, \frac{1}{6}, \frac{1}{12}, \frac{1}{20}$ **15.** $-\frac{1}{3}, -1, 1, \frac{1}{3}$ **17.** $-1, 0, -1, 4$
19. $1, \frac{1}{4}, \frac{1}{9}, \frac{1}{16}$ **21.** $a_n = 2n - 1$ **23.** $a_n = (-1)^{n+1}$
25. $a_n = 2n - 2$ **27.** $a_n = 3n$ **29.** $a_n = 3n + 1$
31. $a_n = (-1)^n 2^{n-1}$ **33.** $a_n = (n - 1)^2$
35. $4, 2, 1, \frac{1}{2}, \frac{1}{4}$
37. \$21,000, \$22,050, \$23,153, \$24,311, \$25,527
39. .9048, .3677, .00004517   **41.** 137,438,953,400

Section 12.2   Warm-ups TFFTFFTTTF

1. 30   **3.** 24   **5.** $\frac{31}{32}$ **7.** 50   **9.** -7 **11.** 0   **13.** $\displaystyle\sum_{i=1}^{6} i$

15. $\sum_{i=1}^{6}(-1)^i(2i-1)$ **17.** $\sum_{i=1}^{6}i^2$ **19.** $\sum_{i=1}^{4}\frac{1}{2+i}$

21. $\sum_{i=1}^{3}\ln(i+1)$ **23.** $\sum_{i=1}^{4}a_i$ **25.** $\sum_{i=1}^{50}x_i$ **27.** $\sum_{i=1}^{n}w_i$

29. $\sum_{j=0}^{4}(j+1)^2$ **31.** $\sum_{j=1}^{6}(2j-3)$ **33.** $\sum_{j=1}^{5}\frac{1}{j+3}$

35. $\sum_{j=0}^{4}x^{2j+5}$ **37.** $\sum_{j=0}^{n-1}x^{j+1}$

39. $x+x^2+x^3+x^4+x^5+x^6$ **41.** $x_0-x_1+x_2-x_3$

43. $x+2x^2+3x^3$ **45.** $\sum_{i=1}^{9}2^{-i}$

Section 12.3   Warm-ups FFFFFFTTTF

1. $a_n=6n-6$ **3.** $a_n=5n+2$ **5.** $a_n=-4n+9$
7. $a_n=-7n+5$ **9.** $a_n=.5n-3.5$
11. $a_n=-.5n-5.5$ **13.** 9, 13, 17, 21, 25
15. 7, 5, 3, 1, −1   **17.** −4, −1, 2, 5, 8
19. −2, −5, −8, −11, −14
21. −7, −11, −15, −19, −23   –
23. 4.5, 5, 5.5, 6, 6.5
25. 1020, 1040, 1060, 1080, 1100   **27.** 51
29. 17   **31.** 4   **33.** 26   **35.** 1176   **37.** 330
39. −481   **41.** 435   **43.** −540   **45.** 150   **47.** −308
49. \$25,000   **51.** 1085

Section 12.4   Warm-ups FFTTTTFTTF

1. $a_n=\frac{1}{3}(3)^{n-1}$ **3.** $a_n=64(\frac{1}{8})^{n-1}$ **5.** $a_n=8(-\frac{1}{2})^{n-1}$
7. $a_n=2(-2)^{n-1}$ **9.** $a_n=-\frac{1}{3}(\frac{3}{4})^{n-1}$ **11.** 2, $\frac{2}{3}$, $\frac{2}{9}$, $\frac{2}{27}$, $\frac{2}{81}$
13. 1, −2, 4, −8, 16   **15.** $\frac{1}{2}$, $\frac{1}{4}$, $\frac{1}{8}$, $\frac{1}{16}$, $\frac{1}{32}$
17. .78, .6084, .4746, .3702, .2887   **19.** 5   **21.** $\frac{1}{3}$
23. $-\frac{1}{9}$ **25.** $\frac{511}{512}$ **27.** $\frac{11}{32}$ **29.** 5115   **31.** .111111
33. $\frac{1}{4}$ **35.** 9   **37.** $\frac{8}{3}$ **39.** $\frac{3}{7}$ **41.** 6   **43.** $\frac{1}{3}$ **45.** $\frac{4}{33}$
47. $\frac{4}{9}$ **49.** \$21,474,836.46   **51.** \$1,146,372.04

Section 12.5   Warm-ups FFFTTTTTTT

1. $r^5+5r^4t+10r^3t^2+10r^2t^3+5rt^4+t^5$
3. $m^3-3m^2n+3mn^2-n^3$ **5.** $x^3+6ax^2+12a^2x+8a^3$
7. $x^8-8x^6+24x^4-32x^2+16$
9. $x^7-7x^6+21x^5-35x^4+35x^3-21x^2+7x-1$
11. $a^{12}-36a^{11}b+594a^{10}b^2-5940a^9b^3$
13. $x^{18}+45x^{16}+900x^{14}+10,500x^{12}$
15. $x^{22}-22x^{21}+231x^{20}-1540x^{19}$

17. $\frac{x^{10}}{1024}+\frac{5x^9y}{768}+\frac{5x^8y^2}{256}+\frac{5x^7y^3}{144}$ **19.** $1287a^8w^5$

21. $-11{,}440m^9n^7$ **23.** $635{,}043{,}840a^{28}b^6$

25. $\sum_{i=0}^{8}\frac{8!}{(8-i)!i!}a^{8-i}m^i$ **27.** $\sum_{i=0}^{5}\frac{5!(-2)^i}{(5-i)!i!}a^{5-i}x^i$

Chapter 12 Review Exercises

1. 1, 8, 27, 64, 125   **3.** 1, 1, −3, 5, −7, 9
5. −1, $-\frac{1}{2}$, $-\frac{1}{3}$ **7.** $\frac{1}{3}$, $\frac{1}{5}$, $\frac{1}{7}$ **9.** 4, 5, 6   **11.** 36

13. 40   **15.** $\sum_{i=1}^{\infty}\frac{1}{2(i+1)}$ **17.** $\sum_{i=1}^{\infty}(i-1)^2$

19. $\sum_{i=1}^{\infty}(-1)^{i+1}x_i$ **21.** $a_n=\frac{n}{3}$ **23.** $a_n=2n$ **25.** 300

27. $\frac{289}{6}$ **29.** 35   **31.** 3, $\frac{3}{2}$, $\frac{3}{4}$, $\frac{3}{8}$ **33.** 1, $\frac{1}{2}$, $\frac{1}{4}$, $\frac{1}{8}$
35. .23, .0023, .000023, .00000023   **37.** $a_n=\frac{1}{2}(6)^{n-1}$
39. $a_n=.7(.1)^{n-1}$ **41.** $\frac{40}{81}$ **43.** .3333333333   **45.** $\frac{3}{8}$
47. 54   **49.** $m^5+5m^4n+10m^3n^2+10m^2n^3+5mn^4+n^5$
51. $a^6-9a^4b+27a^2b^2-27b^3$ **53.** $495x^8y^4$

55. $\sum_{i=0}^{7}\frac{7!}{(7-i)!i!}a^{7-i}w^i$ **57.** Neither   **59.** Arithmetic

61. Arithmetic   **63.** $\frac{1}{\sqrt[3]{180}}$ **65.** $-1+\frac{1}{2}-\frac{1}{6}+\frac{1}{24}-\frac{1}{120}$
67. $a^5+5a^4b+10a^3b^2+10a^2b^3+5ab^4+b^5$ **69.** 26
71. \$12,531.74

Chapter 12 Test

1. −10, −4, 2, 8   **2.** 5, .5, .05, .005
3. −1, $\frac{1}{2}$, $-\frac{1}{6}$, $\frac{1}{24}$ **4.** 1, $\frac{3}{4}$, $\frac{5}{9}$, $\frac{7}{16}$ **5.** $a_n=10-3n$
6. $a_n=-25(-\frac{1}{5})^{n-1}$ **7.** $a_n=(-1)^{n-1}2n$ **8.** $a_n=n^2$
9. $5+7+9+11+13$
10. $5+10+20+40+80+160$
11. $m^4+4m^3q+6m^2q^2+4mq^3+q^4$
12. $1+4+9+16+25+36+49+64+81+100$
13. 750   **14.** $\frac{155}{8}$ **15.** 40,200   **16.** 5   **17.** 2 or −2
18. 11   **19.** $1365r^{11}t^4$ **20.** $-448a^{10}b^3$

Tying It All Together   Chapters 1–12

1. 6   **2.** n^2-3 **3.** $x^2+2xh+h^2-3$ **4.** x^2-2x-2
5. 11   **6.** 6   **7.** $2x^2-7$ **8.** $2x^3-x^2-6x+3$ **9.** 5

10. 3   **11.** $\frac{x+1}{2}$ **12.** x **13.** $-\frac{27}{2}$ **14.** $-\frac{8}{3}$

15. $-\frac{128}{9}$ **16.** 16

17.

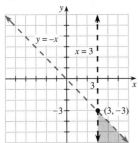

18.

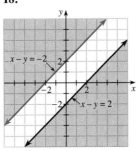

23.

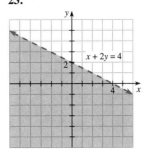

24.

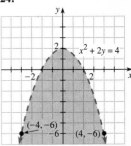

19.

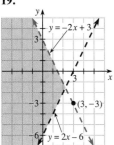

20.

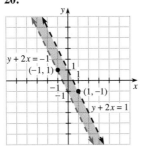

25.

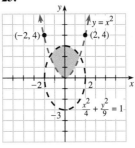

26.

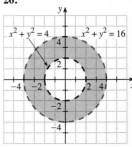

21.

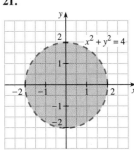

22.

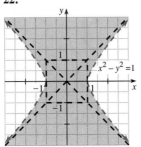

27. $\dfrac{a^2 + b^2}{ab}$ **28.** $\dfrac{y - 3}{y}$ **29.** $\dfrac{10}{(x - 3)(x + 3)(x + 1)}$

30. $\dfrac{2(x^3 - 64)}{x^3 - 16}$ **31.** $\dfrac{a^7}{b^4}$ **32.** 1   **33.** 4   **34.** $\frac{1}{32}$

35. -2 **36.** $\frac{1}{9}$ **37.** $-\frac{1}{8}$ **38.** $\frac{1}{4}$ **39.** $\frac{1}{5}$ **40.** 25

Index